Birkhäuser

Progress in Nonlinear Differential Equations and Their Applications

Volume 84

For further volumes:
http://www.springer.com/series/4889

Massimo Cicognani • Ferruccio Colombini
Daniele Del Santo
Editors

Studies in Phase Space Analysis with Applications to PDEs

Birkhäuser

Editors
Massimo Cicognani
Dipartimento di Matematica
Università di Bologna
Bologna, Italy

Ferruccio Colombini
Dipartimento di Matematica
Università di Pisa
Pisa, Italy

Daniele Del Santo
Dipartimento di Matematica e Geoscienze
Università di Trieste
Trieste, Italy

ISBN 978-1-4614-6347-4 ISBN 978-1-4614-6348-1 (ebook)
DOI 10.1007/978-1-4614-6348-1
Springer New York Heidelberg Dordrecht London

Library of Congress Control Number: 2013931265

Mathematics Subject Classification (2010): 26D10, 35A23, 35A30, 35B30, 35B40, 35B45, 35B60, 35B65, 35J10, 35J75, 35K05, 35K10, 35L15, 35L50, 35L99, 35P25, 35Q30, 35Q40, 35Q41, 35Q53, 35Q55, 35R45, 35S05, 35S30, 42C15, 46E35, 47A40, 47A60, 47G30, 76D03, 81Q05, 81U40

Printed on acid-free paper

Birkhäuser is part of Springer Science+Business Media (www.birkhauser-science.com)

Mais où sont les neiges d'antan?

F. Villon

Preface

This volume is a collection of papers mainly concerning *phase space analysis*, or *microlocal analysis*, and its applications to the theory of *partial differential equations* (PDEs).

A number of papers composing this volume, all written by leading experts in their respective fields, are expanded version of talks given at a meeting held in September 2011 at the *Centro Residenziale Universitario (CEUB)* of Bertinoro, on the hills surrounding Cesena, Italy.

The Bertinoro workshop was the occasion to fix the state of the art in many different aspects of phase space analysis. In fact the results collected here concern general theory of pseudodifferential operators, Hardy-type inequalities, linear and non-linear hyperbolic equations and systems, water-waves equations, Euler–Poisson and Navier–Stokes equations, Schrödinger equations and heat and parabolic equations.

We would like to seize this occasion to thank all the contributors as well as the people who took part in the workshop.

A number of institutions have made it possible to hold the Bertinoro workshop through their financial support. We'd like to list them here: the Italian Ministero dell'Istruzione, dell'Università e della Ricerca by means of the PRIN 2008 project "Phase Space Analysis of PDE's", the Istituto Nazionale di Alta Matematica, the University of Bordeaux 1, the University of Pisa and the Fondazione Cassa di Risparmio di Cesena. We thank all of them for their generosity.

Bologna, Italy — Massimo Cicognani
Pisa, Italy — Ferruccio Colombini
Trieste, Italy — Daniele Del Santo

Contents

List of Contributors

Thomas Alazard
Département de Mathématiques et Applications, École Normale Supérieure et CNRS UMR 8553, 45, rue d'Ulm, F-75230 Paris Cedex 05, France
e-mail: Thomas.Alazard@ens.fr

Jean-Michel Bony
CMLS, École Polytechnique, F-91128 Palaiseau Cedex, France
e-mail: bony@math.polytechnique.fr

Nicolas Burq
Département de Mathématiques, Université Paris-Sud 11 et CNRS, F-91405 Orsay Cedex, France, e-mail: nicolas.burq@math.u-psud.fr

Cristian Cazacu
1. BCAM–Basque Center for Applied Mathematics, Mazarredo, 14, E-48009 Bilbao, Basque Country, Spain
2. Departamento de Matemáticas, Universidad Autónoma de Madrid, E-28049 Madrid, Spain, e-mail: cazacu@bcamath.org

Jean-Yves Chemin
Laboratoire J.-L. Lions UMR 7598, Université Paris VI, 175, rue du Chevaleret, F-75013 Paris, France, e-mail: chemin@ann.jussieu.fr

Elena Cordero
Dipartimento di Matematica, Università di Torino, via Carlo Alberto 10, I-10123 Torino, Italy, e-mail: elena.cordero@unito.it

Raphaël Danchin
LAMA, UMR 8050, Université Paris-Est et Institut Universitaire de France, 61, avenue du Général de Gaulle, F-94010 Créteil Cedex, France
e-mail: raphael.danchin@u-pec.fr

Isabelle Gallagher
Institut de Mathématiques de Jussieu UMR 7586, Université Paris VII, 175, rue du Chevaleret, F-75013 Paris, France, e-mail: gallagher@math.univ-paris-diderot.fr

Todor Gramchev
Dipartimento di Matematica e Informatica, Università di Cagliari, via Ospedale 72, I-09124 Cagliari, Italy, e-mail: todor@unica.it

Nakao Hayashi
Department of Mathematics, Graduate School of Science, Osaka University, Osaka, Toyonaka, 560-0043, Japan, e-mail: nhayashi@math.sci.osaka-u.ac.jp

Jingchi Huang
Academy of Mathematics & Systems Science, Chinese Academy of Sciences, Beijing 100190, P. R. China, e-mail: jchuang@amss.ac.cn

David Lannes
Département de Mathématiques et Applications, École Normale Supérieure, 45, rue d'Ulm, F-75230 Paris Cedex 05, France, e-mail: lannes@ens.fr

Felipe Linares
IMPA, Estrada Dona Castorina 110, Rio de Janeiro 22460-320, RJ Brasil
e-mail: linares@impa.br

Piotr Bogusław Mucha
Instytut Matematyki Stosowanej i Mechaniki, Uniwersytet Warszawski, ul. Banacha 2, 02-097 Warszawa, Poland, e-mail: p.mucha@mimuw.edu.pl

Chloé Mullaert
Laboratoire J.-L. Lions UMR 7598, Université Paris VI, 175, rue du Chevaleret, F-75013 Paris, France, e-mail: cmullaert@ann.jussieu.fr

Takashi Narazaki
Department of Mathematical Sciences, Tokai University, Kitakaname, Kanagawa, 259-1292 Japan, e-mail: narazaki@tokai-u.jp

Pavel I. Naumkin
Centro de Ciencias Matemáticas UNAM, Campus Morelia, AP 61-3 (Xangari), Morelia CP 58089, Michoacán, México, e-mail: pavelni@matmor.unam.mx

Fabio Nicola
Dipartimento di Scienze Matematiche, Politecnico di Torino, corso Duca degli Abruzzi 24, I-10129 Torino, Italy, e-mail: fabio.nicola@polito.it

Tatsuo Nishitani
Department of Mathematics, Osaka University, Machikaneyama 1-1, Toyonaka, 560-0043, Osaka, Japan, e-mail: nishitani@math.sci.osaka-u.ac.jp

Takashi Okaji
Graduate School of Science, Kyoto University, Kyoto 606-8502, Japan
e-mail: okaji@math.kyoto-u.ac.jp

Marius Paicu
Institut de Mathématiques de Bordeaux, Université Bordeaux 1, 351, cours de la Libération, F-33405 Talence Cedex, France
e-mail: marius.paicu@math.u-bordeaux1.fr

Alberto Parmeggiani
Dipartimento di Matematica, Università di Bologna, Piazza di Porta San Donato 5, I-40126 Bologna, Italy, e-mail: alberto.parmeggiani@unibo.it

Vesselin Petkov
Institut de Mathématiques de Bordeaux, Université Bordeaux 1, 351, cours de la Libération, F-33405 Talence, France, e-mail: petkov@math.u-bordeaux1.fr

Michael Reissig
Faculty for Mathematics and Computer Science, TU Bergakademie Freiberg, Prüferstr. 9, D-09596 Freiberg, Germany, e-mail: reissig@math.tu-freiberg.de

Luc Robbiano
Laboratoire de Mathématiques, Université de Versailles Saint-Quentin, 45, avenue des États-Unis, F-78035 Versailles Cedex, France, e-mail: luc.robbiano@uvsq.fr

Luigi Rodino
Dipartimento di Matematica, Università di Torino, via Carlo Alberto 10, I-10123 Torino, Italy, e-mail: luigi.rodino@unito.it

Michael Ruzhansky
Mathematics Department, Imperial College London, Huxley Building, 180 Queen's Gate, London SW7 2AZ, United Kingdom, e-mail: m.ruzhansky@imperial.ac.uk

Jean-Claude Saut
Laboratoire de Mathématiques, UMR 8628, Université Paris-Sud 11 et CNRS, F-91405 Orsay Cedex, France, e-mail: jean-claude.saut@math.u-psud.fr

Nicola Visciglia
Dipartimento di Matematica "L. Tonelli", Università di Pisa, largo Bruno Pontecorvo 5, I-56127 Pisa, Italy, e-mail: viscigli@dm.unipi.it

Ping Zhang
Academy of Mathematics & Systems Science and Hua Loo-Keng Key Laboratory of Mathematics, The Chinese Academy of Sciences, Beijing 100190, P. R. China
e-mail: zp@amss.ac.cn

Enrique Zuazua,
1. BCAM–Basque Center for Applied Mathematics, Mazarredo, 14, E-48009 Bilbao, Basque Country, Spain
2. Ikerbasque, Basque Foundation for Science, Alameda Urquijo 36-5, Plaza Bizkaia, E-48011, Bilbao, Basque Country, Spain, e-mail: zuazua@bcamath.org

Claude Zuily
Département de Mathématiques, Université Paris-Sud 11 et CNRS, F-91405 Orsay Cedex, France, e-mail: claude.zuily@math.u-psud.fr

Chapter 1
The Water-Wave Equations: From Zakharov to Euler

Thomas Alazard, Nicolas Burq, and Claude Zuily

Abstract Starting from the Zakharov/Craig–Sulem formulation of the water-wave equations, we prove that one can define a pressure term and hence obtain a solution of the classical Euler equations. It is proved that these results hold in rough domains, under minimal assumptions on the regularity to ensure, in terms of Sobolev spaces, that the solutions are C^1.

Key words: Cauchy theory, Euler equations, Water-wave system

Mathematics Subject Classification: 35B65, 35B30, 35B60, 35J75, 35S05.

1.1 Introduction

We study the dynamics of an incompressible layer of inviscid liquid, having constant density, occupying a fluid domain with a free surface.

We begin by describing the fluid domain. Hereafter, $d \geq 1$, t denotes the time variable, and $x \in \mathbf{R}^d$ and $y \in \mathbf{R}$ denote the horizontal and vertical variables. We work in a fluid domain with free boundary of the form

$$\Omega = \{ (t,x,y) \in (0,T) \times \mathbf{R}^d \times \mathbf{R} : (x,y) \in \Omega(t) \},$$

where $\Omega(t)$ is the $d+1$-dimensional domain located between two hypersurfaces: a free surface denoted by $\Sigma(t)$ which will be supposed to be a graph and a fixed

T. Alazard
Département de Mathématiques et Applications, École Normale Supérieure et CNRS
UMR 8553, 45, rue d'Ulm, F-75230 Paris Cedex 05, France
e-mail: Thomas.Alazard@ens.fr

N. Burq • C. Zuily (✉)
Département de Mathématiques, Université Paris-Sud 11 et CNRS, F-91405 Orsay Cedex, France
e-mail: nicolas.burq@math.u-psud.fr; claude.zuily@math.u-psud.fr

M. Cicognani et al. (eds.), *Studies in Phase Space Analysis with Applications to PDEs*,
Progress in Nonlinear Differential Equations and Their Applications 84,
DOI 10.1007/978-1-4614-6348-1_1,

bottom Γ. For each time t, one has

$$\Omega(t) = \{(x,y) \in \mathcal{O} : y < \eta(t,x)\},$$

where $\mathcal{O}$ is a given open connected domain and where η is the free surface elevation. We denote by Σ the free surface

$$\Sigma = \{(t,x,y) : t \in (0,T), (x,y) \in \Sigma(t)\},$$

where $\Sigma(t) = \{(x,y) \in \mathbf{R}^d \times \mathbf{R} : y = \eta(t,x)\}$, and we set $\Gamma = \partial\Omega(t) \setminus \Sigma(t)$.

Notice that Γ does not depend on time. Two classical examples are the case of infinite depth ($\mathcal{O} = \mathbf{R}^{d+1}$ so that $\Gamma = \emptyset$) and the case where the bottom is the graph of a function (this corresponds to the case $\mathcal{O} = \{(x,y) \in \mathbf{R}^d \times \mathbf{R} : y > b(x)\}$ for some given function b).

We introduce now a condition which ensures that, at time t, there exists a fixed strip separating the free surface from the bottom:

$$(H_t): \qquad \exists h > 0 : \quad \Gamma \subset \{(x,y) \in \mathbf{R}^d \times \mathbf{R} : y < \eta(t,x) - h\}. \tag{1.1}$$

No regularity assumption will be made on the bottom Γ.

The Incompressible Euler Equation with Free Surface

Hereafter, we use the following notations:

$$\nabla = (\partial_{x_i})_{1\le i\le d}, \quad \nabla_{x,y} = (\nabla, \partial_y), \quad \Delta = \sum_{1\le i\le d} \partial_{x_i}^2, \quad \Delta_{x,y} = \Delta + \partial_y^2.$$

The Eulerian velocity field $v \colon \Omega \to \mathbf{R}^{d+1}$ solves the incompressible Euler equation

$$\partial_t v + v \cdot \nabla_{x,y} v + \nabla_{x,y} P = -g e_y, \quad \operatorname{div}_{x,y} v = 0 \quad \text{in } \Omega,$$

where g is the acceleration due to gravity ($g > 0$) and P is the pressure. The problem is then given by three boundary conditions:

- A kinematic condition (which states that the free surface moves with the fluid)

$$\partial_t \eta = \sqrt{1 + |\nabla\eta|^2}\,(v \cdot n) \quad \text{on } \Sigma, \tag{1.2}$$

  where n is the unit exterior normal to $\Omega(t)$
- A dynamic condition (that expresses a balance of forces across the free surface)

$$P = 0 \quad \text{on } \Sigma \tag{1.3}$$

- The "solid wall" boundary condition at the bottom Γ

$$v \cdot \nu = 0, \tag{1.4}$$

where ν is the normal vector to Γ whenever it exists. In the case of arbitrary bottom, this condition will be implicit and contained in a variational formulation.

The Zakharov/Craig–Sulem Formulation

A popular form of the water-wave system is given by the Zakharov/Craig–Sulem formulation. This is an elegant formulation of the water-wave equations where all the unknowns are evaluated at the free surface only. Let us recall the derivation of this system.

Assume, furthermore, that the motion of the liquid is irrotational. The velocity field v is therefore given by $v = \nabla_{x,y}\Phi$ for some velocity potential $\Phi\colon \Omega \to \mathbf{R}$ satisfying

$$\Delta_{x,y}\Phi = 0 \quad \text{in } \Omega, \qquad \partial_\nu \Phi = 0 \quad \text{on } \Gamma,$$

and the Bernoulli equation

$$\partial_t \Phi + \frac{1}{2}\left|\nabla_{x,y}\Phi\right|^2 + P + gy = 0 \quad \text{in } \Omega. \tag{1.5}$$

Following Zakharov [7], introduce the trace of the potential on the free surface:

$$\psi(t,x) = \Phi(t,x,\eta(t,x)).$$

Notice that since Φ is harmonic, η and Ψ fully determines Φ. Craig and Sulem (see [3]) observe that one can form a system of two evolution equations for η and ψ. To do so, they introduce the Dirichlet-Neumann operator $G(\eta)$ that relates ψ to the normal derivative $\partial_n\Phi$ of the potential by

$$\begin{aligned}(G(\eta)\psi)(t,x) &= \sqrt{1+|\nabla\eta|^2}\,\partial_n\Phi|_{y=\eta(t,x)}\\ &= (\partial_y\Phi)(t,x,\eta(t,x)) - \nabla_x\eta(t,x)\cdot(\nabla_x\Phi)(t,x,\eta(t,x)).\end{aligned}$$

(For the case with a rough bottom, we recall the precise construction later on). Directly from this definition, one has

$$\partial_t\eta = G(\eta)\psi. \tag{1.6}$$

It is proved in [3] (see also the computations in §1.3.6) that the condition $P = 0$ on the free surface implies that

$$\partial_t\psi + g\eta + \frac{1}{2}|\nabla\psi|^2 - \frac{1}{2}\frac{\left(\nabla\eta\cdot\nabla\psi + G(\eta)\psi\right)^2}{1+|\nabla\eta|^2} = 0. \tag{1.7}$$

The system (1.6) and (1.7) is in Hamiltonian form (see [3, 7]), where the Hamiltonian is given by

$$\mathscr{H} = \frac{1}{2}\int_{\mathbf{R}^d} \psi G(\eta)\psi + g\eta^2\,dx.$$

The problem to be considered here is that of the equivalence of the previous two formulations of the water-wave problem. Assume that the Zakharov/Craig–Sulem system has been solved. Namely, assume that, for some $r > 1 + d/2$, $(\eta,\psi) \in C^0(I,H^r(\mathbf{R}^d)\times H^r(\mathbf{R}^d))$ solves (1.6)–(1.7). We would like to show that we have indeed solved the initial system of Euler's equation with free boundary. In particular we have to define the pressure which does not appear in the above system (1.6)–(1.7). To do so, we set

$$B = \frac{\nabla\eta\cdot\nabla\psi + G(\eta)\psi}{1+|\nabla\eta|^2}, \qquad V = \nabla\psi - B\nabla\eta.$$

Then B and V belong to the space $C^0(I,H^{\frac{1}{2}}(\mathbf{R}^d))$. It follows from [1] that (for fixed t) one can define unique variational solutions to the problems

$$\Delta_{x,y}\Phi = 0 \quad \text{in } \Omega, \qquad \Phi|_\Sigma = \psi, \qquad \partial_\nu\Phi = 0 \quad \text{on } \Gamma.$$

$$\Delta_{x,y}Q = 0 \quad \text{in } \Omega, \qquad Q|_\Sigma = g\eta + \frac{1}{2}(B^2+|V|^2), \qquad \partial_\nu Q = 0 \quad \text{on } \Gamma.$$

Then we shall define $P \in \mathscr{D}'(\Omega)$ by

$$P := Q - gy - \frac{1}{2}\left|\nabla_{x,y}\Phi\right|^2$$

and we shall show firstly that P has a trace on Σ which is equal to 0 and secondly that $Q = -\partial_t\Phi$ which will show, according to (1.5), that we have indeed solved Bernoulli's (and therefore Euler's) equation.

These assertions are not straightforward because we are working with solutions of low regularity and we consider general bottoms (namely no regularity assumption is assumed on the bottom). Indeed, the analysis would have been much easier for $r > 2 + d/2$ and a flat bottom.

1.2 Low Regularity Cauchy Theory

Since we are interested in low regularity solutions, we begin by recalling the well-posedness results proved in [2]. These results clarify the Cauchy theory of the water-wave equations as well in terms of regularity indexes for the initial conditions as for the smoothness of the bottom of the domain (namely no regularity assumption is assumed on the bottom).

Recall that the Zakharov/Craig–Sulem system reads

$$\begin{cases} \partial_t\eta - G(\eta)\psi = 0, \\ \partial_t\psi + g\eta + \dfrac{1}{2}|\nabla\psi|^2 - \dfrac{1}{2}\dfrac{\left(\nabla\eta\cdot\nabla\psi + G(\eta)\psi\right)^2}{1+|\nabla\eta|^2} = 0. \end{cases} \tag{1.8}$$

It is useful to introduce the vertical and horizontal components of the velocity

$$B := (v_y)|_{y=\eta} = (\partial_y \Phi)|_{y=\eta}, \quad V := (v_x)|_{y=\eta} = (\nabla_x \Phi)|_{y=\eta}.$$

These can be defined in terms of η and ψ by means of the formulas

$$B = \frac{\nabla\eta\cdot\nabla\psi + G(\eta)\psi}{1+|\nabla\eta|^2}, \qquad V = \nabla\psi - B\nabla\eta. \tag{1.9}$$

Also, recall that the Taylor coefficient $a = -\partial_y P|_\Sigma$ can be defined in terms of $\eta, V, B, and \psi$ only (see §4.3.1 in [5]).

In [2] we proved the following results about low regularity solutions. We refer to the introduction of [2, 4] for references and a short historical survey of the background of this problem.

Theorem 1.1 ([2]). *Let $d \geq 1$ and $s > 1 + d/2$ and consider an initial data (η_0, ψ_0) such that:*

(i) $\eta_0 \in H^{s+\frac{1}{2}}(\mathbf{R}^d), \quad \psi_0 \in H^{s+\frac{1}{2}}(\mathbf{R}^d), \quad V_0 \in H^s(\mathbf{R}^d), \quad B_0 \in H^s(\mathbf{R}^d).$
(ii) The condition (H_0) in (1.1) *holds initially for $t = 0$.*
(iii) There exists a positive constant c such that, for all x in $\mathbf{R}^d$, $a_0(x) \geq c$.

Then there exists $T > 0$ such that the Cauchy problem for (1.8) *with initial data (η_0, ψ_0) has a unique solution:*

$$(\eta, \psi) \in C^0\big([0,T], H^{s+\frac{1}{2}}(\mathbf{R}^d) \times H^{s+\frac{1}{2}}(\mathbf{R}^d)\big),$$

such that

1. *$(V,B) \in C^0\big([0,T], H^s(\mathbf{R}^d) \times H^s(\mathbf{R}^d)\big)$.*
2. *The condition (H_t) in* (1.1) *holds for $t \in [0,T]$ with h replaced by $h/2$.*
3. *$a(t,x) \geq c/2$, for all (t,x) in $[0,T] \times \mathbf{R}^d$.*

In a forthcoming paper we shall prove the following result.

Theorem 1.2. *Assume $\Gamma = \emptyset$. Let $d = 2$ and $s > 1 + \frac{d}{2} - \frac{1}{12}$ and consider an initial data (η_0, ψ_0) such that*

$$\eta_0 \in H^{s+\frac{1}{2}}(\mathbf{R}^d), \quad \psi_0 \in H^{s+\frac{1}{2}}(\mathbf{R}^d), \quad V_0 \in H^s(\mathbf{R}^d), \quad B_0 \in H^s(\mathbf{R}^d).$$

Then there exists $T > 0$ such that the Cauchy problem for (1.8) *with initial data (η_0, ψ_0) has a solution (η, ψ) such that*

$$(\eta, \psi, V, B) \in C^0\big([0,T]; H^{s+\frac{1}{2}}(\mathbf{R}^d) \times H^{s+\frac{1}{2}}(\mathbf{R}^d) \times H^s(\mathbf{R}^d) \times H^s(\mathbf{R}^d)\big).$$

Remark 1.1. *(i)* For the sake of simplicity we stated Theorem 1.2 in dimension $d = 2$ (recall that d is the dimension of the interface). One can prove such a result in any dimension $d \geq 2$, the number $1/12$ being replaced by an index depending on d.

(ii) Notice that in infinite depth ($\Gamma = \emptyset$) the Taylor condition (which is assumption *(iii)* in Theorem 1.1) is always satisfied as proved by Wu ([6]).

Now having solved the system (1.8) in (η,ψ), we have to show that we have indeed solved the initial system in (η,v). This is the purpose of the following section.

There is one point that should be emphasized concerning the regularity. Below we consider solutions (η,ψ) of (1.8) such that

$$(\eta,\psi)\in C^0\big([0,T];H^{s+\frac{1}{2}}(\mathbf{R}^d)\times H^{s+\frac{1}{2}}(\mathbf{R}^d)\big),$$

with the only assumption that $s>\frac{1}{2}+\frac{d}{2}$ (and the assumption that there exists $h>0$ such that the condition (H_t) in (1.1) holds for $t\in[0,T]$). Consequently, the result proved in this note applies to the settings considered in the above theorems.

1.3 From Zakharov to Euler

1.3.1 The Variational Theory

In this paragraph the time is fixed so we will skip it and work in a fixed domain Ω whose top boundary Σ is Lipschitz, i.e, $\eta\in W^{1,\infty}(\mathbf{R}^d)$.

We recall here the variational theory, developed in [1], allowing us to solve the following problem in the case of arbitrary bottom,

$$\Delta\Phi=0 \quad \text{in } \Omega, \quad \Phi|_\Sigma=\psi, \quad \frac{\partial\Phi}{\partial\nu}|_\Gamma=0. \tag{1.10}$$

Notice that Ω is not necessarily bounded below. We proceed as follows.

Denote by $\mathscr{D}$ the space of functions $u\in C^\infty(\Omega)$ such that $\nabla_{x,y}u\in L^2(\Omega)$, and let $\mathscr{D}_0$ be the subspace of functions $u\in\mathscr{D}$ such that u vanishes near the top boundary Σ.

Lemma 1.1 (see Prop 2.2 in [1]). *There exist a positive weight $g\in L^\infty_{loc}(\Omega)$ equal to 1 near the top boundary Σ of Ω and $C>0$ such that for all $u\in\mathscr{D}_0$,*

$$\iint_\Omega g(x,y)|u(x,y)|^2dxdy\le C\iint_\Omega |\nabla_{x,y}u(x,y)|^2dxdy. \tag{1.11}$$

Using this lemma one can prove the following result.

Proposition 1.1 (see page 422 in [1]). *Denote by $H^{1,0}(\Omega)$ the space of functions u on Ω such that there exists a sequence $(u_n)\subset\mathscr{D}_0$ such that*

$$\nabla_{x,y}u_n\to\nabla_{x,y}u \quad \text{in } L^2(\Omega), \quad u_n\to u \quad \text{in } L^2(\Omega,gdxdy),$$

endowed with the scalar product

$$(u,v)_{H^{1,0}(\Omega)}=(\nabla_x u,\nabla_x v)_{L^2(\Omega)}+(\partial_y u,\partial_y v)_{L^2(\Omega)}.$$

Then $H^{1,0}(\Omega)$ is a Hilbert space and (1.11) holds for $u\in H^{1,0}(\Omega)$.

Let $\psi \in H^{\frac{1}{2}}(\mathbf{R}^d)$. One can construct (see below after (1.21)) $\underline{\psi} \in H^1(\Omega)$ such that

$$\operatorname{supp} \underline{\psi} \subset \{(x,y) : \eta(t,x) - h \leq y \leq \eta(x)\}, \quad \underline{\psi}|_\Sigma = \psi.$$

Using Proposition 1.1 we deduce that there exists a unique $u \in H^{1,0}(\Omega)$ such that, for all $\theta \in H^{1,0}(\Omega)$,

$$\iint_\Omega \nabla_{x,y} u(x,y) \cdot \nabla_{x,y}\theta(x,y)dxdy = -\iint_\Omega \nabla_{x,y}\underline{\psi}(x,y) \cdot \nabla_{x,y}\theta(x,y)dxdy.$$

Then to solve the problem (1.10) we set $\Phi = u + \underline{\psi}$.

Remark 1.2. As for the usual Neumann problem the meaning of the third condition in (1.10) is included in the definition of the space $H^{1,0}(\Omega)$. It can be written as in (1.10) if the bottom Γ is sufficiently smooth.

1.3.2 The Main Result

Let us assume that the Zakharov system (1.8) has been solved on $I = (0,T)$, which means that we have found, for $s > \frac{1}{2} + \frac{d}{2}$, a solution

$$(\eta, \psi) \in C^0(\bar{I}, H^{s+\frac{1}{2}}(\mathbf{R}^d) \times H^{s+\frac{1}{2}}(\mathbf{R}^d)),$$

of the system

$$\begin{cases} \partial_t \eta = G(\eta)\psi, \\ \partial \psi = -g\eta - \dfrac{1}{2}|\nabla\psi|^2 + \dfrac{1}{2}\dfrac{(\nabla\psi\cdot\nabla\eta + G(\eta)\psi)^2}{1+|\nabla\eta|^2}. \end{cases} \tag{1.12}$$

Let B, V be defined by (1.9). Then $(B,V) \in C^0(I, H^{s-\frac{1}{2}}(\mathbf{R}^d) \times H^{s-\frac{1}{2}}(\mathbf{R}^d))$.

The above variational theory shows that one can solve (for fixed t) the problem

$$\Delta_{x,y} Q = 0 \quad \text{in } \Omega, \quad Q|_\Sigma = g\eta + \frac{1}{2}(B^2 + |V|^2) \in H^{\frac{1}{2}}(\mathbf{R}^d). \tag{1.13}$$

Here is the main result of this article.

Theorem 1.3. *Let Φ and Q be the variational solutions of the problems* (1.10) *and* (1.13). *Set $P = Q - gy - \frac{1}{2}|\nabla_{x,y}\Phi|^2$. Then $v := \nabla_{x,y}\Phi$ satisfies the Euler system*

$$\partial_t v + (v \cdot \nabla_{x,y})v + \nabla_{x,y} P = -g e_y \quad \text{in } \Omega,$$

together with the conditions

$$\begin{cases} div_{x,y} v = 0, \quad curl_{x,y} v = 0 \quad in\ \Omega, \\ \partial_t \eta = (1+|\nabla\eta|^2)^{\frac{1}{2}} (v\cdot n) \quad on\ \Sigma, \\ P = 0 \quad on\ \Sigma. \end{cases} \tag{1.14}$$

The rest of the paper is devoted to the proof of this result. We proceed in several steps.

1.3.3 Straightening the Free Boundary

First of all if condition (H_t) is satisfied on I, for T small enough, one can find $\eta_* \in L^\infty(\mathbf{R}^d)$ independent of t such that

$$\begin{cases} (i) \quad \nabla_x \eta_* \in H^\infty(\mathbf{R}^d), \quad \|\nabla_x \eta_*\|_{L^\infty(\mathbf{R}^d)} \le C\|\eta\|_{L^\infty(I,H^{s+\frac{1}{2}}(\mathbf{R}^d))}, \\ (ii) \quad \eta(t,x) - h \le \eta_*(x) \le \eta(t,x) - \dfrac{h}{2}, \quad \forall (t,x) \in I \times \mathbf{R}^d, \\ (iii) \quad \Gamma \subset \{(x,y) \in \mathscr{O} : y < \eta_*(x)\}. \end{cases} \tag{1.15}$$

Indeed using the first equation in (1.12) we have

$$\begin{aligned} \|\eta(t,\cdot) - \eta_0\|_{L^\infty(\mathbf{R}^d)} &\le \int_0^t \|G(\eta)\psi(\sigma,\cdot)\|_{H^{s-\frac{1}{2}}(\mathbf{R}^d)} d\sigma \\ &\le TC\big(\|(\eta,\psi)\|_{L^\infty(I,H^{s+\frac{1}{2}}(\mathbf{R}^d)\times H^{s+\frac{1}{2}}(\mathbf{R}^d))}\big). \end{aligned}$$

Therefore taking T small enough we make $\|\eta(t,\cdot) - \eta_0\|_{L^\infty(\mathbf{R}^d)}$ as small as we want. Then we take $\eta_*(x) = -\frac{2h}{3} + e^{-\delta|D_x|}\eta_0$ and writing

$$\eta_*(x) = -\frac{2h}{3} + \eta(t,x) - (\eta(t,x) - \eta_0(x)) + (e^{-\delta|D_x|}\eta_0 - \eta_0(x)),$$

we obtain (1.15).

In what follows we shall set

$$\begin{cases} \Omega_1(t) = \{(x,y) : x \in \mathbf{R}^d, \eta_*(x) < y < \eta(t,x)\}, \\ \Omega_1 = \{(t,x,y) : t \in I, (x,y) \in \Omega_1(t)\}, \quad \Omega_2 = \{(x,y) \in \mathscr{O} : y \le \eta_*(x)\}, \\ \tilde{\Omega}_1 = \{(x,z) : x \in \mathbf{R}^d, z \in (-1,0)\}, \\ \tilde{\Omega}_2 = \{(x,z) \in \mathbf{R}^d \times (-\infty,-1] : (x, z+1+\eta_*(x)) \in \Omega_2\} \\ \tilde{\Omega} = \tilde{\Omega}_1 \cup \tilde{\Omega}_2 \end{cases} \tag{1.16}$$

Following Lannes ([4]), for $t \in I$ consider the map $(x,z) \mapsto (x, \rho(t,x,z))$ from $\tilde{\Omega}$ to $\mathbf{R}^{d+1}$ defined by

$$\begin{cases} \rho(t,x,z) = (1+z)e^{\delta z\langle D_x\rangle}\eta(t,x) - z\eta_*(x) \quad \text{if } (x,z) \in \tilde{\Omega}_1 \\ \rho(t,x,z) = z+1+\eta_*(x) \quad \text{if } (x,z) \in \tilde{\Omega}_2. \end{cases} \tag{1.17}$$

where δ is chosen such that

$$\delta \|\eta\|_{L^\infty(I,H^{s+\frac{1}{2}}(\mathbf{R}^d))} := \delta_0 << 1.$$

Notice that since $s > \frac{1}{2} + \frac{d}{2}$, taking δ small enough and using (1.15) (i) and, (ii), we obtain the estimates

$$\begin{aligned} &(i) \quad \partial_z \rho(t,x,z) \geq \min\left(\frac{h}{3}, 1\right) \quad \forall (t,x,z) \in I \times \tilde{\Omega}, \\ &(ii) \quad \|\nabla_{x,z}\rho\|_{L^\infty(I\times\tilde{\Omega})} \leq C(1 + \|\eta\|_{L^\infty(I,H^{s+\frac{1}{2}}(\mathbf{R}^d))}). \end{aligned} \tag{1.18}$$

It follows from (1.18) (i) that the map $(t,x,z) \mapsto (t,x,\rho(t,x,z))$ is a diffeomorphism from $I \times \tilde{\Omega}$ to Ω which is of class $W^{1,\infty}$.

We denote by κ the inverse map of ρ:

$$\begin{aligned} &(t,x,z) \in I \times \tilde{\Omega}, (t,x,\rho(t,x,z)) = (t,x,y) \\ &\qquad \Longleftrightarrow (t,x,z) = (t,x,\kappa(t,x,y)), (t,x,y) \in \Omega. \end{aligned} \tag{1.19}$$

1.3.4 The Dirichlet-Neumann Operator

Let Φ be the variational solution described above (with fixed t) of the problem

$$\begin{cases} \Delta_{x,y}\Phi = 0 \quad \text{in } \Omega(t), \\ \Phi|_{\Sigma(t)} = \psi(t,\cdot), \\ \partial_\nu \Phi|_\Gamma = 0. \end{cases} \tag{1.20}$$

Let us recall that

$$\Phi = u + \underline{\psi} \tag{1.21}$$

where $u \in H^{1,0}(\Omega(t))$ and $\underline{\psi}$ is an extension of ψ to $\Omega(t)$.

Here is a construction of $\underline{\psi}$. Let $\chi \in C^\infty(\mathbf{R}), \chi(a) = 0$ if $a \leq -1, \chi(a) = 1$ if $a \geq -\frac{1}{2}$. Let $\underline{\tilde{\psi}}(t,x,z) = \chi(z)e^{z\langle D_x\rangle}\psi(t,x)$ for $z \leq 0$. It is classical that $\underline{\tilde{\psi}} \in L^\infty(I, H^1(\tilde{\Omega}))$ if $\psi \in L^\infty(I, H^{\frac{1}{2}}(\mathbf{R}^d))$ and

$$\|\underline{\tilde{\psi}}\|_{L^\infty(I,H^1(\tilde{\Omega}))} \leq C\|\psi\|_{L^\infty(I,H^{\frac{1}{2}}(\mathbf{R}^d))}.$$

Then we set

$$\underline{\psi}(t,x,y) = \underline{\tilde{\psi}}\big(t,x,\kappa(t,x,y)\big). \tag{1.22}$$

Since $\eta \in C^0(I, W^{1,\infty}(\mathbf{R}^d))$ we have $\underline{\psi}(t,\cdot) \in H^1(\Omega(t))$, $\underline{\psi}|_{\Sigma(t)} = \psi$, and

$$\|\underline{\psi}(t,\cdot)\|_{H^1(\Omega(t))} \leq C\big(\|\eta\|_{L^\infty(I,W^{1,\infty}(\mathbf{R}^d))}\big)\|\psi\|_{L^\infty(I,H^{\frac{1}{2}}(\mathbf{R}^d))}.$$

Then we define the Dirichlet-Neumann operator by

$$\begin{aligned} G(\eta)\psi(t,x) &= \sqrt{1+|\nabla\eta|^2}\partial_n\Phi|_\Sigma \\ &= (\partial_y\Phi)(x,\eta(t,x)) - \nabla_x\eta(t,x)\cdot(\nabla_x\Phi)(t,x,\eta(t,x)). \end{aligned} \tag{1.23}$$

It has been shown in [2] (see Sect. 3) that $G(\eta)\psi$ is well defined in $C^0(\bar{I},H^{-\frac{1}{2}}(\mathbf{R}^d))$ if $\eta \in C^0(\bar{I},W^{1,\infty}(\mathbf{R}^d))$ and $\psi \in C^0(\bar{I},H^{\frac{1}{2}}(\mathbf{R}^d))$.

Remark 1.3. Recall that we have set

$$\Omega(t) = \{(x,y)\in\mathscr{O} : y<\eta(t,x)\}, \quad \Omega = \{(t,x,y) : t\in I, (x,y)\in\Omega(t)\}. \tag{1.24}$$

For a function $f \in L^1_{loc}(\Omega)$ if $\partial_t f$ denotes its derivative in the sense of distributions, we have

$$\langle\partial_t f,\varphi\rangle = \lim_{\varepsilon\to 0}\Big\langle\frac{f(\cdot+\varepsilon,\cdot,\cdot)-f(\cdot,\cdot,\cdot)}{\varepsilon},\varphi\Big\rangle, \quad \forall\varphi\in C_0^\infty(\Omega). \tag{1.25}$$

This point should be clarified due to the particular form of the set Ω since we have to show that if $(t,x,y)\in \operatorname{supp}\varphi = K$, then $(t+\varepsilon,x,y)\in\Omega$ for ε sufficiently small independently of the point (t,x,y). This is true. Indeed if $(t,x,y)\in K$, there exists a fixed $\delta>0$ (depending only on K,η) such that $y\le\eta(t,x)-\delta$. Since by (1.12)

$$|\eta(t+\varepsilon,x)-\eta(t,x)| \le \varepsilon\|G(\eta)\psi\|_{L^\infty(I\times\mathbf{R}^d)} \le \varepsilon C$$

where $C = C\big(\|(\eta,\psi)\|_{L^\infty(I,H^{s+\frac{1}{2}}(\mathbf{R}^d)\times H^{s+\frac{1}{2}}(\mathbf{R}^d))}\big)$, we have if $\varepsilon<\frac{\delta}{C}$,

$$y-\eta(t+\varepsilon,x) = y-\eta(t,x)+\eta(t,x)-\eta(t+\varepsilon,x) \le -\delta+\varepsilon C<0.$$

Notice that since $\eta\in C^0(\bar{I},H^{s+\frac{1}{2}}(\mathbf{R}^d)), \partial_t\eta = G(\eta)\psi \in C^0(\bar{I},H^{s-\frac{1}{2}}(\mathbf{R}^d))$, and $s>\frac{1}{2}+\frac{d}{2}$, we have $\rho\in W^{1,\infty}(I\times\tilde{\Omega})$.

The main step in the proof of Theorem 1.3 is the following.

Proposition 1.2. *Let Φ be defined by* (1.20) *and $Q\in H^{1,0}(\Omega(t))$ by* (1.13). *Then for all $t\in I$*

$$(i) \quad \partial_t\Phi(t,\cdot)\in H^{1,0}(\Omega(t)).$$
$$(ii) \quad \partial_t\Phi = -Q \quad in\ \mathscr{D}'(\Omega).$$

This result will be proved in §1.3.6

1.3.5 Preliminaries

If f is a function defined on Ω, we shall denote by $\tilde{f}$ its image by the diffeomorphism $(t,x,z)\mapsto(t,x,\rho(t,x,z))$. Thus we have

$$\tilde{f}(t,x,z) = f(t,x,\rho(t,x,z)) \Leftrightarrow f(t,x,y) = \tilde{f}\big(t,x,\kappa(t,x,y)\big). \tag{1.26}$$

Formally we have the following equalities for $(t,x,y) = (t,x,\rho(t,x,z)) \in \Omega$ and $\nabla = \nabla_x$

$$\begin{cases} \partial_y f(t,x,y) = \dfrac{1}{\partial_z \rho}\partial_z \tilde{f}(t,x,z) \Leftrightarrow \partial_z \tilde{f}(t,x,z) = \partial_z \rho(t,x,\kappa(t,x,y))\partial_y f(t,x,y), \\ \nabla f(t,x,y) = \big(\nabla \tilde{f} - \dfrac{\nabla \rho}{\partial_z \rho}\,\partial_z \tilde{f}\big)(t,x,z) \Leftrightarrow \nabla \tilde{f}(t,x,z) = \big(\nabla f + \nabla \rho\, \partial_y f\big)(t,x,y), \\ \partial_t f(t,x,y) = \big(\partial_t \tilde{f} + \partial_t \kappa(t,x,y)\partial_z \tilde{f}\big)(t,x,\kappa(t,x,y)). \end{cases} \tag{1.27}$$

We shall set in what follows

$$\Lambda_1 = \frac{1}{\partial_z \rho}\partial_z, \quad \Lambda_2 = \nabla_x - \frac{\nabla_x \rho}{\partial_z \rho}\partial_z \tag{1.28}$$

Eventually recall that if u is the function defined by (1.21), we have

$$\iint_{\Omega(t)} \nabla_{x,y} u(t,x,y) \cdot \nabla_{x,y}\theta(x,y)dxdy = -\iint_{\Omega(t)} \nabla_{x,y}\underline{\psi}(t,x,y) \cdot \nabla_{x,y}\theta(x,y)dxdy \tag{1.29}$$

for all $\theta \in H^{1,0}(\Omega(t))$ which implies that for $t \in I$,

$$\|\nabla_{x,y} u(t,\cdot)\|_{L^2(\Omega(t))} \leq C(\|\eta\|_{L^\infty(I,W^{1,\infty}(\mathbf{R}^d))})\|\psi\|_{L^\infty(I,H^{\frac{1}{2}}(\mathbf{R}^d))}. \tag{1.30}$$

Let u be defined by (1.21). Since $(\eta,\psi) \in C^0(\overline{I}, H^{s+\frac{1}{2}}(\mathbf{R}^d) \times H^{s+\frac{1}{2}}(\mathbf{R}^d))$, the elliptic regularity theorem proved in [2] (see Theorem 3.16) shows that

$$\partial_z \tilde{u}, \nabla_x \tilde{u} \in C^0_z([-1,0], H^{s-\frac{1}{2}}(\mathbf{R}^d)) \subset C^0([-1,0] \times \mathbf{R}^d),$$

since $s - \frac{1}{2} > \frac{d}{2}$.

It follows from (1.27) that $\partial_y u$ and $\nabla_x u$ have a trace on Σ and

$$\partial_y u|_\Sigma = \frac{1}{\partial_z \rho(t,x,0)}\partial_z \tilde{u}(t,x,0), \quad \nabla_x u|_\Sigma = \big(\nabla_x \tilde{u} - \frac{\nabla_x \eta}{\partial_z \rho(t,x,0)}\partial_z \tilde{u}\big)(t,x,0).$$

Since $\tilde{u}(t,x,0) = 0$, it follows that

$$\nabla_x u|_\Sigma + (\nabla_x \eta)\partial_y u|_\Sigma = 0$$

from which we deduce, since $\Phi = u + \underline{\psi}$,

$$\nabla_x \Phi|_\Sigma + (\nabla_x \eta)\partial_y \Phi|_\Sigma = \nabla_x \psi. \tag{1.31}$$

On the other hand one has

$$G(\eta)\psi = \big(\partial_y \Phi - \nabla_x \eta \cdot \nabla_x \Phi\big)|_\Sigma. \tag{1.32}$$

It follows from (1.31), (1.32), and (1.9) that we have

$$\nabla_x \Phi|_\Sigma = V, \quad \partial_y \Phi|_\Sigma = B. \tag{1.33}$$

According to (1.13), $P = Q - gy - \frac{1}{2}|\nabla_{x,y}\Phi|^2$ has a trace on Σ and $P|_\Sigma = 0$.

1.3.6 The Regularity Results

The main steps in the proof of Proposition 1.2 are the following.

Lemma 1.2. *Let $\tilde{u}$ be defined by* (1.26) *and κ by* (1.19)*. Then for all $t_0 \in I$ the function $(x,y) \mapsto U(t_0,x,y) := \partial_t \tilde{u}(t_0,x,\kappa(t_0,x,y))$ belongs to $H^{1,0}(\Omega(t_0))$. Moreover there exists a function $\mathscr{F} : \mathbf{R}^+ \to \mathbf{R}^+$ such that*

$$\sup_{t\in I} \iint_{\Omega(t)} |\nabla_{x,y} U(t,x,y)|^2 dxdy \leq \mathscr{F}(\|(\eta,\psi)\|_{L^\infty(I,H^{s+\frac{1}{2}}(\mathbf{R}^d)\times H^{s+\frac{1}{2}}(\mathbf{R}^d))}).$$

Lemma 1.3. *In the sense of distributions on Ω we have the chain rule*

$$\partial_t u(t,x,y) = \partial_t \tilde{u}(t,x,\kappa(t,x,y)) + \partial_t \kappa(t,x,y)\partial_z \tilde{u}(t,x,\kappa(t,x,y)).$$

These lemmas are proved in the next paragraph.

Proof (of Proposition 1.2). According to (1.21) and Lemma 1.3 we have

$$\partial_t \Phi(t,x,y) = \partial_t \tilde{u}(t,x,\kappa(t,x,y)) + \underline{w}(t,x,y) \tag{1.34}$$

where

$$\underline{w}(t,x,y) = \partial_t \kappa(t,x,y)\partial_z \tilde{u}(t,x,\kappa(t,x,y)) + \partial_t \underline{\psi}(t,x,y).$$

According to Lemma 1.2 the first term in the right-hand side of (1.34) belongs to $H^{1,0}(\Omega(t))$. Denoting by $\underline{\tilde{w}}$ the image of $\underline{w}$, if we show that

$$\begin{cases} (i) \quad \underline{\tilde{w}} \in H^1(\mathbf{R}^d \times \mathbf{R}), \\ (ii) \quad \operatorname{supp} \underline{\tilde{w}} \subset \{(x,z) \in \mathbf{R}^d \times (-1,0)\} \\ (iii) \quad w|_\Sigma = -g\eta - \dfrac{1}{2}(B^2 + |V|^2) \end{cases} \tag{1.35}$$

then $\partial_t \Phi$ will be the variational solution of the problem

$$\Delta_{x,y}(\partial_t \Phi) = 0, \quad \partial_t \Phi|_\Sigma = -g\eta - \frac{1}{2}(B^2 + |V|^2).$$

By uniqueness, we deduce from (1.13) that $\partial_t \Phi = -Q$, which completes the proof of Proposition 1.2. Therefore we are left with the proof of (1.35).

Recall that $\underline{\tilde{\psi}}(t,x,z) = \chi(z) e^{z\langle D_x \rangle} \psi(t,x)$. Moreover by Lemma 1.3 we have

$$\widetilde{\partial_t \underline{\psi}} = \Big(\partial_t \underline{\tilde{\psi}} - \frac{\partial_t \rho(t,x,z)}{\partial_z \rho(t,x,z)} \partial_z \underline{\tilde{\psi}}\Big)(t,x,z).$$

Since $\psi \in H^{s+\frac{1}{2}}(\mathbf{R}^d), \partial_t \psi \in H^{s-\frac{1}{2}}(\mathbf{R}^d), \partial_t \eta \in H^{s-\frac{1}{2}}(\mathbf{R}^d)$, the classical properties of the Poisson kernel show that $\partial_t \underline{\tilde{\psi}}$ and $\partial_z \underline{\tilde{\psi}}$ and $\frac{\partial_t \rho}{\partial_z \rho}$ belong to $H^s(\mathbf{R}^d \times (-1,0))$ therefore to $H^1(\mathbf{R}^d \times (-1,0))$ since $s > \frac{1}{2} + \frac{d}{2}$. It follows that the points (i) and (ii) in (1.35) are satisfied by $\widetilde{\partial_t \underline{\psi}}$. Now according to (1.17) $\widetilde{\partial_t \kappa}$ is supported in $\mathbf{R}^d \times (-1,0)$, and it follows from the elliptic regularity that $\widetilde{\partial_t \kappa} \partial_z \tilde{u}$ belongs to $H^1(\mathbf{R}^d \times (-1,0))$. Let us check now point (iii). Since $\partial_t \eta = G(\eta)\psi$ we have

$$\partial_t \underline{\psi}(t,x,y)|_\Sigma = \widetilde{\partial_t \underline{\psi}}|_{z=0} = \partial_t \psi - G(\eta)\psi \cdot \partial_y \underline{\psi}(t,x,y)|_\Sigma. \tag{1.36}$$

On the other hand we have

$$\begin{aligned}
\partial_t \kappa(t,x,y)\partial_z \tilde{u}(t,x,\kappa(t,x,y))|_\Sigma &= \partial_t \kappa(t,x,y)\partial_z \rho(t,x,\kappa(t,x,y))\partial_y u(t,x,y)|_\Sigma \\
&= -\partial_t \rho(t,x,\kappa(t,x,y))\partial_y u(t,x,y)|_\Sigma \\
&= -\partial_t \rho(t,x,\kappa(t,x,y))\big(\partial_y \Phi(t,x,y) - \partial_y \underline{\psi}(t,x,y)\big)|_\Sigma \\
&= -G(\eta)\psi \cdot (B - \partial_y \underline{\psi}(t,x,y)|_\Sigma.
\end{aligned}$$

So using (1.36) we find

$$\underline{w}|_\Sigma = \partial_t \psi - BG(\eta)\psi.$$

It follows from the second equation of (1.12) and from (1.9) that

$$\underline{w}|_\Sigma = -g\eta - \frac{1}{2}(B^2 + |V|^2).$$

This proves the claim (iii) in (1.35) and ends the proof of Proposition 1.2.

1.3.7 Proof of the Lemmas

1.3.7.1 Proof of Lemma 1.2

Recall (see (1.17) and (1.28)) that we have set

$$\begin{cases}
\rho(t,x,z) = (1+z)e^{\delta z \langle D_x \rangle}\eta(t,x) - z\eta_*(x) \quad \text{if } (x,z) \in \tilde{\Omega}_1, \\
\rho(t,x,z) = z + 1 + \eta_*(x) \quad \text{if } (x,z) \in \tilde{\Omega}_2, \\
\Lambda_1(t) = \dfrac{1}{\partial_z \rho(t,\cdot)}\partial_z, \quad \Lambda_2(t) = \nabla_x - \dfrac{\nabla_x \rho(t,\cdot)}{\partial_z \rho(t,\cdot)}\partial_z
\end{cases}$$

and that κ_t has been defined in (1.19).

If we set $\hat{\kappa}_t(x,y) = (x,\kappa(t,x,y))$, then $\hat{\kappa}_t$ is a bijective map from the space $H^{1,0}(\tilde{\Omega})$ (defined as in Proposition 1.1) to the space $H^{1,0}(\Omega(t))$. Indeed near the top boundary ($z \in (-2,0)$), this follows from the classical invariance of the usual space H_0^1 by a $W^{1,\infty}$-diffeomorphism, while, near the bottom, our diffeomorphim is of class H^∞ hence preserves the space $H^{1,0}$.

Now we fix $t_0 \in I$, we take $\varepsilon \in \mathbf{R} \setminus \{0\}$ small enough, and we set for $t \in I$

$$\begin{cases} F(t) = \iint_{\Omega(t)} \nabla_{x,y} u(t,x,y) \cdot \nabla_{x,y} \theta(x,y)\,dxdy \\ H(t) = -\iint_{\Omega(t)} \nabla_{x,y} \underline{\psi}(t,x,y) \cdot \nabla_{x,y} \theta(x,y)\,dxdy \end{cases} \tag{1.37}$$

where $\theta \in H^{1,0}(\Omega(t))$ is chosen as follows.

In $F(t_0+\varepsilon)$ we take

$$\theta_1(x,y) = \frac{u(t_0+\varepsilon,x,y) - \tilde{u}(t_0,x,\kappa(t_0+\varepsilon,x,y))}{\varepsilon} \in H^{1,0}(\Omega(t_0+\varepsilon)).$$

In $F(t_0)$ we take

$$\theta_2(x,y) = \frac{\tilde{u}(t_0+\varepsilon,x,\kappa(t_0,x,y)) - u(t_0,x,y)}{\varepsilon} \in H^{1,0}(\Omega(t_0)).$$

Then in the variables (x,z) we have

$$\begin{aligned} \tilde{\theta}_1(x,z) &= \theta_1(x,\rho(t_0+\varepsilon,x,z)) = \frac{\tilde{u}(t_0+\varepsilon,x,z) - \tilde{u}(t_0,x,z)}{\varepsilon} \\ \tilde{\theta}_2(x,z) &= \theta_2(x,\rho(t_0,x,z)) = \frac{\tilde{u}(t_0+\varepsilon,x,z) - \tilde{u}(t_0,x,z)}{\varepsilon}, \end{aligned} \tag{1.38}$$

so we see that $\tilde{\theta}_1(x,z) = \tilde{\theta}_2(x,z) =: \tilde{\theta}(x,z)$.

It follows from (1.29) that for all $t \in I$, we have $F(t) = H(t)$. Therefore

$$J_\varepsilon(t_0) =: \frac{F(t_0+\varepsilon) - F(t_0)}{\varepsilon} = \mathscr{J}_\varepsilon(t_0) =: \frac{H(t_0+\varepsilon) - H(t_0)}{\varepsilon}.$$

Then after changing variables as in (1.17) we obtain

$$\begin{aligned} J_\varepsilon(t_0) = \frac{1}{\varepsilon} \sum_{j=1}^{2} \iint_{\tilde{\Omega}} \big[&\Lambda_j(t_0+\varepsilon)\tilde{u}(t_0+\varepsilon,x,z)\Lambda_j(t_0+\varepsilon)\tilde{\theta}(x,z)\partial_z\rho(t_0+\varepsilon,x,z) \\ &-\Lambda_j(t_0)\tilde{u}(t_0,x,z)\Lambda_j(t_0)\tilde{\theta}(x,z)\partial_z\rho(t_0,x,z)\big]\,dxdz =: \sum_{j=1}^{2} K_{j,\varepsilon}(t_0). \end{aligned} \tag{1.39}$$

With the notation used in (1.3.7.1) we can write

$$\Lambda_j(t_0+\varepsilon) - \Lambda_j(t_0) = \beta_{j,\varepsilon}(t_0,x,z)\partial_z, \quad j = 1,2. \tag{1.40}$$

Notice that since the function ρ does not depend on t for $z \le -1$ we have $\beta_{j,\varepsilon} = 0$ in this set.

Then we have the following Lemma.

Lemma 1.4. *There exists a nondecreasing function $\mathscr{F} : \mathbf{R}^+ \to \mathbf{R}^+$ such that*

$$\sup_{t_0 \in I} \iint_{\tilde{\Omega}} |\beta_{j,\varepsilon}(t_0,x,z)|^2 dxdz \le \varepsilon^2 \mathscr{F}\left(\|(\eta,\psi)\|_{L^\infty(I,H^{s+\frac{1}{2}}(\mathbf{R}^d)\times H^{s+\frac{1}{2}}(\mathbf{R}^d))}\right).$$

Proof. In the set $\{(x,z) : x \in \mathbf{R}^d, z \in (-1,0)\}$ the most delicate term to deal with is

$$(1) =: \frac{\nabla_x \rho}{\partial_z \rho}(t_0+\varepsilon,x,z) - \frac{\nabla_x \rho}{\partial_z \rho}(t_0,x,z) = \varepsilon \int_0^1 \partial_t \Big(\frac{\nabla_x \rho}{\partial_z \rho}\Big)(t_0+\varepsilon\lambda,x,z)d\lambda.$$

We have

$$\partial_t \Big(\frac{\nabla_x \rho}{\partial_z \rho}\Big) = \frac{\nabla_x \partial_t \rho}{\partial_z \rho} - \frac{(\partial_z \partial_t \rho)\nabla_x \rho}{(\partial_z \rho)^2}.$$

First of all we have $\partial_z \rho \ge \frac{h}{3}$. Now since $s - \frac{1}{2} > \frac{d}{2} \ge \frac{1}{2}$, we can write

$$\begin{aligned}
\|\nabla_x \partial_t \rho(t,\cdot)\|_{L^2(\tilde{\Omega}_1)} &\le 2\|e^{\delta z|D_x|} G(\eta)\psi(t,\cdot)\|_{L^2((-1,0),H^1(\mathbf{R}^d))} \\
&\le C\|G(\eta)\psi(t,\cdot)\|_{H^{\frac{1}{2}}(\mathbf{R}^d)} \le C\|G(\eta)\psi(t,\cdot)\|_{H^{s-\frac{1}{2}}(\mathbf{R}^d)} \\
&\le C\left(\|(\eta,\psi)\|_{L^\infty(I,H^{s+\frac{1}{2}}(\mathbf{R}^d)\times H^{s+\frac{1}{2}}(\mathbf{R}^d))}\right).
\end{aligned} \tag{1.41}$$

On the other hand we have

$$\begin{aligned}
\|\nabla_x \rho(t,\cdot)\|_{L^\infty(\tilde{\Omega}_1)} &\le C\|e^{\delta z|D_x|}\nabla_x \eta(x,\cdot)\|_{L^\infty((-1,0),H^{s-\frac{1}{2}}(\mathbf{R}^d))} + \|\nabla_x \eta_*\|_{L^\infty(\mathbf{R}^d)} \\
&\le C'\|\eta(t,\cdot)\|_{H^{s+\frac{1}{2}}(\mathbf{R}^d)} + \|\nabla_x \eta_*\|_{L^\infty(\mathbf{R}^d)} \le C"\|\eta(t,\cdot)\|_{H^{s+\frac{1}{2}}(\mathbf{R}^d)}
\end{aligned}$$

by (1.15). Eventually since

$$\partial_z \partial_t \rho = e^{\delta z|D_x|} G(\eta)\psi + (1+z)\delta e^{\delta z|D_x|}|D_x| G(\eta)\psi,$$

we have as in (1.41)

$$\|\partial_z \partial_t \rho(t,\cdot)\|_{L^2(\tilde{\Omega})} \le C\left(\|(\eta,\psi)\|_{L^\infty(I,H^{s+\frac{1}{2}}(\mathbf{R}^d)\times H^{s+\frac{1}{2}}(\mathbf{R}^d))}\right). \tag{1.42}$$

Then the Lemma follows.

Thus we can write for $j = 1,2$,

$$
\begin{aligned}
K_{j,\varepsilon}(t_0) &= \sum_{k=1}^{4} \iint_{\tilde{\Omega}_1} A^k_{j,\varepsilon}(t_0,x,z)\,dxdz,\\
A^1_{j,\varepsilon}(t_0,\cdot) &= \Lambda_j(t_0)\Big[\frac{\tilde{u}(t_0+\varepsilon,\cdot)-\tilde{u}(t_0,\cdot)}{\varepsilon}\Big]\Lambda_j(t_0)\tilde{\theta}(\cdot)\partial_z\rho(t_0,\cdot),\\
A^2_{j,\varepsilon}(t_0,\cdot) &= \Big[\frac{\Lambda_j(t_0+\varepsilon)-\Lambda_j(t_0)}{\varepsilon}\Big]\tilde{u}(t_0,\cdot)\Lambda_j(t_0)\tilde{\theta}(\cdot)\partial_z\rho(t_0,\cdot),\\
A^3_{j,\varepsilon}(t_0,\cdot) &= \Lambda_j(t_0+\varepsilon)\tilde{u}(t_0+\varepsilon,\cdot)\Big[\frac{\Lambda_j(t_0+\varepsilon)-\Lambda_j(t_0)}{\varepsilon}\Big]\tilde{\theta}(\cdot)\partial_z\rho(t_0,\cdot),\\
A^4_{j,\varepsilon}(t_0,\cdot) &= \Lambda_j(t_0+\varepsilon)\tilde{u}(t_0+\varepsilon,\cdot)\Lambda_j(t_0+\varepsilon)\tilde{\theta}(\cdot)\Big[\frac{\partial_z\rho(t_0+\varepsilon,\cdot)-\partial_z\rho(t_0,\cdot)}{\varepsilon}\Big].
\end{aligned}
\tag{1.43}
$$

In what follows to simplify the notations we shall set $X=(x,z)\in\tilde{\Omega}$ and we recall that $\Lambda_j(t_0+\varepsilon)-\Lambda_j(t_0)=0$ when $z\le -1$.

First of all, using the lower bound $\partial_z\rho(t_0,X)\ge\frac{h}{3}$, we obtain

$$
\iint_{\tilde{\Omega}} A^1_{j,\varepsilon}(t_0,X)\,dX \ge \frac{h}{3}\left\|\Lambda_j(t_0)\Big[\frac{\tilde{u}(t_0+\varepsilon,\cdot)-\tilde{u}(t_0,\cdot)}{\varepsilon}\Big]\right\|^2_{L^2(\tilde{\Omega})}. \tag{1.44}
$$

Now it follows from (1.40) that

$$
\left|\iint_{\tilde{\Omega}} A^2_{j,\varepsilon}(t_0,X)\,dX\right| \le \sup_{t\in I}\Big\|\frac{\beta_{j,\varepsilon}}{\varepsilon}\Big\|_{L^2(\tilde{\Omega})}\sup_{t\in I}\|\partial_z\tilde{u}(t,\cdot)\|_{L^\infty_z(-1,0,L^\infty(\mathbf{R}^d))}\|\Lambda_j(t_0)\tilde{\theta}\|_{L^2(\tilde{\Omega})}.
$$

Since $s-\frac{1}{2}>\frac{d}{2}$ the elliptic regularity theorem shows that

$$
\begin{aligned}
\sup_{t\in I}\|\partial_z\tilde{u}(t,\cdot)\|_{L^\infty_z(-1,0,L^\infty(\mathbf{R}^d))} &\le \sup_{t\in I}\|\partial_z\tilde{u}(t,\cdot)\|_{L^\infty_z(-1,0,H^{s-\frac{1}{2}}(\mathbf{R}^d))}\\
&\le C\big(\|(\eta,\psi)\|_{L^\infty(I,H^{s+\frac{1}{2}}(\mathbf{R}^d)\times H^{s+\frac{1}{2}}(\mathbf{R}^d))}\big)
\end{aligned}
\tag{1.45}
$$

Using Lemma 1.4 we deduce that

$$
\left|\iint_{\tilde{\Omega}} A^2_{j,\varepsilon}(t_0,X)\,dX\right| \le C\big(\|(\eta,\psi)\|_{L^\infty(I,H^{s+\frac{1}{2}}(\mathbf{R}^d)\times H^{s+\frac{1}{2}}(\mathbf{R}^d))}\big)\|\Lambda_j(t_0)\tilde{\theta}\|_{L^2(\tilde{\Omega})}. \tag{1.46}
$$

Now write

$$
\begin{aligned}
&\iint_{\tilde{\Omega}} A^3_{j,\varepsilon}(t_0,X)\,dX =\\
&\qquad = \iint_{\tilde{\Omega}} \Lambda_j(t_0+\varepsilon)\tilde{u}(t_0+\varepsilon,X)\beta_\varepsilon(t_0+\varepsilon,X)\partial_z\tilde{\theta}(t_0,X)\partial_z\rho(t_0,X)\,dX.
\end{aligned}
$$

By elliptic regularity, $\Lambda_j(t)\tilde{u}$ is bounded in $L^\infty_{t,x,z}$ by a function depending only on $\|(\eta,\psi)\|_{L^\infty(I,H^{s+\frac{1}{2}}(\mathbf{R}^d)\times H^{s+\frac{1}{2}}(\mathbf{R}^d))}$. Therefore we can write

$$\left|\iint_{\tilde{\Omega}} A^3_{j,\varepsilon}(t_0,X)dX\right| \le C(\|(\eta,\psi)\|_{L^\infty(I,H^{s+\frac{1}{2}}(\mathbf{R}^d)\times H^{s+\frac{1}{2}}(\mathbf{R}^d))})\|\partial_z\tilde{\theta}\|_{L^2(\tilde{\Omega})}. \quad (1.47)$$

Since

$$\frac{\partial_z\rho(t_0+\varepsilon,x,z)-\partial_z\rho(t_0,x,z)}{\varepsilon} = \int_0^1 \partial_t\partial_z\rho(t_0+\lambda\varepsilon,x,z)d\lambda$$

(which vanishes when $z\le -1$), we find using (1.45) and (1.42)

$$\left|\iint_{\tilde{\Omega}} A^4_{j,\varepsilon}(t_0,X)dX\right| \le C(\|(\eta,\psi)\|_{L^\infty(I,H^{s+\frac{1}{2}}(\mathbf{R}^d)\times H^{s+\frac{1}{2}}(\mathbf{R}^d))})\|\partial_z\tilde{\theta}\|_{L^2(\tilde{\Omega})}. \quad (1.48)$$

Now we consider

$$\mathscr{J}_\varepsilon = \frac{H(t_0+\varepsilon)-H(t_0)}{\varepsilon}. \quad (1.49)$$

We make the change of variable $(x,z)\to(x,\rho(t_0,x,z))$ in the integral, and we decompose the new integral as in (1.39), (1.43). This gives, with $X=(x,z)$,

$$\mathscr{J}_\varepsilon = \sum_{j=1}^{2}\mathscr{K}_{j,\varepsilon}(t_0), \quad \mathscr{K}_{j,\varepsilon}(t_0) = \sum_{k=1}^{4}\iint_{\tilde{\Omega}}\mathscr{A}^k_{j,\varepsilon}(t_0,X)dX,$$

where $\mathscr{A}^k_{j,\varepsilon}$ has the same form as $-A^k_{j,\varepsilon}$ in (1.43) except the fact that $\tilde{u}$ is replaced by $\underline{\tilde{\psi}}$. Recall that $\underline{\tilde{\psi}}(t,x,z)=\chi(z)e^{z|D_x|}\psi(t,x)$. Now we have

$$\begin{aligned}
\|\Lambda_j\partial_t\underline{\tilde{\psi}}\|_{L^\infty(I,L^2(\tilde{\Omega}))} &\le \mathscr{F}(\|\eta\|_{L^\infty(I,H^{s+\frac{1}{2}}(\mathbf{R}^d))})\|\partial_t\underline{\tilde{\psi}}\|_{L^\infty(I,L^2_z((-1,0),H^1(\mathbf{R}^d)))}\\
&\le \mathscr{F}(\|\eta\|_{L^\infty(I,H^{s+\frac{1}{2}}(\mathbf{R}^d))})\|\partial_t\psi\|_{L^\infty(I,H^{\frac{1}{2}}(\mathbf{R}^d))}\\
&\le \mathscr{F}(\|\eta\|_{L^\infty(I,H^{s+\frac{1}{2}}(\mathbf{R}^d))})\|\partial_t\psi\|_{L^\infty(I,H^{s-\frac{1}{2}}(\mathbf{R}^d))}
\end{aligned}$$

since $s-\frac{1}{2}\ge\frac{1}{2}$. Using the (1.12) on ψ and the fact that $H^{s-\frac{1}{2}}(\mathbf{R}^d)$ is an algebra, we obtain

$$\|\Lambda_j\partial_t\underline{\tilde{\psi}}\|_{L^\infty(I,L^2(\tilde{\Omega}))} \le \mathscr{F}(\|(\eta,\psi)\|_{L^\infty(I,H^{s+\frac{1}{2}}(\mathbf{R}^d))\times H^{s+\frac{1}{2}}(\mathbf{R}^d))}).$$

It follows that we have

$$\left|\iint_{\tilde{\Omega}}\mathscr{A}^1_{j,\varepsilon}(t_0,X)dX\right| \le \mathscr{F}(\|(\eta,\psi)\|_{L^\infty(I,H^{s+\frac{1}{2}}(\mathbf{R}^d)\times H^{s+\frac{1}{2}}(\mathbf{R}^d))})\|\Lambda_j(t_0)\tilde{\theta}\|_{L^2(\tilde{\Omega})}. \quad (1.50)$$

Now since

$$\begin{aligned}
\|\Lambda_j(t_0)\underline{\tilde{\psi}}(t,\cdot)\|_{L^2_z((-1,0),L^\infty(\mathbf{R}^d))} &\le \mathscr{F}(\|\eta\|_{L^\infty(I,H^{s+\frac{1}{2}}(\mathbf{R}^d))})\|\underline{\tilde{\psi}}(t_0,\cdot)\|_{L^2_z((-1,0),H^{\frac{d}{2}+\varepsilon}(\mathbf{R}^d))}\\
&\le \mathscr{F}(\|\eta\|_{L^\infty(I,H^{s+\frac{1}{2}}(\mathbf{R}^d))})\|\psi\|_{L^\infty(I,H^{s+\frac{1}{2}}(\mathbf{R}^d))}
\end{aligned}$$

we can use the same estimates as in (1.46), (1.47), and (1.48) to bound the terms $\mathscr{A}^k_{j,\varepsilon}$ for $k=2,3,4$. We obtain finally

$$\left|\frac{H(t_0+\varepsilon)-H(t_0)}{\varepsilon}\right| \le C\big(\|(\eta,\psi)\|_{L^\infty(I,H^{s+\frac{1}{2}}(\mathbf{R}^d)\times H^{s+\frac{1}{2}}(\mathbf{R}^d))}\big)\sum_{j=1}^{2}\|\Lambda_j(t_0)\tilde{\theta}\|_{L^2(\tilde{\Omega})}. \tag{1.51}$$

Summing up using (1.43), (1.44), (1.46),(1.47),(1.48), and (1.51) we find that, setting

$$\tilde{U}_\varepsilon(t_0,\cdot)=\frac{\tilde{u}(t_0+\varepsilon,\cdot)-\tilde{u}(t_0,\cdot)}{\varepsilon},$$

there exists a nondecreasing function $\mathscr{F}:\mathbf{R}^+\to\mathbf{R}^+$ such that for all $\varepsilon>0$

$$\sum_{j=1}^{2}\sup_{t_0\in I}\|\Lambda_j(t_0)\tilde{U}_\varepsilon(t_0,\cdot)\|_{L^2(\tilde{\Omega})}\le\mathscr{F}\big(\|(\eta,\psi)\|_{L^\infty(I,H^{s+\frac{1}{2}}(\mathbf{R}^d)\times H^{s+\frac{1}{2}}(\mathbf{R}^d))}\big).$$

Since $\tilde{u}(t,\cdot)\in H^{1,0}(\tilde{\Omega})$, the Poincaré inequality ensures that

$$\|\tilde{U}_\varepsilon(t_0,\cdot)\|_{L^2(\tilde{\Omega})}\le C\big(\|(\eta,\psi)\|_{L^\infty(I,H^{s+\frac{1}{2}}(\mathbf{R}^d)\times H^{s+\frac{1}{2}}(\mathbf{R}^d))}\big). \tag{1.52}$$

It follows that we can extract a subsequence $(\tilde{U}_{\varepsilon_k})$ which converges in the weak-star topology of $(L^\infty\cap C^0)(I,H^{1,0}(\tilde{\Omega}))$. But this sequences converge in $\mathscr{D}'(I\times\tilde{\Omega})$ to $\partial_t\tilde{u}$. Therefore $\partial_t\tilde{u}\in C^0(I,H^{1,0}(\tilde{\Omega}))$, and this implies that $\partial_t\tilde{u}(t_0,\cdot,\kappa(t_0,\cdot,\cdot))$ belongs to $H^{1,0}(\Omega(t_0)$ which completes the proof of Lemma 1.2.

1.3.7.2 Proof of Lemma 1.3

Let $\varphi\in C_0^\infty(\Omega)$ and set

$$\begin{aligned}
v_\varepsilon(t,x,y)&=\frac{1}{\varepsilon}[\tilde{u}(t+\varepsilon,x,\kappa(t+\varepsilon,x,y))-\tilde{u}(t+\varepsilon,x,\kappa(t,x,y))],\\
w_\varepsilon(t,x,y)&=\frac{1}{\varepsilon}[\tilde{u}(t+\varepsilon,x,\kappa(t,x,y))-\tilde{u}(t,x,\kappa(t,x,y))],\\
J_\varepsilon&=\iint_\Omega v_\varepsilon(t,x,y)\varphi(t,x,y)dtdxdy,\quad K_\varepsilon=\iint_\Omega w_\varepsilon(t,x,y)\varphi(t,x,y)dtdxdy,\\
I_\varepsilon&=J_\varepsilon+K_\varepsilon.
\end{aligned} \tag{1.53}$$

Let us consider first K_ε. In the integral in y, we make the change of variable $\kappa(t,x,y)=z\Leftrightarrow y=\rho(t,x,z)$. Then setting $\tilde{\varphi}(t,x,z)=\varphi(t,x,\rho(t,x,z))$ and $X=(x,z)\in\tilde{\Omega}$, we obtain

$$K_\varepsilon=\iint_I\int_{\tilde{\Omega}}\frac{\tilde{u}(t+\varepsilon,X)-\tilde{u}(t,X)}{\varepsilon}\tilde{\varphi}(t,X)\partial_z\rho(t,X)dtdX.$$

Since $\rho \in C^1(I \times \tilde{\Omega})$ we have $\tilde{\varphi} \cdot \partial_z \rho \in C_0^0(I \times \tilde{\Omega})$. Now we know that the sequence $\tilde{U}_\varepsilon = \frac{\tilde{u}(\cdot+\varepsilon,\cdot)-\tilde{u}(\cdot,\cdot)}{\varepsilon}$ converges in $\mathscr{D}'(I \times \tilde{\Omega})$ to $\partial_t \tilde{u}$. We use this fact, we approximate $\tilde{\varphi} \cdot \partial_z \rho$ by a sequence in $C_0^\infty(I \times \tilde{\Omega})$, and we use (1.52) to deduce that

$$\lim_{\varepsilon \to 0} K_\varepsilon = \int_I \iint_{\tilde{\Omega}} \partial_t \tilde{u}(t,X) \tilde{\varphi}(t,X) \partial_z \rho(t,X) dt dX.$$

Coming back to the (t,x,y) variables we obtain

$$\lim_{\varepsilon \to 0} K_\varepsilon = \iint_\Omega \partial_t \tilde{u}(t,x,\kappa(t,x,y)) \varphi(t,x,y) dt dx dy. \tag{1.54}$$

Let us look now to J_ε. We cut it into two integrals; in the first we set $\kappa(t+\varepsilon,x,y) = z$, and in the second we set $\kappa(t,x,y) = z$. With $X = (x,z) \in \tilde{\Omega}$ we obtain

$$J_\varepsilon = \frac{1}{\varepsilon} \int_I \iint_{\tilde{\Omega}_1} \tilde{u}(t+\varepsilon,X) \Big(\int_0^1 \frac{d}{d\sigma} \{\varphi(t,x,\rho(t+\varepsilon\sigma,X)) \partial_z \rho(t+\varepsilon\sigma,X)\} d\sigma \Big) dt dX.$$

Differentiating with respect to σ we see easily that

$$J_\varepsilon = \int_I \iint_{\tilde{\Omega}} \tilde{u}(t+\varepsilon,X) \frac{\partial}{\partial z} \Big(\int_0^1 \partial_t \rho(t+\varepsilon\sigma,X) \varphi(t,x,\rho(t+\varepsilon\sigma,X)) d\sigma \Big) dt dX.$$

Since $\tilde{u}$ is continuous in t with values in $L^2(\tilde{\Omega}), \partial_t \rho$ is continuous in (t,x,z) and $\varphi \in C_0^\infty$; we can pass to the limit, and we obtain

$$\lim_{\varepsilon \to 0} J_\varepsilon = \int_I \iint_{\tilde{\Omega}_1} \tilde{u}(t,X) \frac{\partial}{\partial z} \Big(\partial_t \rho(t,X) \varphi(t,x,\rho(t,X)) \Big) dt dX.$$

Now we can integrate by parts. Since, thanks to φ, we have compact support in z we obtain

$$\lim_{\varepsilon \to 0} J_\varepsilon = -\int_I \iint_{\tilde{\Omega}} \partial_z \tilde{u}(t,X) \partial_t \rho(t,X) \varphi(t,x,\rho(t,X)) dt dX.$$

Now since

$$\partial_t \rho(t,X) = -\partial_t \kappa(t,x,y) \partial_z \rho(t,x,z)$$

setting in the integral in z, $\rho(t,X) = y$ we obtain

$$\lim_{\varepsilon \to 0} J_\varepsilon = \iint_\Omega \partial_z \tilde{u}(t,x,\kappa(t,x,y)) \partial_t \kappa(t,x,y) \varphi(t,x,y) dt dx dy. \tag{1.55}$$

Then Lemma 1.3 follows from (1.53)–(1.55).

Acknowledgements Support by the French Agence Nationale de la Recherche, project EDP Dispersives, référence ANR-07-BLAN-0250, is acknowledged.

References

1. Alazard, T., Burq, N., Zuily, C.: On the water-wave equations with surface tension. Duke Math. J. **158**(3), 413–499 (2011)
2. Alazard, T., Burq, N., Zuily, C.: On the Cauchy problem for gravity water waves. Preprint (2012) arxiv:1212.0626
3. Craig, W., Sulem, C.: Numerical simulation of gravity waves. J. Comput. Phys. **108**(1), 73–83 (1993)
4. Lannes, D.: Well-posedness of the water-waves equations. J. Am. Math. Soc. **18**(3), 605–654 (electronic) (2005)
5. Lannes, D.: Water waves: mathematical analysis and asymptotics. to appear
6. Wu, S.: Well-posedness in Sobolev spaces of the full water wave problem in 2-D. Invent. Math. **130**(1), 39–72 (1997)
7. Zakharov, V.E.: Stability of periodic waves of finite amplitude on the surface of a deep fluid. J. Appl. Mech. Tech. Phys. **9**(2), 190–194 (1968)

Chapter 2
On the Characterization of Pseudodifferential Operators (Old and New)

Jean-Michel Bony

Abstract In the framework of the Weyl–Hörmander calculus, under a condition of "geodesic temperance", pseudodifferential operators can be characterized by the boundedness of their iterated commutators. As a corollary, functions of pseudodifferential operators are themselves pseudodifferential. Sufficient conditions are given for the geodesic temperance. In particular, it is valid in the Beals–Fefferman calculus.

Key words: Beals–Fefferman calculus, Functional calculus, Iterated commutators, Pseudodifferential operators, Weyl–Hörmander calculus

2010 Mathematics Subject Classification: 35S05, 47A60, 47G30.

2.1 Introduction

The first historical example of a characterization of pseudodifferential operators is due to Beals [1] for the following class of symbols:

$$S^0_{0,0} = \left\{ a \in C^\infty(\mathbb{R}^n \times \mathbb{R}^{n*}) \middle| \left| \partial_\xi^\alpha \partial_x^\beta a(x,\xi) \right| \leq C_{\alpha,\beta} \right\} .$$

An operator A can be written $A = a(x,D)$ with $a \in S^0_{0,0}$ if and only if A and its iterated commutators with the multiplications by x_j and the derivations $\partial/\partial x_j$ are bounded on L^2.

The aim of this paper is to give an analogous characterization for the more general classes of pseudodifferential operators occurring in the Beals–Fefferman

J.-M. Bony (✉)
CMLS, École Polytechnique, F-91128 Palaiseau Cedex, France
e-mail: bony@math.polytechnique.fr

M. Cicognani et al. (eds.), *Studies in Phase Space Analysis with Applications to PDEs*, Progress in Nonlinear Differential Equations and Their Applications 84,
DOI 10.1007/978-1-4614-6348-1_2,

calculus and the Weyl–Hörmander calculus. Such a characterization has important consequences:

- The Wiener property: if a pseudodifferential operator (of order 0) is invertible as an operator in L^2, its inverse is also a pseudodifferential operator.
- The compatibility with the functional calculus: C^∞ functions of pseudodifferential operator are themselves pseudodifferential.
- It is a good starting point for a general theory of Fourier integral operators. We refer to [4] for that point which will not be developed here.

This paper gives a new presentation, with some simplifications and complements, of the results of [3] where it is shown that the characterization of pseudodifferential operator is valid under an assumption of *geodesic temperance*.

The new point of this paper is the Sect. 2.4 which gives a sufficient condition, easy to check, for the *geodesic temperance*. In the framework of the Beals–Fefferman calculus, this condition is always satisfied and so are its consequences.

2.2 Weyl–Hörmander Calculus

We will denote by small latine letters (such as x) points of the configuration space $\mathbb{R}^n$, by Greek letters (such as ξ) points of its dual space $(\mathbb{R}^n)^*$, and by capital letters ($X=(x,\xi)$) points of the phase space $\mathscr{X}=\mathbb{R}^n\times(\mathbb{R}^n)^*$. The space $\mathscr{X}$ is equipped with the symplectic form σ defined by

$$\sigma(X,Y)=\langle\xi,y\rangle-\langle\eta,x\rangle \quad \text{for} \quad X=(x,\xi) \text{ and } Y=(y,\eta)\,.$$

The Weyl quantization associates to a function or distribution a on $\mathscr{X}$ an operator $a^w(x,D)$ acting in $\mathbb{R}^n$ defined by

$$a^w(x,D)u(x)=\iint e^{i\langle x-y,\xi\rangle}a\left(\tfrac{x+y}{2},\xi\right)u(y)\,dy\,d\xi/(2\pi)^n\,.$$

Considered in a weak sense, for $a\in\mathscr{S}'(\mathscr{X})$, this formula defines $a^w(x,D)$ as an operator mapping the Schwartz space $\mathscr{S}(\mathbb{R}^n)$ into the space $\mathscr{S}'(\mathbb{R}^n)$ of tempered distributions. Conversely, for any such operator A, there is a unique $a\in\mathscr{S}'(\mathscr{X})$, the *symbol* of A, such that $A=a^w(x,D)$.

Definition 2.1. The Hörmander classes of symbol $S(M,g)$ are associated to:

- A Riemannian metric g on $\mathscr{X}$, identified with an application $\left(X\mapsto g_X(\cdot)\right)$, where each g_X is a positive definite quadratic form on $\mathscr{X}$.
- A weight M, i.e. a positive function on $\mathscr{X}$.

They are defined by

$$S(M,g)=\left\{a\in C^\infty(\mathscr{X})\,\Big|\,\left|\partial_{T_1}\dots\partial_{T_k}a(X)\right|\le C_kM(X) \text{ for } k\ge 0 \text{ and } g_X(T_j)\le 1\right\}.$$

We use the notation $\partial_T f(X)=\langle df(X),T\rangle$ for the directional derivatives.

Example 2.1 (Beals–Fefferman classes [2]*).*
For $Q,q \in \mathbb{R}$ and Φ, φ positive functions on $\mathscr{X}$, these classes are defined by

$$a \in S^{Q,q}_{\Phi,\varphi} \Longleftrightarrow \left|\partial_\xi^\alpha \partial_x^\beta a(x,\xi)\right| \le C_{\alpha,\beta}\Phi(x,\xi)^{Q-|\alpha|}\varphi(x,\xi)^{q-|\beta|} .$$

They are actually classes $S(M,g)$ with

$$g_X(\mathrm{d}x,\mathrm{d}\xi) = \frac{\mathrm{d}x^2}{\varphi(X)^2} + \frac{\mathrm{d}\xi^2}{\Phi(X)^2} \quad \text{and} \quad M = \varphi^q \Phi^Q . \tag{2.1}$$

In the particular case

$$\Phi(x,\xi) = (1+|\xi|^2)^{\rho/2} , \ \varphi(x,\xi) = (1+|\xi|^2)^{-\delta/2} ,$$

one recovers the Hörmander class $S^m_{\rho,\delta}$ with $m = Q\rho - q\delta$:

$$a \in S^m_{\rho,\delta} \Longleftrightarrow \left|\partial_\xi^\alpha \partial_x^\beta a(x,\xi)\right| \le C_{\alpha,\beta}(1+|\xi|)^{m-\rho|\alpha|+\delta|\beta|} .$$

2.2.1 Admissible Metrics

Before stating the conditions which guarantee that the classes $S(M,g)$ give rise to a good symbolic calculus, we should recall some well-known properties of quadratic forms in symplectic spaces.

Reduced Form. For each $Y \in \mathscr{X}$, one can choose symplectic coordinates (x',ξ') (depending on Y) such that the quadratic form g_Y takes the diagonal form:

$$g_Y(\mathrm{d}x',\mathrm{d}\xi') = \sum_j \frac{\mathrm{d}x_j'^2 + \mathrm{d}\xi_j'^2}{\lambda_j(Y)} . \tag{2.2}$$

The λ_j are positive and are independent of the particular choice of (x',ξ'). An important invariant, which will characterize the "gain" in the symbolic calculus, is the following:

$$\lambda(Y) = \min_j \lambda_j(Y) .$$

Inverse Metric. It is the Riemannian metric g^σ on $\mathscr{X}$, defined for each Y by $g^\sigma_Y(T) = \sup \frac{\sigma(T,S)^2}{g_Y(S)}$. If $\mathscr{X}$ is identified with its dual via the symplectic form, g^σ_Y is nothing but the inverse quadratic form g_Y^{-1} a priori defined on $\mathscr{X}'$. In the symplectic coordinates above, one has

$$g^\sigma_Y(\mathrm{d}x',\mathrm{d}\xi') = \sum \lambda_j(Y)(\mathrm{d}x_j'^2 + \mathrm{d}\xi_j'^2) .$$

Geometric Mean of g and g^σ. It is a third Riemannian metric $g^\#$ on $\mathscr{X}$. For each Y the quadratic form $g^\#_Y$ is the geometric mean of g_Y and g^σ_Y (the geometric mean of two positive definite quadratic forms is always canonically defined and is easily computed in a basis which diagonalizes the two quadratic forms). In the symplectic coordinates above, one has

$$g^\#_Y(\mathrm{d}x', \mathrm{d}\xi') = \sum(\mathrm{d}x'^2_j + \mathrm{d}\xi'^2_j) .$$

Definition 2.2. The metric g is said *without symplectic eccentricity* (WSE) if for any Y, all the $\lambda_j(Y)$ are equal to $\lambda(Y)$. It is equivalent to saying that $g^\sigma = \lambda^2 g$.

Example 2.2. Metrics of Beals–Fefferman (2.1) are WSE, with $\lambda = \Phi\varphi$.

Definition 2.3. The metric g is said to be *admissible* if it satisfies the following properties:

(i) *Uncertainty principle:* $\lambda(Y) \geq 1$.
(ii) *Slowness:* $\exists C, \quad g_Y(X-Y) \leq C^{-1} \Longrightarrow \frac{g_Y(T)}{g_X(T)} \leq C$.
(iii) *Temperance:* $\exists C,N, \quad \frac{g_Y(T)}{g_X(T)} \leq C(1+g^\sigma_Y(X-Y))^N$.

A weight M is said to be admissible for g (or a g-weight) if it satisfies

(ii′) $\exists C, \quad g_Y(X-Y) \leq C^{-1} \Longrightarrow M(Y)/M(X) \leq C$.
(iii′) $\exists C,N, \quad M(Y)/M(X) \leq C(1+g^\sigma_Y(X-Y))^N$.

As a consequence of (iii), one has

$$(1+g^\sigma_X(X-Y)) \leq C(1+g^\sigma_Y(X-Y))^{N+1} , \tag{2.3}$$

and thus $g_X(T)/g_Y(T)$ is also bounded by the right-hand side of (iii) (with different values of C and N).

Remark 2.1. For an admissible Beals–Fefferman metric (2.1), condition (i) means that $\Phi(X)\varphi(X) \geq 1$ while (ii) and (iii) express bounds of the ratios $(\Phi(Y)/\Phi(X))^{\pm 1}$ and $(\varphi(Y)/\varphi(X))^{\pm 1}$.

The classes $S^m_{\rho,\delta}$ correspond to an admissible metric if $\delta \leq \rho \leq 1$ and $\delta < 1$.

2.2.2 Symbolic Calculus

In this paragraph, g will denote an admissible metric and M a g-weight. Let us denote by $\mathrm{Op}\,S(M,g)$ the class of operators whose symbol belongs to $S(M,g)$ and by # the composition of symbols which corresponds to the composition of operators, i.e.

$$(a_1 \# a_2)^w(x,D) = a_1^w(x,D) \circ a_2^w(x,D) .$$

The following properties are classical [6].

- Operators in $\mathrm{Op}S(M,g)$ map continuously $\mathscr{S}(\mathbb{R}^n)$ into itself, $\mathscr{S}'(\mathbb{R}^n)$ into itself, and, for $M=1$, $L^2(\mathbb{R}^n)$ into itself.
- The formal adjoint of $a^w(x,D)$ is $\overline{a}^w(x,D)$.
- If M_1 and M_2 are two g-weights, then M_1M_2 is also a g-weight and one has $S(M_1,g)\#S(M_2,g)\subset S(M_1M_2,g)$.
- For $a_j\in S(M_j,g)$, one has the asymptotic expansion

$$a_1\#a_2(X)=\sum_{k=0}^{N-1}\frac{1}{k!}\Big\{\big(\tfrac{1}{2\mathrm{i}}\sigma(\partial_Y,\partial_Z)\big)^k a_1(Y)a_2(Z)\Big\}\Big|_{Y=Z=X}+R_N(X)\,,$$

where the kth term in the sum belongs to $S(M_1M_2\lambda^{-k},g)$ and R_N belongs to $S(M_1M_2\lambda^{-N},g)$. The first two terms are the usual product and the Poisson bracket:

$$a_1\#a_2=a_1a_2+\tfrac{1}{2\mathrm{i}}\{a_1,a_2\}+\cdots$$

2.2.3 *Can One Use Only Metrics Without Symplectic Eccentricity?*

As remarked above, Beals–Fefferman metrics are WSE, and one can say that, since the introduction of the Weyl–Hörmander calculus thirty years ago, almost all the metrics which have been used are WSE. It is not uniquely because they are simpler to use, a result of Toft [7] shows that there is a good mathematical reason for this limitation.

Theorem 2.1 (J. Toft).

(i) *If g is an admissible metric, one has*

$$\exists C,N,\quad (g_X(T)/g_Y(T))^{\pm 1}\le C\big(1+g^\#_Y(X-Y)\big)^N. \tag{2.4}$$

(ii) *The metric $g^\#$ is admissible.*

The second point is an easy consequence of the first one. The right-hand side of (2.4), which controls the ratio g_X/g_Y, controls also the ratio of the inverse metric g^σ_X/g^σ_Y and thus the ratio of their geometric mean $g^\#_X/g^\#_Y$, which proves that $g^\#$ is tempered. It controls also the ratio $\lambda(Y)/\lambda(X)$.

Let us introduce a new metric: $\widetilde{g}=\lambda^{-1}g^\#$. The ratio $\widetilde{g}_Y/\widetilde{g}_X$ is then controlled by the right-hand side of (2.4) (with different constants C and N) and a fortiori by a power of $\big(1+\lambda(Y)g^\#_Y(X-Y)\big)$. This proves the temperance of $\widetilde{g}$, because $\widetilde{g}^\sigma=\lambda g^\#$. The slowness being evident, the metric $\widetilde{g}$ is admissible. Moreover, it is WSE: in the coordinates (2.2), one has

$$g_Y(\mathrm{d}x',\mathrm{d}\xi')=\sum_j\frac{\mathrm{d}x_j'^2+\mathrm{d}\xi_j'^2}{\lambda_j(Y)}\quad;\quad \widetilde{g}_Y(\mathrm{d}x',\mathrm{d}\xi')=\sum_j\frac{\mathrm{d}x_j'^2+\mathrm{d}\xi_j'^2}{\lambda(Y)}\,.$$

Let us now compare the g-calculus and the $\widetilde{g}$-calculus.

(i). If M is a g-weight, then M is a $\widetilde{g}$-weight.
(ii). One has then $S(M,g) \subset S(M,\widetilde{g})$.
(iii). Moreover, the Sobolev spaces are the same: $H(M,g) = H(M,\widetilde{g})$.
(iv). Any given $A \in \operatorname{Op} S(M,\widetilde{g})$ maps $H(M_1)$ into $H(M_1/M)$.
(v). The "gain" λ of the symbolic calculus is the same for g and $\widetilde{g}$.

The first point is a consequence of Toft's theorem, the ratio $M(Y)/M(X)$ being also controlled by the right-hand side of (2.4). The second point is evident. The Sobolev space $H(M,g)$ can be defined as the space of $u \in \mathscr{S}'(\mathbb{R}^n)$ such that $a^w(x,D)u \in L^2$ for any $a \in S(M,g)$, which makes (iv) evident. Equivalent definitions can be found in [5], where Theorem 6.9 proves (iii).

It is thus difficult to imagine a situation where it would be more advantageous to use the g-calculus instead of the $\widetilde{g}$-calculus.

2.3 Characterization of Pseudodifferential Operators

2.3.1 *Geodesic Temperance*

Let us denote by $d^\sigma(X,Y)$ the geodesic distance, for the Riemannian metric g^σ, between X and Y.

Definition 2.4. The metric g is said to be *geodesically tempered* if it is tempered and if, moreover, the equivalent following conditions are satisfied:

$$\exists C,N;\ \frac{g_Y(T)}{g_X(T)} \le C\left(1+d^\sigma(X,Y)\right)^N, \tag{2.5}$$

$$\exists C,N;\ C^{-1}\left(1+d^\sigma(X,Y)\right)^{1/N} \le \left(1+g_Y^\sigma(X-Y)\right) \le C\left(1+d^\sigma(X,Y)\right)^N. \tag{2.6}$$

The left inequality in (2.6) is always true. Actually, one has

$$C^{-1}\left(1+g_Y^\sigma(X-Y)\right)^{1/N} \le 1+g^\sigma\text{-length of segment } XY \le C\left(1+g_Y^\sigma(X-Y)\right)^N,$$

which is a simple consequence of (2.3) and thus of the temperance of g. It is clear that the right inequality in (2.6) implies (2.5).

Assume now (2.5), and let $t \in [0,L] \mapsto X(t)$ be a unit-speed geodesic from Y to X. One has

$$g_Y(X-Y)^{1/2} \le \int_0^L g_Y\left(\tfrac{\mathrm{d}X}{\mathrm{d}t}\right)^{1/2}\mathrm{d}t \le C'\int_0^L (1+t)^{N/2}\mathrm{d}t \le C''(1+L)^{N/2+1},$$

which gives the right part of (2.6).

Remark 2.2. The geodesic temperance requires both the temperance and (2.5) (or (2.6)). A metric can satisfy (2.5) without being tempered (example: $e^x dx^2 + e^{-x} d\xi^2$). However, if g satisfies (2.5) and (2.3), it is tempered: the ratio g_X/g_Y is estimated by some power of the g^σ-length of the segment XY which is itself estimated, as remarked above, by a power of $g_Y^\sigma(X-Y)$.

Remark 2.3. There is no known example of a tempered metric g which is not geodesically tempered, but proving that a particular metric is geodesically tempered can be a challenging task. Theorem 2.5 below proves that this is not an issue for the Beals–Fefferman metrics.

2.3.2 Characterization

We shall use the classical notation

$$\operatorname{ad} B \cdot A = [B,A] = B \circ A - A \circ B$$

for the commutator of two operators. The following definition is of interest only for metrics WSE.

Definition 2.5. Let g be a metric WSE. The class $\widehat{S}(\lambda,g)$ is the set of functions $b \in C^\infty(\mathscr{X})$ which satisfy, for any given $\boxed{k \geq 1}$,

$$\exists C_k; \quad \left|\partial_{T_1} \ldots \partial_{T_k} b(X)\right| \leq C_k \lambda(X) \text{ for } g_X(T_j) \leq 1 \,.$$

We refer to [3] for the proof of the following theorem and just add some comments.

Theorem 2.2. *Let g be an admissible metric WSE which is* geodesically tempered. *Then A belongs to* $\operatorname{Op} S(1,g)$ *if and only if the iterated commutators*

$$\operatorname{ad} b_1^w \ldots \operatorname{ad} b_k^w \cdot A\,, \quad k \geq 0,\ b_j \in \widehat{S}(\lambda,g)\,, \tag{2.7}$$

are bounded on L^2.

Remark 2.4. It is easy to see that, for $b \in \widehat{S}(\lambda,g)$ and $a \in S(1,g)$, the Poisson bracket $\{b,a\}$ belongs to $S(1,g)$. The "only if" part of the proof comes from the fact that $b\#a - a\#b$ (a non-local integral expression whose principal part is $-i\{b,a\}$) belongs also to $S(1,g)$.

It would not be sufficient to require the boundedness of the commutators with $b_j \in S(\lambda,g)$. For instance, if g is the Euclidean metric of $\mathscr{X}$, the class $S(1,g)$ is nothing but $S^0_{0,0}$, and one has $\lambda = 1$. For $b_j \in S(\lambda,g) = S(1,g)$ the corresponding operators are bounded on L^2 and (2.7) would be valid for any given A bounded on L^2. On the other hand, $\widehat{S}(\lambda,g)$ contains in that case the functions x_j and ξ_j, and the theorem reduces to the criterium of Beals.

Remark 2.5. One can realize the importance of the geodesic temperance in this way. An essential ingredient in the proof of the "if" part is to prove a "decay of A outside the diagonal of $\mathscr{X}$", namely that given balls $B_Y = \{X \mid g_Y(X-Y) \le r^2\}$ and a family, bounded in $S(1,g)$, of functions α_Y supported in B_Y, one has

$$\|\alpha_Y^w \circ A \circ \alpha_Z^w\|_{\mathscr{L}(L^2)} \le C_N(1 + g_Y^\sigma(B_Y - B_Z))^{-N} ,$$

where $g_Y^\sigma(B_Y - B_Z)$ means $\inf g_Y^\sigma(Y' - Z')$ for $Y' \in B_Y$ and $Z' \in B_Z$.

It turns out that, from the estimate on the commutators with $b^w \in \operatorname{Op}\widehat{S}(\lambda,g)$, one can gain only the variation of b between B_Y and B_Z. But functions in $\widehat{S}(\lambda,g)$ are Lipschitz continuous for g^σ, the variation of b cannot exceed the geodesic distance of the balls and one cannot obtain a better bound than $C_N(1 + d^\sigma(B_Y, B_Z))^{-N}$ in the right-hand side. The geodesic temperance asserts precisely that such a bound can compensate the ratio g_Y/g_Z.

The characterization can be extended to some cases where g is not WSE, the space $\widehat{S}(\lambda,g)$ being accordingly modified (see [3]). However, some extra conditions on g should be added; if not, [3] contains an example where the characterization fails.

Corollary 2.1. *Under the same assumptions on g, let M and M_1 be two g-weights. Then an operator A belongs to $\operatorname{Op}S(M,g)$ if and only if A and its iterated commutators with elements of $\operatorname{Op}\widehat{S}(\lambda,g)$ map continuously $H(M_1,g)$ into $H(M_1/M,g)$.*

Let us choose $B \in \operatorname{Op}S(M_1,g)$ having an inverse $B^{-1} \in S(M_1^{-1},g)$ (see [5, Cor. 6.6.]) and another invertible operator $C \in \operatorname{Op}S(M_1/M,g)$. Then, $A' = CAB^{-1}$ satisfies the assumptions of Theorem 2.2, one has $A' \in \operatorname{Op}S(1,g)$ and thus $A = BA'C^{-1} \in \operatorname{Op}S(M,g)$.

Corollary 2.2 (Wiener property). *Assume g admissible, WSE and geodesically tempered.*

(i) *If $A \in \operatorname{Op}S(1,g)$ is invertible in $\mathscr{L}(L^2)$, then its inverse A^{-1} belongs to $\operatorname{Op}S(1,g)$.*
(ii) *If M and M_1 are two g-weights and if $A \in \operatorname{Op}S(M,g)$ is a bijection from $H(M_1,g)$ onto $H(M_1/M,g)$, then $A^{-1} \in \operatorname{Op}S(M^{-1},g)$.*

For $b_1 \in \widehat{S}(\lambda,g)$ one has $C = \operatorname{ad} b_1^w \cdot A^{-1} = -A^{-1}(\operatorname{ad} b_1^w \cdot A)A^{-1}$ which is bounded on L^2. Next

$$\begin{aligned}\operatorname{ad} b_2^w \cdot C = -(\operatorname{ad} b_2^w \cdot A^{-1})(\operatorname{ad} b_1^w \cdot A)A^{-1} - A^{-1}(\operatorname{ad} b_2^w \cdot \operatorname{ad} b_1^w \cdot A)A^{-1} \\ -A^{-1}(\operatorname{ad} b_1^w \cdot A)(\operatorname{ad} b_2^w \cdot A^{-1})\end{aligned}$$

and the three terms are bounded on L^2. By induction, one gets that all iterated commutators are bounded on L^2 and thus that A^{-1} belongs to $S(1,g)$.

Remark 2.6. The Wiener property is actually valid for some metrics g which are not WSE, including cases where the characterization is not valid. One has just to assume that the metric $\widetilde{g} = \lambda^{-1} g^\#$ of the paragraph 2.2.3 is geodesically tempered. One knows then that $A^{-1} \in \operatorname{Op}S(1,\widetilde{g})$, and proving that its symbol actually belongs to $S(1,g)$ is just a matter of symbolic calculus. The second part of the proof of [5, Thm. 7.6] can be reproduced verbatim.

2.3.3 Functional Calculus

Given a self-adjoint operator A (bounded or unbounded) on L^2, the functional calculus associates to any Borel function f, defined on the spectrum of A, an operator $f(A)$. When f belongs to C^∞, it can be computed via the formula of Helffer-Sjöstrand:

$$f(A) = -\pi^{-1} \iint \overline{\partial}\widetilde{f}(z) R_z \,\mathrm{d}x\,\mathrm{d}y\,, \quad z = x + \mathrm{i}y\,, \tag{2.8}$$

where $R_z = (z - A)^{-1}$ is the resolvent and $\widetilde{f}$ is an almost analytic extension of f.

Theorem 2.3. *Assume g admissible, WSE and geodesically tempered. Let $a \in S(1,g)$ be real valued and let f be a C^∞ function defined in a neighborhood of the spectrum of a^w. Then $f(a^w)$ belongs to* $\mathrm{Op}S(1,g)$.

In that case, $\widetilde{f}$ can be chosen with compact support in $\mathbb{C}$ and the meaning of (2.8) is clear. This theorem is actually a particular case of the following one, where the assumption A self-adjoint with domain $H(M)$ should be thought of as a condition of ellipticity. When $C^{-1}\lambda^{1/N} \le M \le C\lambda^N$, it is equivalent to $1 + |a(X)| \ge C^{-1}M(X)$ and also to the existence of a parametrix $E \in \mathrm{Op}S(M^{-1},g)$ such that $AE - I$ and $EA - I$ belong to $\mathrm{Op}S(\lambda^{-\infty},g)$.

Theorem 2.4. *Under the same assumptions on g, let $M \ge 1$ be a g-weight. Let $a \in S(M,g)$ be real valued such that $A = a^w$ is self-adjoint with domain $H(M,g)$. Let $f \in C^\infty(\mathbb{R})$ be a "symbol of order p", i.e. such that*

$$\left| \frac{\mathrm{d}^k f(t)}{\mathrm{d}t^k} \right| \le C_k (1 + |t|)^{p-k} \quad \textit{for } k \ge 0\,.$$

Then $f(a^w)$ belongs to $\mathrm{Op}S(M^p,g)$. *Moreover, if c is the symbol of $f(a^w)$, one has $c - f \circ a \in S(M^p\lambda^{-2},g)$.*

The result is evident for $1 + A^2$ and, dividing f if necessary by a power of $1+t^2$, one may assume $p < 0$. Now, any negative p can be written $p = \sum_1^N p_j$ with $-1 < p_j < 0$, and one can write $f(t) = \prod_1^N f_j(t)$, where f_j is a symbol of order p_j (it suffices to choose $f_j(t) = (1+t^2)^{p_j/2}$ for $j < N$). One has then $f(a^w) = \prod f_j(a^w)$ and, thanks to the rule of composition of pseudodifferential operators, it suffices to prove the theorem when $-1 < p < 0$ that we assume in the sequel.

One can then choose, for Q large enough,

$$\widetilde{f}(x+iy) = \chi\left(\tfrac{y^2}{1+x^2}\right) \sum_{k=0}^{Q} f^{(k)}(x) \frac{(\mathrm{i}y)^k}{k!}\,,$$

where $\chi \in C^\infty(\mathbb{R})$ satisfies $\chi(s) = 1$ [resp. 0] for $s \le 1/2$ [resp. $s \ge 1$]. One has

$$\left|\overline{\partial}\widetilde{f}(x+iy)\right| \le C(1+|x|)^{p-N-2}|y|^{N+1} \quad \text{for } N \le Q-1, \tag{2.9}$$

and the integral in the right-hand side of (2.8) is thus convergent.

For $\Im z \neq 0$, we know that R_z is a bijection of L^2 onto the domain $H(M)$ and thus, as a consequence of Corollary 2.2, that $R_z \in \mathrm{Op}S(M^{-1})$. Let us denote by r_z its Weyl symbol. We have $\|R_z\|_{\mathscr{L}(L^2)} \leq 1/|\Im z|$. From the resolvent equation $R_z - R_{\mathrm{i}} = (\mathrm{i}-z)R_{\mathrm{i}}R_z$, one gets $\|R_z\|_{\mathscr{L}(L^2,H(M))} \leq C(1+|z|)/|\Im z|$. The iterated commutators can be written

$$\prod (\mathrm{ad}\, b_j^w)\cdot R_z = \sum \pm R_z K_1 R_z \ldots R_z K_p R_z$$

where the sum is finite and each K_l is an iterated commutator of A with some of the b_j. Thus $\|K_l\|_{\mathscr{L}(H(M),L^2)}$ is bounded independently of z and l, and one has

$$\Big\|\prod_1^N (\mathrm{ad}\, b_j^w)\cdot R_z\Big\|_{\mathscr{L}(L^2,H(M))} \leq C\prod_1^N \|b_j\|_{l;\widehat{S}(\lambda,g)} \frac{(1+|z|)^{N+1}}{|\Im z|^{N+1}}$$

$$\Big\|\prod_1^N (\mathrm{ad}\, b_j^w)\cdot R_z\Big\|_{\mathscr{L}(L^2)} \leq C\prod_1^N \|b_j\|_{l;\widehat{S}(\lambda,g)} \frac{(1+|z|)^{N}}{|\Im z|^{N+1}}$$

with $C = C(N)$, $l = l(N)$, denoting by $\|\cdot\|_{k,E}$ the semi-norms of a Frechet space E.

The proof of Theorem 2.2 shows that there exist constants C_k and N_k such that

$$\|r_z\|_{k;S(1,g)} \leq C_k \frac{(1+|z|)^{N_k}}{|\Im z|^{N_k+1}}, \qquad \|r_z\|_{k;S(M^{-1},g)} \leq C_k \frac{(1+|z|)^{N_k+1}}{|\Im z|^{N_k+1}}$$

Let us denote by c the symbol of $f(A)$. Using the estimates above, for $k=0$, one gets from (2.8) and (2.9)

$$\begin{aligned}|c(S)| &\leq C \iint_{|y|<(1+|x|)} \frac{(1+|x|)^N}{|y|^{N+1}} \min\left\{1, \frac{(1+|x|)}{M(S)}\right\} |y|^{N+1} (1+|x|)^{p-N-2}\, \mathrm{d}x\,\mathrm{d}y \\ &\leq C' \int (1+|x|)^{p-1} \min\left\{1, \frac{1+|x|}{M(S)}\right\} \mathrm{d}x \leq C'' M(S)^p .\end{aligned}$$

The estimates of the derivatives $\left(\prod \partial_{T_j}\right) c(S)$ for $g_S(T_j) \leq 1$ are analogous, which ends the proof.

2.4 Sufficient Conditions for the Geodesic Temperance

Theorem 2.5. (i) *Admissible Beals–Fefferman metrics are geodesically tempered.*

(ii) *More generally, let $\mathscr{X} = \mathscr{X}_1 \oplus \mathscr{X}_2 \oplus \cdots \oplus \mathscr{X}_p$ be a decomposition of $\mathscr{X}$ as a vector space, and denote $X = (X_1, \ldots, X_p)$ the components of a vector. Let g be an admissible metric such that g^σ can be written*

$$\begin{aligned}g_X^\sigma(dX) = a_1(X_1,X_2)\Gamma_1(dX_1) &+ a_2(X_1,X_2)\Gamma_2(dX_2) + a_3(X_1,X_2,X_3)\Gamma_3(dX_3) \\ &+ a_4(X_1,X_2,X_3,X_4)\Gamma_4(dX_4) + \cdots + a_p(X)\Gamma_p(dX_p)\,,\end{aligned}$$

where Γ_j is a positive definite quadratic form on $\mathscr{X}_j$ and a_j is a positive function of its arguments. Then g is geodesically tempered.

Remark 2.7. We keep the notations which are of interest for us, but (ii) could be stated for an affine space $\mathscr{X}$ on which a Riemannian metric g^σ is given. The symplectic structure and g itself play no role, the temperance reduces to

$$(g^\sigma_Y(\cdot)/g^\sigma_X(\cdot))^{\pm 1} \leq C(1+g^\sigma_Y(X-Y))^N$$

and one has to prove the right inequality in (2.6).

The particular role played by X_1 and X_2 should be noted. It is only for $X_j, j \geq 3$, that a "triangular structure" of g^σ is required.

2.4.1 Proof of Theorem 2.5 (i)

We have $g^\sigma_X = \Phi(X)^2 \mathrm{d}x^2 + \varphi(X)^2 \mathrm{d}\xi^2$. The temperance and (2.3) can be formulated as follows:

$$\begin{aligned}(g^\sigma_Y(\cdot)/g^\sigma_X(\cdot))^{\pm 1/2} &\leq C \max\{1\,;\Phi(Y)\,|x-y|\,;\varphi(Y)\,|\xi-\eta|\}^N,\\ 1+\Phi(X)\,|x-y|+\varphi(X)\,|\xi-\eta| &\leq C \max\{1\,;\Phi(Y)\,|x-y|\,;\varphi(Y)\,|\xi-\eta|\}^N.\end{aligned} \tag{2.10}$$

The value of $\kappa > 0$, $\varepsilon > 0$ and $R_0 > 1$ will be fixed later, depending only of C and N above. Other constants, such as $C', C'', C_1 \ldots$, may vary from line to line but can be computed depending on C and N.

We have to prove that any given curve $t \in [0,T] \mapsto X(t)$ joining a point (which we take as the origin) to a point of the boundary of $B_x(R) \times B_\xi(R)$ has a length $\geq C'^{-1} R^\delta$, with $\delta > 0$ and C' independent of R. Here,

$$B_x(R) = \{x \,|\, |x| \leq R/\Phi(0)\}\ , \quad B_\xi(R) = \{\xi \,|\, |\xi| \leq R/\varphi(0)\}\ .$$

The result is evident if $R \leq R_0$: in that case, one has $g_X \geq C^{\mathrm{st}} g_0$ for $X \in B_x(R) \times B_\xi(R)$ (with a constant depending on R_0). Thus, we will assume $R \geq R_0$. We may assume that T is the first instant when $X(t)$ reaches the boundary. Exchanging if necessary x and ξ, we may assume that $\Phi(0)\,|x(T)| = R$. Set $R' = \max_{t\in[0,T]} \varphi(0)\,|\xi(t)|$ and let T' be the first instant when R' is reached.

We distinguish two cases.

- *Case I: $R' \leq R^\kappa$.* — Set $Y(t) = (0,\xi(t))$. By temperance, one has $\Phi(Y(t)) \geq C'^{-1} R^{-N\kappa} \Phi(0)$ and, using (2.3),

$$(1+\Phi(X(t))\,|x(t)|) \geq C'^{-1}(1+\Phi(Y(t))\,|x(t)|)^{1/N}\ .$$

Let us consider the length L of the curve between the last instant θ when $\Phi(0)\,|x(t)| = R/2$ and T. For $t \geq \theta$, one has then $\Phi(X(t))/\Phi(0) \geq C'^{-1} R^{-1+1/N-\kappa}$, and

$$L \geq \int_\theta^T \Phi(X(t))\,|x'(t)|\,\mathrm{d}t \geq C'^{-1} R^{-1+1/N-\kappa} \int_\theta^T \Phi(0)\,|x'(t)|\,\mathrm{d}t \geq C'' R^{1/N-\kappa}$$

We may now fix $\kappa = 1/(2N)$ and the result is proved, with $\delta = 1/(2N)$, in the first case.

- *Case II: $R' \geq R^\kappa$.* We distinguish three subcases.

Subcase (IIa). $\Phi(X)/\Phi(0) \geq R^{-1+\varepsilon}$ everywhere in $B_x(R) \times B_\xi(R')$. Then, the length of the curve is greater than

$$\int_0^T \Phi(X(t))\,|x'(t)|\,\mathrm{d}t \geq R^{-1+\varepsilon} \int_0^T \Phi(0)\,|x'(t)|\,\mathrm{d}t \geq R^\varepsilon\,,$$

which ends the proof, with $\delta = \varepsilon$, in this subcase.
Subcase (IIb). $\varphi(X)/\varphi(0) \geq R'^{-1+\varepsilon}$ everywhere in $B_x(R) \times B_\xi(R')$. This is similar; the length is larger than

$$\int_0^{T'} \varphi(X(t))\,|\xi'(t)|\,\mathrm{d}t \geq R'^{\varepsilon} \geq R^{\varepsilon\kappa}\,,$$

and the theorem is proved with $\delta = \kappa\varepsilon$, in this subcase.
Subcase (IIc). It is the remaining case and we will prove that it cannot occur provided that ε and R_0 be conveniently chosen. There should exist $Y_1 = (x_1, \xi_1)$ and $Y_2 = (x_2, \xi_2)$ in $B_x(R) \times B_\xi(R')$ such that

$$\Phi(Y_1)/\Phi(0) \leq R^{-1+\varepsilon} \quad \text{and} \quad \varphi(Y_2)/\varphi(0) \leq R'^{-1+\varepsilon}\,.$$

Let us consider the point $Z = (x_2, \xi_1)$. One has $\Phi(Y_1)\,|x_2 - x_1| \leq 2R^\varepsilon$ and thus

$$\Phi(Z) \leq C' R^{N\varepsilon} \Phi(Y_1) \leq C' R^{-1+(N+1)\varepsilon} \Phi(0)\,. \tag{2.11}$$

Then, assuming $(N+1)\varepsilon < 1$,

$$\Phi(Z)\,|x_2| \leq C' \left(\tfrac{\Phi(0)|x_2|}{R}\right)^{1-(N+1)\varepsilon} (\Phi(0)\,|x_2|)^{(N+1)\varepsilon} \leq C' (\Phi(0)\,|x_2|)^{(N+1)\varepsilon}\,.$$

The same computation, where R is replaced by R', shows that

$$\varphi(Z)\,|\xi_1| \leq C' (\varphi(0)\,|\xi_1|)^{(N+1)\varepsilon}\,.$$

Applying (2.10) between 0 and Z, we get

$$\begin{aligned}(1 + \Phi(0)\,|x_2| + \varphi(0)\,|\xi_1|) &\leq C(1 + \Phi(Z)\,|x_2| + \varphi(Z)\,|\xi_1|)^N \\ &\leq C'(1 + \Phi(0)\,|x_2| + \varphi(0)\,|\xi_1|)^{N(N+1)\varepsilon}\,.\end{aligned}$$

Now, fix $\varepsilon = \frac{1}{2N(N+1)}$. The inequality implies the existence of a constant C_1 such that $(1+\Phi(0)|x_2|+\varphi(0)|\xi_1|) \le C_1$. By temperance, one has $\Phi(0)/\Phi(Z) \le C_2$, which is to compare with (2.11). One gets

$$R^{1-(N+1)\varepsilon} \le C' \frac{\Phi(0)}{\Phi(Z)} \le C' C_2 ,$$

which is impossible for $R \ge R_0$ if we choose, for instance, $R_0 = 2(C'C_2)^2$.

2.4.2 Proof of Theorem 2.5 (ii)

The proof of part (i) is also the proof of the case $p = 2$ (and also of course $p = 1$). As remarked above, the symplectic structure plays no role. Thus, R^n_x, $\mathbb{R}^n_\xi$ and their canonical quadratic forms can be replaced by $\mathscr{X}_1$, $\mathscr{X}_2$, Γ_1 and Γ_2.

For $p > 2$, the theorem is a consequence, by induction, of the following lemma.

Lemma 2.1. *Assume $\mathscr{X} = \mathscr{Y} \oplus \mathscr{Z}$ and let G^σ be a Riemannian metric on $\mathscr{X}$ of the following form*

$$G^\sigma(X, \mathrm{d}X) = g^\sigma(Y, \mathrm{d}Y) + a(Y,Z)\Gamma(\mathrm{d}Z),$$

where g^σ is a Riemannian metric on $\mathscr{Y}$, a is a positive function on $\mathscr{X}$ and Γ is a positive definite quadratic form on $\mathscr{Z}$. Assume the temperance of G^σ and the geodesic temperance of g^σ. Then the geodesic temperance is valid for G^σ.

If we denote by d^σ the geodesic distance for g^σ on $\mathscr{Y}$ and by D^σ the geodesic distance for G^σ on $\mathscr{X}$, there exists thus constants C and N such that

$$\left(G^\sigma_{X_1}/G^\sigma_{X_2}\right)^{\pm 1} \le C\left(1+G^\sigma_{X_1}(X_2-X_1)\right)^N , \tag{2.12}$$

$$\left(g^\sigma_{Y_1}/g^\sigma_{Y_2}\right)^{\pm 1} \le C\left(1+g^\sigma_{Y_1}(Y_2-Y_1)\right)^N , \tag{2.13}$$

$$C^{-1}\left(1+d^\sigma(Y_1,Y_2)\right)^{1/N} \le \left(1+g^\sigma_Y(Y_1-Y_2)\right) \le C\left(1+d^\sigma(Y_1,Y_2)\right)^N . \tag{2.14}$$

Let us consider two points X_0 and X_1 and a curve $[0,1] \ni t \mapsto X(t) = (Y(t), Z(t))$ joining these two points. Let us denote by L the G^σ-length of this curve and set $R^2 = G^\sigma_{X_0}(X_1 - X_0)$. We want to prove that there exist C' and $\delta > 0$, depending just on C and N above, such that $L \ge C'^{-1}R^\delta$. We will assume, as we may, that $R \ge 1$. The value of κ, $0 < \kappa < 1$, will be fixed later, and we distinguish two cases.

- *Case I:* $\forall t$, $g^\sigma_{Y_0}(Y(t) - Y_0)^{1/2} \le R^\kappa/2$. One has then $(a(X_0)\Gamma(Z_1 - Z_0))^{1/2} \ge R/2$. Let us consider the curve $t \mapsto P(t) = (Y_0, Z(t))$. We can apply the case $p = 1$ of the theorem to the metric $a(Y_0, Z)\Gamma(\mathrm{d}Z)$ on the affine space $\{Y_0\} \times \mathscr{Z}$. These metrics depend on Y_0, but they are tempered with the same constants C and N, and thus, they are geodesically tempered with uniform constants. One has thus

$$\int_0^1 \left(a(P(t))\Gamma(\dot{Z}(t))\right)^{1/2} \mathrm{d}t \ge C'^{-1} R^\alpha$$

with $\alpha > 0$ and C' depending just on C and N.

The temperance of G^σ imply $a(X(t))^{1/2} \geq C'^{-1}R^{-N\kappa}a(P(t))^{1/2}$ and thus

$$L \geq \int_0^1 \left(a(X(t))\Gamma(\dot{Z}(t))\right)^{1/2} \mathrm{d}t \geq C'^{-1}R^{-N\kappa} \int_0^1 \left(a(P(t))\Gamma(\dot{Z}(t))\right)^{1/2} \mathrm{d}t$$
$$\geq C''^{-1}R^{\alpha - N\kappa} \ .$$

Fix now $\kappa = \alpha/(2N)$ and the lemma is proved with $\delta = \alpha/2$ in this first case.

- *Case II:* $\exists t_0,\ g^\sigma_{Y_0}(Y(t_0) - Y_0)^{1/2} \geq R^\kappa/2$. Let us consider the curve $[0,t_0] \ni t \mapsto Y(t)$ in $\mathscr{Y}$. By (2.14), one gets

$$L \geq \int_0^{t_0} \left(g^\sigma_{Y(t)}(\dot{Y}(t))\right)^{1/2} \mathrm{d}t \geq C^{-1}(1 + g^\sigma_{Y_0}(Y(t_0) - Y_0))^{1/N} \geq C'^{-1}R^{2\kappa/N} ,$$

which ends the proof of the lemma and of Theorem 2.5.

References

1. Beals, R.: Characterization of pseudodifferential operators and applications. Duke Math. J. **44**(1), 45–57 (1977)
2. Beals, R., Fefferman, C.: Spatially inhomogeneous pseudodifferential operators I. Comm. Pure Appl. Math. **27**, 1–24 (1974)
3. Bony, J.-M.: Caractérisations des opérateurs pseudo-différentiels. In: Séminaire Équations aux Dérivées Partielles 96–97, Les Éditions de l'École Polytechnique, F-91128 Palaiseau cedex, France, Exp. no. XXIII, pp. 17 (1996–1997) ISBN: 2-7302-0396-6
4. Bony, J.-M.: Evolution equations and microlocal analysis. In: Hyperbolic Problems and Related Topics, pp. 17–40. Grad. Ser. Anal. International Press, Somerville (2003)
5. Bony, J.-M., Chemin, J.-Y.: Espaces fonctionnels associés au calcul de Weyl–Hörmander. Bull. Soc. Math. France **122**(1), 77–118 (1994)
6. Hörmander, L.: The Analysis of Linear Partial Differential Operators. Springer, Berlin (1990)
7. Toft, J.: Continuity and Schatten properties for pseudo-differential operators on modulation spaces. In: Modern Trends in Pseudo-Differential Operators. Operational Theory in Advanced Application, vol. 172, pp. 173–206. Birkhäuser, Basel (2007)

Chapter 3
Improved Multipolar Hardy Inequalities

Cristian Cazacu and Enrique Zuazua

Abstract In this paper we prove optimal Hardy-type inequalities for Schrödinger operators with positive multi-singular inverse square potentials of the form

$$A_\lambda := -\Delta - \lambda \sum_{1\leq i<j\leq n} \frac{|x_i - x_j|^2}{|x-x_i|^2|x-x_j|^2}, \quad \lambda > 0.$$

More precisely, we show that A_λ is nonnegative in the sense of L^2 quadratic forms in $\mathbb{R}^N$, if and only if $\lambda \leq (N-2)^2/n^2$, independently of the number n and location of the singularities $x_i \in \mathbb{R}^N$, where $N \geq 3$ denotes the space dimension. This aims to complement some of the results in Bosi et al. (Comm. Pure Appl. Anal. 7:533–562, 2008) obtained by the "expansion of the square" method. Due to the interaction of poles, our optimal result provides a singular quadratic potential behaving like $(n-1)(N-2)^2/(n^2|x-x_i|^2)$ at each pole x_i. Besides, the authors in Bosi et al. (Comm. Pure Appl. Anal. 7:533–562, 2008) showed optimal Hardy inequalities for Schrödinger operators with a finite number of singular poles of the type $B_\lambda := -\Delta - \sum_{i=1}^{n} \lambda/|x-x_i|^2$, up to lower order L^2-reminder terms. By means of the optimal results obtained for A_λ, we also build some examples of bounded domains Ω in which these lower order terms can be removed in $H_0^1(\Omega)$. In this way

C. Cazacu (✉)
BCAM–Basque Center for Applied Mathematics, Mazarredo, 14, E-48009 Bilbao, Basque Country, Spain

Departamento de Matemáticas, Universidad Autónoma de Madrid, E-28049 Madrid, Spain
e-mail: cristicazacu2008@gmail.com

E. Zuazua
BCAM–Basque Center for Applied Mathematics, Mazarredo, 14, E-48009 Bilbao, Basque Country, Spain

Ikerbasque, Basque Foundation for Science, Alameda Urquijo 36-5, Plaza Bizkaia, E-48011, Bilbao, Basque Country, Spain
e-mail: zuazua@bcamath.org

M. Cicognani et al. (eds.), *Studies in Phase Space Analysis with Applications to PDEs*, Progress in Nonlinear Differential Equations and Their Applications 84,
DOI 10.1007/978-1-4614-6348-1_3,

we obtain new lower bounds for the optimal constant in the standard multi-singular Hardy inequality for the operator B_λ in bounded domains. The best lower bounds are obtained when the singularities x_i are located on the boundary of the domain.

Key words: Hardy inequalities, Multipolar potentials, Schrödinger operators

2010 Mathematics Subject Classification: 35J10, 26D10, 46E35, 35Q40, 35J75.

3.1 Introduction

This paper is concerned with a class of Schrödinger operators of the form $-\Delta + V(x)$ with multipolar Hardy-type singular potentials like $V \sim \sum_i \alpha_i/|x-x_i|^2$, $\alpha_i \in \mathbb{R}$, $x_i \in \mathbb{R}^N, N \geq 3$.

The study of such singular potentials is motivated by applications to various fields as molecular physics [26], quantum cosmological models such as the Wheeler–DeWitt equation (see, e.g., [6]), and combustion models [21].

The singularity of inverse square potentials cannot be considered as a lower perturbation of the Laplacian since it has homogeneity -2, being critical from both a mathematical and a physical viewpoint.

Potentials of type $1/|x|^2$ arise, for instance, in Frank et al. [20], where a classification of singular spherical potentials is given in terms of the limit $\lim_{r\to 0} r^2 V(r)$. When the limit is finite and nontrivial, V is said to be a *transition potential.* This potential also arises in point-dipole interactions in molecular physics (see Lévy-Leblond [26]), where the interaction among the poles depends on their relative partitions and the intensity of the singularity in each of them.

Multipolar potentials of type $V = \sum_{i=1}^n \alpha_i/|x-x_i|^2$ are associated with the interaction of a finite number of electric dipoles. They describe molecular systems consisting of n nuclei of unit charge located at a finite number of points $x_1, \ldots, x_n$ and of n electrons. This type of systems is described by the Hartree-Fock model, where Coulomb multi-singular potentials arise in correspondence to the interactions between the electrons and the fixed nuclei; see Catto et al. [10].

Throughout this paper we study the qualitative properties of Schrödinger operators with inverse square potentials V, improving some results already known in the literature. The positivity and coercivity (in the L^2 norm) of such operators are strongly related to Hardy-type inequalities. The first well-known result relies on a 1-d inequality due to Hardy [22] which claims that

$$\forall u \in H_0^1(0,\infty), \quad \int_0^\infty u_x^2 dx > \frac{1}{4}\int_0^\infty \frac{u^2}{x^2} dx, \tag{3.1}$$

where the constant $1/4$ is optimal and not attained. Later on, this inequality was generalized to the multi-d case by Hardy–Littlewood–Polya [23] showing that for any Ω an open subset of $\mathbb{R}^N$, containing the origin, it holds that

$$\forall u \in H_0^1(\Omega), \quad \int_\Omega |\nabla u|^2 dx \geq \frac{(N-2)^2}{4} \int_\Omega \frac{u^2}{|x|^2} dx, \tag{3.2}$$

and the constant $(N-2)^2/4$ is optimal and not attained. The reader interested in the existing literature on the extensions of the classical Hardy inequality (3.2) with a singular potential is referred, in particular, to the following papers and the references therein: [2, 4, 5, 8, 9, 12–14, 19, 25, 29–31].

In the case of a multi-singular potential $V(x) = \sum_{i=1}^n \alpha_i/|x-x_i|^2$ with $\alpha_i \in \mathbb{R}$, where $x_i \in \mathbb{R}^N$ are the singular poles assumed to be fixed; the study of positivity of the quadratic functional

$$\mathscr{D}[u] = S_{\alpha_1,\dots,\alpha_n,x_1,\dots,x_n}[u] := \int_\Omega |\nabla u|^2 dx - \sum_{i=1}^n \alpha_i \int_\Omega \frac{u^2}{|x-x_i|^2} dx \tag{3.3}$$

is much more intricate since the interaction among the poles and their configuration matters.

Among other results, in [17] it was proven that when $\Omega = \mathbb{R}^N$, $\mathscr{D}$ is positive if and only if $\sum_{i=1}^n \alpha_i^+ \leq (N-2)^2/4$ for any configuration of the poles $x_1,\dots,x_n$, where $\alpha^+ = \max\{\alpha, 0\}$. Conversely, if $\sum_{i=1}^n \alpha_i^+ > (N-2)^2/4$, there exist configurations $x_1,\dots,x_n$ for which $\mathscr{D}$ is negative. These results have been improved later on by Bosi, Dolbeault, and Esteban [7] when deriving lower bounds of the spectrum of the operator $-\Delta - \mu \sum_{i=1}^n 1/|x-x_i|^2$, $\mu \in (0, (N-2)^2/4]$, with $x_1, x_2, \dots, x_n \in \mathbb{R}^N$. Roughly speaking, they showed that for any $\mu \in (0, (N-2)^2/4]$ and any configuration $x_1, x_2, \dots, x_n \in \mathbb{R}^N$, $n \geq 2$, there is a nonnegative constant $K_n < \pi^2$ such that

$$u \in C_0^\infty(\mathbb{R}^N), \quad \frac{K_n + (n+1)\mu}{d^2} \int_{\mathbb{R}^N} u^2 dx + \int_{\mathbb{R}^N} |\nabla u|^2 dx - \mu \sum_{i=1}^n \int_{\mathbb{R}^N} \frac{u^2}{|x-x_i|^2} dx \geq 0, \tag{3.4}$$

where d denotes $d := \min_{i \neq j} |x_i - x_j|/2$. The original proof of (3.4) in [7] employs a partition of unity technique, the so-called IMS (for Ismagilov, Morgan-Simon, Sigal, see [27, 28]), localizing the singular Schrödinger operator. Inequality (3.4) emphasizes that we can reach the critical singular mass $(N-2)^2/(4|x-x_i|^2)$ at any singular pole x_i to the price of adding a lower order term in L^2-norm.

To simplify the notations, here and throughout the paper when writing $\int \cdot dx$ we denote the integral over $\mathbb{R}^N$. Besides, using the so-called "expansion of the square" method, the authors in [7] proved the following inequality without lower order terms

$$\int |\nabla u|^2 dx \geq \frac{(N-2)^2}{4n} \sum_{i=1}^n \int \frac{u^2}{|x-x_i|^2} dx + \frac{(N-2)^2}{4n^2} \sum_{1 \leq i < j \leq n} \int \frac{|x_i - x_j|^2}{|x-x_i|^2 |x-x_j|^2} u^2 dx, \tag{3.5}$$

for any $u \in H^1(\mathbb{R}^N)$ and any set of poles $x_1, x_2, \dots, x_n \in \mathbb{R}^N$, $n \geq 2$. Let us denote the singular potentials in (3.5) by

$$V_i(x) := \frac{1}{|x-x_i|^2}, \quad V_{ij}(x) := \frac{|x_i - x_j|^2}{|x-x_i|^2 |x-x_j|^2}, \quad \forall i, j \in \{1, \dots, n\}, \quad i \neq j. \tag{3.6}$$

Observe that both potentials in (3.6) have a quadratic singularity at each pole x_i, i.e.

$$\lim_{x\to x_i} V_i(x)|x-x_i|^2 = 1, \quad \lim_{x\to x_i} V_{ij}(x)|x-x_i|^2 = 1, \quad \forall i,j\in\{1,\dots,n\}, \quad i\neq j. \tag{3.7}$$

Moreover, due to the symmetry we notice that for any $i\in\{1,\dots,n\}$ we have the asymptotic formula as $x\to x_i$:

$$\sum_{1\le i<j\le n} V_{ij}(x) = \frac{1}{|x-x_i|^2}\Big[\sum_{\substack{j=1\\ j\neq i}}^{n} \frac{|x_i-x_j|^2}{|x-x_j|^2} + O(|x-x_i|^2)\Big] \sim \frac{n-1}{|x-x_i|^2}, \quad \text{as } x\to x_i. \tag{3.8}$$

Therefore, due to (3.8) we remark that the total mass arising at a singular pole x_i in (3.5) is proportional to

$$\frac{(N-2)^2}{4n}\sum_{i=1}^{n} V_i(x) + \frac{(N-2)^2}{4n^2}\sum_{1\le i<j\le n} V_{ij}(x) \sim \frac{(N-2)^2}{4}\frac{2n-1}{n^2}\frac{1}{|x-x_i|^2}, \tag{3.9}$$

$$\text{as } x\to x_i, \quad \forall i\in\{1,\dots,n\}.$$

Note however that the multiplicative factor in each singularity in (3.9) is smaller than the optimal one that (3.4) yields for $\mu=(N-2)^2/4$. This is so because in (3.5) no other corrected terms are added.

We also mention the articles [1, 7, 16–18] and the references therein for other inequalities with multipolar singularities.

It is also worth mentioning the literature on Hardy-type inequalities of different nature than those studied in this paper. In particular, in [24] (see also the references therein) the authors investigated the so-called Hardy inequalities for m-dimensional particles of the form

$$\sum_{j=1}^{N}\int_{\mathbb{R}^{mN}} |\nabla_j u|^2 dx \ge \mathscr{C}(m,N)\sum_{1\le i<j\le N}\int_{\mathbb{R}^{mN}} \frac{|u|^2}{r_{ij}^2}dx, \tag{3.10}$$

giving explicit lower bounds for the best constant $\mathscr{C}(m,N)$ when applied to test functions $u\in H^1(\mathbb{R}^{mN})$. In (3.10), we have denoted $x=(x_1,\dots,x_N)$ with $x_i=(x_{i,1},\dots,x_{i,m})\in\mathbb{R}^m$, $r_{ij}=\sqrt{\sum_{k=1}^{m}(x_{i,k}-x_{j,k})^2}$ for all $i,j\in\{1,\dots,N\}$ and ∇_j for the gradient associated to the j-th particle. Roughly speaking, the singularities in (3.10) occur in uncountable sets given by the diagonals $x_i=x_j$. The optimality of (3.10) is still an open question excepting the case $m=1$ for which $\mathscr{C}(1,N)=1/2$ provided the test functions u vanish on diagonals $x_i=x_j$.

In this paper we develop new optimal Hardy-type inequalities with multipolar potentials.

In Sect. 3.2 we present some general strategies to handle the Hardy inequalities.

In Sect. 3.3 we complement and improve some results in [7] related to inequality (3.5). Our proofs use convenient transformations involving the product of the fundamental solutions E_i of the Laplacian at the poles x_i, $i \in \{1,\ldots,n\}$. In Theorem 3.1 of Sect. 3.3, we give an optimal inequality for the operator $A_\lambda = -\Delta - \lambda \sum_{1\leq i<j\leq n} V_{ij}(x)$, $\lambda > 0$, showing a better singular behavior of the potential at each pole x_i than pointed out in (3.5)–(3.9). This allows to show the existence of bounded domains in which, for the bipolar Hardy inequality, the L^2-reminder term in (3.4) can be removed. For this to be done, the best situation seems to be the case in which the singularities are localized on the boundary of the domain, as emphasized in Sect. 3.4, Proposition 3.1.

In Sect. 3.5 we end up with some further comments and open questions.

3.2 Preliminaries: Some Strategies to Prove Hardy-Type Inequalities

There are several techniques for proving Hardy inequalities in smooth domains (including the whole space) which are all interlinked by the following integral identity.

Assume $\Omega \subset \mathbb{R}^N$, $N \geq 2$, is an open set and let $x_1,\ldots,x_n \in \overline{\Omega}$. We also consider a distribution $\varphi \in D'(\Omega)$ such that $\varphi(x) > 0$ in $\Omega \setminus \{x_1,\ldots,x_n\}$ and $\varphi \in C^2(\Omega \setminus \{x_1,x_2,\ldots x_n\})$. Then it holds that

$$\int_\Omega \left[|\nabla u|^2 + \frac{\Delta\varphi}{\varphi}u^2\right] dx = \int_\Omega \left|\nabla u - \frac{\nabla\varphi}{\varphi}u\right|^2 dx = \int_\Omega \varphi^2 |\nabla(u\varphi^{-1})|^2 dx,$$
$$\forall u \in C_0^1(\Omega \setminus \{x_1,x_2,\ldots x_n\}). \tag{3.11}$$

The proof of (3.11) can be done using integration by parts. In particular, (3.11) can be extended to test functions $u \in H_0^1(\Omega)$ since $C_0^1(\Omega \setminus \{x_1,\ldots,x_n\})$ is dense in $H_0^1(\Omega)$, see, e.g., [15].

The identity (3.11) could be extended to more general classes of distributions φ depending on the applications that we have in mind. Here we are interested in applications to Hardy inequalities with multipolar potentials located at the poles $x_1,\ldots,x_n$, with $x_i \neq x_j$ for all $i \neq j$ and $i,j \in \{1,\ldots,n\}$.

Various aspects of the identities involved in (3.11) have been used in the literature to prove and analyze Hardy inequalities in different contexts. But, as far as we know, (3.11) has not been stated explicitly as it stands before.

Identity (3.11) could be directly applied to obtain Hardy inequalities with potentials of the form $-\Delta\varphi/\varphi$, i.e.

$$\int_\Omega |\nabla u|^2 dx \geq \int_\Omega \left(-\frac{\Delta\varphi}{\varphi}\right) u^2 dx, \quad \forall u \in H_0^1(\Omega). \tag{3.12}$$

In order to derive inequalities for a concrete potential $V = V(x) \in L^1_{loc}(\Omega)$, one needs to look for a corresponding φ such that

$$-\frac{\Delta\varphi}{\varphi} \geq V(x), \quad \forall x \in \Omega. \tag{3.13}$$

Some of the existing techniques to prove Hardy-type inequalities use "the expansion of the square" method (e.g., [7]) or suitable functional transformations (e.g., [3, 8]). In view of (3.11), all these techniques are actually equivalent, and the problem can always be reduced to checking pointwise inequalities for a potential V and a corresponding φ as in (3.13).

Optimality. For a general φ satisfying (3.11), we cannot say anything about the optimality of (3.12). To argue in that sense, next we give a counterexample by means of the standard Hardy inequality. Assume $\Omega = \mathbb{R}^N$, $N \geq 3$ and let $\lambda < \lambda_\star := (N-2)^2/4$. Then we consider

$$\varphi = |x|^{-(N-2)/2+\sqrt{\lambda_\star-\lambda}}$$

and observe that $\varphi > 0$ in $\mathbb{R}^N$, $\varphi \in C^\infty(\mathbb{R}^N \setminus \{0\})$ and therefore φ satisfies the identity (3.11) before. Then for such φ, inequality (3.12) becomes

$$\int |\nabla u|^2 dx \geq \lambda \int \frac{u^2}{|x|^2} dx, \quad \forall u \in H^1(\mathbb{R}^N),$$

which is not optimal as follows from (3.2).

3.3 Multipolar Hardy Inequalities

Assume $N \geq 3$ and consider n poles $x_1, \dots x_n \in \mathbb{R}^N$, $n \geq 2$, such that $x_i \neq x_j$ for any $i \neq j$, and $i, j \in \{1, 2, \dots, n\}$. In the sequel we improve the result (3.5) by Bosi et al. [7]. The main result of this section is as follows:

Theorem 3.1. *It holds that*

$$\int |\nabla u|^2 dx \geq \frac{(N-2)^2}{n^2} \sum_{1 \leq i < j \leq n} \int \left| \frac{x - x_i}{|x - x_i|^2} - \frac{x - x_j}{|x - x_j|^2} \right|^2 u^2 dx, \quad \forall u \in H^1(\mathbb{R}^N), \tag{3.14}$$

or equivalently

$$\int |\nabla u|^2 dx \geq \frac{(N-2)^2}{n^2} \sum_{1 \leq i < j \leq n} \int \frac{|x_i - x_j|^2}{|x - x_i|^2 |x - x_j|^2} u^2 dx, \quad \forall u \in H^1(\mathbb{R}^N). \tag{3.15}$$

Moreover, the constant $(N-2)^2/n^2$ is optimal.

In the sequel, we prove Theorem 3.1 applying identity (3.11) before.

Proof of Theorem 3.1.

By density arguments it is sufficient to prove (3.14) for any function $u \in C_0^1(\mathbb{R}^N \setminus \{x_1, \dots, x_n\})$. Then, according to (3.11)–(3.13), it is enough to find a proper φ satisfying

$$-\frac{\Delta\varphi}{\varphi} \geq \frac{(N-2)^2}{n^2} \sum_{1\leq i<j\leq n} \left| \frac{x-x_i}{|x-x_i|^2} - \frac{x-x_j}{|x-x_j|^2} \right|^2, \quad \forall x \in \mathbb{R}^N \setminus \{x_1, \dots, x_n\}.$$

Let us choose

$$\varphi = E^{1/n} = \prod_{i=1}^{n} E_i^{1/n}, \tag{3.16}$$

where $E = \prod_{i=1}^{n} E_i$ and E_i is the fundamental solution of the Laplacian at the singular pole x_i, $i \in \{1, \dots, n\}$, i.e.

$$E_i = \frac{|x-x_i|^{2-N}}{\omega_N (N-2)}. \tag{3.17}$$

Here ω_N denotes the $(N-1)$-Hausdorff measure of the unit sphere S^{N-1} in $\mathbb{R}^N$. Note that φ chosen in (3.16) verifies the integrability conditions to validate the identity (3.11). On the other hand, we have

$$\nabla E = \Big(\sum_{i=1}^{n} \frac{\nabla E_i}{E_i}\Big) E, \quad \forall x \in \mathbb{R}^N \setminus \{x_1, \dots, x_n\}. \tag{3.18}$$

Due to the fact that $-\Delta E_i = \delta_{x_i}$ for all $i \in \{1, \dots, n\}$ we obtain

$$\begin{aligned} \Delta E &= \Big(\sum_{i=1} \frac{\Delta E_i}{E_i} + 2 \sum_{1\leq i<j\leq n} \frac{\nabla E_i \cdot \nabla E_j}{E_i E_j}\Big) E \\ &= 2\Big(\sum_{1\leq i<j\leq n} \frac{\nabla E_i \cdot \nabla E_j}{E_i E_j}\Big) E, \quad \forall x \in \mathbb{R}^N \setminus \{x_1, \dots, x_n\}. \end{aligned} \tag{3.19}$$

Combining (3.16)–(3.19) we notice that φ satisfies precisely the equation

$$-\Delta\varphi - \frac{(N-2)^2}{n^2} \sum_{1\leq i<j\leq n} \left| \frac{x-x_i}{|x-x_i|^2} - \frac{x-x_j}{|x-x_j|^2} \right|^2 \varphi = 0, \quad \forall x \in \mathbb{R}^N \setminus \{x_1, \dots, x_n\}. \tag{3.20}$$

Then, identity (3.11) becomes

$$\begin{aligned} &\int \Big[|\nabla u|^2 - \frac{(N-2)^2}{n^2} \sum_{1\leq i<j\leq n} \left| \frac{x-x_i}{|x-x_i|^2} - \frac{x-x_j}{|x-x_j|^2} \right|^2 u^2 \Big] dx \\ &\qquad = \int \left| \nabla u - \frac{\nabla(E^{1/n})}{E^{1/n}} u \right|^2 dx = \int |\nabla(u E^{-1/n})|^2 E^{2/n} dx \geq 0. \end{aligned} \tag{3.21}$$

This concludes the proof of (3.14).

Optimality of the constant.

Next we complete the proof of Theorem 3.1 by showing the optimality of the constant $(N-2)^2/n^2$ in (3.14).

According to (3.21), we actually showed that for all $u \in H^1(\mathbb{R}^N)$ we have

$$\int |\nabla u|^2 dx - \frac{(N-2)^2}{n^2} \sum_{1\leq i<j\leq n} \int \left| \frac{x-x_i}{|x-x_i|^2} - \frac{x-x_j}{|x-x_j|^2} \right|^2 u^2 dx \\ = \int |\nabla (uE^{-1/n})|^2 E^{2/n} dx. \quad (3.22)$$

Here $B_r(x) \subset \mathbb{R}^N$, for some fixed $r > 0$ and $x \in \mathbb{R}^N$, denotes the ball of radius r centered at x.

For $\varepsilon > 0$ aimed to be small ($\varepsilon < \min\{1, d/2\}$), we consider the cutoff functions $\theta_\varepsilon \in C_0(\mathbb{R}^N)$ defined by

$$\theta_\varepsilon(x) = \begin{cases} 0, & |x-x_i| \leq \varepsilon^2, \quad \forall i \in \{1,\dots,n\}, \\ \frac{\log |x-x_i|/\varepsilon^2}{\log 1/\varepsilon}, & \varepsilon^2 \leq |x-x_i| \leq \varepsilon, \quad \forall i \in \{1,\dots,n\}, \\ 1, & x \in B_{1/\varepsilon}(0) \setminus \cup_{i=1}^n B_\varepsilon(x_i), \\ \varepsilon\left(\frac{2}{\varepsilon} - |x|\right), & 1/\varepsilon \leq |x| \leq 2/\varepsilon, \\ 0, & |x| \geq 2/\varepsilon. \end{cases} \quad (3.23)$$

Then we consider the sequence $\{u_\varepsilon\}_{\varepsilon>0}$ defined by

$$u_\varepsilon := E^{1/n}\theta_\varepsilon, \quad \varepsilon > 0,$$

which belongs to $C_0(\mathbb{R}^N)$ $(\subset H^1(\mathbb{R}^N))$ since θ_ε belongs to $C_0(\mathbb{R}^N)$ and is supported far from the poles x_i.

In the sequel we show that $\{u_\varepsilon\}_{\varepsilon>0}$ is an approximating sequence for $(N-2)^2/n^2$, that is,

$$\lim_{\varepsilon\searrow 0} \frac{\int |\nabla u_\varepsilon|^2 dx}{\sum_{1\leq i<j\leq n} \int \left| \frac{x-x_i}{|x-x_i|^2} - \frac{x-x_j}{|x-x_j|^2} \right|^2 u_\varepsilon^2 dx} = \frac{(N-2)^2}{n^2}. \quad (3.24)$$

Firstly, we can easily notice that there exists a constant $C > 0$ depending on d (uniformly in ε) such that

$$\sum_{1\leq i<j\leq n} \int \left| \frac{x-x_i}{|x-x_i|^2} - \frac{x-x_j}{|x-x_j|^2} \right|^2 u_\varepsilon^2 dx > C, \quad \forall \varepsilon > 0. \quad (3.25)$$

On the other hand, taking into account where $\nabla\theta_\varepsilon$ is supported, we split in two parts the right-hand side of (3.22) as

$$\begin{aligned}
&\int |\nabla(u_\varepsilon E^{-1/n})|^2 E^{2/n} dx \\
&\quad = \sum_{i=1}^{n} \int_{B_\varepsilon(x_i)\setminus B_{\varepsilon^2}(x_i)} |\nabla \theta_\varepsilon|^2 E^{2/n} dx + \int_{B_{2/\varepsilon}(0)\setminus B_{1/\varepsilon}(0)} |\nabla \theta_\varepsilon|^2 E^{2/n} dx \\
&\qquad := I_1 + I_2.
\end{aligned} \tag{3.26}$$

Next we obtain

$$I_1 = \frac{1}{\omega_N^2 (N-2)^2} \sum_{i=1}^{n} \int_{B_\varepsilon(x_i)\setminus B_{\varepsilon^2}(x_i)} \frac{1}{\log^2(1/\varepsilon)} \frac{1}{|x-x_i|^2} \prod_{j=1}^{n} |x-x_j|^{2(2-N)/n} dx. \tag{3.27}$$

Since

$$|x-x_j| \geq \frac{d}{2}, \quad \forall x \in B_\varepsilon(x_i), \quad \forall j \neq i, \quad \forall i,j \in \{1,\dots,n\}, \tag{3.28}$$

from (3.27) we deduce that

$$\begin{aligned}
I_1 &\leq \frac{\left(\frac{d}{2}\right)^{2(n-1)(2-N)/n}}{\omega_N^2 (N-2)^2} \frac{1}{\log^2(1/\varepsilon)} \sum_{i=1}^{n} \int_{B_\varepsilon(x_i)\setminus B_{\varepsilon^2}(x_i)} |x-x_i|^{2(2-N)/n-2} dx \\
&= \frac{n\left(\frac{d}{2}\right)^{2(n-1)(2-N)/n}}{\omega_N^2 (N-2)^2} \frac{1}{\log^2(1/\varepsilon)} \int_{\varepsilon^2}^{\varepsilon} r^{N-1} \int_{S^{N-1}} r^{2(2-N)/n-2} d\sigma dr \\
&= \frac{n\left(\frac{d}{2}\right)^{2(n-1)(2-N)/n}}{\omega_N (N-2)^2} \frac{1}{\log^2(1/\varepsilon)} \int_{\varepsilon^2}^{\varepsilon} r^{(N-2)(1-2/n)-1} dr.
\end{aligned} \tag{3.29}$$

From (3.27) and (3.29) we obtain that

$$I_1 = \begin{cases} O\left(\frac{1}{\log(1/\varepsilon)}\right), & n=2, \\ O(\varepsilon^{(N-2)(1-2/n)}), & n \geq 3. \end{cases} \tag{3.30}$$

Taking $\varepsilon > 0$ small enough such that $\varepsilon < 1/2m$, where $m = \max_{i=1,\dots,n} |x_i|$, it holds

$$|x-x_i| \geq \frac{1}{2\varepsilon}, \quad \forall x \in B_{2/\varepsilon}(0) \setminus B_{1/\varepsilon}(0), \quad \forall i \in \{1,\dots,n\}. \tag{3.31}$$

Due to (3.31) we obtain

$$\begin{aligned}
I_2 &= \frac{1}{\omega_N^2 (N-2)^2} \int_{B_{2/\varepsilon}(0)\setminus B_{1/\varepsilon}(0)} \varepsilon^2 \prod_{i=1}^{n} |x-x_i|^{2(2-N)/n} dx \\
&\leq \frac{1}{\omega_N^2 (N-2)^2} \int_{B_{2/\varepsilon}(0)\setminus B_{1/\varepsilon}(0)} \varepsilon^2 \prod_{i=1}^{n} \left(\frac{1}{2\varepsilon}\right)^{2(2-N)/n} dx \\
&= \frac{1}{\omega_N^2 (N-2)^2} \left(\frac{1}{2}\right)^{2(2-N)} \int_{B_{2/\varepsilon}(0)\setminus B_{1/\varepsilon}(0)} \varepsilon^{2(N-1)} dx
\end{aligned}$$

$$= \frac{2^{2(N-2)}}{\omega_N (N-2)^2} \varepsilon^{2(N-1)} \int_{1/\varepsilon}^{2/\varepsilon} r^{N-1} dr$$
$$= O(\varepsilon^{N-2}), \quad \text{as } \varepsilon \to 0. \tag{3.32}$$

In conclusion, according to (3.26), (3.30), and (3.32) we get

$$\lim_{\varepsilon \searrow 0} \int |\nabla(u_\varepsilon E^{-1/n})|^2 E^{2/n} dx = 0, \quad \forall n \geq 2. \tag{3.33}$$

Combining (3.22), (3.25), and (3.33) we end up with the optimality of $(N-2)^2/n^2$ as in (3.24), and the proof of Theorem 3.1 is complete.

□

Remark 3.1.
Our optimal result in Theorem 3.1 provides an inequality with a positive singular quadratic potential which behaves asymptotically like

$$\frac{(N-2)^2}{n^2} \sum_{1 \leq i < j \leq n} V_{ij}(x) \sim \frac{(N-2)^2}{4} \frac{4n-4}{n^2} \frac{1}{|x-x_i|^2} \qquad \text{as } x \to x_i, \quad \forall i \in \{1, \ldots, n\}, \tag{3.34}$$

at each pole x_i. In particular, for any $n \geq 2$, Theorem 3.1 represents an improvement of (3.5), in the sense that the multiplication factor in (3.34) which corresponds to the quadratic singularity is larger than that one obtained in inequality (3.5) as emphasized in (3.9).

Remark 3.2.
The proof of (3.5) in [7] was obtained by expanding the square

$$\int \left| \nabla u + \alpha \sum_{i=1}^{n} \frac{x-x_i}{|x-x_i|^2} u \right|^2 dx \geq 0, \quad \alpha \in \mathbb{R}, \tag{3.35}$$

which gives

$$0 \leq \int |\nabla u|^2 dx + [n\alpha^2 - (N-2)\alpha] \sum_{i=1}^{n} \int \frac{u^2}{|x-x_i|^2} dx$$
$$- \alpha^2 \sum_{1 \leq i < j \leq n} \int \frac{|x_i - x_j|^2}{|x-x_i|^2 |x-x_j|^2} u^2 dx. \tag{3.36}$$

More precisely, (3.5) is a consequence of (3.36) when $\alpha = (N-2)/(2n)$. Moreover, we remark that the expansion (3.36) also applies to derive the inequality of Theorem 3.1 with a different choice of α, that is, $\alpha = (N-2)/n$.

The quadratic term in (3.21) is given by the formula

$$\int \left| \nabla u - \frac{\nabla(E^{1/n})}{E^{1/n}} u \right|^2 dx = \int \left| \nabla u + \frac{N-2}{n} \sum_{i=1}^{n} \frac{x-x_0}{|x-x_0|^2} u \right|^2 dx,$$

which motivates the use of the "expansion of the square" emphasized above for $\alpha = (N-2)/n$. This was not observed in [7]. In fact we got to this point indirectly as a consequence of the direct application of identity (3.11).

Remark 3.3.
Adimurthi et al. proved in particular in [3] that, whenever E satisfies $-\Delta E = \sum_{1\leq i\leq n} \delta_{x_i}$ for some given poles $x_1, \ldots, x_n \in \mathbb{R}^N$, the following inequality holds

$$\int |\nabla u|^2 dx \geq \frac{1}{4} \int \left|\frac{\nabla E}{E}\right|^2 u^2 dx, \quad \forall u \in C_0^\infty(\mathbb{R}^N). \tag{3.37}$$

A direct application of (3.37) in the context of multipolar Hardy inequalities would consist on taking $E = E_1 + \ldots + E_n$. If $N \geq 3$, we then get (cf. [1])

$$\int |\nabla u|^2 dx \geq \frac{(N-2)^2}{4} \int \left|\frac{\sum_{i=1}^n (x-x_i)|x-x_i|^{-N}}{\sum_{i=1}^n |x-x_i|^{2-N}}\right|^2 u^2 dx, \quad \forall u \in H^1(\mathbb{R}^N). \tag{3.38}$$

Observe that the potential V in (3.38), given by

$$V_{N,n,x_1,\ldots,x_n}(x) := \left|\frac{\sum_{i=1}^n (x-x_i)|x-x_i|^{-N}}{\sum_{i=1}^n |x-x_i|^{2-N}}\right|^2, \tag{3.39}$$

is nonnegative and moreover has a quadratic singularity at each pole x_i. More precisely, $V_{N,n,x_1,\ldots,x_n}$ satisfies

$$V_{N,n,x_1,\ldots,x_n}(x) = \frac{1}{|x-x_i|^2} + O(|x-x_i|^{N-4}), \text{ as } x \to x_i, \quad \forall i \in \{1, \ldots, n\}, \tag{3.40}$$

respectively

$$V_{N,n,x_1,\ldots,x_n}(x) = O(\frac{1}{|x|^2}), \quad \text{as } |x| \to \infty. \tag{3.41}$$

From (3.40) and (3.41) we can easily deduce that

$$V_{N,n,x_1,\ldots,x_n}(x) = \sum_{i=1}^n \frac{1}{|x-x_i|^2} + O(1), \quad \forall x \in \mathbb{R}^N, \tag{3.42}$$

where $O(1)$ denotes a changing sign quantity, uniformly bounded in $\mathbb{R}^N$. For $N \geq 4$, the identification (3.42) shows that inequality (3.38) allows to deduce an inequality in the spirit of (3.4) in which the same critical singular potential is obtained, paying the prize of adding a lower order term in L^2-norm. The multiplication factor of the lower order term obtained through (3.38) remains to be compared with that one which corresponds to (3.4).

On the contrary, inequality (3.38) does not allow to get optimal results as in Theorem 3.1 when removing the corrected lower order terms in L^2-norm.

We point out that the key role for showing Theorem 3.1 was played by identity (3.11) applying for suitable distributions involving the product of the fundamental solutions of the Laplacian at each singular pole x_i. This allows to prove optimal Hardy inequalities for singular quadratic potentials of the form

$$W(x) = \sum_{i=1}^{n} \frac{\lambda_i(x)}{|x - x_i|^2}, \quad \forall x \in \mathbb{R}^N,$$

where

$$\lambda_i(x) > 0 \text{ in } \mathbb{R}^N, \quad \lim_{x \to x_i} \lambda_i(x) = (n-1)\frac{(N-2)^2}{2n^2}, \quad \forall i \in \{1, \dots, n\}.$$

The weights λ_i in Theorem 3.1 are given by $\lambda_i(x) = (N-2)^2/(2n^2) \sum_{\substack{j=1 \\ j \neq i}}^{n} |x_i - x_j|^2 / |x - x_j|^2$.

As we mentioned before, the potential $V_{N,n,x_1,\dots x_n}$ cannot be compared with W on any bounded connected domain Ω with $x_1, \dots, x_n \in \overline{\Omega}$. Indeed, next we emphasize this in two concrete examples.

Firstly, for $N = 3$, $n = 2$, we consider the singular poles $0, x_0 \in \overline{\Omega} \subset \mathbb{R}^3$, and we obtain $V_{3,2,0,x_0}(x_0/2) = 0$ while $W(x_0/2) > 0$.

Secondly, let us consider a configuration with three singular poles $x_1, x_2, x_3 \in \mathbb{R}^3$ determining an equilateral triangle such that

$$|x_1| = |x_2| = |x_3| > 0, \quad x_1 + x_2 + x_3 = 0$$

and let $\Omega \subset \mathbb{R}^3$ be a connected bounded open set with $x_1, x_2, x_3 \in \overline{\Omega}$. Then $V_{3,3,x_1,x_2,x_3}(0) = 0$ while $W(0) > 0$.

3.4 New Bounds for the Bipolar Hardy Inequality in Bounded Domains

We now present some consequences of the previous multipolar Hardy inequality in Theorem 3.1 to bounded domains in $H_0^1(\Omega)$.

In this subsection we present some applications of Theorem 3.1 to bounded domains in the case of a bipolar potential

$$V(x) = \frac{1}{|x - x_1|^2} + \frac{1}{|x - x_2|^2}, \tag{3.43}$$

for some $x_1, x_2 \in \mathbb{R}^N$, $N \geq 3$ with $x_1 \neq x_2$. In consequence, we derive new lower bounds for the bipolar Hardy inequality, which turn out to be optimal in the case where the poles are located on the boundary of the domain.

We have seen that Theorem 3.1 provides an inequality involving a bipolar potential which behaves asymptotically like

$$\frac{(N-2)^2}{4}V_{12}(x) \sim \frac{(N-2)^2}{4}\frac{1}{|x-x_i|^2}, \qquad \text{as } x \to x_i, \quad \forall i \in \{1,2\}. \tag{3.44}$$

On the other hand, inequality (3.5) provides a bipolar potential with a weaker quadratic singularity which is asymptotically given by

$$\frac{(N-2)^2}{8}(V_1+V_2) + \frac{(N-2)^2}{16}V_{12}(x) \sim \frac{3(N-2)^2}{16}\frac{1}{|x-x_i|^2}, \tag{3.45}$$
$$\text{as } x \to x_i, \quad \forall i \in \{1,2\}.$$

Theorem 3.1 may give better lower bounds than inequality (3.5) for the Hardy inequality with the bipolar potential V as in (3.43). The main results of this section are as follows:

As a consequence of Theorem 3.1, we have

Proposition 3.1. *Assume* $0 \le \alpha, \beta \le 1$.
For any $x_1 \neq x_2$ *and for all* $u \in C_0^\infty(B_{r(x_1,x_2)}(C(x_1,x_2)))$*, we have*

$$\int_{B_{r(x_1,x_2)}(C(x_1,x_2))} |\nabla u|^2 dx \ge \frac{(N-2)^2}{4}\int_{B_{r(x_1,x_2)}(C(x_1,x_2))}\left[\frac{\alpha}{|x-x_1|^2}+\frac{\beta}{|x-x_2|^2}\right]u^2 dx, \tag{3.46}$$

where $B_{r(x_1,x_2)}(C(x_1,x_2))$ *is the ball centered at the point*

$$C(x_1,x_2) = \frac{\beta}{\alpha+\beta}x_1 + \frac{\alpha}{\alpha+\beta}x_2,$$

of radius

$$r(x_1,x_2) = \frac{\sqrt{\alpha+\beta-\alpha\beta}}{\alpha+\beta}|x_1-x_2|,$$

as shown in Fig. 3.1.

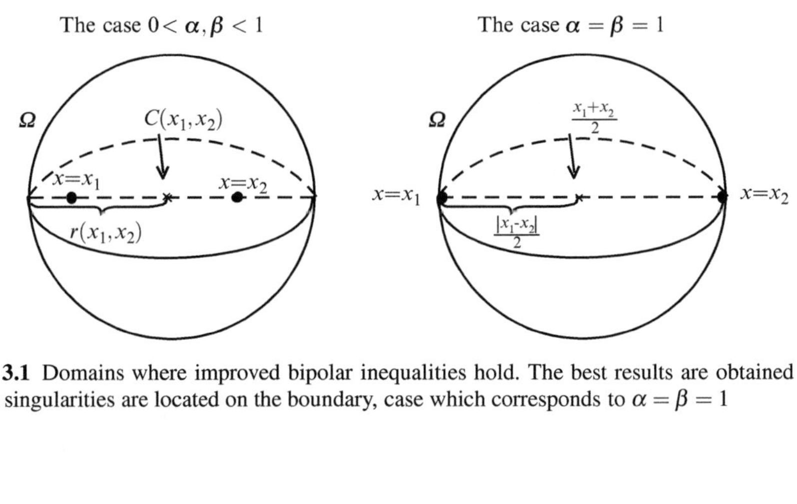

Fig. 3.1 Domains where improved bipolar inequalities hold. The best results are obtained when both singularities are located on the boundary, case which corresponds to $\alpha=\beta=1$

As a consequence of inequality (3.5) and Proposition 3.1 we have

Proposition 3.2. *Assume* $0 \le \alpha, \beta \le 1$.
For any $x_1 \neq x_2$ *and for all* $u \in C_0^\infty(B_{r(x_1,x_2)}(C(x_1,x_2)))$, *we have*

$$\int_{B_{r(x_1,x_2)}(C(x_1,x_2))} |\nabla u|^2 dx \ge \Big[\frac{(N-2)^2}{8} + \frac{(N-2)^2}{16}\alpha\Big] \int_{B_{r(x_1,x_2)}(C(x_1,x_2))} \frac{u^2}{|x-x_1|^2} dx + \Big[\frac{(N-2)^2}{8} + \frac{(N-2)^2}{16}\beta\Big] \int_{B_{r(x_1,x_2)}(C(x_1,x_2))} \frac{u^2}{|x-x_2|^2} dx, \tag{3.47}$$

where $B_{r(x_1,x_2)}(C(x_1,x_2))$ *is the ball centered at the point*

$$C(x_1,x_2) = \frac{\beta}{\alpha+\beta} x_1 + \frac{\alpha}{\alpha+\beta} x_2,$$

of radius

$$r(x_1,x_2) = \frac{\sqrt{\alpha+\beta-\alpha\beta}}{\alpha+\beta} |x_1 - x_2|,$$

as shown in Fig. 3.1.

Remark 3.4. The constraints $\alpha, \beta \le 1$ impose to the singular poles x_1, x_2 to belong to $\in \overline{B_{r(x_1,x_2)}(C(x_1,x_2))}$.

Remark 3.5. We observe that for α, β getting closer to 1, the result of Proposition 3.1 is better than the one of Proposition 3.2.

Next we prove only Proposition 3.1 since the proof of Proposition 3.2 follows the same steps.

Proof of Proposition 3.1.

Let us consider an open bounded subset $\Omega \subset \mathbb{R}^N$, $N \ge 3$. Applying Theorem 3.1 we have that

$$\int_\Omega |\nabla u|^2 dx \ge \frac{(N-2)^2}{4} \int_\Omega \Big| \frac{x-x_1}{|x-x_1|^2} - \frac{x-x_2}{|x-x_2|^2} \Big|^2 u^2 dx, \quad \forall u \in H_0^1(\Omega). \tag{3.48}$$

In the sequel, we are seeking for domains $\Omega \subset \mathbb{R}^N$ such that $x_1, x_2 \in \overline{\Omega}$ and

$$\Big| \frac{x-x_1}{|x-x_1|^2} - \frac{(x-x_2)}{|x-x_2|^2} \Big|^2 \ge \frac{\alpha}{|x-x_1|^2} + \frac{\beta}{|x-x_2|^2}, \quad \forall x \in \overline{\Omega}. \tag{3.49}$$

Using the identity $2(x-x_1)(x-x_2) = |x-x_1|^2 + |x-x_2|^2 - |x_1-x_2|^2$, then (3.49) is equivalent to

$$|x_1 - x_2|^2 \ge \alpha |x-x_2|^2 + \beta |x-x_1|^2, \quad \forall x \in \overline{\Omega}. \tag{3.50}$$

Expanding the squares in (3.50) and dividing by $\alpha+\beta$ we obtain

$$\frac{1-\beta}{\alpha+\beta}|x_1|^2+\frac{1-\alpha}{\alpha+\beta}|x_2|^2-\frac{2}{\alpha+\beta}x_1\cdot x_2\geq |x|^2-2x\cdot\Big(\frac{\alpha}{\alpha+\beta}x_2+\frac{\beta}{\alpha+\beta}x_1\Big). \tag{3.51}$$

Coupling the squares we rewrite (3.51) as

$$\begin{aligned}\frac{1-\beta}{\alpha+\beta}|x_1|^2+\frac{1-\alpha}{\alpha+\beta}|x_2|^2-\frac{2}{\alpha+\beta}x_1\cdot x_2+\Big|\frac{\alpha}{\alpha+\beta}x_2+\frac{\beta}{\alpha+\beta}x_1\Big|^2&\\ \geq\Big|x-\Big(\frac{\alpha}{\alpha+\beta}x_2+\frac{\beta}{\alpha+\beta}x_1\Big)\Big|^2.&\end{aligned} \tag{3.52}$$

After some computations on the left-hand side of (3.52) we get

$$\frac{\alpha+\beta-\alpha\beta}{(\alpha+\beta)^2}|x_1-x_2|^2\geq\Big|x-\Big(\frac{\alpha}{\alpha+\beta}x_2+\frac{\beta}{\alpha+\beta}x_1\Big)\Big|^2,\quad\forall x\in\overline{\Omega}. \tag{3.53}$$

Due to this, the proof is finished by identifying properly the set Ω. □

We notice that, as far as α and β get closer to 1, the poles $x=x_1$, respectively, $x=x_2$, are pushed to the boundary of the domain as drawn in Fig. 3.1. Indeed, if $\alpha=\beta=1$, then x_1 and x_2 are located on the boundary of $B_{r(x_1,x_2)}(C(x_1,x_2))=B_{|x_1-x_2|/2}((x_1+x_2)/2)$. Moreover, we will have the nontrivial inequality

$$\int_{B_{\frac{|x_1-x_2|}{2}}\left(\frac{x_1+x_2}{2}\right)}|\nabla u|^2dx\geq\frac{(N-2)^2}{4}\int_{B_{\frac{|x_1-x_2|}{2}}\left(\frac{x_1+x_2}{2}\right)}\Big[\frac{1}{|x-x_1|^2}+\frac{1}{|x-x_2|^2}\Big]u^2dx \tag{3.54}$$

for all $u\in C_0^1(B_{|x_1-x_2|/2}((x_1+x_2)/2))$.

As we said before, inequality (3.54) is new and not trivial. In fact, it provides an improved result in higher dimensions as follows.

Applying Hardy inequalities with boundary singularities (see, e.g., [11]), we have that

$$\int_{B_{\frac{|x_1-x_2|}{2}}\left(\frac{x_1+x_2}{2}\right)}|\nabla u|^2dx\geq\frac{N^2}{4}\int_{B_{\frac{|x_1-x_2|}{2}}\left(\frac{x_1+x_2}{2}\right)}\frac{u^2}{|x-x_1|^2}dx$$

and

$$\int_{B_{\frac{|x_1-x_2|}{2}}\left(\frac{x_1+x_2}{2}\right)}|\nabla u|^2dx\geq\frac{N^2}{4}\int_{B_{\frac{|x_1-x_2|}{2}}\left(\frac{x_1+x_2}{2}\right)}\frac{u^2}{|x-x_2|^2}dx,$$

the constant $N^2/4$ being optimal in both cases. Thus

$$\int_{B_{\frac{|x_1-x_2|}{2}}\left(\frac{x_1+x_2}{2}\right)}|\nabla u|^2dx\geq\frac{N^2}{8}\int_{B_{\frac{|x_1-x_2|}{2}}\left(\frac{x_1+x_2}{2}\right)}\Big[\frac{1}{|x-x_1|^2}+\frac{1}{|x-x_2|^2}\Big]u^2dx.$$

Note that inequality (3.54) is better for $N\geq 7$ since

$$\frac{(N-2)^2}{4} \geq \frac{N^2}{8}, \quad \forall N \geq 7.$$

3.5 Further Comments and Open Problems

- In the spirit of (3.11) we may look for potentials V with an infinite number of singularities for which the Hardy inequality holds true. Besides, as such V are given as an infinite series one needs to make sure that they are well-defined. For instance, a potential of the form

$$V(x) = \sum_{(i,j,k)\in\mathbb{Z}^3} \frac{1}{|x_1-i|^2+|x_2-j|^2+|x_3-k|^2}, \quad x=(x_1,x_2,x_3)\in\mathbb{R}^3,$$

  diverges at every point $x\in\mathbb{R}^3$ and therefore the corresponding Hardy inequality does not make sense.
  The issue of potentials with an infinite number of singularities will be the subject of a forthcoming work.

- The optimality of the result in Proposition 3.1 remains to be analyzed.
- Identity (3.11) could be also applied for other choices of φ that we performed in Theorem 3.1. In particular, if we choose $\varphi = E^{1/2} = \prod_{i=1}^n E_i^{1/2}$, then we deduce the following inequality

$$\int |\nabla u|^2 dx$$

$$\geq \frac{(N-2)^2}{4}\sum_{i=1}^n \int \frac{u^2}{|x-x_i|^2}dx - \frac{(N-2)^2}{2}\sum_{1\leq i<j\leq n}\int \frac{(x-x_i)(x-x_j)}{|x-x_i|^2|x-x_j|^2}u^2 dx, \tag{3.55}$$

  for all $u\in C_0^\infty(\mathbb{R}^N)$. Next we may wonder the question if we can reprove inequality (3.4) through (3.55) since the potential in (3.55) has a critical quadratic singularity at any pole $x_i\in\mathbb{R}^N$. Firstly, we note that this is possible for the subcritical case $\mu < (N-2)^2/4$ which allows to get lower order terms in L^2-norm. However, (3.55) does not provide better L^2 lower order term than (3.4) does.
  In the critical case $\mu = (N-2)^2/4$ we cannot obtain L^2-reminder terms from (3.55) unless the lower order term in inequality (3.55) has positive sign in a small neighborhood of the singular poles x_i. More precisely, the question to answer is whether for any configuration of the poles $x_1,\dots,x_n$, there exists $\varepsilon>0$ small enough such that

$$-\sum_{1\leq i<j\leq n}\frac{(x-x_i)(x-x_j)}{|x-x_i|^2|x-x_j|^2} \geq 0, \quad \forall x\in \cup_{i=1}^n B_\varepsilon(x_i)? \tag{3.56}$$

  Unfortunately, (3.56) is not true. We give a counterexample below (Fig. 3.2).

Let us consider a configuration of three poles x_1, x_2, and x_3 determining an equilateral triangle with the vertices at x_i, $i \in \{1,2,3\}$ such that

$$|x_i - x_j| = l > 0, \quad \forall i \neq j, \quad \forall i, j \in \{1,2,3\}.$$

Given $\varepsilon > 0$, we also consider $x_\varepsilon \in \mathbb{R}^3$ located on the line determined by x_1 and x_3 such that $|x_\varepsilon - x_1| = \varepsilon$, $|x_\varepsilon - x_3| = \varepsilon + l$ (as in Fig. 3.2). Then we have

$$|x_\varepsilon - x_2|^2 = \varepsilon^2 + l^2 + \varepsilon l, \quad |x_\varepsilon - x_3| = (\varepsilon + l)^2.$$

In view of this, we can easily obtain that

$$-\sum_{1 \leq i < j \leq 3} \frac{(x_\varepsilon - x_i)(x_\varepsilon - x_j)}{|x_\varepsilon - x_i|^2 |x_\varepsilon - x_j|^2} < 0,$$

for $\varepsilon > 0$ small enough, fact which contradicts (3.56).

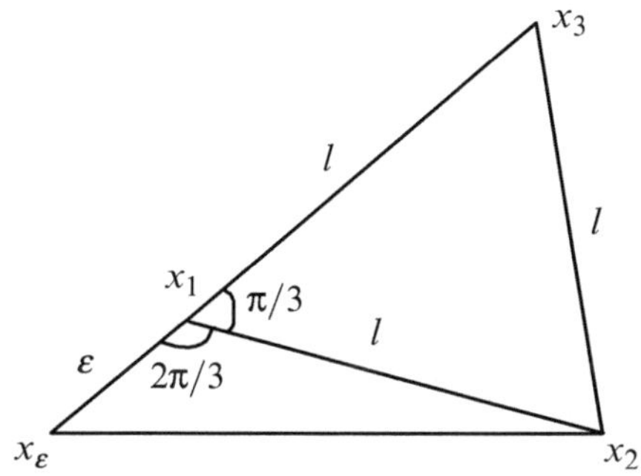

Fig. 3.2 Counterexample to (3.56)

More general, one can show that there is no configuration $x_1, \ldots, x_n$ for which (3.56) is true. The condition (3.56) is violated at the singular poles x_{k_i}, with $\{k_i \mid i \in \{1, \ldots, n\}\} \subset \{1, \ldots, n\}$, which are located on the boundary of the smallest convex set containing all the poles $x_1, \ldots, x_n$.

Acknowledgments Partially supported by the Grant MTM2011-29306-C02-00 of the MICINN (Spain), project PI2010-04 of the Basque Government, the ERC Advanced Grant FP7-246775 NUMERIWAVES, the ESF Research Networking Program OPTPDE, the grant PN-II-ID-PCE-2011-3-0075 of CNCS-UEFISCDI Romania, and a doctoral fellowship from UAM (Universidad Autónoma de Madrid)

References

1. Adimurthi: Best constants and Pohozaev identity for Hardy-Sobolev type operators, to appear in Communications in Contemporary Mathematics
2. Adimurthi, Sandeep, K.: Existence and non-existence of the first eigenvalue of the perturbed Hardy-Sobolev operator. Proc. Roy. Soc. Edinburgh Sect. A **132**, 1021–1043 (2002)

3. Adimurthi, Sekar, A.: Role of the fundamental solution in Hardy-Sobolev-type inequalities. Proc. Roy. Soc. Edinburgh Sect. A **136**, 1111–1130 (2006)
4. Azorero, J.G., Peral, I.: Hardy inequalities and some critical elliptic and parabolic problems. J. Differ. Equat. **144**, 441–476 (1998)
5. Barbatis, G., Filippas, S., Tertikas, A.: A unified approach to improved L^p Hardy inequalities with best constants. Trans. Am. Math. Soc. **356**, 2169–2196 (2004)
6. Berestycki, H., Esteban, M.J.: Existence and bifurcation of solutions for an elliptic degenerate problem. J. Different. Equat. **134**, 1–25 (1997)
7. Bosi, R., Dolbeault, J., Esteban, M.J.: Estimates for the optimal constants in multipolar Hardy inequalities for Schrödinger and Dirac operators. Comm. Pure Appl. Anal. **7**, 533–562 (2008)
8. Brezis, H., Vázquez, J.L.: Blow-up solutions of some nonlinear elliptic problems. Rev. Mat. Univ. Complut. Madrid **10**, 443–469 (1997)
9. Caffarelli, L., Kohn, R., Nirenberg, L.: First order interpolation inequalities with weights. Compos. Math. **53**, 259–275 (1984)
10. Catto, I., Le Bris, C., Lions, P.-L.: On the thermodynamic limit for Hartree-Fock type models. Ann. Inst. H. Poincaré Anal. Non Linéaire **18**, 687–760 (2001)
11. Cazacu, C.: On Hardy inequalities with singularities on the boundary. C. R. Acad. Sci. Paris, Ser. I, **349**, 273–277 (2011)
12. Dolbeault, J., Duoandikoetxea, J., Esteban, M.J., Vega, L.: Hardy-type estimates for Dirac operators. Ann. Sci. École Norm. Sup. (4) **40**, 885–900 (2007)
13. Duyckaerts, T.: A singular critical potential for the Schrödinger operator. Can. Math. Bull. **50**, 35–47 (2007)
14. Duyckaerts, T.: Inégalités de résolvante pour l'opérateur de Schrödinger avec potentiel multipolaire critique. Bull. Soc. Math. Fr. **134**, 201–239 (2006)
15. Evans, L.C., Gariepy, R.F.: Measure Theory and Fine Properties of Functions. CRC, Boca Raton (1992)
16. Felli, V., Marchini, E.M., Terracini, S.: On Schrödinger operators with multipolar inverse-square potentials. J. Funct. Anal. **250**, 265–316 (2007)
17. Felli, V., Terracini, S.: Elliptic equations with multi-singular inverse-square potentials and critical nonlinearity. Comm. Part. Differ. Equat. **31**, 469–495 (2006)
18. Felli, V., Terracini, S.: Nonlinear Schrödinger equations with symmetric multi-polar potentials. Calc. Var. Part. Differ. Equat. **27**, 25–58 (2006)
19. Filippas, S., Maz'ya, V., Tertikas, A.: On a question of Brezis and Marcus. Calc. Var. Part. Differ. Equat. **25**, 491–501 (2006)
20. Frank, W.M., Land, D.J., Spector, R.M.: Singular potentials. Rev. Modern Phys. **43**, 36–98 (1971)
21. Gel'fand, I.M.: Some problems in the theory of quasi-linear equations. Uspehi Mat. Nauk **14**, 87–158 (1959)
22. Hardy, G.H.: An inequality between integrals. Messenger Math. **54**, 150–156 (1925)
23. Hardy, G.H., Littlewood, J.E., Pólya, G.: Inequalities. Cambridge University Press, Cambridge (1988)
24. Hoffmann-Ostenhof, M., Hoffmann-Ostenhof, T., Laptev, A., Tidblom, J.: Many-particle Hardy inequalities. J. Lond. Math. Soc. (2) **77**, 99–114 (2008)
25. Krejčiřík, D., Zuazua, E.: The Hardy inequality and the heat equation in twisted tubes. J. Math. Pures Appl. (9) **94**, 277–303 (2010)
26. Lévy-Leblond, J.M.: Electron capture by polar molecules. Phys. Rev. **153**, 1–4 (1967)
27. Morgan, J.D.: Schrödinger operators whose operators have separated singularities. J. Operat. Theor. **1**, 109–115 (1979)
28. Simon, B.: Semiclassical analysis of low lying eigenvalues. I. Nondegenerate minima: asymptotic expansions. Ann. Inst. H. Poincaré Sect. A (N.S.) **38**, 295–308 (1983)
29. Tertikas, A.: Critical phenomena in linear elliptic problems. J. Funct. Anal. **154**, 42–66 (1998)
30. Tintarev, K., Fieseler, K.H.: Concentration Compactness. Imperial College Press, London (2007)
31. Vázquez, J.L., Zuazua, E.: The Hardy inequality and the asymptotic behaviour of the heat equation with an inverse-square potential. J. Funct. Anal. **173**, 103–153 (2000)

Chapter 4
The Role of Spectral Anisotropy in the Resolution of the Three-Dimensional Navier–Stokes Equations

Jean-Yves Chemin, Isabelle Gallagher, and Chloé Mullaert

Abstract We present different classes of initial data to the three-dimensional, incompressible Navier–Stokes equations, which generate a global in time, unique solution though they may be arbitrarily large in the endpoint function space in which a fixed-point argument may be used to solve the equation locally in time. The main feature of these initial data is an anisotropic distribution of their frequencies. One of those classes is taken from Chemin and Gallagher (Trans. Am. Math. Soc. 362(6):2859–2873, 2010) and Chemin et al. (J. für die reine und angewandte Mathematik, to appear), and another one is new.

Key words: Anisotropy, Navier–Stokes equations

2010 Mathematics Subject Classification: 35Q30, 76D03.

4.1 Introduction

In this article, we are interested in the construction of global smooth solutions which cannot be obtained in the framework of small data. Let us recall what the incompressible Navier–Stokes (with constant density) is:

J.-Y. Chemin • C. Mullaert
Laboratoire J.-L. Lions UMR 7598, Université Paris VI, 175, rue du Chevaleret, F-75013 Paris, France
e-mail: chemin@ann.jussieu.fr; cmullaert@ann.jussieu.fr

I. Gallagher (✉)
Institut de Mathématiques de Jussieu UMR 7586, Université Paris VII, 175, rue du Chevaleret, F-75013 Paris, France
e-mail: gallagher@math.univ-paris-diderot.fr

M. Cicognani et al. (eds.), *Studies in Phase Space Analysis with Applications to PDEs*, Progress in Nonlinear Differential Equations and Their Applications 84,
DOI 10.1007/978-1-4614-6348-1_4,

$$(NS)\begin{cases}\partial_t u + u\cdot\nabla u - \Delta u = -\nabla p & \text{in} \quad \mathbb{R}^+\times\mathbb{R}^3\\ \operatorname{div} u = 0\\ u_{|t=0} = u_0.\end{cases}$$

In all this paper $x=(x_h,x_3)=(x_1,x_2,x_3)$ will denote a generic point of $\mathbb{R}^3$, and we shall write $u=(u^h,u^3)=(u^1,u^2,u^3)$ for a vector field on $\mathbb{R}^3=\mathbb{R}^2_h\times\mathbb{R}_v$. We also define the horizontal differentiation operators $\nabla^h \stackrel{\text{def}}{=} (\partial_1,\partial_2)$ and $\operatorname{div}_h \stackrel{\text{def}}{=} \nabla^h\cdot$, as well as $\Delta_h \stackrel{\text{def}}{=} \partial_1^2+\partial_2^2$.

First, let us recall the history of global existence results for small data. In his seminal work [13], J. Leray proved in 1934 that if $\|u_0\|_{L^2}\|\nabla u_0\|_{L^2}$ is small enough, then there exists a global regular solution of (NS). Then in [6], H. Fujita and T. Kato proved in 1964 that if

$$\|u_0\|_{\dot H^{\frac12}} \stackrel{\text{def}}{=} \Big(\int_{\mathbb{R}^3}|\xi|\,|\widehat{u}_0(\xi)|^2 d\xi\Big)^{\frac12}$$

is small enough, then there exists a unique global solution in the space

$$C_b(\mathbb{R}^+;\dot H^{\frac12})\cap L^4(\mathbb{R}^+;\dot H^1).$$

After works of many authors on this question (see in particular [2, 9, 11, 15]), the optimal norm to express the smallness of the initial data was found on 2001 by Koch and Tataru in [12]. This is the BMO^{-1} norm. We are not going to define precisely this norm here. Let us simply notice that this norm is in between two Besov norms which can be easily defined. More precisely we have

$$\|u_0\|_{\dot B^{-1}_{\infty,\infty}} \lesssim \|u_0\|_{BMO^{-1}} \lesssim \|u_0\|_{\dot B^{-1}_{\infty,2}} \quad \text{with}$$

$$\|u_0\|_{\dot B^{-1}_{\infty,\infty}} \stackrel{\text{def}}{=} \sup_{t>0} t^{\frac12}\|e^{t\Delta}u_0\|_{L^\infty} \quad \text{and} \quad \|u_0\|_{\dot B^{-1}_{\infty,2}} \stackrel{\text{def}}{=} \|e^{t\Delta}u_0\|_{L^2(\mathbb{R}^+;L^\infty)}.$$

First of all, let us mention that $\dot H^{\frac12}$ is continuously embedded in $\dot B^{-1}_{\infty,2}$. To have a more precise idea of what these spaces mean, let us observe that the space $\dot B^{-1}_{\infty,\infty}$ we shall denote by $\dot C^{-1}$ from now on, contains all the derivatives of order 1 of bounded functions. Let us give some examples. If we consider a divergence-free vector field of the type

$$u_{\varepsilon,0}(x)=\frac{1}{\varepsilon}\cos\Big(\frac{x_3}{\varepsilon}\Big)\big(-\partial_2\varphi(x),\partial_1\varphi(x),0\big)$$

for some given function φ in the Schwartz class of $\mathbb{R}^3$, then we have

$$\|u_{\varepsilon,0}\|_{\dot B^{-1}_{\infty,2}} \sim \|u_{\varepsilon,0}\|_{\dot C^{-1}} \sim 1 \quad \text{and} \quad \|u_{\varepsilon,0}\|_{\dot H^{\frac12}} \sim \varepsilon^{-\frac32}.$$

Another example which will be a great interest for this paper is the case when

$$u_{\varepsilon,0}(x)=\varphi_0(\varepsilon x_3)\big(-\partial_2\varphi(x_h),\partial_1\varphi(x_h),0\big).$$

As claimed by Proposition 1.1 of [4], we have, for small enough ε,

$$\|u_{\varepsilon,0}\|_{\dot{C}^{-1}} \geq \frac{1}{2}\|\varphi\|_{\dot{C}^{-1}(\mathbb{R}^2)}\|\varphi_0\|_{L^\infty(\mathbb{R})}. \tag{4.1}$$

In this paper, we are going to consider the initial data, the regularity of which will be (at least) $\dot{H}^{\frac{1}{2}}$. Our interest is focused on the size of the initial data measured in the $\dot{C}^{-1}$ norm.

Let us define $\mathcal{G}$ the set of divergence-free vector fields in $\dot{H}^{\frac{1}{2}}(\mathbb{R}^3)$ generating global smooth solutions to (NS) and let us recall some known results about the geometry of $\mathcal{G}$.

First of all, Fujita-Kato' theorem [6] can be interpreted as follows: the set $\mathcal{G}$ contains a ball of positive radius. Next let us assume that $\mathcal{G}$ is not the whole space $\dot{H}^{\frac{1}{2}}$ (in other words, we assume that an initial data exists which generates singularities in finite time). Then there exists a critical radius ρ_c such that if u_0 is an initial data such that $\|u_0\|_{\dot{H}^{\frac{1}{2}}} < \rho_c$, then u_0 generates a global regular solution and for any $\rho > \rho_c$, there exists an initial data of $\dot{H}^{\frac{1}{2}}$ norm ρ which generates a singularity at finite time. Using the theory of profiles introduced in the context of Navier–Stokes equations by the second author (see [7]), Rusin and Sverak prove in [14] that the set (where $\mathcal{G}^c$ denotes the complement of $\mathcal{G}$ in $\dot{H}^{\frac{1}{2}}$)

$$\mathcal{G}^c \cap \left\{u_0 \in \dot{H}^{\frac{1}{2}} \,/\, \|u_0\|_{\dot{H}^{\frac{1}{2}}} = \rho_c\right\}$$

is nonempty and compact up to dilations and translations.

In collaboration with P. Zhang, the first two authors prove in [5] that any point u_0 of $\mathcal{G}$ is at the center of an interval I included in $\mathcal{G}$ and such that the length of I measured in the $\dot{C}^{-1}$ norm is arbitrary large. In other words for any u_0 in $\mathcal{G}$, there exist arbitrary large (in $\dot{C}^{-1}$) perturbations of this initial data that generate global solutions. As we shall see, the perturbations are strongly anisotropic.

Our aim is to give a new point of view about the important role played by anisotropy in the resolution of the Cauchy problem for (NS).

The first result we shall present shows that as soon as enough anisotropy is present in the initial data (where the degree of anisotropy is given by the norm of the data only), then it generates a global unique solution. A similar result can be found in [1, Theorem 1].

Theorem 4.1. *A constant c_0 exists which satisfies the following. If $(u_{\varepsilon,0})_{\varepsilon>0}$ is a family of divergence-free vector fields in $\dot{H}^{\frac{1}{2}}$ satisfying*

$$\forall \xi \in \operatorname{Supp} \widehat{u}_{\varepsilon,0}, \quad \textit{either} \quad |\xi_h| \leq \varepsilon|\xi_3| \quad \text{or} \quad |\xi_3| \leq \varepsilon|\xi_h|, \tag{4.2}$$

then, if $\varepsilon^4\|u_{\varepsilon,0}\|_{\dot{H}^{\frac{1}{2}}}$ is less than c_0, $u_{\varepsilon,0}$ belongs to $\mathcal{G}$.

Let us remark that this result has little to do with the precise structure of the equations: as will appear clearly in its proof in Sect. 4.2, it can actually easily be recast as a small data theorem, the smallness being measured in anisotropic Sobolev spaces. It is therefore of a different nature than the next Theorems 4.2 and 4.3, whose proofs on the contrary rely heavily on the structure of the nonlinearity (more precisely on the fact that the two-dimensional equations are globally well-posed).

The next theorem shows that as soon as the initial data has slow variations in one direction, then it generates a global solution, which, roughly speaking, corresponds to the case when the support in Fourier space of the initial data lies in the region where $|\xi_3| \leq \varepsilon|\xi_h|$. Furthermore, one can add to any initial data in $\mathcal{G}$ any such slowly varying data, and the superposition still generates a global solution (provided the variation is slow enough and the profile vanishes at zero).

Theorem 4.2 ([4, 5]). *Let $v_0^h = (v_0^1, v_0^2)$ be a horizontal, smooth divergence-free vector field on $\mathbb{R}^3$ (i.e., v_0^h is in $L^2(\mathbb{R}^3)$ as well as all its derivatives), belonging, as well as all its derivatives, to $L^2(\mathbb{R}_{x_3}; \dot{H}^{-1}(\mathbb{R}^2))$; let w_0 be a smooth divergence-free vector field on $\mathbb{R}^3$. Then, there exists a positive ε_0 depending on norms of v_0^h and w_0 such that, if $\varepsilon \leq \varepsilon_0$, then the following initial data belongs to $\mathcal{G}$:*

$$v_{\varepsilon,0}(x) \stackrel{\text{def}}{=} (v_0^h + \varepsilon w_0^h, w_0^3)(x_1, x_2, \varepsilon x_3).$$

If moreover $v_0^h(x_1,x_2,0) = w_0^3(x_1,x_2,0) = 0$ for all $(x_1,x_2) \in \mathbb{R}^2$ and if u_0 belongs to $\mathcal{G}$, then there exists a positive number ε_0' depending on u_0 and on norms of v_0^h and w_0 such that if $\varepsilon \leq \varepsilon_0'$, the following initial data belongs to $\mathcal{G}$:

$$u_{\varepsilon,0} \stackrel{\text{def}}{=} u_0 + v_{\varepsilon,0}.$$

One can assume that v_0^h and w_0^3 have frequency supports in a given ring of $\mathbb{R}^3$, so that (4.2) holds. Nevertheless, Theorem 4.1 does not apply since $v_{\varepsilon,0}$ is of the order of $\varepsilon^{-\frac{1}{2}}$ in $\dot{H}^{\frac{1}{2}}$. Actually the proof of Theorem 4.2 is deeper than that of Theorem 4.1, as it uses the structure of the quadratic term in (NS). The proof of Theorem 4.2 may be found in [4, 5], we shall not give it here. Note that Inequality (4.1) implies that $v_{\varepsilon,0}$ may be chosen arbitrarily large in $\dot{C}^{-1}$.

One formal way to translate the above result is that the vertical frequencies of the initial data $v_{\varepsilon,0}$ are actually very small, compared with the horizontal frequencies. The following theorem gives a statement in terms of frequency sizes, in the spirit of Theorem 4.1. However as already pointed out, Theorem 4.1 again does not apply because the initial data is too large in $\dot{H}^{\frac{1}{2}}$. Notice also that the assumption made in the statement of Theorem 4.2 that the profile should vanish at $x_3 = 0$ is replaced here by a smallness assumption in $L^2(\mathbb{R}^2)$.

Theorem 4.3. *Let $(v_{\varepsilon,0})_\varepsilon$ be a family of smooth divergence-free vector field, uniformly bounded in the space $L^\infty(\mathbb{R}; \dot{H}^s(\mathbb{R}^2))$ for all $s \geq -1$, such that $(\sqrt{\varepsilon}\, v_{\varepsilon,0})_\varepsilon$ is uniformly bounded in the space $L^2(\mathbb{R}_{x_3}; \dot{H}^s(\mathbb{R}^2))$ for $s \geq -1$ and satisfying*

$$\forall \varepsilon \in]0,1[, \quad \forall \xi \in \operatorname{Supp} \widehat{v}_{\varepsilon,0}, \quad |\xi_3| \leq \varepsilon|\xi_h|.$$

Then there exists a positive number ε_0 such that for all $\varepsilon \leq \varepsilon_0$, the data $v_{\varepsilon,0}$ belongs to $\mathcal{G}$.

Moreover if u_0 belongs to $\mathscr{G}$, then there are positive constants c_0 and ε_0' such that if

$$\|v_{\varepsilon,0}(\cdot,0)\|_{L^2(\mathbb{R}^2_h)} \leq c_0,$$

then for all $\varepsilon \leq \varepsilon_0'$, the following initial data belongs to $\mathscr{G}$:

$$u_{\varepsilon,0} \stackrel{\text{def}}{=} u_0 + v_{\varepsilon,0}\,.$$

Let us remark that as in [4], the data $v_{\varepsilon,0}$ may be arbitrarily large in $\dot{C}^{-1}$. Note that Theorems 4.2 and 4.3, though of similar type, are not comparable (unless one imposes the spectrum of the initial profiles in Theorem 4.2 to be included in a ring of $\mathbb{R}^3$, in which case the result follows from Theorem 4.3).

The paper is organized as follows. In the second section, we introduce anisotropic Sobolev spaces, and as a warm up, we prove Theorem 4.1.

The rest of the paper is devoted to the proof of Theorem 4.3. In the third section, we define an (global) approximated solution and prove estimates on this approximated solutions and prove Theorem 4.3.

The last section is devoted to the proof of a propagation result for a linear transport-diffusion equation we admit in the preceding section. Let us point out that we make the choice not to use the technology anisotropic paradifferential calculus and to present an elementary proof.

4.2 Preliminaries: Notation and Anisotropic Function Spaces

In this section we recall the definition of the various function spaces we shall be using in this paper, namely, anisotropic Lebesgue and Sobolev spaces.

We denote by $L^p_hL^q_v$ (resp. $L^q_v(L^p_h)$) the space $L^p(\mathbb{R}^2_h;L^q(\mathbb{R}_v))$ (resp. $L^q(\mathbb{R}_v;L^p(\mathbb{R}^2_h))$ equipped with the norm

$$\|f\|_{L^p_hL^q_v} \stackrel{\text{def}}{=} \left(\int_{\mathbb{R}^2_h}\left(\int_{\mathbb{R}_v}|f(x_h,x_3)|^q\,dx_3\right)^{\frac{p}{q}}dx_h\right)^{\frac{1}{p}}$$

and similarly $\dot{H}^{s,\sigma}$ is the space $\dot{H}^s(\mathbb{R}^2;\dot{H}^\sigma(\mathbb{R}))$ with

$$\|f\|_{\dot{H}^{s,\sigma}} \stackrel{\text{def}}{=} \left(\int_{\mathbb{R}^3}|\xi_h|^{2s}|\xi_3|^{2\sigma}|\widehat{f}(\xi_h,\xi_3)|^2\,d\xi_h d\xi_3\right)^{\frac{1}{2}}$$

where $\widehat{f} = \mathscr{F}f$ is the Fourier transform of f. Note that $\dot{H}^{s,\sigma}$ is a Hilbert space as soon as $s<1$ and $\sigma<1/2$. We define also

$$\|f\|_{\dot{H}^{s_1,s_2,s_3}} \stackrel{\text{def}}{=} \left(\int_{\mathbb{R}^3}|\xi_1|^{2s_1}|\xi_2|^{2s_2}|\xi_3|^{2s_3}|\widehat{f}(\xi_1,\xi_2,\xi_3)|^2\,d\xi_1 d\xi_2 d\xi_3\right)^{\frac{1}{2}}.$$

This is a Hilbert space if all s_j are less than $1/2$. Finally we shall often use the spaces $L^p_v\dot H^s_h = L^p(\mathbb{R}_v;\dot H^s(\mathbb{R}^2_h))$. Let us notice that $L^2_v\dot H^s_h = \dot H^{s,0}$. The following result, proved by D. Iftimie in [10], is the basis of the proof of Theorem 4.1.

Theorem 4.4. *There is a constant ε_0 such that the following result holds. Let $(s_i)_{1\le i\le 3}$ be such that $s_1+s_2+s_3=1/2$ and $-1/2<s_i<1/2$. Then any divergence-free vector field of norm smaller than ε_0 in $\dot H^{s_1,s_2,s_3}$ generates a global smooth solution to (NS).*

This theorem implies obviously the following corollary, since $\dot H^{s,\frac12-s}$ is continuously embedded in $\dot H^{\frac s2,\frac s2,\frac12-s}$ as soon as $0<s<1/2$. More precisely, we have that the space $\dot H^{s,\frac12-s}$ is the space $\dot H^{s,0,\frac12-s}\cap \dot H^{0,s,\frac12-s}$.

Corollary 4.1. *There is a constant ε_0 such that the following result holds. Let s be given in $]0,1/2[$. Then any divergence-free vector field of norm smaller than ε_0 in $\dot H^{s,\frac12-s}$ generates a global smooth solution to* (NS).

Proof (Proof of Theorem 4.1). Let us decompose u_0 into two parts, namely, we write $u_0=v_0+w_0$, with

$$v_0 \stackrel{\text{def}}{=} \mathscr{F}^{-1}\big(\mathbf{1}_{|\xi_h|\le\varepsilon|\xi_3|}\widehat{u}_0(\xi)\big) \quad\text{and}\quad w_0 \stackrel{\text{def}}{=} \mathscr{F}^{-1}\big(\mathbf{1}_{|\xi_3|\le\varepsilon|\xi_h|}\widehat{u}_0(\xi)\big).$$

Let $0<s<1/2$ be given. On the one hand we have

$$\|v_0\|^2_{\dot H^{s,\frac12-s}} = \int_{|\xi_3|\le\varepsilon|\xi_h|} |\xi_h|^{2s}|\xi_3|^{1-2s}|\widehat{u}_0(\xi)|^2\,d\xi$$

hence, since $s<1/2$,

$$\begin{aligned}\|v_0\|^2_{\dot H^{s,\frac12-s}} &\le \varepsilon^{1-2s}\int |\xi_h|\,|\widehat{u}_0(\xi)|^2\,d\xi\\ &\le \varepsilon^{1-2s}\|u_0\|^2_{\dot H^{\frac12}}.\end{aligned}$$

Identical computations give, since $s>0$,

$$\begin{aligned}\|w_0\|^2_{\dot H^{s,\frac12-s}} &= \int_{|\xi_h|\le\varepsilon|\xi_3|} |\xi_h|^{2s}|\xi_3|^{1-2s}|\widehat{u}_0(\xi)|^2\,d\xi\\ &\le \varepsilon^{2s}\int |\xi_3|\,|\widehat{u}_0(\xi)|^2\,d\xi\\ &\le \varepsilon^{2s}\|u_0\|^2_{\dot H^{\frac12}}.\end{aligned}$$

To conclude we can choose $s=1/4$, which gives

$$\|u_0\|_{\dot H^{\frac14,\frac14}} \le \varepsilon^{\frac14}\|u_0\|_{\dot H^{\frac12}}.$$

Then, the result follows by the well-posedness of (NS) in $\dot H^{\frac14,\frac14}$ given by Corollary 4.1.

Remark 4.1. The proof of Theorem 4.1 does not use the special structure of the nonlinear term in (NS) as it reduces to checking that the initial data is small in an adequate scale-invariant space.

4.3 Proof of Theorem 4.3

In this section we shall prove the second part of Theorem 4.3: we consider an initial data $u_0 + v_{\varepsilon,0}$ satisfying the assumptions of the theorem and we prove that for $\varepsilon > 0$ small enough, it generates a global, unique solution to (NS). It will be clear from the proof that in the case when $u_0 \equiv 0$ (which amounts to the first part of Theorem 4.3), the assumption that $v_{\varepsilon,0}(x_h, 0)$ is small in $L^2(\mathbb{R}^2)$ is not necessary. Thus the proof of the whole of Theorem 4.3 will be obtained.

4.3.1 Decomposition of the Initial Data

The first step of the proof consists in decomposing the initial data as follows.

Proposition 4.1. *Let $v_{\varepsilon,0}$ be a divergence-free vector field satisfying*

$$\forall \varepsilon \in]0,1[\,, \quad \forall \xi \in \operatorname{Supp} \widehat{v}_{\varepsilon,0}\,, \quad |\xi_3| \leq \varepsilon |\xi_h|\,.$$

Then there exist two divergence-free vector fields $(\overline{v}^h_{\varepsilon,0}, 0)$, and $w_{\varepsilon,0}$ the spectrum of which is included in that of $v_{\varepsilon,0}$, and such that

$$v_{\varepsilon,0} = (\overline{v}^h_{\varepsilon,0}, 0) + w_{\varepsilon,0} \quad \textit{with} \quad \left|\widehat{w}^h_{\varepsilon,0}\right| \leq \varepsilon \left|\widehat{w}^3_{\varepsilon,0}\right|.$$

Proof. Let $\mathbb{P}_h \stackrel{\text{def}}{=} \mathrm{Id} - \nabla_h \Delta_h^{-1} \mathrm{div}_h$ be the Leray projector onto horizontal divergence-free vector fields and define

$$\overline{v}^h_{\varepsilon,0} \stackrel{\text{def}}{=} \mathbb{P}_h v^h_{\varepsilon,0} \quad \text{and} \quad w_{\varepsilon,0} \stackrel{\text{def}}{=} v_{\varepsilon,0} - (\overline{v}^h_{\varepsilon,0}, 0)\,. \tag{4.3}$$

The estimate on $w_{\varepsilon,0}$ simply comes from the fact that obviously

$$\widehat{w}^h_{\varepsilon,0}(\xi) = \frac{\xi_h \cdot \widehat{v}^h_{\varepsilon,0}}{|\xi_h|^2}\, \xi_h\,,$$

and therefore since $v_{\varepsilon,0}$ is divergence free and using the spectral assumption we find

$$|\widehat{w}^h_{\varepsilon,0}(\xi)| \leq \frac{|\xi_h \cdot \widehat{v}^h_{\varepsilon,0}|}{|\xi_h|} = \frac{|\xi_3 \widehat{v}^3_{\varepsilon,0}|}{|\xi_h|} \leq \varepsilon |\widehat{v}^3_{\varepsilon,0}| = \varepsilon |\widehat{w}^3_{\varepsilon,0}(\xi)|\,.$$

That proves the proposition.

4.3.2 Construction of an Approximate Solution and End of the Proof of Theorem 4.3

The construction of the approximate solution follows closely the ideas of [4, 5]. We write indeed

$$v_\varepsilon^{app} \stackrel{\text{def}}{=} (\overline{v}_\varepsilon^h, 0) + w_\varepsilon \quad \text{and} \quad u_\varepsilon^{app} \stackrel{\text{def}}{=} u + v_\varepsilon^{app},$$

where u is the global unique solution associated with u_0 and $\overline{v}_\varepsilon^h$ solves the two-dimensional Navier–Stokes equations for each given x_3:

$$(\text{NS2D})_{x_3} \quad \begin{cases} \partial_t \overline{v}_\varepsilon^h + \overline{v}_\varepsilon^h \cdot \nabla^h \overline{v}_\varepsilon^h - \Delta_h \overline{v}_\varepsilon^h = -\nabla^h \overline{p}_\varepsilon \quad \text{in} \quad \mathbb{R}^+ \times \mathbb{R}^2 \\ \operatorname{div}_h \overline{v}_\varepsilon^h = 0 \\ \overline{v}_{\varepsilon|t=0}^h = \overline{v}_{\varepsilon,0}^h(\cdot, x_3), \end{cases}$$

while w_ε solves the linear transport-diffusion-type equation

$$(\text{T}) \quad \begin{cases} \partial_t w_\varepsilon + \overline{v}_\varepsilon^h \cdot \nabla^h w_\varepsilon - \Delta w_\varepsilon = -\nabla q_\varepsilon \quad \text{in} \quad \mathbb{R}^+ \times \mathbb{R}^3 \\ \operatorname{div} w_\varepsilon = 0 \\ w_{\varepsilon|t=0} = w_{\varepsilon,0} \,. \end{cases}$$

Those vector fields satisfy the following bounds (see Paragraph 4.3.3 for a proof).

Lemma 4.1. *Under the assumptions of Theorem 4.3, the family u_ε^{app} is uniformly bounded in $L^2(\mathbb{R}^+; L^\infty(\mathbb{R}^3))$, and $\nabla u_\varepsilon^{app}$ is uniformly bounded in $L^2(\mathbb{R}^+; L_v^\infty L_h^2)$.*

Now define u_ε the solution associated with the initial data $u_0 + v_{\varepsilon,0}$, which a priori has a finite life span, depending on ε. Consider

$$R_\varepsilon \stackrel{\text{def}}{=} u_\varepsilon - u_\varepsilon^{app},$$

which satisfies the following property (see Paragraph 4.3.4 for a proof).

Lemma 4.2. *For any positive δ there exists $\varepsilon(\delta)$ and $c(\delta)$ such that if*

$$\varepsilon \le \varepsilon(\delta) \quad \textit{and if} \quad \|v_{\varepsilon,0}(\cdot,0)\|_{L_h^2} \le c(\delta),$$

then the vector field $R^\varepsilon \stackrel{\text{def}}{=} u_\varepsilon - u_\varepsilon^{app}$ satisfies the equation

$$(\text{E}_\varepsilon) \begin{cases} \partial_t R_\varepsilon + R_\varepsilon \cdot \nabla R_\varepsilon - \Delta R_\varepsilon + u_\varepsilon^{app} \cdot \nabla R_\varepsilon + R_\varepsilon \cdot \nabla u_\varepsilon^{app} = F_\varepsilon - \nabla \widetilde{q}_\varepsilon \\ \operatorname{div} R_\varepsilon = 0 \\ R_{\varepsilon|t=0} = 0 \end{cases}$$

with $\|F_\varepsilon\|_{L^2(\mathbb{R}^+; \dot{H}^{-\frac{1}{2}}(\mathbb{R}^3))} \le \delta$.

Assuming those two lemmas to be true, the end of the proof of Theorem 4.3 follows very easily using the method given in [4, Sect. 2]: an energy estimate in $\dot H^{\frac12}(\mathbb{R}^3)$ on (E_ε), using the fact that the forcing term is as small as needed and that the initial data is zero, gives that R_ε is unique, and uniformly bounded in $L^\infty(\mathbb{R}^+;\dot H^{\frac12})\cap L^2(\mathbb{R}^+;\dot H^{\frac32})$. Since the approximate solution is also unique and globally defined, Theorem 4.3 is proved.

4.3.3 *Proof of the Estimates on the Approximate Solution (Lemma 4.1)*

As noted in [5, Appendix B], the global solution u associated with $u_0\in\dot H^{\frac12}$ belongs to $L^2(\mathbb{R}^+;L^\infty(\mathbb{R}^3))$, and ∇u belongs to $L^2(\mathbb{R}^+;L^\infty_v L^2_h)$. So we just need to study v_ε^{app}, which we shall do in two steps: first $\overline{v}_\varepsilon^h$, then w_ε.

4.3.3.1 Estimates on $\overline{v}_\varepsilon^h$

Due to the spectral assumption on $\overline{v}_{\varepsilon,0}^h$, it is easy to see that

$$\begin{aligned}\forall\alpha=(\alpha_h,\alpha_3)\in\mathbb{N}^2\times\mathbb{N},\quad &\varepsilon^{\frac12-\alpha_3}\partial^\alpha\overline{v}_{\varepsilon,0}^h \quad\text{is uniformly bounded in } L^2_v\dot H^s_h,\\ \text{and}\quad &\varepsilon^{-\alpha_3}\partial^\alpha\overline{v}_{\varepsilon,0}^h \quad\text{is uniformly bounded in } L^\infty_v\dot H^s_h.\end{aligned}$$

Indeed the definition of $\overline{v}_{\varepsilon,0}^h$ given in (4.3) and the spectral assumption as well as the a priori bounds on $v_{\varepsilon,0}$ give directly the first result. To prove the second result one uses first the Gagliardo-Nirenberg inequality:

$$\|\partial^\alpha\overline{v}_{\varepsilon,0}^h\|^2_{L^\infty_v\dot H^s_h}\le\|\partial^\alpha\overline{v}_{\varepsilon,0}^h\|_{L^2_v\dot H^s_h}\|\partial_3\partial^\alpha\overline{v}_{\varepsilon,0}^h\|_{L^2_v\dot H^s_h}$$

and then the same arguments. The proof of [4, Lemma 3.1 and Corollary 3.1] enables us to infer from those bounds the following result.

Proposition 4.2. *Under the assumptions of Theorem 4.3, for all real numbers $s>-1$ and all $\alpha=(\alpha_h,\alpha_3)\in\mathbb{N}^2\times\mathbb{N}$, there is a constant $\overline{C}$ such that the vector field $\overline{v}_\varepsilon^h$ satisfies the following bounds:*

$$\begin{aligned}&\|\partial^\alpha\overline{v}_\varepsilon^h(t)\|^2_{L^\infty_v\dot H^s_h}+\sup_{x_3\in\mathbb{R}}\int_0^t\|\partial^\alpha\nabla^h\overline{v}_\varepsilon^h(t')\|^2_{\dot H^s_h}\,dt'\\ &\qquad+\varepsilon\Big(\|\partial^\alpha\nabla^h\overline{v}_\varepsilon^h(t)\|^2_{L^2_v\dot H^s_h}+\int_0^t\|\partial^\alpha\overline{v}_\varepsilon^h(t')\|^2_{L^2(\mathbb{R}^3)}\,dt'\Big)\le\overline{C}\varepsilon^{2\alpha_3}.\end{aligned}$$

4.3.3.2 Estimates on w_ε

The definition of $w_{\varepsilon,0}$ given in (4.3), along with the spectral assumption on $(v_{\varepsilon,0})_{\varepsilon>0}$, leads to

$$\forall\varepsilon\in]0,1[,\quad \forall\xi\in\operatorname{Supp}\widehat{w}_{\varepsilon,0},\quad |\xi_3|\le\varepsilon|\xi_h|\quad\text{and}\quad \big|\widehat{w}^h_{\varepsilon,0}(\xi)\big|\le\varepsilon\big|\widehat{w}^3_{\varepsilon,0}(\xi)\big|.$$

The proof of the following result is technical and postponed to Sect. 4.4.

Proposition 4.3. *Under the assumptions of Theorem 4.3, w^3_ε and $\varepsilon^{-1}w^h_\varepsilon$ are uniformly bounded in the space $L^\infty(\mathbb{R}^+;L^\infty_vL^2_h)\cap L^2(\mathbb{R}^+;L^\infty_v\dot{H}^s_h)$ for all $s\ge 0$. Moreover $\varepsilon^{\frac12-\alpha_3}\partial^\alpha w_\varepsilon$ is uniformly bounded in $L^\infty(\mathbb{R}^+;L^2_v\dot{H}^s_h)\cap L^2(\mathbb{R}^+;L^2_v\dot{H}^s_h)$ for all $s\ge 0$ and all $\alpha=(\alpha_h,\alpha_3)\in\mathbb{N}^2\times\mathbb{N}$.*

The Gagliardo-Nirenberg inequality and Sobolev embeddings lead to Lemma 4.1.

4.3.4 Proof of the Estimates on the Remainder (Lemma 4.2)

Subtracting the equation on u^{app}_ε from the equation on u, one finds directly that

$$F_\varepsilon=(\partial_3^2\overline{v}^h_\varepsilon,\partial_3\overline{p}_\varepsilon)+w_\varepsilon\cdot\nabla v^{app}_\varepsilon+u\cdot\nabla v^{app}_\varepsilon+v^{app}_\varepsilon\cdot\nabla u,$$

which we decompose into $F_\varepsilon=G_\varepsilon+H_\varepsilon$ with

$$G_\varepsilon\stackrel{\text{def}}{=}(\partial_3^2\overline{v}^h_\varepsilon,\partial_3\overline{p}_\varepsilon)+w_\varepsilon\cdot\nabla v^{app}_\varepsilon\quad\text{and}\quad H_\varepsilon\stackrel{\text{def}}{=}u\cdot\nabla v^{app}_\varepsilon+v^{app}_\varepsilon\cdot\nabla u.$$

Lemma 4.2 follows from the two following propositions.

Proposition 4.4. *There is a positive constant C such that for all ε in $]0,1[$,*

$$\|G_\varepsilon\|_{L^2(\mathbb{R}^+;\dot{H}^{-\frac12}(\mathbb{R}^3))}\le C\varepsilon^{\frac12}.$$

Proof. Let us start by splitting G_ε in three parts: $G_\varepsilon=G^1_\varepsilon+G^2_\varepsilon+G^3_\varepsilon$ with

$$G^1_\varepsilon\stackrel{\text{def}}{=}(\partial_3^2\overline{v}^h_\varepsilon,0),\ G^2_\varepsilon\stackrel{\text{def}}{=}(0,\partial_3\overline{p}_\varepsilon),\quad\text{and}\quad G^3_\varepsilon\stackrel{\text{def}}{=}w_\varepsilon\cdot\nabla v^{app}_\varepsilon.$$

On the one hand we have obviously

$$\|G^1_\varepsilon\|_{L^2(\mathbb{R}^+;\dot{H}^{-\frac12}(\mathbb{R}^3))}\le\|\partial_3\overline{v}^h_\varepsilon\|_{L^2(\mathbb{R}^+;\dot{H}^{\frac12}(\mathbb{R}^3))}.$$

Proposition 4.2 applied with $\alpha=(0,1)$, $\alpha=(0,2)$, and $\alpha=(\alpha_h,1)$ with $|\alpha_h|=1$ gives

$$\int_0^t\|\partial_3\overline{v}_\varepsilon(t',\cdot)\|^2_{L^2}dt'\lesssim\varepsilon\quad\text{and}\quad\int_0^t\|\partial_3\nabla\overline{v}_\varepsilon(t',\cdot)\|^2_{L^2}dt'\lesssim\varepsilon.$$

By interpolation, we infer that

$$\|G^1_\varepsilon\|_{L^2(\mathbb{R}^+;\dot H^{-\frac12}(\mathbb{R}^3))} \lesssim \varepsilon^{\frac12}. \tag{4.4}$$

To estimate G^2_ε we use the fact that

$$-\Delta_h \overline{p}_\varepsilon = \sum_{j,k=1}^{2} \partial_j\partial_k(\overline{v}^j_\varepsilon \overline{v}^k_\varepsilon),$$

and since $(-\Delta_h)^{-1}\partial_j\partial_k$ is a Fourier multiplier of order 0 for each (j,k) in $\{1,2\}^2$, we get

$$\|G^2_\varepsilon\|_{L^2(\mathbb{R}^+;\dot H^{-\frac12}(\mathbb{R}^3))} \lesssim \sum_{j,k=1}^{2} \|\overline{v}^j_\varepsilon \partial_3 \overline{v}^k_\varepsilon\|_{L^2(\mathbb{R}^+;\dot H^{-\frac12}(\mathbb{R}^3))}.$$

As $L^2_v \dot H_h^{-\frac12} \hookrightarrow \dot H^{-\frac12}(\mathbb{R}^3)$, we get

$$\|G^2_\varepsilon\|_{L^2(\mathbb{R}^+;\dot H^{-\frac12}(\mathbb{R}^3))} \lesssim \sum_{j,k=1}^{2} \|\overline{v}^j_\varepsilon \partial_3 \overline{v}^k_\varepsilon\|_{L^2(\mathbb{R}^+;L^2_v \dot H_h^{-\frac12})}.$$

Using the Sobolev embedding $L_h^{\frac43} \hookrightarrow \dot H_h^{-\frac12}$ and Hölder's inequality gives

$$\begin{aligned}\|G^2_\varepsilon\|_{L^2(\mathbb{R}^+;\dot H^{-\frac12}(\mathbb{R}^3))} &\lesssim \sum_{j,k=1}^{2} \|\overline{v}^j_\varepsilon \partial_3 \overline{v}^k_\varepsilon\|_{L^2(\mathbb{R}^+;L^2_v L_h^{\frac43})} \\ &\le C\|\overline{v}^h_\varepsilon\|_{L^\infty(\mathbb{R}^+;L^\infty_v L_h^{\frac83})} \|\partial_3 \overline{v}^h_\varepsilon\|_{L^2(\mathbb{R}^+;L^2_v L_h^{\frac83})}\end{aligned},$$

so the Sobolev embedding $\dot H_h^{\frac14} \hookrightarrow L_h^{\frac83}$ gives finally

$$\|G^2_\varepsilon\|_{L^2(\mathbb{R}^+;\dot H^{-\frac12}(\mathbb{R}^3))} \lesssim C\|\overline{v}^h_\varepsilon\|_{L^\infty(\mathbb{R}^+;L^\infty_v \dot H_h^{\frac14})} \|\partial_3 \overline{v}^h_\varepsilon\|_{L^2(\mathbb{R}^+;L^2_v \dot H_h^{\frac14})}.$$

The result follows again from Proposition 4.2: choosing $s = 1/4$ and $\alpha = 0$ we get that $\overline{v}^h_\varepsilon$ is uniformly bounded in $L^\infty(\mathbb{R}^+;L^\infty_v \dot H_h^{\frac14})$, while $s = -3/4$ and $\alpha = (\alpha_h, 1)$ with $|\alpha_h| = 1$ gives

$$\|\partial_3 \overline{v}^h_\varepsilon\|_{L^2(\mathbb{R}^+;L^2_v \dot H_h^{\frac14})} \lesssim \varepsilon^{\frac12}.$$

We infer finally that

$$\|G^2_\varepsilon\|_{L^2(\mathbb{R}^+;\dot H^{-\frac12}(\mathbb{R}^3))} \lesssim \varepsilon^{\frac12}. \tag{4.5}$$

To end the proof of the proposition let us estimate G^3_ε. We simply use two-dimensional product laws, which gives

$$\begin{aligned}\|G^3_\varepsilon\|_{L^2(\mathbb{R}^+;\dot H^{-\frac12}(\mathbb{R}^3))} &= \|w_\varepsilon\cdot\nabla v^{app}_\varepsilon\|_{L^2(\mathbb{R}^+;\dot H^{-\frac12}(\mathbb{R}^3))}\\ &\lesssim \|w^h_\varepsilon\|_{L^\infty(\mathbb{R}^+;L^\infty_v\dot H^{\frac14}_h)}\|\nabla^h v^{app}_\varepsilon\|_{L^2(\mathbb{R}^+;L^2_v\dot H^{\frac14}_h)}\\ &\quad+\|w^3_\varepsilon\|_{L^\infty(\mathbb{R}^+;L^\infty_v\dot H^{\frac14}_h)}\|\partial_3 v^{app}_\varepsilon\|_{L^2(\mathbb{R}^+;L^2_v\dot H^{\frac14}_h)}\lesssim\varepsilon^{\frac12},\end{aligned}$$

due to Propositions 4.2 and 4.3. Together with Inequalities (4.4) and (4.5) that proves Proposition 4.4.

Proposition 4.5. *Let $\delta>0$ be given. There are positive constants $\varepsilon(\delta)$ and $c(\delta)$ such that if $\varepsilon\le\varepsilon(\delta)$ and if $\|v_{\varepsilon,0}(\cdot,0)\|_{L^2_h}\le c(\delta)$, then*

$$\|H_\varepsilon\|_{L^2(\mathbb{R}^+;\dot H^{-\frac12}(\mathbb{R}^3))}\le\delta.$$

Proof. First, we approximate H_ε, and then we estimate this approximation.

Using [8, Theorem 2.1] we get

$$\lim_{t\to\infty}\|u(t,.)\|_{\dot H^{\frac12}(\mathbb{R}^3)}=0$$

so we can approximate u in $L^\infty(\mathbb{R}^+,\dot H^{\frac12})$: for all $\eta>0$, there exists an integer N, real numbers $(t_j)_{0\le j\le N}$ and smooth, compactly supported, divergence-free functions $(\varphi_j)_{1\le j\le N}$ such that

$$\widetilde u_\eta(t)\stackrel{\text{def}}{=}\sum_{j=1}^N 1_{[t_{j-1},t_j]}(t)\,\varphi_j$$

is uniformly bounded in $L^\infty(\mathbb{R}^+,\dot H^{\frac12})\cap L^2(\mathbb{R}^+,\dot H^{\frac32})$ and satisfies

$$\|u-\widetilde u_\eta\|_{L^\infty(\mathbb{R}^+,\dot H^{\frac12}(\mathbb{R}^3))}\le\eta\,. \tag{4.6}$$

We split H_ε into two contributions

$$H_\varepsilon=H_{\varepsilon,\eta}+(\widetilde u_\eta-u)\cdot\nabla v^{app}_\varepsilon+v^{app}_\varepsilon\cdot\nabla(\widetilde u_\eta-u)$$

with $H_{\varepsilon,\eta}\stackrel{\text{def}}{=}\widetilde u_\eta\cdot\nabla v^{app}_\varepsilon+v^{app}_\varepsilon\cdot\nabla\widetilde u_\eta\,.$

As v^{app}_ε and $\widetilde u_\eta-u$ are divergence-free vector fields,

$$H_\varepsilon-H_{\varepsilon,\eta}=\operatorname{div}\big((\widetilde u_\eta-u)\otimes v^{app}_\varepsilon+v^{app}_\varepsilon\otimes(\widetilde u_\eta-u)\big)\,.$$

Thanks to [5, Lemma 3.3] we get

$$\|H_\varepsilon-H_{\varepsilon,\eta}\|_{\dot H^{-\frac12}}\lesssim\|\widetilde u_\eta-u\|_{\dot H^{\frac12}}\big(\|\nabla^h v^{app}_\varepsilon\|_{L^\infty_vL^2_h}+\|v^{app}_\varepsilon\|_{L^\infty}+\|\partial_3 v^{app}_\varepsilon\|_{L^2_v\dot H^{\frac12}_h}\big)$$

and Proposition 4.2 along with (4.6) lead to

$$\|H_\varepsilon - H_{\varepsilon,\eta}\|_{L^2(\mathbb{R}^+,\dot H^{-\frac12}(\mathbb{R}^3))} \lesssim \eta\,.$$

It remains to estimate $H_{\varepsilon,\eta} = \widetilde{u}_\eta \cdot \nabla v_\varepsilon^{app} + v_\varepsilon^{app} \cdot \nabla \widetilde{u}_\eta$. By Propositions 4.2 and 4.3 we have

$$\begin{aligned}\|\widetilde{u}_\eta^3 \partial_3 v_\varepsilon^{app}\|_{L^2(\mathbb{R}^+,\dot H^{-\frac12}(\mathbb{R}^3))} &\lesssim \|\widetilde{u}_\eta^3\|_{L^\infty(\mathbb{R}^+,\dot H^{\frac12}(\mathbb{R}^3))} \|\partial_3 v_\varepsilon^{app}\|_{L^2(\mathbb{R}^+,\dot H^{\frac12}(\mathbb{R}^3))}\\ &\lesssim \|\widetilde{u}_\eta^3\|_{L^\infty(\mathbb{R}^+,\dot H^{\frac12}(\mathbb{R}^3))} \varepsilon^{\frac12}.\end{aligned}$$

Since $\widetilde{u}_\eta$ is uniformly bounded in $L^\infty(\mathbb{R}^+,\dot H^{\frac12}(\mathbb{R}^3))$, we infer that

$$\lim_{\varepsilon\to0} \|\widetilde{u}_\eta^3 \partial_3 v_\varepsilon^{app}\|_{L^2(\mathbb{R}^+,\dot H^{-\frac12}(\mathbb{R}^3))} = 0\,.$$

Lemma 3.4 of [5] claims that

$$\|ab\|_{\dot H^{-\frac12}} \le C\|a\|_{L^2_v\dot H^{\frac12}_h} \|b(\cdot,0)\|_{L^2_h} + C\|x_3 a\|_{L^2} \|\partial_3 b\|_{L^\infty_v \dot H^{\frac12}_h}\,.$$

So we get

$$\|\widetilde{u}_\eta^h \cdot \nabla^h v_\varepsilon^{app}\|_{\dot H^{-\frac12}} \lesssim \|\widetilde{u}_\eta^h\|_{L^2_v\dot H^{\frac12}_h} \|\nabla^h v_\varepsilon^{app}(\cdot,0)\|_{L^2_h} + \|x_3 \widetilde{u}_\eta^h\|_{L^2} \|\partial_3 \nabla^h v_\varepsilon^{app}\|_{L^\infty_v\dot H^{\frac12}_h}$$

and

$$\|v_\varepsilon^{app} \cdot \nabla \widetilde{u}_\eta\|_{\dot H^{-\frac12}} \lesssim \|\nabla \widetilde{u}_\eta\|_{L^2_v\dot H^{\frac12}_h} \|v_\varepsilon^{app}(\cdot,0)\|_{L^2_h} + \|x_3 \nabla \widetilde{u}_\eta\|_{L^2} \|\partial_3 v_\varepsilon^{app}\|_{L^\infty_v\dot H^{\frac12}_h}\,.$$

Propositions 4.2 and 4.3 lead to

$$\lim_{\varepsilon\to0} \int_{\mathbb{R}+} \|x_3 \widetilde{u}_\eta^h(t)\|^2_{L^2(\mathbb{R}^3)} \|\partial_3 \nabla^h v_\varepsilon^{app}(t)\|^2_{L^\infty_v \dot H^{\frac12}_h(\mathbb{R}^3)}\, dt = 0$$

and

$$\lim_{\varepsilon\to0} \int_{\mathbb{R}+} \|x_3 \nabla \widetilde{u}_\eta(t)\|^2_{L^2(\mathbb{R}^3)} \|\partial_3 v_\varepsilon^{app}(t)\|^2_{L^\infty_v \dot H^{\frac12}_h(\mathbb{R}^3)}\, dt = 0\,.$$

Now we recall that $\widetilde{u}_\eta$ is uniformly bounded in $L^\infty(\mathbb{R}^+,\dot H^{\frac12}) \cap L^2(\mathbb{R}^+,\dot H^{\frac32})$; hence, $\widetilde{u}_\eta$ is uniformly bounded in $L^\infty(\mathbb{R}^+,L^2_v\dot H^{\frac12}_h)$ and $\nabla\widetilde{u}_\eta$ is uniformly bounded in $L^2(\mathbb{R}^+, L^2_v\dot H^{\frac12}_h)$. So in order to conclude we just have to estimate

$$\|v_\varepsilon^{app}(\cdot,0)\|_{L^\infty(\mathbb{R}^+,L^2_h(\mathbb{R}^2))} + \|\nabla^h v_\varepsilon^{app}(\cdot,0)\|_{L^2(\mathbb{R}^+,L^2_h(\mathbb{R}^2))}\,.$$

This is done in the following proposition, which concludes the proof of Proposition 4.5.

Proposition 4.6. *For all $\delta > 0$ there are positive constants $\varepsilon(\delta)$ and $c(\delta)$ such that for all $0 < \varepsilon \le \varepsilon(\delta)$, if $\|u_{\varepsilon,0}(\cdot,0)\|_{L^2_h} \le c(\delta)$ then*

$$\|v^{app}_\varepsilon(\cdot,0)\|_{L^\infty(\mathbb{R}^+,L^2_h(\mathbb{R}^2))} + \|\nabla^h v^{app}_\varepsilon(\cdot,0)\|_{L^2(\mathbb{R}^+,L^2_h(\mathbb{R}^2))} \le \delta\,.$$

Proof. First, we estimate $\overline{v}^h_\varepsilon$ and w^h_ε. For all $\varepsilon > 0$, an energy estimate in L^2_h gives

$$\frac{1}{2}\|\overline{v}^h_\varepsilon(t,\cdot,0)\|^2_{L^2_h} + \int_0^t \|\nabla^h \overline{v}^h_\varepsilon(t',\cdot,0)\|^2_{L^2_h} dt' = \frac{1}{2}\|v_{\varepsilon,0}(\cdot,0)\|^2_{L^2_h}\,. \tag{4.7}$$

Then, for all $\delta > 0$ there is a constant $c(\delta)$ such that if $\|v_{\varepsilon,0}(\cdot,0)\|_{L^2_h} \le c(\delta)$ then

$$\|\overline{v}^h_\varepsilon(\cdot,0)\|_{L^\infty(\mathbb{R}^+,L^2_h(\mathbb{R}^2))} + \|\nabla^h \overline{v}^h_\varepsilon(\cdot,0)\|_{L^2(\mathbb{R}^+,L^2_h(\mathbb{R}^2))} \le \delta\,.$$

Moreover, by Proposition 4.3 we have

$$\|w^h_\varepsilon(\cdot,0)\|_{L^\infty(\mathbb{R}^+,L^2_h)} + \|\nabla^h w^h_\varepsilon(\cdot,0)\|_{L^2(\mathbb{R}^+,L^2_h)} \lesssim \varepsilon\,.$$

It remains to estimate w^3_ε. According to Proposition 4.3, w_ε and $\nabla^h w_\varepsilon$ are uniformly bounded respectively in $L^\infty(\mathbb{R}^+, L^\infty_v \dot{H}_h^{-\frac{1}{2}})$ and $L^2(\mathbb{R}^+, L^\infty_v \dot{H}_h^{-\frac{1}{2}})$, so we shall get the result by proving that for all $\delta > 0$ there are positive constants $\varepsilon(\delta)$ and $c(\delta)$ such that if $\varepsilon \le \varepsilon(\delta)$ and $\|u_{\varepsilon,0}(\cdot,0)\|_{L^2_h} \le c(\delta)$ then

$$\|w^3_\varepsilon(\cdot,0)\|_{L^\infty(\mathbb{R}^+,\dot{H}_h^{\frac{1}{2}})} + \|\nabla^h w^3_\varepsilon(\cdot,0)\|_{L^2(\mathbb{R}^+,\dot{H}_h^{\frac{1}{2}})} \le \delta\,.$$

Recall that w^3_ε satisfies

$$\begin{cases} \partial_t w^3_\varepsilon + \overline{v}^h_\varepsilon \cdot \nabla^h w^3_\varepsilon - \Delta_h w^3_\varepsilon = \partial_3^2 w^3_\varepsilon - \partial_3 q_\varepsilon \\ w^3_{\varepsilon|t=0} = w^3_{\varepsilon,0}\,. \end{cases}$$

Define $T_\varepsilon \stackrel{\text{def}}{=} \partial_3^2 w^3_\varepsilon - \partial_3 q_\varepsilon$. An energy estimate in $\dot{H}_h^{\frac{1}{2}}$ gives

$$\begin{aligned} &\|w^3_\varepsilon(t,0)\|^2_{\dot{H}_h^{\frac{1}{2}}} + \int_0^t \|\nabla^h w^3_\varepsilon(t',0)\|^2_{\dot{H}_h^{\frac{1}{2}}} dt' \\ &\lesssim \|w^3_{\varepsilon,0}(\cdot,0)\|^2_{\dot{H}_h^{\frac{1}{2}}} + \|T_\varepsilon(\cdot,0)\|^2_{L^2(\mathbb{R}^+,\dot{H}_h^{-\frac{1}{2}})} + \int_0^t \left|\langle \overline{v}^h_\varepsilon \cdot \nabla^h w^3_\varepsilon, w^3_\varepsilon\rangle_{\dot{H}_h^{\frac{1}{2}}}\right|(t',0)dt'\,. \end{aligned} \tag{4.8}$$

Using [3, Lemma 1.1] we get for each fixed x_3

$$\left|\langle \overline{v}^h_\varepsilon \cdot \nabla^h w^3_\varepsilon, w^3_\varepsilon\rangle_{\dot{H}_h^{\frac{1}{2}}}(x_3)\right| \lesssim \|\nabla^h \overline{v}^h_\varepsilon(x_3)\|_{L^2_h} \|\nabla^h w^3_\varepsilon(x_3)\|_{\dot{H}_h^{\frac{1}{2}}} \|w^3_\varepsilon(x_3)\|_{\dot{H}_h^{\frac{1}{2}}}\,.$$

In particular, using (4.7), we get

$$\int_0^t \left|\langle \overline{v}_\varepsilon^h\cdot\nabla^h w_\varepsilon^3, w_\varepsilon^3\rangle_{\dot H_h^{\frac12}}(t',0)\right|dt'$$
$$\lesssim \|\nabla^h\overline{v}_\varepsilon^h(\cdot,0)\|_{L^2(\mathbb{R}^+,L_h^2)}\|\nabla^h w_\varepsilon^3(\cdot,0)\|_{L^2(\mathbb{R}^+,\dot H_h^{\frac12})}\|w_\varepsilon^3(\cdot,0)\|_{L^\infty(\mathbb{R}^+,\dot H_h^{\frac12})}$$
$$\lesssim \|\overline{v}_{\varepsilon,0}^h(\cdot,0)\|_{L_h^2}\|\nabla^h w_\varepsilon^3(\cdot,0)\|_{L^2(\mathbb{R}^+,\dot H_h^{\frac12})}\|w_\varepsilon^3(\cdot,0)\|_{L^\infty(\mathbb{R}^+,\dot H_h^{\frac12})}$$

Then we infer that

$$\int_0^t \left|\langle \overline{v}_\varepsilon^h\cdot\nabla^h w_\varepsilon^3, w_\varepsilon^3\rangle_{\dot H_h^{\frac12}}(t',\cdot,0)\right|dt' \lesssim \|u_{\varepsilon,0}(\cdot,0)\|_{L_h}$$
$$\times\left(\|\nabla^h w_\varepsilon^3(\cdot,0)\|^2_{L^2(\mathbb{R}^+,\dot H_h^{\frac12})}+\|w_\varepsilon^3(\cdot,0)\|^2_{L^\infty(\mathbb{R}^+,\dot H_h^{\frac12})}\right).$$

Plugging this inequality into (4.8) we obtain that there is a constant C such that

$$\|w_\varepsilon^3(\cdot,0)\|^2_{L^\infty(\mathbb{R}^+,\dot H_h^{\frac12})}+\left(1-C\|u_{\varepsilon,0}(\cdot,0)\|_{L_h^2}\right)\|\nabla^h w_\varepsilon^3(\cdot,0)\|^2_{L^2(\mathbb{R}^+,\dot H_h^{\frac12})}$$
$$\lesssim \|w_{\varepsilon,0}^3(\cdot,0)\|^2_{\dot H_h^{\frac12}}+\|T_\varepsilon(\cdot,0)\|^2_{L^2(\mathbb{R}^+,\dot H_h^{-\frac12})}$$
$$\lesssim \|w_{\varepsilon,0}^3(\cdot,0)\|_{L_h^2}\|w_{\varepsilon,0}^3(\cdot,0)\|_{\dot H_h^1}+\|T_\varepsilon(\cdot,0)\|^2_{L^2(\mathbb{R}^+,\dot H_h^{-\frac12})}.$$

As $w_{\varepsilon,0}$ is uniformly bounded in $L_v^\infty\dot H_h^1$, it remains to prove that

$$\lim_{\varepsilon\to0}\|T_\varepsilon(\cdot,0)\|_{L^2(\mathbb{R}^+,\dot H_h^{-\frac12})}=0.$$

As $\partial_3^2 w_\varepsilon^3=-\partial_3\mathrm{div}_h w_\varepsilon^h$, we get

$$\|\partial_3^2 w_\varepsilon^3(\cdot,0)\|_{L^2(\mathbb{R}^+,\dot H_h^{-\frac12})}\le\|\partial_3\nabla^h w_\varepsilon^h(\cdot,0)\|_{L^2(\mathbb{R}^+,\dot H_h^{-\frac12})}\le\|\partial_3\nabla^h w_\varepsilon^h\|_{L^2(\mathbb{R}^+,L_v^\infty\dot H_h^{-\frac12})}.$$

The bounds on w_ε given in Proposition 4.3 along with the Gagliardo-Nirenberg inequality lead to

$$\|\partial_3^2 w_\varepsilon^3(\cdot,0)\|_{L^2(\mathbb{R}^+,\dot H_h^{-\frac12})}\le\|\partial_3\nabla^h w_\varepsilon^h\|^{\frac12}_{L^2(\mathbb{R}^+,L_v^2\dot H_h^{-\frac12})}\|\partial_3^2\nabla^h w_\varepsilon^h\|^{\frac12}_{L^2(\mathbb{R}^+,L_v^2\dot H_h^{-\frac12})}$$
$$\lesssim\varepsilon^2.$$

Now let us turn to the pressure term. Recall that

$$-\Delta q_\varepsilon=\mathrm{div}N_\varepsilon\,,\quad\text{with}\quad N_\varepsilon\overset{\mathrm{def}}{=}\overline{v}_\varepsilon^h\cdot\nabla^h w_\varepsilon=\mathrm{div}_h\left(\overline{v}_\varepsilon^h\otimes w_\varepsilon\right)$$

since $\overline{v}_\varepsilon^h$ is divergence free. To estimate $\partial_3 q_\varepsilon(\cdot,0)$ we use Gagliardo-Nirenberg inequality, according to which it suffices to estimate $\partial_3 q_\varepsilon$ in L_v^2 and in $\dot H_v^1$.

Since $(-\Delta)^{-1}\mathrm{div}_h\,\mathrm{div}$ is a zero-order Fourier multiplier, we have

$$\|\partial_3 q_\varepsilon\|_{L^2(\mathbb{R}^+,H^{-\frac12,1})} \lesssim \|\partial_3(\overline{v}_\varepsilon^h \otimes w_\varepsilon)\|_{L^2(\mathbb{R}^+,H^{-\frac12,1})}\cdot$$

On the one hand we write

$$\|w_\varepsilon \partial_3 \overline{v}_\varepsilon^h\|_{L^2(\mathbb{R}^+,L^2_v\dot H_h^{-\frac12})} \lesssim \|w_\varepsilon\|_{L^2(\mathbb{R}^+,L^2_v\dot H_h^{\frac12})} \|\partial_3 \overline{v}_\varepsilon^h\|_{L^\infty(\mathbb{R}^+,L^\infty_v L^2_h)} \lesssim \varepsilon^{\frac12}$$

by Propositions 4.2 and 4.3, and similarly,

$$\|\overline{v}_\varepsilon^h \partial_3 w_\varepsilon\|_{L^2(\mathbb{R}^+,L^2_v\dot H_h^{-\frac12})} \lesssim \|\partial_3 w_\varepsilon\|_{L^2(\mathbb{R}^+,L^2_v\dot H_h^{\frac12})} \|\overline{v}_\varepsilon^h\|_{L^\infty(\mathbb{R}^+,L^\infty_v L^2_h)} \lesssim \varepsilon^{\frac12}\,.$$

In the same way we find that

$$\|\partial_3(\overline{v}_\varepsilon^h \otimes w_\varepsilon)\|_{L^2(\mathbb{R}^+,\dot H^{-\frac12,1})} \lesssim \varepsilon^{\frac32}\,.$$

This ends the proof of Proposition 4.6.

4.4 Estimates on the Linear Transport-Diffusion Equation

In this appendix we shall prove Proposition 4.3. It turns out to be convenient to rescale w_ε. Thus we define the vector field

$$W_\varepsilon(t,x) \stackrel{\text{def}}{=} \Big(\frac{w_\varepsilon^h}{\varepsilon}, w_\varepsilon^3\Big)(t,x_h,\varepsilon^{-1}x_3)$$

which satisfies

$$\begin{cases} \partial_t W_\varepsilon + \overline{V}_\varepsilon^h \cdot \nabla^h W_\varepsilon - \Delta_h W_\varepsilon - \varepsilon^2\,\partial_3^2 W_\varepsilon = -\left(\nabla^h Q_\varepsilon, \varepsilon^2\,\partial_3 Q_\varepsilon\right) \\ \mathrm{div}\,W_\varepsilon = 0 \\ W_\varepsilon\,(0,.) = W_{\varepsilon,0} \end{cases}$$

where

$$\overline{V}_\varepsilon^h(t,x) \stackrel{\text{def}}{=} \overline{v}_\varepsilon^h(t,x_h,\varepsilon^{-1}x_3) \quad \text{and} \quad Q_\varepsilon(t,x) \stackrel{\text{def}}{=} \varepsilon^{-1} q_\varepsilon(t,x_h,\varepsilon^{-1}x_3)\,.$$

Note that thanks to Proposition 4.2, the vector field $\partial^\alpha \overline{V}_\varepsilon^h$ is uniformly bounded in the space $L^\infty(\mathbb{R}^+,L^2_v\dot H_h^s)\cap L^2(\mathbb{R}^+,L^2_v\dot H_h^{s+1})$ for each $\alpha\in\mathbb{N}^3$ and any $s>-1$ and hence also in $L^\infty(\mathbb{R}^+,L^\infty_v\dot H_h^s)\cap L^2(\mathbb{R}^+,L^\infty_v\dot H_h^{s+1})$.

Similarly we have defined

$$W_{\varepsilon,0}(x) \stackrel{\text{def}}{=} \Big(\frac{w^h_{\varepsilon,0}}{\varepsilon}, w^3_{\varepsilon,0}\Big)(x_h, \varepsilon^{-1}x_3),$$

and by construction it is bounded in $\dot{H}^s(\mathbb{R}^3)$ for all $s \geq -1$.

Proposition 4.3 is a corollary of the next statement.

Proposition 4.7. *Under the assumptions of Theorem 4.3, the following results hold.*

1. *For all $s > -1$ and all $\alpha \in \mathbb{N}^3$, $\partial^\alpha W_\varepsilon$ is bounded in $L^\infty(\mathbb{R}^+, L^2_v\dot{H}^s_h) \cap L^2(\mathbb{R}^+, L^2_v\dot{H}^{s+1}_h)$; in particular $\partial^\alpha W_\varepsilon$ is bounded in $L^\infty(\mathbb{R}^+, L^\infty_v\dot{H}^s_h) \cap L^2(\mathbb{R}^+, L^\infty_v\dot{H}^{s+1}_h)$.*
2. *For all $\alpha \in \mathbb{N}^3$, $\partial^\alpha W_\varepsilon$ is bounded in $L^2(\mathbb{R}^+, L^2)$, hence in particular in $L^2(\mathbb{R}^+, L^\infty_v L^2_h)$.*

Proof. Let us start by proving the first statement of the proposition. We notice that it is enough to prove the result for $s \in]-1,1[$, and we shall argue by induction on α.

- Let us start by considering the case $\alpha = 0$. An energy estimate in $L^2_v\dot{H}^s_h$ on the equation satisfied by W_ε gives

$$\frac{1}{2}\frac{\mathrm{d}}{\mathrm{d}t}\|W_\varepsilon\|^2_{L^2_v\dot{H}^s_h} + \|\nabla^h W_\varepsilon\|^2_{L^2_v\dot{H}^s_h} + \varepsilon^2\|\partial_3 W_\varepsilon\|^2_{L^2_v\dot{H}^s_h}$$
$$= -\langle \overline{V}^h_\varepsilon \cdot \nabla^h W_\varepsilon, W_\varepsilon\rangle_{L^2_v\dot{H}^s_h} - \langle \nabla^h Q_\varepsilon, W^h_\varepsilon\rangle_{L^2_v\dot{H}^s_h} - \langle \varepsilon^2\,\partial_3 Q_\varepsilon, W^3_\varepsilon\rangle_{L^2_v\dot{H}^s_h}.$$

For the nonlinear term we have, by [3, Lemma 1.1], and for each given t and x_3,

$$\big|\langle \overline{V}^h_\varepsilon \cdot \nabla^h W_\varepsilon, W_\varepsilon\rangle_{\dot{H}^s_h}(t,x_3)\big| \lesssim \|\nabla^h \overline{V}^h_\varepsilon(t,x_3)\|_{L^2_h}\|\nabla^h W_\varepsilon(t,x_3)\|_{\dot{H}^s_h}\|W_\varepsilon(t,x_3)\|_{\dot{H}^s_h}$$
$$\leq \frac{1}{4}\|\nabla^h W_\varepsilon(t,x_3)\|^2_{\dot{H}^s_h} + C\|\nabla^h \overline{V}^h_\varepsilon(t,x_3)\|^2_{L^2_h}\|W_\varepsilon(t,x_3)\|^2_{\dot{H}^s_h}$$

so after integration over x_3, we find

$$\frac{1}{2}\frac{\mathrm{d}}{\mathrm{d}t}\|W_\varepsilon\|^2_{L^2_v\dot{H}^s_h} + \frac{3}{4}\|\nabla^h W_\varepsilon\|^2_{L^2_v\dot{H}^s_h} + \varepsilon^2\|\partial_3 W_\varepsilon\|^2_{L^2_v\dot{H}^s_h}$$
$$\leq C\|\nabla^h \overline{V}^h_\varepsilon\|^2_{L^\infty_v L^2_h}\|W_\varepsilon\|^2_{L^2_v\dot{H}^s_h} - \langle \nabla^h Q_\varepsilon, W^h_\varepsilon\rangle_{L^2_v\dot{H}^s_h} - \langle \varepsilon^2\,\partial_3 Q_\varepsilon, W^3_\varepsilon\rangle_{L^2_v\dot{H}^s_h}.$$

Now let us study the pressure term. As W_ε is a divergence-free vector field we have

$$-\langle \nabla^h Q_\varepsilon, W^h_\varepsilon\rangle_{L^2_v\dot{H}^s_h} - \langle \varepsilon^2\,\partial_3 Q_\varepsilon, W^3_\varepsilon\rangle_{L^2_v\dot{H}^s_h} = (\varepsilon^2 - 1)\langle \nabla^h Q_\varepsilon, W^h_\varepsilon\rangle_{L^2_v\dot{H}^s_h}.$$

We claim that

$$\big|\langle \nabla^h Q_\varepsilon(t), W^h_\varepsilon(t)\rangle_{L^2_v\dot{H}^s_h}\big| \leq \frac{1}{4}\|\nabla^h W_\varepsilon(t)\|^2_{L^2_v\dot{H}^s_h} + C_\varepsilon(t)\|W_\varepsilon(t)\|^2_{L^2_v\dot{H}^s_h} \tag{4.9}$$

where C_ε is uniformly bounded in $L^1(\mathbb{R}^+)$. Assuming that claim to be true, we infer (up to changing C_ε) that

$$\frac{\mathrm{d}}{\mathrm{d}t}\|W_\varepsilon(t)\|^2_{L^2_v\dot{H}^s_h} + \|\nabla^h W_\varepsilon(t)\|^2_{L^2_v\dot{H}^s_h} + \varepsilon^2\|\partial_3 W_\varepsilon(t)\|^2_{L^2_v\dot{H}^s_h} \lesssim C_\varepsilon(t)\|W_\varepsilon(t)\|^2_{L^2_v\dot{H}^s_h}\,.$$

Thanks to Gronwall's lemma this gives

$$\|W_\varepsilon(t)\|^2_{L^2_v\dot{H}^s_h} + \int_0^t \|\nabla^h W_\varepsilon(t')\|^2_{L^2_v\dot{H}^s_h}\,dt' \lesssim \|W_{\varepsilon,0}\|^2_{L^2_v\dot{H}^s_h}\,,$$

and the conclusion of Proposition 4.7 (4.7), for $\alpha = 0$ and $-1 < s < 1$, comes from the a priori bounds on $W_{\varepsilon,0}$. It remains to prove the claim (4.9). For all real numbers r, we have

$$\big|\langle \nabla^h Q_\varepsilon(t), W^h_\varepsilon(t)\rangle_{L^2_v\dot{H}^s_h}\big| \leq \|\nabla^h Q_\varepsilon(t)\|_{L^2_v\dot{H}^r_h}\|W^h_\varepsilon(t)\|_{L^2_v\dot{H}^{2s-r}_h}\,.$$

As W_ε is a divergence-free vector field we can write

$$\mathrm{div}\,(\overline{V}^h_\varepsilon\cdot\nabla^h W_\varepsilon) = -\Delta_h Q_\varepsilon - \varepsilon^2\,\partial_3^2 Q_\varepsilon\,.$$

Then we define

$$M^h_\varepsilon \stackrel{\mathrm{def}}{=} \overline{V}^h_\varepsilon\cdot\nabla^h W^h_\varepsilon + \partial_3(W^3_\varepsilon\overline{V}^h_\varepsilon),$$

and using the fact that $\overline{V}^h_\varepsilon$ is divergence free, we have

$$\mathrm{div}(\overline{V}^h_\varepsilon\cdot\nabla^h W_\varepsilon) = \mathrm{div}_h M^h_\varepsilon\,.$$

It follows that

$$Q_\varepsilon = (-\Delta_h - \varepsilon^2\,\partial_3^2)^{-1}\mathrm{div}_h M^h_\varepsilon\,, \tag{4.10}$$

and since $\nabla^h(-\Delta_h - \varepsilon^2\,\partial_3^2)^{-1}\mathrm{div}_h$ is a zero-order Fourier multiplier mapping $L^2_v\dot{H}^r_h$ to itself with norm 1, we infer that for all real numbers r,

$$\|\nabla^h Q_\varepsilon\|_{L^2_v\dot{H}^r_h} \leq \|M^h_\varepsilon\|_{L^2_v\dot{H}^r_h}\,,$$

and therefore,

$$\big|\langle \nabla^h Q_\varepsilon(t), W^h_\varepsilon(t)\rangle_{L^2_v\dot{H}^s_h}\big| \leq \|M^h_\varepsilon(t)\|_{L^2_v\dot{H}^r_h}\|W^h_\varepsilon(t)\|_{L^2_v\dot{H}^{2s-r}_h}\,. \tag{4.11}$$

We can estimate $\|M^h_\varepsilon\|_{L^2_v\dot{H}^r_h}$ as follows, thanks to the divergence-free condition on W_ε:

$$\begin{aligned}\|M^h_\varepsilon\|_{L^2_v\dot{H}^r_h} &\leq \|\overline{V}^h_\varepsilon\cdot\nabla^h W_\varepsilon\|_{L^2_v\dot{H}^r_h} + \|\partial_3(W^3_\varepsilon\overline{V}^h_\varepsilon)\|_{L^2_v\dot{H}^r_h}\\ &\leq \|\overline{V}^h_\varepsilon\cdot\nabla^h W_\varepsilon\|_{L^2_v\dot{H}^r_h} + \|W^3_\varepsilon\,\partial_3\overline{V}^h_\varepsilon\|_{L^2_v\dot{H}^r_h} + \|\overline{V}^h_\varepsilon\,\mathrm{div}_h W^h_\varepsilon\|_{L^2_v\dot{H}^r_h}\,.\end{aligned}$$

Thanks to two-dimensional product laws, if $-1 < r < 0$, then we get

$$\|\overline{V}^h_\varepsilon \cdot \nabla^h W_\varepsilon\|_{L^2_v \dot H^r_h} + \|\overline{V}^h_\varepsilon \mathrm{div}_h W^h_\varepsilon\|_{L^2_v \dot H^r_h} \lesssim \|\overline{V}^h_\varepsilon\|_{L^\infty_v \dot H_h^{\frac12}} \|\nabla^h W_\varepsilon\|_{L^2_v \dot H_h^{r+\frac12}}$$

and

$$\|W^3_\varepsilon \partial_3 \overline{V}^h_\varepsilon\|_{L^2_v \dot H^r_h} \lesssim \|\nabla \overline{V}^h_\varepsilon\|_{L^\infty_v L^2_h} \|W^3_\varepsilon\|_{L^2_v \dot H^{r+1}_h} \cdot$$

So if $-1 < r < 0$, then

$$\|M^h_\varepsilon\|_{L^2_v \dot H^r_h} \lesssim \|\overline{V}^h_\varepsilon\|_{L^\infty_v \dot H_h^{\frac12}} \|\nabla^h W_\varepsilon\|_{L^2_v \dot H_h^{r+\frac12}} + \|\nabla \overline{V}^h_\varepsilon\|_{L^\infty_v L^2_h} \|W^3_\varepsilon\|_{L^2_v \dot H^{r+1}_h}, \tag{4.12}$$

and this leads to (4.9) for $-1 < s < 1$, due to the following computations.

- If $0 < s < 1$, we choose $r = s - 1$ to get

  $$\|M^h_\varepsilon\|_{L^2_v \dot H^{s-1}_h} \lesssim \|\overline{V}^h_\varepsilon\|_{L^\infty_v \dot H_h^{\frac12}} \|\nabla^h W_\varepsilon\|_{L^2_v \dot H_h^{s-\frac12}} + \|\nabla \overline{V}^h_\varepsilon\|_{L^\infty_v L^2_h} \|W^3_\varepsilon\|_{L^2_v \dot H^s_h},$$

  so by (4.11) with $r = s - 1$, we infer that

  $$\begin{aligned} \big|\langle \nabla^h Q_\varepsilon, W^h_\varepsilon\rangle_{L^2_v \dot H^s_h}\big| \lesssim{}& \|\overline{V}^h_\varepsilon\|_{L^\infty_v \dot H_h^{\frac12}} \|\nabla^h W_\varepsilon\|_{L^2_v \dot H_h^{s-\frac12}} \|\nabla^h W^h_\varepsilon\|_{L^2_v \dot H^s_h} \\ &+ \frac18 \|\nabla^h W^h_\varepsilon\|^2_{L^2_v \dot H^s_h} + C\|\nabla \overline{V}^h_\varepsilon\|^2_{L^\infty_v L^2_h} \|W^3_\varepsilon\|^2_{L^2_v \dot H^s_h} \cdot \end{aligned} \tag{4.13}$$

  We then use the interpolation inequality

  $$\|\overline{V}^h_\varepsilon\|_{L^\infty_v \dot H_h^{\frac12}} \|\nabla^h W_\varepsilon\|_{L^2_v \dot H_h^{s-\frac12}} \lesssim \|\overline{V}^h_\varepsilon\|^{\frac12}_{L^\infty_v L^2_h} \|\nabla^h \overline{V}^h_\varepsilon\|^{\frac12}_{L^\infty_v L^2_h} \|W_\varepsilon\|^{\frac12}_{L^2_v \dot H^s_h} \|\nabla^h W_\varepsilon\|^{\frac12}_{L^2_v \dot H^s_h}$$

  along with the convexity inequality $ab \leq \frac34 a^{4/3} + \frac14 b^4$, to get

  $$\begin{aligned} \|\nabla^h W_\varepsilon\|_{L^2_v \dot H^s_h} \|\overline{V}^h_\varepsilon\|_{L^\infty_v \dot H_h^{\frac12}} \|\nabla^h W_\varepsilon\|_{L^2_v \dot H_h^{s-\frac12}} \leq{}& \frac18 \|\nabla^h W_\varepsilon\|^2_{L^2_v \dot H^s_h} \\ &+ C\|\overline{V}^h_\varepsilon\|^2_{L^\infty_v L^2_h} \|\nabla \overline{V}^h_\varepsilon\|^2_{L^\infty_v L^2_h} \|W_\varepsilon\|^2_{L^2_v \dot H^s_h} \cdot \end{aligned}$$

  It remains to define

  $$C_\varepsilon(t) \stackrel{\text{def}}{=} C\|\nabla \overline{V}^h_\varepsilon(t)\|^2_{L^\infty_v L^2_h} \big(1 + \|\overline{V}^h_\varepsilon(t)\|^2_{L^\infty_v L^2_h}\big) \tag{4.14}$$

  to obtain from (4.13) that

  $$\big|\langle \nabla^h Q_\varepsilon(t), W^h_\varepsilon(t)\rangle_{L^2_v \dot H^s_h}\big| \leq \frac14 \|\nabla^h W_\varepsilon(t)\|^2_{L^2_v \dot H^s_h} + C_\varepsilon(t) \|W_\varepsilon(t)\|^2_{L^2_v \dot H^s_h} \cdot$$

Notice that C_ε belongs to $L^1(\mathbb{R}^+)$ thanks to the uniform bounds on $\overline{V}^h_\varepsilon$ derived above from Proposition 4.2.

- If $s=0$, we choose $r=-\frac{1}{2}$ and hence by (4.11) and (4.12),

$$\left|\langle\nabla^h Q_\varepsilon, W^h_\varepsilon\rangle_{L^2}\right| \lesssim \|W^h_\varepsilon\|_{L^2_v\dot H_h^{\frac12}}\big(\|\overline{V}^h_\varepsilon\|_{L^\infty_v\dot H_h^{\frac12}}\|\nabla^h W_\varepsilon\|_{L^2}+\|\nabla\overline{V}^h_\varepsilon\|_{L^\infty_v L^2_h}\|W^3_\varepsilon\|_{L^2_v\dot H_h^{\frac12}}\big).$$

By interpolation we infer that

$$\begin{aligned}\left|\langle\nabla^h Q_\varepsilon, W^h_\varepsilon\rangle_{L^2}\right| \lesssim{}& \|W^h_\varepsilon\|^{\frac12}_{L^2}\|\nabla^h W^h_\varepsilon\|^{3/2}_{L^2}\|\overline{V}^h_\varepsilon\|^{\frac12}_{L^\infty_v L^2_h}\|\nabla\overline{V}^h_\varepsilon\|^{\frac12}_{L^\infty_v L^2_h}\\ &+\|W_\varepsilon\|_{L^2}\|\nabla^h W_\varepsilon\|_{L^2}\|\nabla\overline{V}^h_\varepsilon\|_{L^\infty_v L^2_h}.\end{aligned}$$

The convexity inequality $ab\le\frac34 a^{4/3}+\frac14 b^4$ implies that

$$\begin{aligned}&\|W^h_\varepsilon\|^{\frac12}_{L^2}\|\nabla^h W^h_\varepsilon\|^{3/2}_{L^2}\|\overline{V}^h_\varepsilon\|^{\frac12}_{L^\infty_v L^2_h}\|\nabla\overline{V}^h_\varepsilon\|^{\frac12}_{L^\infty_v L^2_h}\\ &\qquad\le\frac18\|\nabla^h W_\varepsilon\|^2_{L^2}+C\|W^h_\varepsilon\|^2_{L^2}\|\overline{V}^h_\varepsilon\|^2_{L^\infty_v L^2_h}\|\nabla\overline{V}^h_\varepsilon\|^2_{L^\infty_v L^2_h}\end{aligned}\tag{4.15}$$

and

$$\|W_\varepsilon\|_{L^2}\|\nabla^h W_\varepsilon\|_{L^2}\|\nabla\overline{V}^h_\varepsilon\|_{L^\infty_v L^2_h}\le\frac18\|\nabla^h W_\varepsilon\|^2_{L^2}+C\|W_\varepsilon\|^2_{L^2}\|\nabla\overline{V}^h_\varepsilon\|^2_{L^\infty_v L^2_h}.\tag{4.16}$$

With the above choice (4.14) for C_ε we obtain

$$\left|\langle\nabla^h Q_\varepsilon(t), W^h_\varepsilon(t)\rangle_{L^2}\right|\le\frac14\|\nabla^h W_\varepsilon(t)\|^2_{L^2}+C_\varepsilon(t)\|W_\varepsilon(t)\|^2_{L^2}.$$

- Finally if $-1<s<0$, we proceed slightly differently. We recall that

$$\mathrm{div}_h M^h_\varepsilon=-\Delta_h Q_\varepsilon-\varepsilon^2\partial_3^2 Q_\varepsilon,$$

and as W_ε is divergence free, we have

$$M^h_\varepsilon=\overline{V}^h_\varepsilon\cdot\nabla^h W^h_\varepsilon-\overline{V}^h_\varepsilon\mathrm{div}_h W^h_\varepsilon+W^3_\varepsilon\,\partial_3\overline{V}^h_\varepsilon.$$

Defining

$$M^h_{\varepsilon,1}\stackrel{\text{def}}{=}\mathrm{div}_h(\overline{V}^h_\varepsilon\otimes W^h_\varepsilon-W^h_\varepsilon\otimes\overline{V}^h_\varepsilon)\quad\text{and}\quad M^h_{\varepsilon,2}\stackrel{\text{def}}{=}W_\varepsilon\cdot\nabla\overline{V}^h_\varepsilon,$$

we can split $M^h_\varepsilon=M^h_{\varepsilon,1}+M^h_{\varepsilon,2}$ and estimate each term differently.

Since $\nabla^h(-\Delta_h-\varepsilon^2\partial_3^2)\mathrm{div}_h$ is a zero-order Fourier multiplier mapping $L^2_v\dot H^s_h$ to itself with norm 1,

$$\big|\langle\nabla^h Q_\varepsilon, W_\varepsilon^h\rangle_{L^2_v\dot H^s_h}\big| \le \|M^h_{\varepsilon,1}\|_{L^2_v\dot H^{s-1}_h}\|W^h_\varepsilon\|_{L^2_v\dot H^{s+1}_h} + \|M^h_{\varepsilon,2}\|_{L^2_v\dot H^s_h}\|W^h_\varepsilon\|_{L^2_v\dot H^s_h}\,.$$

Using two-dimensional product laws we obtain

$$\|M^h_{\varepsilon,1}\|_{L^2_v\dot H^{s-1}_h} \lesssim \|\overline{V}^h_\varepsilon W_\varepsilon\|_{L^2_v\dot H^s_h} \lesssim \|\overline{V}^h_\varepsilon\|_{L^\infty_v\dot H^{\frac12}_h}\|W_\varepsilon\|_{L^2_v\dot H^{s+\frac12}_h}$$

and

$$\|M^h_{\varepsilon,2}\|_{L^2_v\dot H^s_h} \lesssim \|W_\varepsilon\cdot\nabla\overline{V}^h_\varepsilon\|_{L^2_v\dot H^s_h} \lesssim \|\nabla\overline{V}^h_\varepsilon\|_{L^\infty_v L^2}\|W_\varepsilon\|_{L^2_v\dot H^{s+1}_h}\,.$$

Therefore, we get

$$\begin{aligned}|\langle\nabla^h Q_\varepsilon, W_\varepsilon^h\rangle_{L^2_v\dot H^s_h}| \le{}& \|\overline{V}^h_\varepsilon\|_{L^\infty_v\dot H^{\frac12}_h}\|W_\varepsilon\|_{L^2_v\dot H^{s+\frac12}_h}\|\nabla^h W^h_\varepsilon\|_{L^2_v\dot H^s_h}\\ &+\|\nabla\overline{V}^h_\varepsilon\|_{L^\infty_v L^2}\|\nabla^h W_\varepsilon\|_{L^2_v\dot H^s_h}\|W^h_\varepsilon\|_{L^2_v\dot H^s_h}\,.\end{aligned}\tag{4.17}$$

Then we use the interpolation inequality

$$\|W_\varepsilon\|_{L^2_v\dot H^{s+\frac12}_h}\|\nabla^h W^h_\varepsilon\|_{L^2_v\dot H^s_h} \lesssim \|W_\varepsilon\|^{\frac12}_{L^2_v\dot H^s_h}\|\nabla^h W^h_\varepsilon\|^{3/2}_{L^2_v\dot H^s_h}$$

along with the convexity inequalities $ab\le\frac34 a^{4/3}+\frac14 b^4$ and $ab\le\frac12 a^2+\frac12 b^2$, to infer that again with the choice (4.14) for C_ε,

$$\big|\langle\nabla^h Q_\varepsilon(t), W_\varepsilon^h(t)\rangle_{L^2_v\dot H^s_h}\big| \le \frac14\|\nabla^h W_\varepsilon(t)\|^2_{L^2_v\dot H^s_h} + C_\varepsilon(t)\|W_\varepsilon(t)\|^2_{L^2_v\dot H^s_h}\,.$$

The first result of the proposition is therefore proved in the case when $\alpha=0$ and $-1<s<1$.

- To go further in the induction process, let $k\in\mathbb N$ be given and suppose the result proved for all $\alpha\in\mathbb N^3$ such that $|\alpha|\le k$, still for $-1<s<1$. Now consider $\alpha\in\mathbb N^3$ such that $|\alpha|=k+1$. The vector field $\partial^\alpha W_\varepsilon$ solves

$$\partial_t\partial^\alpha W_\varepsilon + \partial^\alpha(\overline{V}^h_\varepsilon\cdot\nabla^h W_\varepsilon) - \Delta_h\partial^\alpha W_\varepsilon - \varepsilon^2\partial_3^2\partial^\alpha W_\varepsilon = -(\nabla^h\partial^\alpha Q_\varepsilon, \varepsilon^2\partial_3\partial^\alpha Q_\varepsilon)\,.$$

An energy estimate in $L^2_v\dot H^s_h$ gives

$$\begin{aligned}\frac12\frac{\mathrm d}{\mathrm dt}\|\partial^\alpha W_\varepsilon\|^2_{L^2_v\dot H^s_h} + \langle\partial^\alpha(\overline{V}^h_\varepsilon\cdot\nabla^h W_\varepsilon),\partial^\alpha W_\varepsilon\rangle_{L^2_v\dot H^s_h} + \|\nabla^h\partial^\alpha W_\varepsilon\|^2_{L^2_v\dot H^s_h}\\ = -\langle\nabla^h\partial^\alpha Q_\varepsilon,\partial^\alpha W^h_\varepsilon\rangle_{L^2_v\dot H^s_h} - \varepsilon^2\langle\partial_3\partial^\alpha Q_\varepsilon,\partial^\alpha W^3_\varepsilon\rangle_{L^2_v\dot H^s_h}\,.\end{aligned}$$

We split $\langle\partial^\alpha(\overline{V}^h_\varepsilon\cdot\nabla^h W_\varepsilon),\partial^\alpha W_\varepsilon\rangle_{L^2_v\dot H^s_h}$ into two contributions:

$$\langle\overline{V}^h_\varepsilon\cdot\nabla^h\partial^\alpha W_\varepsilon,\partial^\alpha W_\varepsilon\rangle_{L^2_v\dot H^s_h} + \sum_{0<\beta\le\alpha} C_\beta\langle\partial^\beta\overline{V}^h_\varepsilon\cdot\nabla^h\partial^{\alpha-\beta}W_\varepsilon,\partial^\alpha W_\varepsilon\rangle_{L^2_v\dot H^s_h}\,.\tag{4.18}$$

The first term in (4.18) satisfies, as in [3, Lemma 1.1]

$$\begin{aligned}|\langle \overline{V}^h_\varepsilon \cdot \nabla^h \partial^\alpha W_\varepsilon, \partial^\alpha W_\varepsilon\rangle_{\dot H^s_h}| &\lesssim \|\nabla^h \overline{V}^h_\varepsilon\|_{L^2_h} \|\nabla^h \partial^\alpha W_\varepsilon\|_{\dot H^s_h} \|\partial^\alpha W_\varepsilon\|_{\dot H^s_h}\\ &\le \frac{1}{4}\|\nabla^h \partial^\alpha W_\varepsilon\|^2_{\dot H^s_h} + C\|\nabla^h \overline{V}^h_\varepsilon\|^2_{L^2_h}\|\partial^\alpha W_\varepsilon\|^2_{\dot H^s_h}\end{aligned}$$

so

$$|\langle \overline{V}^h_\varepsilon \cdot \nabla^h \partial^\alpha W_\varepsilon, \partial^\alpha W_\varepsilon\rangle_{L^2_v \dot H^s_h}| \le \frac{1}{4}\|\nabla^h \partial^\alpha W_\varepsilon\|^2_{L^2_v \dot H^s_h} + C\|\nabla^h \overline{V}^h_\varepsilon\|^2_{L^\infty_v L^2_h}\|\partial^\alpha W_\varepsilon\|^2_{L^2_v \dot H^s_h} \cdot$$

For the remaining terms in (4.18), as $\overline{V}^h_\varepsilon$ is a horizontal, divergence-free vector field, two-dimensional product laws give

$$\begin{aligned}|\langle \partial^\beta \overline{V}^h_\varepsilon \cdot \nabla^h \partial^{\alpha-\beta} W_\varepsilon, \partial^\alpha W_\varepsilon\rangle_{\dot H^s_h}| &= \big|\langle \mathrm{div}_h\big(\partial^\beta \overline{V}^h_\varepsilon \otimes \partial^{\alpha-\beta} W_\varepsilon\big), \partial^\alpha W_\varepsilon\rangle_{\dot H^s_h}\big|\\ &\lesssim \|\partial^\beta \overline{V}^h_\varepsilon \otimes \partial^{\alpha-\beta} W_\varepsilon\|_{\dot H^s_h}\|\nabla^h \partial^\alpha W_\varepsilon\|_{\dot H^s_h}\\ &\lesssim \|\partial^\beta \overline{V}^h_\varepsilon\|_{\dot H_h^{\frac{s+1}{2}}}\|\partial^{\alpha-\beta} W_\varepsilon\|_{\dot H_h^{\frac{s+1}{2}}}\|\nabla^h \partial^\alpha W_\varepsilon\|_{\dot H^s_h}\end{aligned}$$

so

$$\begin{aligned}&|\langle \partial^\beta \overline{V}^h_\varepsilon \cdot \nabla^h \partial^{\alpha-\beta} W_\varepsilon, \partial^\alpha W_\varepsilon\rangle_{L^2_v \dot H^s_h}|\\ &\qquad\le \frac{1}{4}\|\nabla^h \partial^\alpha W_\varepsilon\|^2_{L^2_v \dot H^s_h} + C\|\partial^\beta \overline{V}^h_\varepsilon\|^2_{L^\infty_v \dot H_h^{\frac{s+1}{2}}}\|\partial^{\alpha-\beta} W_\varepsilon\|^2_{L^2_v \dot H_h^{\frac{s+1}{2}}} \cdot\end{aligned}$$

Then we get

$$\begin{aligned}&\frac{1}{2}\frac{\mathrm{d}}{\mathrm{d}t}\|\partial^\alpha W_\varepsilon\|^2_{L^2_v \dot H^s_h} + \frac{1}{2}\|\nabla^h \partial^\alpha W_\varepsilon\|^2_{L^2_v \dot H^s_h} + \varepsilon^2\|\partial_3 \partial^\alpha W_\varepsilon\|^2_{L^2_v \dot H^s_h}\\ &\quad\lesssim \|\nabla^h \overline{V}^h_\varepsilon\|^2_{L^\infty_v L^2_h}\|\partial^\alpha W_\varepsilon\|^2_{L^2_v \dot H^s_h} + \big|\langle \nabla^h \partial^\alpha Q_\varepsilon, \partial^\alpha W^h_\varepsilon\rangle_{L^2_v \dot H^s_h} - \varepsilon^2\langle \partial_3 \partial^\alpha Q_\varepsilon, \partial^\alpha W^3_\varepsilon\rangle_{L^2_v \dot H^s_h}\big|\\ &\qquad + C\sum_{0<\beta\le\alpha}\|\partial^\beta \overline{V}^h_\varepsilon\|^2_{L^\infty_v \dot H_h^{\frac{s+1}{2}}}\|\partial^{\alpha-\beta} W_\varepsilon\|^2_{L^2_v \dot H_h^{\frac{s+1}{2}}} \cdot\end{aligned}$$

Now let us estimate the pressure term. We recall that

$$-\langle \nabla^h \partial^\alpha Q_\varepsilon, \partial^\alpha W^h_\varepsilon\rangle_{L^2_v \dot H^s_h} - \langle \varepsilon^2 \partial_3 \partial^\alpha Q_\varepsilon, \partial^\alpha W^3_\varepsilon\rangle_{L^2_v \dot H^s_h} = (\varepsilon^2 - 1)\langle \nabla^h \partial^\alpha Q_\varepsilon, \partial^\alpha W^h_\varepsilon\rangle_{L^2_v \dot H^s_h}$$

and we claim that

$$|\langle \nabla^h \partial^\alpha Q_\varepsilon, \partial^\alpha W^h_\varepsilon\rangle_{L^2_v \dot H^s_h}(t)| \le \frac{1}{4}\|\nabla^h \partial^\alpha W_\varepsilon(t)\|^2_{L^2_v \dot H^s_h} + C_{1,\varepsilon}(t) + C_{2,\varepsilon}(t)\|\partial^\alpha W_\varepsilon(t)\|^2_{L^2_v \dot H^s_h} \tag{4.19}$$

with $C_{1,\varepsilon}$ and $C_{2,\varepsilon}$ uniformly bounded in $L_1(\mathbb{R}^+)$. By the induction assumption (noticing that $(s+1)/2+\alpha-1<\alpha$), we deduce that

$$\sum_{0<\beta\leq\alpha}\|\partial^\beta\overline{V}^h_\varepsilon\|^2_{L^\infty_v\dot H_h^{\frac{s+1}{2}}}\|\partial^{\alpha-\beta}W_\varepsilon\|^2_{L^2_v\dot H_h^{\frac{s+1}{2}}}$$

is uniformly bounded in $L^1(\mathbb{R}^+)$ so up to changing $C_{1,\varepsilon}$ and $C_{2,\varepsilon}$ we get

$$\frac{\mathrm{d}}{\mathrm{d}t}\|\partial^\alpha W_\varepsilon(t)\|^2_{L^2_v\dot H^s_h}+\|\nabla^h\partial^\alpha W_\varepsilon(t)\|^2_{L^2_v\dot H^s_h}\leq C_{1,\varepsilon}(t)+C_{2,\varepsilon}(t)\|\partial^\alpha W_\varepsilon(t)\|^2_{L^2_v\dot H^s_h}.$$

Using Gronwall's lemma in turn this implies that

$$\|\partial^\alpha W_\varepsilon(t)\|^2_{L^2_v\dot H^s_h}+\int_0^t\|\nabla^h\partial^\alpha W_\varepsilon(t')\|^2_{L^2_v\dot H^s_h}\mathrm{d}t'\lesssim\|\partial^\alpha W_{\varepsilon,0}\|^2_{L^2_v\dot H^s_h},$$

and the bounds on $W_{\varepsilon,0}$ conclude the proof if $-1<s<1$. It remains to prove the estimate (4.19) on the pressure term. We shall adapt the computations of the case $\alpha=0$. We define

$$N_{\varepsilon,\alpha,\beta}\stackrel{\mathrm{def}}{=}\partial^\beta\overline{V}^h_\varepsilon\cdot\nabla^h\partial^{\alpha-\beta}W^h_\varepsilon+\partial_3(\partial^{\alpha-\beta}W^3_\varepsilon\partial^\beta\overline{V}^h_\varepsilon),$$

and recalling (4.10) we get, since $\nabla^h(-\Delta_h-\varepsilon^2\partial_3^2)^{-1}\mathrm{div}_h$ is a Fourier multiplier of order 0 mapping $L^2_v\dot H^r_h$ to itself with norm 1,

$$\langle\nabla^h\partial^\alpha Q_\varepsilon,\partial^\alpha W^h_\varepsilon\rangle_{L^2_v\dot H^s_h}\lesssim\sum_{0\leq\beta\leq\alpha}\|N_{\varepsilon,\alpha,\beta}\|_{L^2_v\dot H_h^{r_\beta}}\|\partial^\alpha W^h_\varepsilon(t,\cdot)\|_{L^2_v\dot H_h^{2s-r_\beta}}$$

where r_β is any real number. Then we define

$$(*)_{\alpha,\beta}:=\|N_{\varepsilon,\alpha,\beta}\|_{L^2_v\dot H_h^{r_\beta}}\|\partial^\alpha W^h_\varepsilon(t,\cdot)\|_{L^2_v\dot H_h^{2s-r_\beta}}.$$

The term $(*)_{\alpha,0}$ can be treated as was done for $\alpha=0$, changing W^h_ε into $\partial^\alpha W^h_\varepsilon$. So we have, as in the proof of (4.9),

$$|(*)_{\alpha,0}|\leq\frac{1}{8}\|\nabla^h\partial^\alpha W_\varepsilon\|^2_{L^2_v\dot H^s_h}+C\|\partial^\alpha W_\varepsilon\|^2_{L^2_v\dot H^s_h}\|\nabla^h\partial^\alpha\overline{V}^h_\varepsilon\|^2_{L^\infty_vL^2_h}\big(1+\|\partial^\alpha\overline{V}^h_\varepsilon\|^2_{L^\infty_vL^2_h}\big).\tag{4.20}$$

For the others terms we have the following estimates.

- If $0<s<1$, we choose $r_\beta=s-1$ like in the case $\alpha=0$, and as in (4.13) we obtain

$$\sum_{0<\beta\leq\alpha}|(*)_{\alpha,\beta}|\leq\frac{1}{8}\|\nabla^h\partial^\alpha W_\varepsilon\|^2_{L^2_v\dot H^s_h}+C\sum_{0<\beta\leq\alpha}\|\partial^\beta\overline{V}^h_\varepsilon\|^2_{L^\infty_v\dot H_h^{\frac12}}\|\nabla^h\partial^{\alpha-\beta}W_\varepsilon\|^2_{L^2_v\dot H_h^{s-\frac12}}$$
$$+C\sum_{0<\beta\leq\alpha}\|\nabla\partial^\beta\overline{V}^h_\varepsilon\|^2_{L^\infty_vL^2_h}\|\partial^{\alpha-\beta}W^3_\varepsilon\|^2_{L^2_v\dot H^s_h}.$$

Then we define, recalling (4.20),

$$C_{1,\varepsilon} \stackrel{\text{def}}{=} C \sum_{0<\beta\leq\alpha} \|\partial^\beta \overline{V}_\varepsilon^h\|^2_{L^\infty_v \dot{H}_h^{\frac{1}{2}}} \|\nabla^h \partial^{\alpha-\beta} W_\varepsilon\|^2_{L^2_v \dot{H}_h^{s-\frac{1}{2}}} + C \sum_{0<\beta\leq\alpha} \|\nabla \partial^\beta \overline{V}_\varepsilon^h\|^2_{L^\infty_v L^2_h} \|\partial^{\alpha-\beta} W_\varepsilon^3\|^2_{L^2_v \dot{H}_h^s}$$

and

$$C_{2,\varepsilon} \stackrel{\text{def}}{=} C \|\nabla^h \partial^\alpha \overline{V}_\varepsilon^h\|^2_{L^\infty_v L^2_h} \big(1 + \|\partial^\alpha \overline{V}_\varepsilon^h\|^2_{L^\infty_v L^2_h}\big)$$

to get

$$\sum_{0\leq\beta\leq\alpha} \big|(*)_{\alpha,\beta}\big| \leq \frac{1}{4} \|\nabla^h \partial^\alpha W_\varepsilon\|^2_{L^2_v \dot{H}_h^s} + C_{1,\varepsilon} + C_{2,\varepsilon} \|\partial^\alpha W_\varepsilon\|^2_{L^2_v \dot{H}_h^s} \cdot$$

Note that the families $(C_{1,\varepsilon})_{\varepsilon>0}$ and $(C_{2,\varepsilon})_{\varepsilon>0}$ are bounded in $L^1(\mathbb{R}^+)$ thanks to the induction assumption and Proposition 4.2.

- If $s = 0$ then following the steps leading to (4.15), (4.16) we choose $r_\beta = -1/2$ and write

$$\big|(*)_{\alpha,\beta}\big| \lesssim \|\partial^\alpha W_\varepsilon^h\|_{L^2_v \dot{H}_h^{\frac{1}{2}}} \Big(\|\partial^\beta \overline{V}_\varepsilon^h\|_{L^\infty_v \dot{H}_h^{\frac{1}{2}}} \|\nabla^h \partial^{\alpha-\beta} W_\varepsilon\|_{L^2} + \|\nabla \partial^\beta \overline{V}_\varepsilon^h\|_{L^\infty_v L^2_h} \|\partial^{\alpha-\beta} W_\varepsilon^3\|_{L^2_v \dot{H}_h^{\frac{1}{2}}}\Big)$$

so, by interpolation, we get

$$\big|(*)_{\alpha,\beta}\big| \lesssim \|\partial^\alpha W_\varepsilon^h\|^{\frac{1}{2}}_{L^2} \|\partial^\alpha \nabla^h W_\varepsilon^h\|^{\frac{1}{2}}_{L^2} \|\partial^\beta \overline{V}_\varepsilon^h\|_{L^\infty_v \dot{H}_h^{\frac{1}{2}}} \|\nabla^h \partial^{\alpha-\beta} W_\varepsilon\|_{L^2} + \|\partial^\alpha W_\varepsilon^h\|^{\frac{1}{2}}_{L^2} \|\partial^\alpha \nabla^h W_\varepsilon^h\|^{\frac{1}{2}}_{L^2} \|\nabla \partial^\beta \overline{V}_\varepsilon^h\|_{L^\infty_v L^2_h} \|\partial^{\alpha-\beta} W_\varepsilon^3\|_{L^2_v \dot{H}_h^{\frac{1}{2}}} \cdot$$

When $\beta > 0$, the convexity inequality $abc \leq \frac{1}{4}a^4 + \frac{1}{4}b^4 + \frac{1}{2}c^2$ leads to

$$\sum_{0<\beta\leq\alpha} \big|(*)_{\alpha,\beta}\big| \leq \frac{1}{8} \|\nabla^h \partial^\alpha W_\varepsilon\|^2_{L^2} + C \sum_{0<\beta\leq\alpha} \|\partial^\alpha W_\varepsilon^h\|^2_{L^2} \big(\|\partial^\beta \overline{V}_\varepsilon^h\|^4_{L^\infty_v \dot{H}_h^{\frac{1}{2}}} + \|\nabla \partial^\beta \overline{V}_\varepsilon^h\|^4_{L^\infty_v L^2_h}\big) + C \sum_{0<\beta\leq\alpha} \big(\|\nabla^h \partial^{\alpha-\beta} W_\varepsilon\|^2_{L^2} + \|\nabla^h \partial^{\alpha-\beta} W_\varepsilon^3\|^2_{L^2_v \dot{H}_h^{-\frac{1}{2}}}\big) \cdot$$

We define

$$C_{1,\varepsilon} \stackrel{\text{def}}{=} C \sum_{0<\beta\leq\alpha} \big(\|\nabla^h \partial^{\alpha-\beta} W_\varepsilon\|^2_{L^2} + \|\nabla^h \partial^{\alpha-\beta} W_\varepsilon^3\|^2_{L^2_v \dot{H}_h^{-\frac{1}{2}}}\big)$$

and

$$C_{2,\varepsilon} \stackrel{\text{def}}{=} C\|\nabla^h\partial^\alpha\overline{V}_\varepsilon^h\|_{L_v^\infty L_h^2}^2(1+\|\partial^\alpha\overline{V}_\varepsilon^h\|_{L_v^\infty L_h^2}^2)$$
$$+C\sum_{0<\beta\leq\alpha}(\|\partial^\beta\overline{V}_\varepsilon^h\|^4_{L_v^\infty\dot{H}_h^{\frac{1}{2}}}+\|\nabla\partial^\beta\overline{V}_\varepsilon^h\|^4_{L_v^\infty L_h^2})$$

to get when $s=0$ and recalling (4.20),

$$\sum_{0<\beta\leq\alpha}\big|(*)_{\alpha,\beta}\big|\leq\frac{1}{4}\|\nabla^h\partial^\alpha W_\varepsilon\|_{L^2}^2+C_{1,\varepsilon}+C_{2,\varepsilon}\|\partial^\alpha W_\varepsilon\|_{L^2}^2\,.$$

Again note that the families $(C_{1,\varepsilon})_{\varepsilon>0}$ and $(C_{2,\varepsilon})_{\varepsilon>0}$ are bounded in $L^1(\mathbb{R}^+)$ thanks to the induction assumption and Proposition 4.2.

∘ If $-1<s<0$ then following the computations leading to (4.17), we write

$$\big|(*)_{\alpha,\beta}\big|\lesssim\|\partial^\beta\overline{V}_\varepsilon^h\|_{L_v^\infty\dot{H}_h^{\frac{1}{2}}}\|\partial^{\alpha-\beta}W_\varepsilon\|_{L_v^2\dot{H}_h^{s+\frac{1}{2}}}\|\nabla\partial^\alpha W_\varepsilon^h\|_{L_v^2\dot{H}_h^s}$$
$$+\|\nabla\partial^\beta\overline{V}_\varepsilon^h\|_{L_v^\infty L^2}\|\nabla\partial^{\alpha-\beta}W_\varepsilon\|_{L_v^2\dot{H}_h^s}\|\partial^\alpha W_\varepsilon^h\|_{L_v^2\dot{H}_h^s}$$

so

$$\sum_{0<\beta\leq\alpha}\big|(*)_{\alpha,\beta}\big|\leq\frac{1}{4}\|\nabla^h\partial^\alpha W_\varepsilon\|_{L_v^2\dot{H}_h^s}^2+\sum_{\beta<\alpha}\|\partial^\beta\overline{V}_\varepsilon^h\|^2_{L_v^\infty\dot{H}_h^{\frac{1}{2}}}\|\partial^{\alpha-\beta}W_\varepsilon\|^2_{L_v^2\dot{H}_h^{s+\frac{1}{2}}}$$
$$+\sum_{\beta<\alpha}\|\nabla\partial^\beta\overline{V}_\varepsilon^h\|_{L_v^\infty L^2}\|\nabla\partial^{\alpha-\beta}W_\varepsilon\|_{L_v^2\dot{H}_h^s}\|\partial^\alpha W_\varepsilon^h\|_{L_v^2\dot{H}_h^s}\,.$$

In this case, we define

$$C_{1,\varepsilon}\stackrel{\text{def}}{=}C\sum_{0<\beta\leq\alpha}\|\partial^\beta\overline{V}_\varepsilon^h\|^2_{L_v^\infty\dot{H}_h^{\frac{1}{2}}}\|\partial^{\alpha-\beta}W_\varepsilon\|^2_{L_v^2\dot{H}_h^{s+\frac{1}{2}}}\,.$$

and

$$C_{2,\varepsilon} \stackrel{\text{def}}{=} C\|\nabla^h\partial^\alpha\overline{V}_\varepsilon^h\|_{L_v^\infty L_h^2}^2(1+\|\partial^\alpha\overline{V}_\varepsilon^h\|_{L_v^\infty L_h^2}^2)$$
$$+C\sum_{0<\beta\leq\alpha}\|\nabla\partial^\beta\overline{V}_\varepsilon^h\|_{L_v^\infty L^2}\|\nabla\partial^{\alpha-\beta}W_\varepsilon\|_{L_v^2\dot{H}_h^s}$$

which as before are bounded in $L^1(\mathbb{R}^+)$, and we obtain, recalling (4.20),

$$\sum_{0\leq\beta\leq\alpha}(*)_{\alpha,\beta}\leq\frac{1}{4}\|\nabla^h\partial^\alpha W_\varepsilon\|^2_{L_v^2\dot{H}_h^s}+C_{1,\varepsilon}+C_{2,\varepsilon}\|\partial^\alpha W_\varepsilon\|^2_{L_v^2\dot{H}_h^s}\,.$$

The first part of the proposition is proved.

Now let us turn to the second part. As noted above, for all $\alpha \in \mathbb{N}^3$, $\partial^\alpha W_\varepsilon$ satisfies

$$\partial_t \partial^\alpha W_\varepsilon + \partial^\alpha(\overline{V}_\varepsilon^h \cdot \nabla^h W_\varepsilon) - \Delta_h \partial^\alpha W_\varepsilon - \varepsilon^2 \partial_3^2 \partial^\alpha W_\varepsilon = -\partial^\alpha(\nabla^h Q_\varepsilon, \varepsilon^2 \partial_3 Q_\varepsilon)\,.$$

Defining

$$g_\varepsilon \stackrel{\text{def}}{=} \overline{V}_\varepsilon^h \cdot \nabla^h W_\varepsilon + (\nabla^h Q_\varepsilon, \varepsilon^2 \partial_3 Q_\varepsilon)\,,$$

an energy estimate in $L^2_v \dot{H}_h^{-1}$ gives

$$\begin{aligned}\frac{1}{2}\|\partial^\alpha W_\varepsilon(t)\|^2_{L^2_v \dot{H}_h^{-1}} + \int_0^t \|\partial^\alpha W_\varepsilon(t')\|^2_{L^2} \mathrm{d}t' &\le \frac{1}{2}\|\partial^\alpha W_{\varepsilon,0}\|^2_{L^2_v \dot{H}_h^{-1}} \\ &+ \int_0^t \big|\langle \partial^\alpha g_\varepsilon, \partial^\alpha W_\varepsilon\rangle_{L^2_v \dot{H}_h^{-1}}(t')\big|\, dt'\,. \end{aligned} \tag{4.21}$$

We define $K_\varepsilon(t) \stackrel{\text{def}}{=} \sup\limits_{0\le t' \le t} \|\partial^\alpha W_\varepsilon(t')\|_{\dot{H}_h^{-1}}$, so that

$$\frac{1}{2}K_\varepsilon^2(t) \le \frac{1}{2}\|\partial^\alpha W_{\varepsilon,0}\|^2_{L^2_v \dot{H}_h^{-1}} + K_\varepsilon(t) \int_0^t \|\partial^\alpha g_\varepsilon(t')\|_{L^2_v \dot{H}_h^{-1}}\, dt'\,.$$

This implies that

$$\frac{1}{4}K_\varepsilon^2(t) \le \frac{1}{2}\|\partial^\alpha W_{\varepsilon,0}\|^2_{L^2_v \dot{H}_h^{-1}} + \|\partial^\alpha g_\varepsilon\|^2_{L^1(\mathbb{R}^+, L^2_v \dot{H}_h^{-1})}\,. \tag{4.22}$$

But according to (4.21) we know that

$$\int_0^t \|\partial^\alpha W_\varepsilon(t')\|^2_{L^2}\, dt' \le \frac{1}{2}\|\partial^\alpha W_{\varepsilon,0}\|^2_{L^2_v \dot{H}_h^{-1}} + K_\varepsilon(t) \int_0^t \|\partial^\alpha g_\varepsilon(t')\|_{L^2_v \dot{H}_h^{-1}}\, dt'\,,$$

so with (4.22) we infer that

$$\int_0^t \|\partial^\alpha W_\varepsilon(t')\|^2_{L^2} dt' \lesssim \|\partial^\alpha W_{\varepsilon,0}\|^2_{L^2_v \dot{H}_h^{-1}} + \|\partial^\alpha g_\varepsilon\|^2_{L^1(\mathbb{R}^+, L^2_v \dot{H}_h^{-1})}\,.$$

It remains to estimate $\|\partial^\alpha g_\varepsilon\|_{L^1(\mathbb{R}^+, L^2_v \dot{H}_h^{-1})}$. As $\overline{V}_\varepsilon^h$ is a divergence-free vector field, we have

$$\begin{aligned}\|\partial^\alpha(\overline{V}_\varepsilon^h \cdot \nabla^h W_\varepsilon)\|_{L^1(\mathbb{R}^+, L^2_v \dot{H}_h^{-1})} &\le \|\partial^\alpha(\overline{V}_\varepsilon^h \otimes W_\varepsilon)\|_{L^1(\mathbb{R}^+, L^2)} \\ &\lesssim \sum_{0\le\beta\le\alpha} \|\partial^\beta \overline{V}_\varepsilon^h\|_{L^2(\mathbb{R}^+; L^2_v \dot{H}_h^{1/2})} \|\partial^{\alpha-\beta} W_\varepsilon\|_{L^2(\mathbb{R}^+; L^\infty_v \dot{H}_h^{1/2})}\end{aligned} \tag{4.23}$$

which gives the expected bound due to Proposition 4.7 (4.7) proved above. On the other hand, we recall that as computed in (4.10),

$$\Delta_h Q_\varepsilon - \varepsilon^2 \partial_3^2 Q_\varepsilon = \mathrm{div}_h\big(\overline{V}_\varepsilon^h \cdot \nabla^h W_\varepsilon^h + \partial_3(W_\varepsilon^3 \overline{V}_\varepsilon^h)\big)\,.$$

So since $(\Delta_h - \varepsilon^2 \partial_3^2)^{-1}\nabla_h \mathrm{div}_h$ and $(\Delta_h - \varepsilon^2 \partial_3^2)^{-1}\varepsilon\partial_3 \mathrm{div}_h$ are zero-order Fourier multipliers, the same estimates give the expected a priori bound on $(\nabla^h Q_\varepsilon, \varepsilon^2 \partial_3 Q_\varepsilon)$, and the result follows.

References

1. Bahouri, H., Gallagher, I.: On the stability in weak topology of the set of global solutions to the Navier–Stokes equations. (submitted) http://front.math.ucdavis.edu/1109.4043
2. Cannone, M., Meyer, Y., Planchon, F.: Solutions autosimilaires des équations de Navier–Stokes. Séminaire sur les Équations aux Dérivées Partielles de l'École Polytechnique 1993–1994, Exp. No. VIII, École Polytech., Palaiseau, pp. 12, (1994)
3. Chemin, J.-Y.: Remarques sur l'existence pour le système de Navier–Stokes incompressible. SIAM J. Math. Anal. **23**, 20–28 (1992)
4. Chemin, J.-Y., Gallagher, I.: Large, global solutions to the Navier–Stokes equations, slowly varying in one direction. Trans. Am. Math. Soc. **362**(6), 2859–2873 (2010)
5. Chemin, J.-Y., Gallagher, I., Zhang, P.: On large perturbations to global solutions of the 3-D incompressible Navier–Stokes equations, Journal für die reine und angewandte Mathematik, de Gruyter.
6. Fujita, H., Kato, T.: On the Navier–Stokes initial value problem I. Arch. Ration. Mech. Anal. **16**, 269–315 (1964)
7. Gallagher, I.: Profile decomposition for solutions of the Navier–Stokes equations. Bulletin de la Société Mathématique de France **129**, 285–316 (2001)
8. Gallagher, I., Iftimie, D., Planchon, F.: Asymptotics and stability for global solutions to the Navier–Stokes equations. Annales de l'Institut fourier **53**, 2075–2083 (2003)
9. Giga, Y., Miyakawa, T.: Solutions in L^r of the Navier–Stokes initial value problem. Arch. Ration. Mech. Anal. **89**, 267–281 (1985)
10. Iftimie, D.: Resolution of the Navier–Stokes equations in anisotropic spaces. Revista Matematica Ibero-americana **15**, 1–36 (1999)
11. Kato, T.: Strong L^p-solutions of the Navier–Stokes equation in $\mathbb{R}^m$ with applications to weak solutions. Mathematische Zeitschrift **187** 471–480 (1984)
12. Koch, H., Tataru, D.: Well–posedness for the Navier–Stokes equations. Adv. Math. **157**, 22–35 (2001)
13. Leray, J.: Essai sur le mouvement d'un liquide visqueux emplissant l'espace. Acta Math. **63**, 193–248 (1933)
14. Rusin, W., Šverák, V.: Minimal initial data for potential NavierâĂ"Stokes singularities. J. Funct. Anal. **260**, 879–891 (2011)
15. Weissler, F.: The Navier–Stokes initial value problem in L^p. Arch. Ration. Mech. Anal. **74**, 219–230 (1980)

Chapter 5
Schrödinger Equations in Modulation Spaces

Elena Cordero, Fabio Nicola, and Luigi Rodino

Abstract We consider the linear propagator for Schrödinger equations with variable coefficients in $\mathbb{R}^d$. We show that it is bounded on some spaces arising in time–frequency analysis, known as modulation spaces. This generalizes recent results where the case of constant coefficients was considered.

Key words: Fourier integral operators, Modulation spaces, Schrödinger equations, Short-time Fourier transform

2010 Mathematics Subject Classification: 35S30, 47G30, 42C15.

5.1 Introduction

Modulation spaces, denoted by $\mathscr{M}^{p,q}$, $1 \le p,q \le \infty$, were introduced by Feichtinger in 1983 and nowadays used in many fields of analysis [20–22, 27, 44]. For heuristic purposes, functions in $\mathscr{M}^{p,q}$ may be regarded as functions which are locally in $\mathscr{F}L^q$, i.e., Fourier transforms of functions in L^q, and decay at infinity-like functions in L^p. One writes $\mathscr{M}^p$ for $\mathscr{M}^{p,p}$, $1 \le p \le \infty$, and we have $\mathscr{M}^2 = L^2$. We address to the next Sect. 5.1 for definitions, given in terms of the short-time Fourier transform, and related properties. For a comprehensive presentation, see [20, 27].

E. Cordero (✉) • L. Rodino
Dipartimento di Matematica, Università di Torino, via Carlo Alberto 10, I-10123 Torino, Italy
e-mail: elena.cordero@unito.it; luigi.rodino@unito.it

F. Nicola
Dipartimento di Scienze Matematiche, Politecnico di Torino, corso Duca degli Abruzzi 24, I-10129 Torino, Italy
e-mail: fabio.nicola@polito.it

M. Cicognani et al. (eds.), *Studies in Phase Space Analysis with Applications to PDEs*, Progress in Nonlinear Differential Equations and Their Applications 84,
DOI 10.1007/978-1-4614-6348-1_5,

Modulation spaces turned out to be the appropriate function spaces for many problems in time–frequency analysis, with relevant applications to the theory of signals and related issues in numerical analysis. Besides the above-mentioned contributions, see, for example, [9, 13, 17, 23–26].

More recently, modulation spaces have been used in different problems for partial differential equations, cf. [2–4, 10–12, 16, 45, 46], and pseudodifferential operators [28, 29, 34, 35, 37, 40–42]. Our attention will be here limited to linear equations of Schrödinger type, with emphases on boundedness properties of the propagators. Consider as a basic example the free particle Cauchy problem

$$\begin{cases} i\partial_t u + \Delta u = 0 \\ u(0,x) = u_0(x), \end{cases} \tag{5.1}$$

with $x \in \mathbb{R}^d$, $d \geq 1$. The solution $u(t,x)$ is given by the propagator

$$\begin{aligned} u(t,x) &= e^{it\Delta} u_0(x) \\ &= \int_{\mathbb{R}^d} e^{2\pi i(x\eta - 2\pi t|\eta|^2)} \widehat{u_0}(\eta) d\eta, \end{aligned} \tag{5.2}$$

where we normalize the Fourier transform as $\widehat{u_0}(\eta) = \int_{\mathbb{R}^d} e^{-2\pi i x\eta} u_0(x)\, dx$. From [3, Theorem 1] we have for every fixed $t \in \mathbb{R}$

$$e^{it\Delta} : \mathscr{M}^{p,q} \to \mathscr{M}^{p,q}, \quad 1 \leq p,q \leq \infty. \tag{5.3}$$

So, in modulation spaces there is no loss of regularity of $u(t,x)$, $t \neq 0$, with respect to the initial datum $u_0(x)$, whereas in the Lebesgue spaces L^p, $p \neq 2$, the initial regularity is lost because of the strong oscillations of the Fourier multiplier $e^{4\pi^2 it|\eta|^2}$.

Beside other possible applications, the continuity of the map (5.3) provides a good frame to numerical computation of the solutions; cf. [16].

In this paper we shall give a version of (5.3) for more general Schrödinger equations. Consider the Cauchy problem

$$\begin{cases} i\partial_t u + Au = 0 \\ u(0,x) = u_0(x), \end{cases} \tag{5.4}$$

where $A = a(x,D)$ is a second-order pseudodifferential operator, acting on the variables $x \in \mathbb{R}^d$. In the Kohn-Nirenberg quantization,

$$a(x,D)f(x) = \int_{\mathbb{R}^d} e^{2\pi i x\xi} a(x,\xi)\hat{f}(\xi)d\xi. \tag{5.5}$$

We assume that $a(x,D)$ is self-adjoint, with poly-homogeneous symbol $a(x,\xi)$ belonging to the classes of Shubin [39] and Helffer [32]. Namely we have

$$a(x,\xi) \sim \sum_{j=0}^{\infty} a_{2-2j}(x,\xi), \tag{5.6}$$

where $a_{2-2j} \in \mathscr{C}^\infty(\mathbb{R}^{2d} \setminus \{0\})$ is (positively) homogeneous of order $2-2j$ with respect to (x,ξ), i.e. $a_{2-2j}(\lambda x, \lambda \xi) = \lambda^{2-2j} a_{2-2j}(x,\xi)$, for $\lambda > 0$. The principal symbol $a_2(x,\xi)$ is real-valued since A is self-adjoint. The problem (5.4) is forward and backward well posed. In fact, the corresponding evolution operator e^{itA}, acting from $\mathscr{S}(\mathbb{R}^d)$ into $\mathscr{S}(\mathbb{R}^d)$, extends to L^2-isometries [32].

Theorem 5.1. *Under the preceding assumptions, for every fixed $t \in \mathbb{R}$, the operator e^{itA} extends to a continuous map*

$$e^{itA} : \mathscr{M}^p \to \mathscr{M}^p, \quad 1 \le p \le \infty. \tag{5.7}$$

For $p \neq q$, boundedness in $\mathscr{M}^{p,q}$ fails in general, as will be seen below. The proof of Theorem 5.1 and related properties will be given in what follows by combining two kinds of results. In fact, on the one hand, we may represent e^{itA}, $t \in]-T,T[$, $T > 0$ sufficiently small, as a Fourier integral operator (FIO), see, for example, [32, 33]:

$$\begin{aligned}(e^{itA}u_0)(t,x) &= (F_t u_0)(t,x) \\ &= \int_{\mathbb{R}^d} e^{2\pi i \Phi(t,x,\eta)} \sigma(t,x,\eta) \widehat{u_0}(\eta) d\eta + (R_t u_0)(t,x),\end{aligned} \tag{5.8}$$

where σ is in the same class of symbols as a of order 0, depending smoothly on t and $\Phi \in \mathscr{C}^\infty(]-T,T[\times(\mathbb{R}^{2d} \setminus \{0\}))$ is real-valued, smooth for $(x,\eta) \neq (0,0)$, satisfying $\Phi(t,\lambda x, \lambda \eta) = \lambda^2 \Phi(t,x,\eta)$ for $\lambda > 0$, $|\det \partial^2_{x,\eta} \Phi(x,\eta)| > 0$. The operator R_t in (5.8) is regularizing, i.e., $R_t : \mathscr{S}'(\mathbb{R}^d) \to \mathscr{S}(\mathbb{R}^d)$. For the sake of clarity and for later reference in this paper, we shall outline the standard construction of F_t in Sect. 5.3.

On the other hand, previous results of the authors [11, 15] show that the FIOs in the right-hand side of (5.8), for fixed t, map continuously M^p to M^p, $1 \le p \le \infty$. This will be recalled in Sect. 5.4, where we shall also review some more general continuity properties of FIOs. A combination of the results of Sects. 5.3 and 5.4 gives therefore Theorem 5.1.

In Sect. 5.5, a further analysis of the eikonal equation of Sect. 5.3, relating $a(x,\xi)$ and $\Phi(t,x,\eta)$, will allow to give sufficient condition on $a(x,\xi)$, granting continuity of e^{itA} on $\mathscr{M}^{p,q}$, $p \neq q$, and on Wiener amalgam spaces. In particular, we shall recapture (5.3).

To end this introduction, we would like to outline the study of a more general situation, namely, the case when the symbol $a(x,\xi)$ satisfies the $S^0_{0,0}$-type estimates

$$|\partial_x^\alpha \partial_\xi^\beta a(x,\xi)| \le c_{\alpha,\beta}, \quad \text{for } |\alpha| + |\beta| \ge 2. \tag{5.9}$$

If we assume that $A = a(x,D)$ is self-adjoint, the operator e^{itA} is still defined as an L^2-isometry, for any $t \in \mathbb{R}$. The classical procedure in Sect. 5.3 cannot be applied in this general case, because there is no symbolic calculus for $S^0_{0,0}$-type operators. However, alternative representations of e^{itA} have been given by Tataru [43] and Bony [5, 6]; see also Asada–Fujiwara [1]. In particular, a combination of the Bargman representation of Tataru [43] and a result in the recent paper [14] allows us to write e^{itA}, $t \in]-T,T[$ in the FIO form (5.8) for suitable Φ and σ of $S^0_{0,0}$-type. Our results of Sect. 5.4 apply to this situation as well, and we may conclude that Theorem 5.1 remains valid for such general $A = a(x,D)$. Details of the proof will be given in a subsequent paper.

Notation. We write $t^2 = t \cdot t$, for $t \in \mathbb{R}^d$ and $xy = x \cdot y$ for the scalar product on $\mathbb{R}^d$.

The Schwartz class is denoted by $\mathscr{S}(\mathbb{R}^d)$ and the space of tempered distributions by $\mathscr{S}'(\mathbb{R}^d)$. We use the brackets $\langle f,g \rangle$ to denote the extension to $\mathscr{S}(\mathbb{R}^d) \times \mathscr{S}'(\mathbb{R}^d)$ of the inner product $\langle f,g \rangle = \int f(t)\overline{g(t)}dt$ on $L^2(\mathbb{R}^d)$. The Fourier transform is normalized to be $\hat{f}(\eta) = \mathscr{F}f(\eta) = \int f(t)e^{-2\pi i t\eta}dt$.

Throughout this paper, we shall use the notation $A \lesssim B$ to indicate $A \leq cB$ for a suitable constant $c > 0$, whereas $A \asymp B$ if $A \leq cB$ and $B \leq kA$ for suitable $c, k > 0$.

5.2 Modulation Spaces and Time–Frequency Analysis

At the basis of the time–frequency analysis there are the linear operators of translation and modulation (so-called time–frequency shifts) given by

$$T_x f(t) = f(t-x) \quad \text{and} \quad M_\eta f(t) = e^{2\pi i \eta t} f(t). \tag{5.10}$$

These occur in the following time–frequency representation: Let g be a nonzero window function in the Schwartz class $\mathscr{S}(\mathbb{R}^d)$; then the short-time Fourier transform (STFT) of a signal $f \in L^2(\mathbb{R}^d)$ with respect to the window g is given by

$$V_g f(x,\eta) = \langle f, M_\eta T_x g \rangle = \int_{\mathbb{R}^d} f(t)\,\overline{g(t-x)}\,e^{-2\pi i \eta t}\,dt. \tag{5.11}$$

We have $V_g f \in L^2(\mathbb{R}^{2d})$. This definition can be extended to every pair of dual topological vector spaces, whose duality, denoted by $\langle \cdot, \cdot \rangle$, extends the inner product on $L^2(\mathbb{R}^d)$. For instance, it may be adapted to the framework of tempered distributions.

We briefly explain the meaning of the "time–frequency" representation. If $f(t)$ represents a signal varying in time, its Fourier transform $\hat{f}(\eta)$ shows the distribution of its frequency η, but its magnitude $|\hat{f}(\eta)|$ alone does not give information about "when" these frequencies appear. To overcome this problem, one may choose a nonnegative window function g well localized around the origin. Then, the information of the signal f at the instant x can be obtained by shifting the window g till the instant x under consideration is reached and by computing the Fourier transform of the product $f(x)g(t-x)$ that localizes f around the instant time x.

The STFT $V_g f$ is defined on many pairs of Banach spaces. For instance, it maps $L^2(\mathbb{R}^d) \times L^2(\mathbb{R}^d)$ into $L^2(\mathbb{R}^{2d})$ and $\mathscr{S}(\mathbb{R}^d) \times \mathscr{S}(\mathbb{R}^d)$ into $\mathscr{S}(\mathbb{R}^{2d})$. Furthermore, it can be extended to a map from $\mathscr{S}'(\mathbb{R}^d) \times \mathscr{S}(\mathbb{R}^d)$ into $\mathscr{S}'(\mathbb{R}^{2d})$.

Once the analysis of the signal f is terminated, we can reconstruct the original signal f by a suitable inversion procedure. Namely, the reproducing formula related to the STFT, for every pairs of windows $\varphi_1, \varphi_2 \in \mathscr{S}(\mathbb{R}^d)$ with $\langle \varphi_1, \varphi_2 \rangle \neq 0$, becomes

$$\int_{\mathbb{R}^{2d}} V_{\varphi_1} f(x,\omega) M_\omega T_x \varphi_2 \, dx d\omega = \langle \varphi_2, \varphi_1 \rangle f. \tag{5.12}$$

In this paper we shall mainly concerned with the discretized version of (5.11), namely, the so-called Gabor frames. For $\alpha, \beta > 0$, $g \in L^2(\mathbb{R}^d)$, the set of time–frequency shifts $\mathscr{G}(g,\alpha,\beta) = \{g_{m,n} := M_n T_m g\}$, with $(m,n) \in \alpha\mathbb{Z}^d \times \beta\mathbb{Z}^d$, is a *Gabor frame* if there exist positive constants $A, B > 0$, such that

$$A\|f\|_{L^2} \leq \sum_{m,n} |\langle f, T_m M_n g\rangle|^2 \leq B\|f\|_{L^2}, \quad \forall f \in L^2(\mathbb{R}^d). \tag{5.13}$$

As a counterpart of (5.12), if condition (5.13) is satisfied, we may reconstruct the original signal from the Gabor coefficients $\langle f, g_{m,n}\rangle$ in terms of the so-called dual window γ, using a discrete analog of (5.12).

5.2.1 Modulation Spaces

Modulation spaces were introduced by Feichtinger in [20] (for their basic properties we also refer to [27, Chap. 11–13] and the original literature quoted there), and their norms are a measure of the joint time–frequency distribution of $f \in \mathscr{S}'$.

For the quantitative description of decay properties, we use weight functions on the time–frequency plane. In the sequel v will always be a continuous, strictly positive, even, submultiplicative weight function (in short, a submultiplicative weight); hence, $v(0) = 1$ up to a multiplicative factor, $v(z) = v(-z)$, and $v(z_1 + z_2) \leq v(z_1)v(z_2)$ for all $z, z_1, z_2 \in \mathbb{R}^{2d}$. A positive weight function μ on $\mathbb{R}^{2d}$ belongs to $\mathscr{M}_v$, that is, is *v-moderate* if $\mu(z_1+z_2) \leq Cv(z_1)\mu(z_2)$ for all $z_1, z_2 \in \mathbb{R}^{2d}$.

For our investigation of FIOs we assume $v \in \mathscr{S}'(\mathbb{R}^{2d})$, and we shall mostly use the polynomial weights defined by

$$v_s(z) = v_s(x,\eta) = \langle z\rangle^s = (1+|x|^2+|\eta|^2)^{s/2}, \quad z = (x,\eta) \in \mathbb{R}^{2d}.$$

Given a nonzero window $g \in \mathscr{S}(\mathbb{R}^d)$, $\mu \in \mathscr{M}_v$, and $0 < p, q \leq \infty$, the *modulation space* $M^{p,q}_\mu(\mathbb{R}^d)$ consists of all tempered distributions $f \in \mathscr{S}'(\mathbb{R}^d)$ such that $V_g f \in L^{p,q}_\mu(\mathbb{R}^{2d})$ (weighted mixed-norm spaces). The norm on $M^{p,q}_\mu$ is

$$\|f\|_{M^{p,q}_\mu} = \|V_g f\|_{L^{p,q}_\mu} = \left(\int_{\mathbb{R}^d}\left(\int_{\mathbb{R}^d}|V_g f(x,\eta)|^p \mu(x,\eta)^p \, dx\right)^{q/p} d\eta\right)^{1/p} \tag{5.14}$$

(with obvious changes when $p = \infty$ or $q = \infty$). If $p = q$, we write M^p_μ instead of $M^{p,p}_\mu$, and if $\mu(z) \equiv 1$ on $\mathbb{R}^{2d}$, then we write $M^{p,q}$ and M^p for $M^{p,q}_\mu$ and $M^{p,p}_\mu$, respectively.

Then $M^{p,q}_\mu(\mathbb{R}^d)$ is a Banach space for $1 \leq p, q \leq \infty$ whose definition is always independent of the choice of the window g. Moreover, if $\mu \in \mathscr{M}_v$, $1 \leq p, q \leq \infty$, and $g \in M^1_v \setminus \{0\}$, then $\|V_g f\|_{L^{p,q}_\mu}$ is an equivalent norm for $M^{p,q}_\mu(\mathbb{R}^d)$ (see [27, Theorem 11.3.7]):

$$\|f\|_{M_m^{p,q}} \asymp \|V_g f\|_{L_\mu^{p,q}}.$$

The class of modulation spaces contains the following well-known function spaces:
Weighted L^2-spaces: $M^2_{\langle x\rangle^s}(\mathbb{R}^d) = L^2_s(\mathbb{R}^d) = \{f : f(x)\langle x\rangle^s \in L^2(\mathbb{R}^d)\}$, $s \in \mathbb{R}$
Sobolev spaces: $M^2_{\langle \eta\rangle^s}(\mathbb{R}^d) = H^s(\mathbb{R}^d) = \{f : \hat{f}(\eta)\langle \eta\rangle^s \in L^2(\mathbb{R}^d)\}$, $s \in \mathbb{R}$
Shubin–Sobolev spaces [7, 39]: $M^2_{v_s(x,\eta)}(\mathbb{R}^d) = L^2_s(\mathbb{R}^d) \cap H^s(\mathbb{R}^d) = Q_s(\mathbb{R}^d)$, $s \geq 0$
Feichtinger's algebra: $M^1(\mathbb{R}^d) = S_0(\mathbb{R}^d)$

The Schwartz class and the space of tempered distributions are characterized as $\mathscr{S}(\mathbb{R}^d) = \bigcap_{s\geq 0} M^1_{v_s}(\mathbb{R}^d)$ and $\mathscr{S}'(\mathbb{R}^d) = \bigcup_{s\geq 0} M^\infty_{v_{-s}}(\mathbb{R}^d)$, respectively.

In the sequel we shall mainly work with the spaces denoted by $\mathscr{M}_\mu^{p,q}(\mathbb{R}^d)$ that are the closure of the Schwartz space $\mathscr{S}(\mathbb{R}^d)$ with respect to the norm of $M_\mu^{p,q}(\mathbb{R}^d)$. Hence $\mathscr{M}_\mu^{p,q}(\mathbb{R}^d) = M_\mu^{p,q}(\mathbb{R}^d)$ if $p,q < \infty$. Moreover we set $\mathscr{M}_\mu^{p}(\mathbb{R}^d) = \mathscr{M}_\mu^{p,p}(\mathbb{R}^d)$.

Among the properties of modulation spaces, we list the following results:

Lemma 5.1. *We have*

(i) $M^{p_1,q_1} \hookrightarrow M^{p_2,q_2}$, *if* $p_1 \leq p_2$ *and* $q_1 \leq q_2$.
(ii) If $1 \leq p,q < \infty$, *then* $(M^{p,q})' = M^{p',q'}$, *where* p', q' *are dual exponents to* p, q, *respectively.*
(iii) For $1 \leq p,q,p_i,q_i \leq \infty$, $i = 1,2$, *with* $q_1 < \infty$ *or* $q_2 < \infty$, *and*

$$\frac{1}{p} = \frac{1-\theta}{p_1} + \frac{\theta}{p_2}, \quad \frac{1}{q} = \frac{1-\theta}{q_1} + \frac{\theta}{q_2},$$

we have

$$[M^{p_1,q_1}, M^{p_2,q_2}]_\theta = M^{p,q},$$

where $[\cdot,\cdot]_\theta$ *denotes the complex interpolation space.The same holds if every modulation space is replaced by the closure of the Schwartz space into itself.*
(iv) If $1 \leq p_i,q_i \leq \infty$, $i = 1,2,3$, *with*

$$\frac{1}{p_1} + \frac{1}{p_2} = 1 + \frac{1}{p_3}, \quad \frac{1}{q_1} + \frac{1}{q_2} = \frac{1}{q_3}$$

then

$$M^{p_1,q_1} * M^{p_2,q_2} \hookrightarrow M^{p_3,q_3}.$$

Wiener Amalgam Spaces. We briefly recall the definition and the main properties of Wiener amalgam spaces. We refer to [18, 19, 31] for details.

Let $g \in \mathscr{C}_0^\infty$ be a test function that satisfies $\|g\|_{L^2} = 1$. We will refer to g as a window function. Let B one of the following Banach spaces: $L^p, \mathscr{F}L^p$, $L^{p,q}$, L^pL^q, $1 \leq p,q \leq \infty$. Let C be one of the following Banach spaces: L^p, $L^{p,q}$, L^pL^q, $1 \leq p,q \leq \infty$. For any given tempered distribution f which is locally in B (i.e., $gf \in B$, $\forall g \in \mathscr{C}_0^\infty$), we set $f_B(x) = \|fT_xg\|_B$.

The *Wiener amalgam space* $W(B,C)$ with local component B and global component C is defined as the space of all tempered distributions f locally in B such that

$f_B \in C$. Endowed with the norm $\|f\|_{W(B,C)} = \|f_B\|_C$, $W(B,C)$ is a Banach space. Moreover, different choices of $g \in \mathscr{C}_0^\infty$ generate the same space and yield equivalent norms.

If $B = \mathscr{F}L^1$ (the Fourier algebra), the space of admissible windows for the Wiener amalgam spaces $W(\mathscr{F}L^1, C)$ can be enlarged to the so-called Feichtinger algebra $W(\mathscr{F}L^1, L^1)$. Recall that the Schwartz class $\mathscr{S}$ is dense in $W(\mathscr{F}L^1, L^1)$.

The main properties of Wiener amalgam spaces are the following:

Lemma 5.2. *Let B_i, C_i, $i = 1,2,3$ be Banach spaces such that $W(B_i, C_i)$ are well defined. Then,*

(1) Convolution: *If $B_1 * B_2 \hookrightarrow B_3$ and $C_1 * C_2 \hookrightarrow C_3$, we have*

$$W(B_1, C_1) * W(B_2, C_2) \hookrightarrow W(B_3, C_3). \tag{5.15}$$

(2) Inclusions: *If $B_1 \hookrightarrow B_2$ and $C_1 \hookrightarrow C_2$,*

$$W(B_1, C_1) \hookrightarrow W(B_2, C_2).$$

Moreover, the inclusion of B_1 into B_2 need only hold "locally" and the inclusion of C_1 into C_2 "globally." In particular, for $1 \le p_i, q_i \le \infty$, $i = 1,2$, we have

$$p_1 \ge p_2 \, and \, q_1 \le q_2 \Longrightarrow W(L^{p_1}, L^{q_1}) \hookrightarrow W(L^{p_2}, L^{q_2}). \tag{5.16}$$

(3) Complex interpolation: *For $0 < \theta < 1$, we have*

$$[W(B_1, C_1), W(B_2, C_2)]_{[\theta]} = W\left([B_1, B_2]_{[\theta]}, [C_1, C_2]_{[\theta]}\right),$$

if C_1 or C_2 has absolutely continuous norm. The same holds if every Wiener amalgam space is replaced by the closure of the Schwartz space into itself.

(4) Duality: *If B', C' are the topological dual spaces of the Banach spaces B, C, respectively, and the space of test functions $\mathscr{C}_0^\infty$ is dense in both B and C, then*

$$W(B,C)' = W(B', C'). \tag{5.17}$$

(5) Pointwise products: *If $B_1 \cdot B_2 \hookrightarrow B_3$ and $C_1 \cdot C_2 \hookrightarrow C_3$, we have*

$$W(B_1, C_1) \cdot W(B_2, C_2) \hookrightarrow W(B_3, C_3). \tag{5.18}$$

Modulation spaces and Wiener amalgam spaces are closely related: for $p = q$, we have

$$\|f\|_{W(\mathscr{F}L^p, L^p)} = \left(\int_{\mathbb{R}^d} \int_{\mathbb{R}^d} |V_g f(x, \eta)|^p \, dx d\eta\right)^{1/p} \asymp \|f\|_{M^p}. \tag{5.19}$$

More generally, from comparing the definitions of $M^{p,q}$ and $W(\mathscr{F}L^p, L^q)$, it is obvious that $M^{p,q} = \mathscr{F}W(\mathscr{F}L^p, L^q)$.

5.2.2 Gabor Frames [23]

Fix a function $g \in L^2(\mathbb{R}^d)$ and a lattice $\Lambda = \alpha\mathbb{Z}^d \times \beta\mathbb{Z}^d$ for $\alpha, \beta > 0$. For $(m,n) \in \Lambda$, define $g_{m,n} := M_n T_m g$. The set of time–frequency shifts $\mathscr{G}(g,\alpha,\beta) = \{g_{m,n}, (m,n) \in \Lambda\}$ is called Gabor system. Associated to $\mathscr{G}(g,\alpha,\beta)$, we define the coefficient operator C_g, which maps functions to sequences as follows:

$$(C_g f)_{m,n} = (C_g^{\alpha,\beta} f)_{m,n} := \langle f, g_{m,n} \rangle, \quad (m,n) \in \Lambda, \tag{5.20}$$

the synthesis operator

$$D_g c = D_g^{\alpha,\beta} c = \sum_{(m,n)\in\Lambda} c_{m,n} T_m M_n g, \quad c = \{c_{m,n}\}_{(m,n)\in\Lambda}$$

and the Gabor frame operator

$$S_g f = S_g^{\alpha,\beta} f := D_g S_g f = \sum_{(m,n)\in\Lambda} \langle f, g_{m,n} \rangle g_{m,n}. \tag{5.21}$$

The set $\mathscr{G}(g,\alpha,\beta)$ is called a Gabor frame for the Hilbert space $L^2(\mathbb{R}^d)$ if S_g is a bounded and invertible operator on $L^2(\mathbb{R}^d)$. Equivalently, C_g is bounded from $L^2(\mathbb{R}^d)$ to $l^2(\alpha\mathbb{Z}^d \times \beta\mathbb{Z}^d)$ with closed range, i.e., $\|f\|_{L^2} \asymp \|C_g f\|_{l^2}$, cf. (5.13). If $\mathscr{G}(g,\alpha,\beta)$ is a Gabor frame for $L^2(\mathbb{R}^d)$, then the so-called *dual window* $\gamma = S_g^{-1} g$ is well defined and the set $\mathscr{G}(\gamma,\alpha,\beta)$ is a frame (the so-called canonical dual frame of $\mathscr{G}(g,\alpha,\beta)$). Every $f \in L^2(\mathbb{R}^d)$ possesses the frame expansion

$$f = \sum_{(m,n)\in\Lambda} \langle f, g_{m,n} \rangle \gamma_{m,n} = \sum_{(m,n)\in\Lambda} \langle f, \gamma_{m,n} \rangle g_{m,n} \tag{5.22}$$

with unconditional convergence in $L^2(\mathbb{R}^d)$ and norm equivalence:

$$\|f\|_{L^2} \asymp \|C_g f\|_{l^2} \asymp \|C_\gamma f\|_{l^2}.$$

This result is contained in [27, Proposition 5.2.1]. In particular, if $\gamma = g$ and $\|g\|_{L^2} = 1$, the frame is called *normalized tight* Gabor frame, and the expansion (5.22) reduces to

$$f = \sum_{(m,n)\in\Lambda} \langle f, g_{m,n} \rangle g_{m,n}. \tag{5.23}$$

If we ask for more regularity on the window g, then the previous result can be extended to modulation spaces, as shown below [23, 30].

Theorem 5.2. *Let $\mu \in \mathscr{M}_v$, $\mathscr{G}(g,\alpha,\beta)$ be a frame for $L^2(\mathbb{R}^d)$, with lattice $\Lambda = \alpha\mathbb{Z}^d \times \beta\mathbb{Z}^d$ and $g \in \mathscr{S}$. Define $\tilde{\mu} = \mu_{|\Lambda}$.*

(i) For every $1 \le p,q \le \infty$, $C_g : M_\mu^{p,q} \to l_{\tilde{\mu}}^{p,q}$ and $D_g : l_{\tilde{\mu}}^{p,q} \to M_\mu^{p,q}$ continuously, and if $f \in M_\mu^{p,q}$, then the Gabor expansions (5.22) converge unconditionally in $M_\mu^{p,q}$ for $1 \le p,q < \infty$ and weak-M_μ^∞ unconditionally if $p = \infty$ or $q = \infty$.*

(ii) The following norms are equivalent on $M^{p,q}_{\mu}$*:*

$$\|f\|_{M^{p,q}_{\mu}} \asymp \|C_g f\|_{l^{p,q}_{\tilde{\mu}}}. \tag{5.24}$$

5.3 Classical Fourier Integral Operators in $\mathbb{R}^d$

As first step of the proof of Theorem 5.1, we write e^{itA} in the form of FIO

$$(F_t u_0)(t,x) = \int_{\mathbb{R}^d} e^{2\pi i \Phi(t,x,\eta)} \sigma(t,x,\eta) \widehat{u_0}(\eta) d\eta. \tag{5.25}$$

Following Helffer [32, Chap. III], we assume that F_t is a classical FIO, which in our context means

$$|\partial_t^\gamma \partial_x^\alpha \partial_\eta^\beta \sigma(t,x,\eta)| \le c_{\alpha,\beta,\gamma}, \quad \forall (\alpha,\beta,\gamma) \in \mathbb{N}^d \times \mathbb{N}^d \times \mathbb{N} \tag{5.26}$$

and $\sigma \sim \sum_{j=0}^{\infty} \sigma_{-2j}$, with $\sigma_{-2j}(t,\lambda x,\lambda \eta) = \lambda^{-2j} \sigma_{-2j}(t,x,\eta)$.

As for $\Phi(t,x,\eta)$, we take it as solution of the eikonal equation [32, page 142]

$$\begin{cases} 2\pi \partial_t \Phi - a_2(x, \nabla_x \Phi) = 0 \\ \Phi(0,x,\eta) = x\eta, \end{cases} \tag{5.27}$$

the factor 2π depending on our normalization of Fourier transform. The motivation of this choice of Φ in (5.25) will be clear in a moment. Because of the homogeneity of $a_2(x,\xi)$ and $\Phi(0,x,\eta)$, the function $\Phi(t,x,\eta)$ is homogeneous of order 2 in (x,η), well-defined and smooth for $(x,\eta) \neq (0,0)$, $t \in]-T,T[$, with $T > 0$ sufficiently small. Moreover, $\Phi(t,x,\eta)$ is real-valued since the principal symbol $a_2(x,\xi)$ is real-valued. In view of the initial condition in (5.27) we also have

$$|\det \partial^2_{x,\eta} \Phi(t,x,\eta)| \ge c > 0, \quad (t,x,\eta) \in]-T,T[\times (\mathbb{R}^{2d} \setminus \{0\}), \tag{5.28}$$

after possibly shrinking $T > 0$.

Let us recall the following standard formula for composition of FIOs and pseudodifferential operators $A = a(x,D)$ adapted to our Situation:

Proposition 5.1. *The composition* AF_t *is again an FIO of the type* (5.25)*, with the same phase function* $\Phi(t,x,\eta)$*:*

$$AF_t u_0(t,x) = \int_{\mathbb{R}^d} e^{2\pi i \Phi(t,x,\eta)} b(t,x,\eta) \widehat{u_0}(\eta) d\eta, \tag{5.29}$$

where

$$b(t,x,\eta) \sim \sum_{\alpha \in \mathbb{N}^d} b_\alpha(t,x,\eta) \tag{5.30}$$

with

$$b_\alpha(t,x,\eta) = \frac{1}{\alpha!}\partial_\xi^\alpha a(x,\nabla_x\Phi(t,x,\eta))D_y^\alpha\left(e^{i\psi}\sigma(t,y,\eta)\right)_{|y=x} \tag{5.31}$$

and

$$\psi = \psi(t,x,y,\eta) = \Phi(t,y,\eta) - \Phi(t,x,\eta) - \langle y-x, \nabla_x\Phi(t,x,\eta)\rangle. \tag{5.32}$$

In particular if $a \sim \sum_{j=0}^\infty a_{2-2j}$, $\sigma \sim \sum_{j=0}^\infty \sigma_{-2j}$, we may rearrange the asymptotic expansion of $b(t,x,\eta)$ according to the order of homogeneity of the terms: $b \sim \sum_{j=0}^\infty b_{2-2j}$. We have in particular

$$b_2(t,x,\eta) = a_2(x,\nabla_x\Phi(t,x,\eta))\sigma_0(t,x,\eta), \tag{5.33}$$

$$\begin{aligned} b_0(t,x,\eta) &= a_2(x,\nabla_x\Phi(t,x,\eta))\sigma_{-2}(t,x,\eta) \\ &\quad + \sum_{j=1}^d \partial_{\xi_j} a_2(x,\partial_x\Phi(t,x,\eta))\partial_{x_j}\sigma_0(t,x,\eta). \end{aligned} \tag{5.34}$$

With $\Phi(t,x,\eta)$ as in (5.27), we want to find $\sigma(t,x,\eta)$ in (5.25) such that $e^{itA} = F_t + R_t$, with R_t regularizing. To this end, we try to verify

$$(i\partial_t + A)F_t = 0 \quad \text{for} \quad t \in]-T,T[, \quad F_0 = I. \tag{5.35}$$

The initial condition $F_0 = I$ is satisfied if in (5.25), we impose $\sigma(0,x,\eta) = 1$, taking also into account the initial condition in (5.27). By applying Proposition 5.1 and letting ∂_t act under the integral sign, we have $(i\partial_t + A)F_t = G_t$, where G_t is an FIO with same phase Φ and symbol

$$i\partial_t\sigma(t,x,\eta) - 2\pi\partial_t\Phi(t,x,\eta)\sigma(t,x,\eta) + b(t,x,\eta). \tag{5.36}$$

Arguing first formally, we consider the asymptotic expansion of all the terms in (5.36), we impose the sum vanishes, and we solve with respect to σ_{-2j}. In view of (5.33) and (5.34), the first two equations, collecting the terms of order 2 and 0, are

$$(-2\pi\partial_t\Phi + a_2(x,\nabla_x\Phi))\sigma_0(t,x,\eta) = 0 \tag{5.37}$$

$$(-2\pi\partial_t\Phi + a_2(x,\nabla_x\Phi))\sigma_{-2}(t,x,\eta) + L\sigma_0(t,x,\eta) = 0 \tag{5.38}$$

with

$$L = \partial_t + \sum_{j=1}^d \partial_{\xi_j} a_2(x,\nabla_x\Phi(t,x,\eta))\partial_{x_j}. \tag{5.39}$$

In view of our choice of Φ as solution of the eikonal equation (5.27), the identity (5.37) is satisfied and (5.38) reduces to

$$L\sigma_0(t,x,\eta) = 0, \tag{5.40}$$

to which we add the initial condition $\sigma_0(0,x,\eta) = 1$. Since the vector field L in (5.39) is transversal to the manifold $t = 0$, we may solve and obtain a solution $\sigma_0(t,x,\eta)$, homogeneous of order 0 with respect to (x,η). By using (5.30)–(5.32), similar arguments allow to determine $\sigma_{-2j}(t,x,\eta)$, $j > 1$, to which we impose the initial condition $\sigma_{-2j}(0,x,\eta) = 0$. See Helffer [32, Chap. 3.1] for details.

In this way we solved formally (5.35). To pass to a true solution, we construct $\sigma(t,x,\eta)$ satisfying (5.26) with $\sigma \sim \sum_{j=0}^{\infty} \sigma_{-2j}$, according to standard proceedings. We may as well obtain a smooth phase function, cutting off Φ in a neighborhood of the origin in $\mathbb{R}^{2d}$. This produces errors in the right-hand side of (5.35), which however are smooth families of regularizing operators. By a simple argument (see Helffer [32, Proposition 3.1.1]), we may conclude $e^{itA} = F_t + R_t$, with R_t regularizing, for $t \in]-T,T[$.

5.4 Boundedness of FIOs on Modulation Spaces and Wiener Amalgam Spaces

Let us first recall the basic results in this framework, contained in [11, 15]. There we focused on FIOs of the type

$$Tf(x) = \int_{\mathbb{R}^d} e^{2\pi i \Phi(x,\eta)} \sigma(x,\eta) \hat{f}(\eta) d\eta \tag{5.41}$$

for $f \in \mathscr{S}(\mathbb{R}^d)$.

The phase function $\Phi(x,\eta)$ fulfills the following properties:

(i) $\Phi \in \mathscr{C}^\infty(\mathbb{R}^{2d})$.
(ii) There exist constants $C_\alpha > 0$ such that

$$|\partial^\alpha \Phi(x,\eta)| \leq C_\alpha, \quad \forall \alpha \in \mathbb{N}^{2d},\ |\alpha| \geq 2. \tag{5.42}$$

(iii) There exists $\delta > 0$ such that

$$\left| \det \left(\frac{\partial^2 \Phi}{\partial x \partial \eta} \Big|_{(x,\eta)} \right) \right| \geq \delta \quad \forall (x,\eta) \in \mathbb{R}^{2d}. \tag{5.43}$$

If we set

$$\begin{cases} y = \nabla_\eta \Phi(x,\eta) \\ \xi = \nabla_x \Phi(x,\eta), \end{cases} \tag{5.44}$$

and solve with respect to (x,ξ), we obtain a mapping χ, defined by $(x,\xi) = \chi(y,\eta)$, which is a smooth bi-Lipschitz canonical transformation. This means that

- χ is a smooth diffeomorphism on $\mathbb{R}^{2d}$.
- both χ and χ^{-1} are uniformly Lipschitz continuous.
- χ preserves the symplectic form, i.e.,

$$dx \wedge d\xi = dy \wedge d\eta.$$

Indeed, under the above assumptions, the global inversion function theorem (see, e.g., [36]) allows us to solve the first equation in (5.44) with respect to x for fixed η, and substituting in the second equation yields the smooth map χ. The bounds on the derivatives of χ, which give the Lipschitz continuity, follow from the expression for the derivatives of an inverse function combined with the bounds in (i) and (ii). The symplectic nature of the map χ is classical; see, e.g. [8]. Similarly, solving the second equation in (5.44) with respect to η, one obtains the map χ^{-1} with the desired properties.

The main results concerning the boundedness of such operators on modulation spaces are as follows (see [11, 15]): As symbol class we choose the so-called Sjöstrand class $M^{\infty,1}(\mathbb{R}^{2d})$, which contains Hörmander's class $S^0_{0,0}$ and so allows us to consider rougher symbols. For the sake of clarity, we present only the simplified unweighted version of the issues.

Theorem 5.3. *Consider a phase Φ satisfying* (i), (ii), *and* (iii) *and a symbol $\sigma \in M^{\infty,1}(\mathbb{R}^{2d})$:*

(i) *Then the corresponding Fourier integral operator T extends to a bounded operator on $\mathcal{M}^p(\mathbb{R}^d)$, $1 \le p \le \infty$.*
(ii) *If the phase $\Phi(x,\eta)$ satisfies in addition*

$$\sup_{x,x',\eta \in \mathbb{R}^d} \left|\nabla_x \Phi(x,\eta) - \nabla_x \Phi(x',\eta)\right| < \infty, \tag{5.45}$$

then the operator T extends to a bounded operator on $\mathcal{M}^{p,q}(\mathbb{R}^d)$ for every $1 \le p,q \le \infty$.
(iii) *If $1 \le q < p \le \infty$ and the phase $\Phi(x,\eta)$ satisfies, for some $\delta > 0$,*

$$\left|\det\left(\left.\frac{\partial^2 \Phi}{\partial x \partial x}\right|_{(x,\eta)}\right)\right| \ge \delta \quad \forall (x,\eta) \in \mathbb{R}^{2d}, \tag{5.46}$$

then the operator T extends to a bounded operator from $\mathcal{M}^{p,q}(\mathbb{R}^d)$ into $\mathcal{M}^{q,p}(\mathbb{R}^d)$.

Observe that, in general, without further assumptions on the phase or weights on the symbol class (see below), if we simply choose $\sigma \in M^{\infty,1}(\mathbb{R}^{2d})$, then the corresponding operator T may not be bounded on $\mathcal{M}^{p,q}(\mathbb{R}^d)$, with $p \ne q$. A simple example is provided by the pointwise multiplication operator

$$Af(x) = e^{i\pi|x|^2} f(x), \quad x \in \mathbb{R}^d$$

which is an FIO of the type above, having phase $\Phi(x,\eta) = x\eta + \frac{|x|^2}{2}$ and symbol $\sigma \equiv 1 \in M^{\infty,1}(\mathbb{R}^{2d})$ which is bounded on $\mathcal{M}^{p,q}(\mathbb{R}^d)$ if and only if $p = q$; see [15, Proposition 7.1].

So, we add a further decay condition on the symbol, represented by suitable weights on the symbol class. Namely, for $s_1, s_2 \in \mathbb{R}$, we define the weight function $v_{s_1,s_2}(x,\eta) := \langle x\rangle^{s_1}\langle\eta\rangle^{s_2}$, $(x,\eta) \in \mathbb{R}^{2d}$. Then,

Theorem 5.4. *Consider a phase Φ satisfying* (i), (ii), *and* (iii) *and a symbol $\sigma \in M^{\infty,1}_{v_{s_1,s_2}\otimes 1}(\mathbb{R}^{2d})$, $s_1, s_2 \in \mathbb{R}$.*

(*i*) *Let $1 \le q < p \le \infty$. If $s_1 > d\left(\frac{1}{q}-\frac{1}{p}\right)$, $s_2 \ge 0$, T extends to a bounded operator on $\mathscr{M}^{p,q}(\mathbb{R}^d)$.*

(*ii*) *Let $1 \le p < q \le \infty$. If $s_1 \ge 0$, $s_2 > d\left(\frac{1}{p}-\frac{1}{q}\right)$, T extends to a bounded operator on $\mathscr{M}^{p,q}(\mathbb{R}^d)$.*

In all cases,

$$\|Tf\|_{\mathscr{M}^{p,q}} \lesssim \|\sigma\|_{M^{\infty,1}_{v_{s_1,s_2}\otimes 1}} \|f\|_{\mathscr{M}^{p,q}}. \tag{5.47}$$

Theorem 5.3 has the following counterpart in the framework of Wiener amalgam spaces. The first item is trivial since $\mathscr{W}(\mathscr{F}L^p, L^p) = \mathscr{M}^p$.

Theorem 5.5. *Consider a phase Φ satisfying* (i), (ii), *and* (iii) *and a symbol $\sigma \in M^{\infty,1}(\mathbb{R}^{2d})$.*

(*i*) *Then the corresponding Fourier integral operator T extends to a bounded operator on $\mathscr{W}(\mathscr{F}L^p, L^p)(\mathbb{R}^d)$, $1 \le p \le \infty$.*

(*ii*) *If the phase $\Phi(x,\eta)$ satisfies in addition*

$$\sup_{x,\eta,\eta' \in \mathbb{R}^d} \left|\nabla_\eta \Phi(x,\eta) - \nabla_\eta \Phi(x,\eta')\right| < \infty, \tag{5.48}$$

then the operator T extends to a bounded operator on $\mathscr{W}(\mathscr{F}L^p, L^q)$, for every $1 \le p,q \le \infty$.

(*iii*) *If $1 \le q < p \le \infty$ and the phase $\Phi(x,\eta)$ satisfies, for some $\delta > 0$,*

$$\left|\det\left(\frac{\partial^2 \Phi}{\partial\eta\partial\eta}\Big|_{(x,\eta)}\right)\right| \ge \delta \quad \forall (x,\eta) \in \mathbb{R}^{2d}, \tag{5.49}$$

then the corresponding FIO T extends to a bounded operator $\mathscr{W}(\mathscr{F}L^p, L^q) \to \mathscr{W}(\mathscr{F}L^q, L^p)$.

Similarly, the issues of Theorem 5.4 can be rephrased for amalgam spaces as follows:

Theorem 5.6. *Consider a phase Φ and a symbol σ as in Theorem 5.4.*

(*i*) *Let $1 \le q < p \le \infty$. If $s_1 > d\left(\frac{1}{q}-\frac{1}{p}\right)$, $s_2 \ge 0$, T extends to a bounded operator on $\mathscr{W}(\mathscr{F}L^p, L^q)$.*

(*ii*) *Let $1 \le p < q \le \infty$. If $s_1 \ge 0$, $s_2 > d\left(\frac{1}{p}-\frac{1}{q}\right)$, T extends to a bounded operator on $\mathscr{W}(\mathscr{F}L^p, L^q)$.*

In all cases,

$$\|Tf\|_{\mathscr{W}(\mathscr{F}L^p,L^q)} \lesssim \|\sigma\|_{M^{\infty,1}_{v_{s_1,s_2}\otimes 1}} \|f\|_{\mathscr{W}(\mathscr{F}L^p,L^q)}. \tag{5.50}$$

For the benefit of the reader, we shall give a sketch of the proof of Theorem 5.3, item (i), first appeared in [15], and of Theorem 5.5, item (ii), which is new. For simplicity, we limit ourselves to $\sigma \in S^0_{0,0}$.

First, let us recall the key issue for the proof: an almost diagonalization result for FIOs as above, with respect to Gabor frames. For clarity, we shall only consider a normalized tight frame $\mathscr{G}(g,\alpha,\beta)$, with $g \in \mathscr{S}(\mathbb{R}^d)$.

Theorem 5.7. *([15, Thm. 3.3]) Consider a phase function Φ satisfying (i) and (ii), a symbol σ in $S^0_{0,0}(\mathbb{R}^{2d})$, and a function $g \in \mathscr{S}(\mathbb{R}^d)$. Then for every $N \in \mathbb{N}$, there exists a constant $C_N > 0$ such that*

$$|\langle T g_{m,n}, g_{m',n'}\rangle| \le C_N \langle \chi(m,n) - (m',n')\rangle^{-2N}, \tag{5.51}$$

where χ is the canonical transformation generated by Φ.

This result shows that the matrix representation of an FIO with respect a Gabor frame is well organized. More precisely, if $\sigma \in S^0_{0,0}$, then the Gabor matrix of T is highly concentrated along the graph of χ.

Further, the continuity on modulation spaces uses the Schur's test [15, Lemma 4.1]:

Lemma 5.3. *Consider a lattice Λ and an operator K defined on sequences as*

$$(Kc)_\lambda = \sum_{\nu\in\Lambda} K_{\lambda,\nu} c_\nu,$$

where

$$\sup_{\nu\in\Lambda} \sum_{\lambda\in\Lambda} |K_{\lambda,\nu}| < \infty, \quad \sup_{\lambda\in\Lambda} \sum_{\nu\in\Lambda} |K_{\lambda,\nu}| < \infty.$$

Then K is continuous on $l^p(\Lambda)$ for every $1 \le p \le \infty$ and moreover maps the space $c_0(\Lambda)$ of sequences vanishing at infinity into itself.

We can now sketch the proof of Theorem 5.3, item (i).

Proof of Theorem 5.3 (i). We first prove the theorem in the case $p < \infty$.

For $T = C_g \circ T_{m',n',m,n} \circ D_g$, the following diagram is commutative:

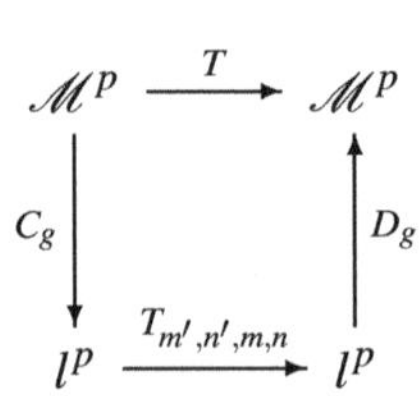

where T is viewed as an operator with dense domain $\mathscr{S}(\mathbb{R}^d)$. Whence, it is enough to prove the continuity of the infinite matrix $T_{m',n',m,n}$ from l^p into l^p.

This follows from Schur's test if we prove that, upon setting

$$K_{m',n',m,n} = T_{m',n',m,n},$$

we have

$$K_{m',n',m,n} \in l^\infty l^1, \tag{5.52}$$

and

$$K_{m',n',m,n} \in l^\infty l^1. \tag{5.53}$$

In view of (5.51) we have

$$|K_{m',n',m,n}| \lesssim \langle \chi(m,n) - (m',n') \rangle^{-2N}, \tag{5.54}$$

for every integer $N > 0$. Now, the last quotient in (5.54) is bounded, so we deduce (5.52).

Finally, since χ is a bi-Lipschitz function we have

$$|\chi(m,n) - (m',n')| \asymp |(m,n) - \chi^{-1}(m',n')| \tag{5.55}$$

so that (5.53) follows as well.

The case $p = \infty$ follows analogously by using the last part of the statement of Lemma 5.3. □

Proof of Theorem 5.5 (*ii*). The result can be obtained from Theorem 5.3, item (*ii*) as follows: Assume, for simplicity, that $\sigma \in S^0_{0,0}$. Conjugating with the Fourier transform yields the operator

$$\tilde{T} f(x) = \mathscr{F} \circ T \circ \mathscr{F}^{-1} f(x)$$

Since $\mathscr{M}^{p,q} = \mathscr{F}^{-1}\mathscr{W}(\mathscr{F}L^p, L^q)$, it suffices to prove that $\tilde{T}$ extends to a bounded operator on $\mathscr{M}^{p,q}$. By duality and an explicit computation, this is equivalent to verifying that the operator

$$\tilde{T}^* f(x) = \int_{\mathbb{R}^d} e^{-2\pi i \Phi(\eta,x)} \overline{\sigma(\eta,x)} \hat{f}(\eta)\, d\eta$$

extends to a bounded operator on $\mathscr{M}^{p',q'}$. Since $\sigma \in S^0_{0,0}$ implies that $\sigma^*(x,\eta) = \overline{\sigma(\eta,x)} \in S^0_{0,0}$, the operator $\tilde{T}^*$ has a phase satisfying (5.15) by the assumption (5.48), and the desired conclusion follows. □

To end this section, we shall prove the main result of this paper, which uses the issues explained above.

Proof of Theorem 5.1. Using Sect. 5.3, we can write $e^{itA} = F_t + R_t$, where the FIO F_t is defined in (5.25) and R_t is regularizing.

Observe that the phase $\Phi(t,x,\eta)$ trivially satisfies the assumptions (i), (ii), and (iii) for every $t \in]-T,T[$. Moreover the symbol σ satisfies (5.26), hence in particular is in $S^0_{0,0}(\mathbb{R}^{2d}) \subset M^{\infty,1}(\mathbb{R}^{2d})$ with respect to the variables (x,η), for every $t \in]-T,T[$. So Theorem 5.3, item (i), assures the continuity of F_t on $\mathcal{M}^p$ for every $1 \leq p \leq \infty$. Since the remainder R_t is regularizing, it maps every $\mathcal{M}^p(\mathbb{R}^d) \subset \mathcal{S}'(\mathbb{R}^d)$ into $\mathcal{S}(\mathbb{R}^d) \subset \mathcal{M}^p(\mathbb{R}^d)$; this gives the continuity of the operator e^{itA} on $\mathcal{M}^p(\mathbb{R}^d)$ for every $t \in]-T,T[$. Next, we want to enlarge the time interval $]-T,T[$. We start with the right-hand side and take a time $t_0 < T$, close to T. Then we set $\tilde{u}_0(x) := e^{it_0A}u_0(x)$ and take $\tilde{u}_0(x)$ as initial datum in (5.4) so that we can extend the solution on the right-hand side to a time $T' > T$. By iterating this argument both on the left- and on the right-hand side of the interval $]-T,T[$, one obtains a solution for every $t \in \mathbb{R}$, being T independent of the initial datum. □

5.5 Other Results

In order to apply the other issues in Sect. 5.4 to the propagator e^{itA}, we have to relate more precisely the properties of the principal symbol $a_2(x,\xi)$ and the phase function $\Phi(t,x,\eta)$.

Lemma 5.4. *Let $\Phi(t,x,\eta)$ be the solution of the eikonal equation* (5.27). *We have*

$$\Phi(t,x,\eta) = x\eta + (2\pi)^{-1}ta_2(x,\eta) + t^2\tilde{\Phi}(t,x,\eta), \tag{5.56}$$

where $\tilde{\Phi}(t,x,\eta) \in \mathcal{C}^\infty(]-T,T[\times(\mathbb{R}^{2d}\setminus\{0\}))$ is homogeneous of order 2 *in (x,η).*

Proof. From Taylor's formula for $\Phi(t,x,\eta)$, with respect to the variable t at $t=0$, we have

$$\Phi(t,x,\eta) = \Phi(0,x,\eta) + t\partial_t\Phi(0,x,\eta) + t^2\tilde{\Phi}(t,x,\eta),$$

for some function $\tilde{\Phi} \in \mathcal{C}^\infty(]-T,T[\times(\mathbb{R}^{2d}\setminus\{0\}))$. On the other hand, it follows from (5.27) that $\Phi(0,x,\eta) = x\eta$, $\nabla_x\Phi(0,x,\eta) = \eta$,

$$2\pi\partial_t\Phi(0,x,\eta) = a_2(x,\nabla_x\Phi(0,x,\eta)) = a_2(x,\eta).$$

Moreover, since $\Phi(t,x,\eta)$ and $a_2(t,x,\eta)$ are homogeneous of order 2 with respect to (x,η), so is $\tilde{\Phi}(t,x,\eta)$. The lemma is proved. □

We observe that the phase condition (5.45) is satisfied when

$$\Phi(t,x,\eta) = x\eta + (2\pi)^{-1}ta_2(\eta),$$

that is, the principal symbol a_2 does not depend on x. One then has the following result:

Theorem 5.8. *Assume that the symbol* $a(x,\xi)$ *enjoys* (5.6) *and its principal symbol*

$$a_2(x,\xi) = a_2(\xi), \quad \text{for every } (x,\xi) \in \mathbb{R}^{2d}. \tag{5.57}$$

Then the propagator e^{itA} *is bounded on* $\mathcal{M}^{p,q}$, *for every* $1 \le p,q \le \infty$ *and for every* $t \in \mathbb{R}$.

Observe that if the total symbol $a(x,\xi) = a_2(\xi)$, then the propagator e^{itA} reduces to a Fourier multiplier and the boundedness on $\mathcal{M}^{p,q}$ was already proved in [3, Theorem 11]; see also [38].

Similarly, using Lemma 5.4, we observe that, under the assumptions of the previous theorem and with the assumption (5.57) replaced by

$$\left| \det \left(\frac{\partial^2 a_2}{\partial x \partial x}\Big|_{(x,\eta)} \right) \right| \ge \delta \quad \forall (x,\eta) \in \mathbb{R}^{2d}, \tag{5.58}$$

we obtain the boundedness from $\mathcal{M}^{p,q}(\mathbb{R}^d)$ into $\mathcal{M}^{q,p}(\mathbb{R}^d)$. Similar assumptions give the boundedness on Wiener amalgam spaces as well.

References

1. Asada, K., Fujiwara, D.: On some oscillatory transformation in $L^2(\mathbb{R}^n)$. Jpn. J. Math. **4**, 299–361 (1978)
2. Bényi, A., Grafakos, L., Gröchenig, K., Okoudjou, K.: A class of Fourier multipliers for modulation spaces. Appl. Comput. Harmonic Anal. **19**, 131–139 (2005)
3. Bényi, A., Gröchenig, K.: Okoudjou, K.A., Rogers, L.G.: Unimodular Fourier multipliers for modulation spaces. J. Funct. Anal. **246**(2), 366–384 (2007)
4. Bényi, A., Okoudjou, K.A.: Local well-posedness of nonlinear dispersive equations on modulation spaces. Bull. Lond. Math. Soc. **41**(3), 549–558 (2009)
5. Bony, J.: Opérateurs intégraux de Fourier et calcul de Weyl-Hörmander (cas d'une métrique symplectique). (French) [Fourier integral operators and the Weyl-Hörmander calculus (case of a symplectic metric)] *Journées "Équations aux Dérivées Partielles" (Saint-Jean-de-Monts, 1994)*, Exp. No. IX, 14 pp., *École Polytech., Palaiseau*, 1994.
6. Bony, J.: Evolution equations and generalized Fourier integral operators. In: Advances in Phase Space Analysis of Partial Differential Equations. Progress in Nonlinear Differential Equations and Their Applications, vol. 78, pp. 59–72. Birkhäuser Boston Inc., Boston (2009)
7. Boggiatto, P., Cordero, E., Gröchenig, K.: Generalized Anti-Wick operators with symbols in distributional Sobolev spaces. Integr. Equat. Operat. Theor. **48**, 427–442 (2004)
8. Carathéodory, C.: Variationsrechnung und partielle Differentialglichungen erster Ordnung. Teubner, Berlin (1935); Leipzig (1956). English transl.: Holden-Day, San Francisco (1965)
9. Cordero, E., Nicola, F.: Metaplectic representation on Wiener amalgam spaces and applications to the Schrödinger equation. J. Funct. Anal. **254**, 506–534 (2008)
10. Cordero, E., Nicola, F.: Remarks on Fourier multipliers and applications to the wave equation. J. Math. Anal. Appl. **353**(2), 583–591 (2009)
11. Cordero, E., Nicola, F.: Boundedness of Schrödinger type propagators on modulation spaces. J. Fourier Anal. Appl. **16**, 311–339 (2010)
12. Cordero, E., Zucco, D.: The Cauchy Problem for the vibrating plate equation in modulation spaces. J. Pseudo-Differ. Operat. Appl. **2**, 343–354 (2011). DOI: 10.1007/s11868-011-0032-7

13. Cordero, E., Feichtinger, H., Luef, F.: Banach gelfand triples for gabor analysis. In: Pseudo-Differential Operators. Lecture Notes in Mathematics, vol. 1949, pp. 1–33. Springer, Berlin (2008)
14. Cordero, E., Gröchenig, K., Nicola, F., Rodino, L.: The Wiener property for a class of Fourier Integral Operators. Journal de mathématiques pures et appliquées **99** (2013), pp. 219–233 DOI:10.1016/j.matpur.2012.06.012
15. Cordero, E., Nicola, F., Rodino, L.: Time-frequency analysis of Fourier integral operators. Commun. Pure Appl. Anal. **9**(1), 1–21 (2010)
16. Cordero, E., Nicola, F., Rodino, L.: Sparsity of Gabor representation of Schrödinger propagators. Appl. Comput. Harmon. Anal. **26**(3), 357–370 (2009)
17. Cordero, E., Nicola, F., Rodino, L.: Boundedness of Fourier integral operators on $\mathscr{F}L^p$ spaces. Trans. Am. Math. Soc. **361**(11), 6049–6071 (2009)
18. Feichtinger, H.G.: Banach spaces of distributions of Wiener's type and interpolation. In: Proceedings of the Conference on Functional Analysis and Approximation, Oberwolfach August 1980. International Series in Numerical Mathematics, vol. 69, pp. 153–165. Birkhäuser, Boston (1981)
19. Feichtinger, H.G.: Banach convolution algebras of Wiener's type. In: Proceedings of the Conference on "Function, Series, Operators", Budapest August 1980. Colloquia Mathematical Society János Bolyai, vol. 35, pp. 509–524. North-Holland, Amsterdam (1983)
20. Feichtinger, H.G.: Modulation spaces on locally compact abelian groups. Technical Report, University Vienna, 1983, and also In: Krishna, M., Radha, R., Thangavelu, S. (eds.) Wavelets and Their Applications, pp. 99–140. Allied Publishers, New Delhi (2003)
21. Feichtinger, H.G.: Atomic characterizations of modulation spaces through Gabor-type representations. Rocky Mountain J. Math. **19**, 113–126 (1989). Proceedings of the Conference on Constructive Function Theory .
22. Feichtinger, H.G.: Generalized amalgams, with applications to Fourier transform. Can. J. Math. **42**(3), 395–409 (1990)
23. Feichtinger, H.G., Gröchenig, K.: Gabor frames and time–frequency analysis of distributions. J. Funct. Anal. **146**(2), 464–495 (1997)
24. Feichtinger, H.G., Gröchenig, K.: Banach spaces related to integrable group representations and their atomic decompositions I. J. Funct. Anal. **86**(2), 307–340 (1989)
25. Feichtinger, H.G., Gröchenig, K.H.: Banach spaces related to integrable group representations and their atomic decompositions II. Monatsh. Math. **108**, 129–148 (1989)
26. Galperin, Y.V., Samarah, S.: Time-frequency analysis on modulation spaces $M_m^{p,q}$, $0 < p$, $q \leq \infty$. Appl. Comp. Harmonic Anal. **16**, 1–18 (2004)
27. Gröchenig, K.: Foundations of Time–Frequency Analysis. Birkhäuser, Boston (2001)
28. Gröchenig, K., Heil, C.: Modulation spaces and pseudodifferential operators. Integr. Equat. Oper. Theor. **34**, 439–457 (1999)
29. Gröchenig, K., Heil, C.: Counterexamples for boundedness of pseudodifferential operators. Osaka J. Math. **41**, 681–691 (2004)
30. Gröchenig, K., Leinert, M.: Wiener's lemma for twisted convolution and Gabor frames. J. Am. Math. Soc. **17**, 1–18 (2004)
31. Heil, C.: An introduction to weighted Wiener amalgams. In: Krishna, M., Radha, R., Thangavelu, S. (eds.) Wavelets and Their Applications, pp. 183–216. Allied Publishers Private Limited, New Delhi (2003)
32. Helffer, B.: Théorie Spectrale pour des Operateurs Globalement Elliptiques. Astérisque, Société Mathématique de France (1984)
33. Helffer, B., Robert, D.: Comportement Asymptotique Precise du Spectre d'Operateurs Globalement Elliptiques dans $\mathbb{R}^d$. Sem. Goulaouic-Meyer-Schwartz 1980–81. École Polytechnique, Exposé II (1980)
34. Kobayashi, M., Sugimoto, M., Tomita, N.: Trace ideals for pseudo-differential operators and their commutators with symbols in α-modulation spaces. J. Anal. Math. **107**, 141–160 (2009)
35. Kobayashi, M., Sugimoto, M., Tomita, N.: On the L^2-boundedness of pseudo-differential operators and their commutators with symbols in α-modulation spaces. J. Math. Anal. Appl. **350**(1), 157–169 (2009)

36. Krantz, S.G., Parks, H.R.: The Implicit Function Theorem. Birkhäuser Boston Inc., Boston (2002)
37. Labate, D.: Pseudodifferential operators on modulation spaces. J. Math. Anal. Appl. **262**, 242–255 (2001)
38. Miyachi, A., Nicola, F., Rivetti, S., Tabacco, A., Tomita, N.: Estimates for unimodular Fourier multipliers on modulation spaces. Proc. Am. Math. Soc. **137**, 3869–3883 (2009)
39. Shubin, M.A.: Pseudodifferential Operators and Spectral Theory, 2nd edn. Springer, Berlin (2001). Translated from the 1978 Russian original by Stig I. Andersson
40. Sjöstrand, J.: An algebra of pseudodifferential operators. Math. Res. Lett. **1**, 185–192 (1994)
41. Sugimoto, M., Tomita, N.: A counterexample for boundedness of pseudo-differential operators on modulation spaces. Proc. Am. Math. Soc. **136**(5), 1681–1690 (2008)
42. Sugimoto, M., Tomita, N.: Boundedness properties of pseudo-differential and Calderón-Zygmund operators on modulation spaces. J. Fourier Anal. Appl. **14**(1), 124–143 (2008)
43. Tataru, D.: Phase space transforms and microlocal analysis. Phase Space Analysis of Partial Differential Equations, vol. II, pp. 505–524. Pubbl. Cent. Ric. Mat. Ennio Giorgi, Scuola Norm. Sup., Pisa (2004)
44. Triebel, H.: Modulation spaces on the Euclidean n-spaces. Z. Anal. Anwendungen **2**, 443–457 (1983)
45. Wang, B., Huang, C.: Frequency-uniform decomposition method for the generalized BO, KdV and NLS equations. J. Differ. Equat. **239**(1), 213–250 (2007)
46. Wang, B., Lifeng, Z., Boling, G.: Isometric decomposition operators, function spaces $E^{\lambda}_{p,q}$ and applications to nonlinear evolution equations. J. Funct. Anal. **233**(1), 1–39 (2006)

Chapter 6
New Maximal Regularity Results for the Heat Equation in Exterior Domains, and Applications

Raphaël Danchin and Piotr Bogusław Mucha

Abstract This paper is dedicated to the proof of new maximal regularity results involving Besov spaces for the heat equation in the half-space or in bounded or exterior domains of $\mathbb{R}^n$. We strive for *time independent*, a priori estimates in regularity spaces of type $L_1(0,T;X)$ where X stands for some *homogeneous* Besov space. In the case of bounded domains, the results that we get are similar to those of the whole space or of the half-space. For exterior domains, we need to use mixed Besov norms in order to get a control on the low frequencies. Those estimates are crucial for proving global-in-time results for nonlinear heat equations in a critical functional framework.

Key words: Besov spaces, Exterior domain, Heat equation, Maximal regularity, L^1 regularity

2010 Mathematics Subject Classification: 35K05, 35K10, 35B65.

6.1 Introduction

We are concerned with the proof of maximal regularity estimates for the heat equation with Dirichlet boundary conditions, namely,

---

R. Danchin (✉)
LAMA, UMR 8050, Université Paris-Est et Institut Universitaire de France,
61, avenue du Général de Gaulle, F-94010 Créteil Cedex, France
e-mail: raphael.danchin@u-pec.fr

P.B. Mucha
Instytut Matematyki Stosowanej i Mechaniki, Uniwersytet Warszawski,
ul. Banacha 2, 02-097 Warszawa, Poland
e-mail: p.mucha@mimuw.edu.pl

M. Cicognani et al. (eds.), *Studies in Phase Space Analysis with Applications to PDEs*, Progress in Nonlinear Differential Equations and Their Applications 84,
DOI 10.1007/978-1-4614-6348-1_6,

$$\begin{array}{lll} u_t - \nu\Delta u = f & \text{in} & (0,T)\times\Omega, \\ u = 0 & \text{at} & (0,T)\times\partial\Omega, \\ u = u_0 & \text{on} & \Omega \end{array} \tag{6.1}$$

in various domains Ω of $\mathbb{R}^n$ ($n \geq 2$).

We are interested in L_1-in-time estimates for the solutions to (6.1) with a gain of two full spatial derivatives with respect to the data, that is,

$$\|u_t, \nu\nabla^2 u\|_{L_1(0,T;X)} \leq C\big(\|u_0\|_X + \|f\|_{L_1(0,T;X)}\big) \tag{6.2}$$

with a constant C *independent of* T.

Such time-independent estimates are of importance not only for the heat semi-group theory but also in the applications. Typically, they are crucial for proving global existence and uniqueness statements for nonlinear heat equations with small data in a critical functional framework. Moreover, the fact that two full derivatives may be gained with respect to the source term allows to consider not only the $-\Delta$ operator but also small perturbations of it. In addition, we shall see below that it is possible to choose X in such a way that the constructed solution u is L_1 in time with values in the set of Lipschitz functions. Hence, if the considered nonlinear heat equation determines the velocity field of some fluid, then this velocity field admits a unique Lipschitzian flow for all time. The model may thus be reformulated equivalently in Lagrangian variables (see, e.g., our recent work [4] in the slightly different context of incompressible flows). This is obviously of interest to investigate free boundary problems.

Let us recall however that estimates such as (6.2) are false if X is any reflexive Banach space, hence, in particular, if X is a Lebesgue or Sobolev space (see, e.g., [6]). On the other hand, it is well known that (6.2) holds true in the whole space $\mathbb{R}^n$ if X is a homogeneous Besov space *with third index* 1. Let us be more specific. Let us fix some homogeneous Littlewood–Paley decomposition $(\dot\Delta_j)_{j\in\mathbb{Z}}$ (see the definition in the next section) and denote by $(e^{\alpha\Delta})_{\alpha>0}$ the heat semigroup over $\mathbb{R}^n$. Then it is well known (see, e.g., [1]) that there exist two constants c and C such that for all $j\in\mathbb{Z}$ and $\alpha\in\mathbb{R}^+$, one has

$$\|e^{\alpha\Delta}\dot\Delta_j h\|_{L_p(\mathbb{R}^n)} \leq Ce^{-c\alpha 2^{2j}}\|\dot\Delta_j h\|_{L_p(\mathbb{R}^n)}. \tag{6.3}$$

Hence if u satisfies (6.1), then one may write

$$\dot\Delta_j u(t) = e^{\nu t\Delta}\dot\Delta_j u_0 + \int_0^t e^{\nu(t-\tau)\Delta}\dot\Delta_j f\,d\tau.$$

Therefore, taking advantage of (6.3), we discover that

$$\|\dot\Delta_j u(t)\|_{L_p(\mathbb{R}^n)} \leq C\bigg(e^{-c\nu t 2^{2j}}\|\dot\Delta_j u_0\|_{L_p(\mathbb{R}^n)} + \int_0^t e^{-c\nu(t-\tau)2^{2j}}\|\dot\Delta_j f\|_{L_p(\mathbb{R}^n)}\,d\tau\bigg),$$

whence

$$\begin{aligned}\|\dot\Delta_j u\|_{L_\infty(0,T;L_p(\mathbb{R}^n))} &+ \nu 2^{2j}\|\dot\Delta_j u\|_{L_1(0,T;L_p(\mathbb{R}^n))} \\ &\leq C\big(\|\dot\Delta_j u_0\|_{L_p(\mathbb{R}^n)} + \|\dot\Delta_j f\|_{L_1(0,T;L_p(\mathbb{R}^n))}\big).\end{aligned}$$

Multiplying the inequality by 2^{js} and summing up over j, we thus eventually get for some absolute constant C independent of ν and T :

$$\|u\|_{L_\infty(0,T;\dot{B}^s_{p,1}(\mathbb{R}^n))} + \|u_t, \nu\nabla^2 u\|_{L_1(0,T;\dot{B}^s_{p,1}(\mathbb{R}^n))} \leq C(\|f\|_{L_1(0,T;\dot{B}^s_{p,1}(\mathbb{R}^n))} + \|u_0\|_{\dot{B}^s_{p,1}(\mathbb{R}^n)}), \tag{6.4}$$

where the homogeneous Besov semi-norm that is used in the above inequality is defined by

$$\|u\|_{\dot{B}^s_{p,1}(\mathbb{R}^n)} := \sum_{j\in\mathbb{Z}} 2^{sj}\|\dot{\Delta}_j u\|_{L_p(\mathbb{R}^n)}.$$

From this and the definition of homogeneous Besov space $\dot{B}^s_{p,1}(\mathbb{R}^n)$ (see Sect. 6.2), we easily deduce the following classical result:

Theorem 6.1. *Let $p \in [1,\infty]$ and $s \in \mathbb{R}$. Let $f \in L_1(0,T;\dot{B}^s_{p,1}(\mathbb{R}^n))$ and $u_0 \in \dot{B}^s_{p,1}(\mathbb{R}^n)$. Then* (6.1) *with $\Omega = \mathbb{R}^n$ has a unique solution u in*

$$\mathscr{C}([0,T);\dot{B}^s_{p,1}(\mathbb{R}^n)) \quad \textit{with} \quad \partial_t u, \nabla^2 u \in L_1(0,T;\dot{B}^s_{p,1}(\mathbb{R}^n))$$

and (6.4) *is satisfied.*

The present paper is mainly devoted to generalizations of Theorem 6.1 to the half-space, bounded or exterior domains (that is, the complement of bounded simply connected domains), and applications to the global solvability of nonlinear heat equations.

Proving maximal regularity estimates for general domains essentially relies on Theorem 6.1 and localization techniques. More precisely, after localizing the equation, thanks to a suitable resolution of unity, one has to estimate "interior terms" with support that do not intersect the boundary of Ω and "boundary terms" the support of which meets $\partial\Omega$. In order to prove *interior estimates*, that is, bounds for the interior terms, it suffices to resort to the theorem in the whole space, Theorem 6.1, for those interior terms that satisfy (6.1) (with suitable data) once extended by zero onto the whole space. In contrast, the extension of the boundary terms by zero does not satisfy (6.1) on $\mathbb{R}^n$. However, performing a change of variable reduces their study to that of (6.1) on the half-space $\mathbb{R}^n_+$. Therefore, proving maximal regularity estimates in general domains mainly relies on such estimates on $\mathbb{R}^n$ and on $\mathbb{R}^n_+$. As a matter of fact, we shall see that the latter case stems from the former, by symmetrization, *provided s is close enough to* 0. In the case of a general domain, owing to change of variables and localization, however, we shall obtain (6.4) either *up to low-order terms* or with a *time-dependent* constant C. In a bounded domain, it turns out that Poincaré inequality (or equivalently the fact that the Dirichlet Laplacian operator has eigenvalues bounded away from 0) allows to prove an exponential decay which is sufficient to cancel out those lower-order terms. In the case of an exterior domain, that decay turns out to be only algebraic (at most $t^{-n/2}$ in dimension n). As a consequence, absorbing the lower-order terms will enforce us to use mixed Besov norms and to assume that $n \geq 3$.

The paper unfolds as follows. The basic tools for our analysis (Besov spaces on domains, product estimates, embedding results) are presented in the next section. In Sect. 6.3 we prove maximal regularity estimates similar to those of Theorem 6.1 first in the half-space and next in exterior or bounded domains. As an application, in the last section, we establish global existence results for nonlinear heat equations with small data in a critical functional framework.

6.2 Tools

In this section, we introduce the main functional spaces and (harmonic analysis) tools that will be needed in this paper.

6.2.1 Besov Spaces on the Whole Space

Throughout we fix a smooth nonincreasing radial function $\chi:\mathbb{R}^n\to[0,1]$ supported in $B(0,1)$ and such that $\chi\equiv 1$ on $B(0,1/2)$ and set $\varphi(\xi):=\chi(\xi/2)-\chi(\xi)$. Note that this implies that φ is valued in $[0,1]$, supported in $\{1/2\le r\le 2\}$, and that

$$\sum_{k\in\mathbb{Z}}\varphi(2^{-k}\xi)=1\quad\text{for all}\quad \xi\neq 0. \tag{6.5}$$

Then we introduce the homogeneous Littlewood–Paley decomposition $(\dot\Delta_k)_{k\in\mathbb{Z}}$ over $\mathbb{R}^n$ by setting

$$\dot\Delta_k u:=\varphi(2^{-k}D)u=\mathscr{F}^{-1}\big(\varphi(2^{-k}\cdot)\mathscr{F}u\big).$$

Above $\mathscr{F}$ stands for the Fourier transform on $\mathbb{R}^n$. We also define the low-frequency cutoff $\dot S_k:=\chi(2^{-k}D)$.

In order to define Besov spaces on $\mathbb{R}^n$, we first introduce the following homogeneous semi-norms and nonhomogeneous Besov norms (for all $s\in\mathbb{R}$ and $(p,r)\in[1,\infty]^2$):

$$\|u\|_{\dot B^s_{p,r}(\mathbb{R}^n)}:=\big\|2^{sk}\|\dot\Delta_k u\|_{L_p(\mathbb{R}^n)}\big\|_{\ell_r(\mathbb{Z})}$$
$$\|u\|_{B^s_{p,r}(\mathbb{R}^n)}:=\big\|2^{sk}\|\dot\Delta_k u\|_{L_p(\mathbb{R}^n)}\big\|_{\ell_r(\mathbb{N})}+\|\dot S_0 u\|_{L_p(\mathbb{R}^n)}.$$

The nonhomogeneous Besov space $B^s_{p,r}(\mathbb{R}^n)$ is the set of tempered distributions u such that $\|u\|_{B^s_{p,r}(\mathbb{R}^n)}$ is finite. Following [1], we define the homogeneous Besov space $\dot B^s_{p,r}(\mathbb{R}^n)$ as

$$\dot B^s_{p,r}(\mathbb{R}^n)=\Big\{u\in\mathscr{S}'_h(\mathbb{R}^n)\,:\,\|u\|_{\dot B^s_{p,r}(\mathbb{R}^n)}<\infty\Big\},$$

where $\mathscr{S}'_h(\mathbb{R}^n)$ stands for the set of tempered distributions u over $\mathbb{R}^n$ such that for all smooth compactly supported function θ over $\mathbb{R}^n$, we have $\lim_{\lambda\to+\infty}\theta(\lambda D)u=0$ in $L_\infty(\mathbb{R}^n)$. Note that any distribution $u\in\mathscr{S}'_h(\mathbb{R}^n)$ satisfies $u=\sum_{k\in\mathbb{Z}}\dot\Delta_k u$ in $\mathscr{S}'_h(\mathbb{R}^n)$.

We shall make an extensive use of the following result (see the proof in, e.g., [1, 5]):

Proposition 6.1. *Let $b^s_{p,r}$ denote $\dot B^s_{p,r}(\mathbb{R}^n)$ or $B^s_{p,r}(\mathbb{R}^n)$. Then the following a priori estimates hold true:*

- *For any $s>0$,*
$$\|uv\|_{b^s_{p,r}}\lesssim\|u\|_{L^\infty}\|v\|_{b^s_{p,r}}+\|v\|_{L^\infty}\|u\|_{b^s_{p,r}}.$$
- *For any $s>0$ and $t>0$,*
$$\|uv\|_{b^s_{p,r}}\lesssim\|u\|_{L^\infty}\|v\|_{b^s_{p,r}}+\|v\|_{b^{-t}_{\infty,r}}\|u\|_{b^{s+t}_{p,\infty}}.$$
- *For any $t>0$ and $s>-n/p'$,*
$$\|uv\|_{b^s_{p,r}}\lesssim\|u\|_{L^\infty}\|v\|_{b^s_{p,r}}+\|u\|_{b^{n/p'}_{p',\infty}}\|v\|_{b^s_{p,r}}+\|v\|_{b^{-t}_{\infty,r}}\|u\|_{b^{s+t}_{p,\infty}}.$$
- *For any $q>1$ and $1-n/q\le s\le 1$,*
$$\|uv\|_{b^0_{q,1}}\lesssim\|u\|_{b^s_{n,1}}\|v\|_{b^{1-s}_{q,1}}.$$

As obviously a smooth compactly supported function belongs to any space $\dot B^{n/p}_{p,1}(\mathbb{R}^n)$ with $1\le p\le\infty$, and to any Besov space $B^\sigma_{p,1}(\mathbb{R}^n)$, we deduce from the previous proposition and embedding that (see the proof in [5]):

Corollary 6.1. *Let θ be in $\mathscr{C}^\infty_c(\mathbb{R}^n)$. Then $u\mapsto\theta u$ is a continuous mapping of $b^s_{p,r}(\mathbb{R}^n)$*

- *For any $s\in\mathbb{R}$ and $1\le p,r\le\infty$, if $b^s_{p,r}(\mathbb{R}^n)=B^s_{p,r}(\mathbb{R}^n)$.*
- *For any $s\in\mathbb{R}$ and $1\le p,r\le\infty$ satisfying*

$$-n/p'<s<n/p\quad\big(-n/p<s\le n/p\ \text{if}\ r=1,\quad -n/p'\le s<n/p\ \text{if}\ r=\infty\big)\tag{6.6}$$

if $b^s_{p,r}(\mathbb{R}^n)=\dot B^s_{p,r}(\mathbb{R}^n)$.

The following proposition allows us to compare the spaces $B^s_{p,r}(\mathbb{R}^n)$ and $\dot B^s_{p,r}(\mathbb{R}^n)$ for compactly supported functions[1] (see the proof in [5]):

Proposition 6.2. *Let $1\le p,r\le\infty$, and $s>-n/p'$ (or $s\ge -n/p'$ if $r=\infty$). Then for any compactly supported distribution f, we have*

$$f\in B^s_{p,r}(\mathbb{R}^n)\iff f\in\dot B^s_{p,r}(\mathbb{R}^n)$$

[1] Without any support assumption, it is obvious that if $s>0$, then we have $\|\cdot\|_{\dot B^s_{p,r}(\mathbb{R}^n)}\lesssim\|\cdot\|_{B^s_{p,r}(\mathbb{R}^n)}$ and that, if $s<0$, then $\|\cdot\|_{\dot B^s_{p,r}(\mathbb{R}^n)}\gtrsim\|\cdot\|_{B^s_{p,r}(\mathbb{R}^n)}$.

and there exists a constant $C = C(s,p,r,n,K)$ (with $K = \mathrm{Supp}\ f$) such that

$$C^{-1}\|f\|_{\dot{B}^s_{p,r}(\mathbb{R}^n)} \le \|f\|_{B^s_{p,r}(\mathbb{R}^n)} \le C\|f\|_{\dot{B}^s_{p,r}(\mathbb{R}^n)}.$$

The following lemma will be useful for boundary estimates (see the proof in [5]):

Lemma 6.1. *Let Z be a Lipschitz diffeomorphism on $\mathbb{R}^n$ with DZ and DZ^{-1} bounded, $(p,r) \in [1,\infty]^2$, and s a real number satisfying $-1+1/p < s < 1/p$.*

- *If in addition Z is measure preserving, then the linear map $u \mapsto u \circ Z$ is continuous on $\dot{B}^s_{p,r}(\mathbb{R}^n)$.*
- *In the general case, the map $u \mapsto u \circ Z$ is continuous on $\dot{B}^s_{p,r}(\mathbb{R}^n)$ provided in addition $J_{Z^{-1}} \in \dot{B}^{n/p'}_{p',\infty} \cap L_\infty$ with $J_{Z^{-1}} := |\det DZ^{-1}|$.*

6.2.2 Besov Spaces on Domains

We aim at extending the definition of homogeneous Besov spaces to general domains. We proceed by restriction as follows[2]:

Definition 6.1. For $s \in \mathbb{R}$ and $1 \le p,q \le \infty$, we define the homogeneous Besov space $\dot{B}^s_{p,q}(\Omega)$ over Ω as the restriction (in the distributional sense) of $\dot{B}^s_{p,q}(\mathbb{R}^n)$ on Ω, that is,

$$\varphi \in \dot{B}^s_{p,q}(\Omega) \Longleftrightarrow \varphi = \psi_{|\Omega} \quad \text{for some} \quad \psi \in \dot{B}^s_{p,q}(\mathbb{R}^n).$$

We then set

$$\|\varphi\|_{\dot{B}^s_{p,q}(\Omega)} := \inf_{\psi_{|\Omega}=\varphi} \|\psi\|_{\dot{B}^s_{p,q}(\mathbb{R}^n)}.$$

The embedding, duality, and interpolation properties of these Besov spaces may be deduced from those on $\mathbb{R}^n$. As regards duality, we shall use repeatedly the following result:

Proposition 6.3. *If $-1+1/p < s < 1/p$ (with $1 \le p,r < \infty$), then the space $\dot{B}^{-s}_{p',r'}(\Omega)$ may be identified with the dual space of $\dot{B}^s_{p,r}(\Omega)$; in the limit case $r = \infty$, then $\dot{B}^{-s}_{p',1}(\Omega)$ may be identified with the dual space of the completion of $\mathcal{C}^\infty_c(\mathbb{R}^n)$ for $\|\cdot\|_{\dot{B}^s_{p,\infty}(\Omega)}$. Furthermore, without any condition over (s,p,r), we have*

$$\left|\int_\Omega uv\,dx\right| \le C\|u\|_{\dot{B}^s_{p,r}(\Omega)}\|v\|_{\dot{B}^{-s}_{p',r'}(\Omega)}.$$

Similarly, some product laws for Besov spaces on $\mathbb{R}^n$ may be extended to the domain case. We shall use the last inequality of Proposition 6.1 and also the following result that is proved in [5]:

[2] Nonhomogeneous Besov spaces on domains may be defined by the same token.

Proposition 6.4. *Let $b^s_{p,r}(\Omega)$ denote $\dot B^s_{p,r}(\Omega)$ or $B^s_{p,r}(\Omega)$ and Ω be any domain of $\mathbb{R}^n$. Then for any $p\in[1,\infty]$, s such that $-n/p'<s<n/p$ (or $-n/p'<s\le n/p$ if $r=1$ or $-n/p'\le s<n/p$ if $r=\infty$), the following inequality holds true:*

$$\|uv\|_{b^s_{p,r}(\Omega)}\le C\|u\|_{b^{n/q}_{q,1}(\Omega)}\|v\|_{b^s_{p,r}(\Omega)}\quad \textit{with}\ \ q=\min(p,p').$$

A very useful feature of Besov spaces is their interpolation properties. We refer to the books [2, 14] for the proof of the following statement:

Proposition 6.5. *Let $b^s_{p,q}$ denote $B^s_{p,q}$ or $\dot B^s_{p,q}$, $s\in\mathbb{R}$, $p\in[1,\infty)$, and $q\in[1,\infty]$. The real interpolation of Besov spaces gives the following statement if $s_1\neq s_2$:*

$$\left(b^{s_1}_{p,q_1}(\Omega),b^{s_2}_{p,q_2}(\Omega)\right)_{\theta,q}=b^s_{p,q}(\Omega)$$

with $s=\theta s_2+(1-\theta)s_1$ and $\frac{1}{p}=\frac{\theta}{p_2}+\frac{1-\theta}{p_1}$.

Moreover, if $s_1\neq s_2$, $t_1\neq t_2$ and if $T: b^{s_1}_{p_1,q_1}(\Omega)+b^{s_2}_{p_2,q_2}(\Omega)\to b^{t_1}_{k_1,l_1}(\Omega)+b^{t_2}_{k_2,l_2}(\Omega)$ is a linear map, bounded from $b^{s_1}_{p_1,q_1}(\Omega)$ to $b^{t_1}_{k_1,l_1}(\Omega)$ and from $b^{s_2}_{p_2,q_2}(\Omega)$ to $b^{t_2}_{k_2,l_2}(\Omega)$, then for any $\theta\in(0,1)$, the map T is also bounded from $b^s_{p,q}(\Omega)$ to $b^t_{k,q}(\Omega)$ with

$$s=\theta s_2+(1-\theta)s_1,\quad t=\theta t_2+(1-\theta)t_1,\quad \frac{1}{p}=\frac{\theta}{p_2}+\frac{1-\theta}{p_1},\quad \frac{1}{k}=\frac{\theta}{k_2}+\frac{1-\theta}{k_1}.$$

The following composition estimate will be of constant use in the last section of this paper.

Proposition 6.6. *Let $f:\mathbb{R}^r\to\mathbb{R}$ be a C^1 function such that $f(\mathbf{0})=0$ and, for some $m\ge 1$ and $K\ge 0$,*

$$|df(\mathbf{u})|\le K|\mathbf{u}|^{m-1}\quad \textit{for all}\ \ \mathbf{u}\in\mathbb{R}^r. \tag{6.7}$$

Then for all $s\in(0,1)$ and $1\le p,q\le\infty$, there exists a constant C so that

$$\|f(\mathbf{u})\|_{\dot B^s_{p,q}(\Omega)}\le CK\|\mathbf{u}\|^{m-1}_{L_\infty(\Omega)}\|u\|_{\dot B^s_{p,q}(\Omega)}. \tag{6.8}$$

Proof. The proof relies on the characterization of the norm of $\dot B^s_{p,q}(\Omega)$ by finite differences, namely,[3]

$$\|f(\mathbf{u})\|_{\dot B^s_{p,q}(\Omega)}=\left(\int_\Omega\left(\int_\Omega\frac{|f(\mathbf{u}(y))-f(\mathbf{u}(x))|^p}{|y-x|^{n+sp}}\,dy\right)^{\frac{q}{p}}dx\right)^{\frac{1}{q}}. \tag{6.9}$$

[3] Here we just consider the case $q<\infty$ to shorten the presentation.

Now the mean value formula implies that

$$f(\mathbf{u}(y)) - f(\mathbf{u}(x)) = \left(\int_0^1 df(\mathbf{u}(x) + t(\mathbf{u}(y) - \mathbf{u}(x)))\,dt\right) \cdot (\mathbf{u}(y) - \mathbf{u}(x)).$$

Hence using the growth assumption (6.7),

$$|f(\mathbf{u}(y)) - f(\mathbf{u}(x))| \le K\left(\int_0^1 |\mathbf{u}(x) + t(\mathbf{u}(y) - \mathbf{u}(x))|^{m-1}\,dt\right)|\mathbf{u}(y) - \mathbf{u}(x)|. \quad (6.10)$$

Therefore we get

$$|f(\mathbf{u}(y)) - f(\mathbf{u}(x))| \le K\|\mathbf{u}\|_{L_\infty(\Omega)}^{m-1}|\mathbf{u}(y) - \mathbf{u}(x)|.$$

Inserting this latter inequality in (6.9), we readily get (6.8). □

In [3, 5], we proved that:

Proposition 6.7. *Let Ω be the half-space or a bounded or exterior domain with C^1 boundary. For all $1 \le p, q < \infty$, and $-1 + 1/p < s < 1/p$, we have*

$$B^s_{p,q}(\Omega) = \overline{\mathscr{C}^\infty_c(\Omega)}^{\|\cdot\|_{B^s_{p,q}(\Omega)}}. \quad (6.11)$$

Remark 6.1. In any C^1 domain Ω and for $0 < s < n/p$ the space $\dot{B}^s_{p,q}(\Omega)$ embeds in $\dot{B}^0_{m,q}(\Omega)$ with $1/m = 1/p - s/n$. Therefore, if $q \le \min(2,m)$, it also embeds in the Lebesgue space $L_m(\Omega)$. So finally if $s \in (0, \frac{1}{p})$ and $q \le \min(2,m)$ with m as above, then Proposition 6.7 allows us to redefine the space $\dot{B}^s_{p,q}(\Omega)$ by

$$\dot{B}^s_{p,q}(\Omega) = \overline{\mathscr{C}^\infty_c(\Omega)}^{\|\cdot\|_{\dot{B}^s_{p,q}(\Omega)}}. \quad (6.12)$$

Remark 6.2. In particular under the above hypotheses, both classes of Besov spaces admit trivial extension by zero onto the whole space. Combining with Proposition 6.2, we deduce that

$$B^s_{p,q}(\Omega) = \dot{B}^s_{p,q}(\Omega) \quad \text{if} \quad -1 + 1/p < s < 1/p \quad \text{and} \quad \Omega \text{ is } \textit{bounded}.$$

Note also that, for obvious reasons, the above density result does not hold true if $q = \infty$, for the strong topology. However, it holds for the weak $*$ topology.

6.3 A Priori Estimates for the Heat Equation

This section is the core of the paper. Here we prove generalizations of Theorem 6.1 to more general domains. First we consider the half-space case, and then we consider the exterior and bounded cases. We shall mainly focus on the unbounded case which is more tricky and just indicate at the end of this section what has to be changed in the bounded domain case.

6.3.1 The Heat Equation in the Half-Space

The purpose of this paragraph is to extend Theorem 6.1 to the half-space case $\mathbb{R}^n_+$, namely,

$$\begin{aligned} u_t - \nu\Delta u = f & \text{ in } (0,T)\times\mathbb{R}^n_+,\\ u|_{x_n=0} = 0 & \quad\text{on } (0,T)\times\partial\mathbb{R}^n_+,\\ u|_{t=0} = u_0 & \quad\text{on } \mathbb{R}^n_+. \end{aligned} \tag{6.13}$$

Theorem 6.2. *Let $p\in[1,\infty)$ and $s\in(-1+1/p,1/p)$. Assume that f belongs to $L_1(0,T;\dot{B}^s_{p,1}(\mathbb{R}^n_+))$ and that u_0 is in $\dot{B}^s_{p,1}(\mathbb{R}^n_+)$. Then (6.13) has a unique solution u satisfying*

$$u\in\mathscr{C}([0,T);\dot{B}^s_{p,1}(\mathbb{R}^n_+)),\quad u_t,\nabla^2 u\in L_1(0,T;\dot{B}^s_{p,1}(\mathbb{R}^n_+))$$

and the following estimate is valid:

$$\begin{aligned}\|u\|_{L_\infty(0,T;\dot{B}^s_{p,1}(\mathbb{R}^n_+))} + \|u_t,\nu\nabla^2 u\|_{L_1(0,T;\dot{B}^s_{p,1}(\mathbb{R}^n_+))}\\ \le C(\|f\|_{L_1(0,T;\dot{B}^s_{p,1}(\mathbb{R}^n_+))} + \|u_0\|_{\dot{B}^s_{p,1}(\mathbb{R}^n_+)}),\end{aligned} \tag{6.14}$$

where C is an absolute constant with no dependence on ν and T.

Proof. We argue by symmetrization. Let $\widetilde{u}_0$ and $\widetilde{f}$ be the antisymmetric extensions over $\mathbb{R}^n$ to the data u_0 and f. Then, given our assumptions over s and Proposition 6.7, one may assert that $\widetilde{u}_0\in\dot{B}^s_{p,1}(\mathbb{R}^n)$, $\widetilde{f}\in L_1(0,T;\dot{B}^s_{p,1}(\mathbb{R}^n))$ and that, in addition,

$$\|\widetilde{u}_0\|_{\dot{B}^s_{p,1}(\mathbb{R}^n)}\approx\|u_0\|_{\dot{B}^s_{p,1}(\mathbb{R}^n_+)}\quad\text{and}\quad\|\widetilde{f}\|_{L_1(0,T;\dot{B}^s_{p,1}(\mathbb{R}^n))}\approx\|f\|_{L_1(0,T;\dot{B}^s_{p,1}(\mathbb{R}^n_+))}.$$

Let $\widetilde{u}$ be the solution given by Theorem 6.1. As this solution is unique in the corresponding functional framework, the symmetry properties of the data ensure that $\widetilde{u}$ is antisymmetric with respect to $\{x_n=0\}$. As a consequence, it vanishes over $\{x_n=0\}$. Hence the restriction u of $\widetilde{u}$ to the half-space satisfies (6.13). In addition:

- $\widetilde{u}_t$ coincides with the antisymmetric extension of u_t.
- $\nabla^2_{x'}\widetilde{u}$ coincides with the antisymmetric extension of $\nabla^2_{x'}u$.
- $\nabla_{x'}\partial_{x_n}\widetilde{u}$ coincides with the symmetric extension of $\nabla_{x'}\partial_{x_n}u$.
- $\partial^2_{x_n,x_n}\widetilde{u}=(\Delta-\Delta_{x'})\widetilde{u}$ hence coincides with $\widetilde{u}_t-\widetilde{f}-\Delta_{x'}\widetilde{u}$.

Hence one may conclude that

$$\begin{aligned}\|u\|_{L_\infty(0,T;\dot{B}^s_{p,1}(\mathbb{R}^n_+))} + \|u_t,\nu\nabla^2 u\|_{L_1(0,T;\dot{B}^s_{p,1}(\mathbb{R}^n_+))}\\ \le\|\widetilde{u}\|_{L_\infty(0,T;\dot{B}^s_{p,1}(\mathbb{R}^n))} + \|\widetilde{u}_t,\nu\nabla^2\widetilde{u}\|_{L_1(0,T;\dot{B}^s_{p,1}(\mathbb{R}^n))}.\end{aligned}$$

This implies (6.14). □

Remark 6.3. The case of nonhomogeneous boundary conditions, where u equals some given h at the boundary, reduces to the homogeneous case: it is only a matter of assuming that h admits some extension $\widetilde{h}$ over $(0,T)\times\mathbb{R}^n_+$ so that $\widetilde{h}_t-\nu\Delta\widetilde{h}\in L_1(0,T;\dot{B}^s_{p,1}(\mathbb{R}^n_+))$.

6.3.2 The Exterior Domain Case

Here we extend Theorem 6.1 to the case where Ω is an exterior domain (that is, the complement of a bounded simply connected domain). Here is our main statement:

Theorem 6.3. *Let Ω be a smooth exterior domain of $\mathbb{R}^n$ with $n\geq 3$. Let $1<q\leq p<\infty$ with $q<n/2$. Let $-1+1/p<s<1/p$ and $-1+1/q<s'<1/q-2/n$. Let*

$$u_0\in\dot{B}^s_{p,1}\cap\dot{B}^{s'}_{q,1}(\Omega)\quad\text{and}\quad f\in L_1(0,T;\dot{B}^s_{p,1}\cap\dot{B}^{s'}_{q,1}(\Omega)).$$

Then there exists a unique solution u to (6.1) such that

$$u\in\mathscr{C}([0,T];\dot{B}^s_{p,1}\cap\dot{B}^{s'}_{q,1}(\Omega)),\qquad u_t,\nabla^2u\in L_1(0,T;\dot{B}^s_{p,1}\cap B^{s'}_{q,1}(\Omega))$$

and the following inequality is satisfied:

$$\|u\|_{L_\infty(0,T;\dot{B}^s_{p,1}\cap\dot{B}^{s'}_{q,1}(\Omega))}+\|u_t,\nu\nabla^2u\|_{L_1(0,T;\dot{B}^s_{p,1}\cap\dot{B}^{s'}_{q,1}(\Omega))}\leq C\big(\|u_0\|_{\dot{B}^s_{p,1}\cap\dot{B}^{s'}_{q,1}(\Omega)}+\|f\|_{L_1(0,T;\dot{B}^s_{p,1}\cap\dot{B}^{s'}_{q,1}(\Omega))}\big),\tag{6.15}$$

where the constant C is independent of T and ν.

Proving this theorem relies on the following statement (that is of independent interest and holds in any dimension $n\geq 2$) and on lower-order estimates (see Lemma 6.2 below) which will enable us to remove the time dependency.

Theorem 6.4. *Let Ω be a C^2 exterior domain of $\mathbb{R}^n$ with $n\geq 2$. Let $1<p<\infty$, $-1+1/p<s<1/p$, $f\in L_1(0,T;\dot{B}^s_{p,1}(\Omega))$, and $u_0\in\dot{B}^s_{p,1}(\Omega)$. Then (6.1) has a unique solution u such that*

$$u\in\mathscr{C}([0,T];\dot{B}^s_{p,1}(\Omega)),\qquad \partial_t u,\nabla^2u\in L_1(0,T;\dot{B}^s_{p,1}(\Omega))$$

and the following estimate is valid:

$$\|u\|_{L_\infty(0,T;\dot{B}^s_{p,1}(\Omega))}+\|u_t,\nu\nabla^2u\|_{L_1(0,T;\dot{B}^s_{p,1}(\Omega))}\leq Ce^{CT\nu}\big(\|u_0\|_{\dot{B}^s_{p,1}(\Omega)}+\|f\|_{L_1(0,T;\dot{B}^s_{p,1}(\Omega))}\big),\tag{6.16}$$

where the constant C depends only on s, p, and Ω.

Additionally if K is a compact subset of Ω such that $\mathrm{dist}(\partial\Omega,\Omega\setminus K)>0$, *there holds*

$$\|u\|_{L_\infty(0,T;\dot{B}^s_{p,1}(\Omega))}+\|u_t,\nu\nabla^2 u\|_{L_1(0,T;\dot{B}^s_{p,1}(\Omega))}$$
$$\leq C\big(\|u_0\|_{\dot{B}^s_{p,1}(\Omega)}+\|f\|_{L_1(0,T;\dot{B}^s_{p,1}(\Omega))}+\nu\|u\|_{L_1(0,T;\dot{B}^s_{p,1}(K))}\big), \tag{6.17}$$

where C is as above.

Proof. We suppose that we have a smooth enough solution and focus on the proof of the estimates. We shall do it in three steps: first we prove interior estimates, next boundary estimates and finally global estimates after summation.

Note that performing the following change of unknown and data:

$$u_{new}(t,x)=\nu u_{old}(\nu^{-1}t,x),\quad u_{0,new}(x)=\nu u_{0,old}(x),\quad f_{new}(t,x)=f_{old}(\nu^{-1}t,x)$$

reduces the study to the case $\nu=1$. So we shall make this assumption in all that follows.

Throughout we fix some covering $(B(x^\ell,\lambda))_{1\leq\ell\leq L}$ of K by balls of radius λ and take some neighborhood $\Omega^0\subset\Omega$ of $\mathbb{R}^n\setminus K$ such that $d(\Omega^0,\partial\Omega)>0$. We assume in addition that the first M balls do not intersect K while the last $L-M$ balls are centered at some point of $\partial\Omega$.

Let $\eta^0:\mathbb{R}^n\to[0,1]$ be a smooth function supported in Ω^0 and with value 1 on a neighborhood of $\Omega\setminus K$. Then we consider a subordinate partition of unity $(\eta^\ell)_{1\leq\ell\leq L}$ such that:

1. $\sum_{0\leq l\leq L}\eta^\ell=1$ on Ω.
2. $\|\nabla^k\eta^\ell\|_{L_\infty(\mathbb{R}^n)}\leq C_k\lambda^{-k}$ for $k\in\mathbb{N}$ and $1\leq\ell\leq L$.
3. Supp $\eta^\ell\subset B(x^\ell,\lambda)$.

We also introduce another smooth function $\widetilde{\eta}^0$ supported in K and with value 1 on Supp $\nabla\eta^0$ and smooth functions $\widetilde{\eta}^1,\cdots,\widetilde{\eta}^L$ with compact support in $B(x^\ell,\lambda)$ and such that $\widetilde{\eta}^\ell\equiv 1$ on Supp η^ℓ.

Note that for $\ell\in\{1,\cdots,L\}$, the bounds for the derivatives of η^ℓ together with the fact that $|\text{Supp }\nabla\eta^\ell|\approx\lambda^n$ and Proposition 6.5 implies that for $k=0,1$ and any $q\in[1,\infty]$, we have

$$\|\nabla\eta^\ell\|_{\dot{B}^{k+n/q}_{q,1}(\mathbb{R}^n)}\lesssim\lambda^{-1-k}. \tag{6.18}$$

The same holds for the functions $\widetilde{\eta}^\ell$.

First Step: The Interior Estimate

The vector field $U^0:=u\eta^0$ satisfies the following modification of (6.1):

$$\begin{aligned} &U^0_t-\Delta U^0=\eta^0 f-2\nabla\eta^0\cdot\nabla u-u\Delta\eta^0 &&\text{in } (0,T)\times\mathbb{R}^n,\\ &U^0|_{t=0}=u_0\eta^0 &&\text{on } \mathbb{R}^n.\end{aligned} \tag{6.19}$$

Theorem 6.1 thus yields the following estimate:

$$\|U^0\|_{L_\infty(0,T;\dot B^s_{p,1}(\mathbb{R}^n))}+\|U^0_t,\nabla^2U^0\|_{L_1(0,T;\dot B^s_{p,1}(\mathbb{R}^n))}\lesssim\|\eta^0 f\|_{L_1(0,T;\dot B^s_{p,1}(\mathbb{R}^n))}$$
$$+\|\nabla\eta^0\cdot\nabla u\|_{L_1(0,T;\dot B^s_{p,1}(\mathbb{R}^n))}+\|u\Delta\eta^0\|_{L_1(0,T;\dot B^s_{p,1}(\mathbb{R}^n))}+\|\eta^0u_0\|_{\dot B^s_{p,1}(\mathbb{R}^n)}.$$

Let us emphasize that as $\nabla\eta^0\cdot\nabla u$ and $u\Delta\eta^0$ are compactly supported; we may replace the homogeneous norms by nonhomogeneous ones in the first two terms. As a consequence, because the function $\nabla\eta^0$ is in $\mathscr{C}^\infty_c(\mathbb{R}^n)$ and $\widetilde\eta^0\equiv 1$ on Supp $\nabla\eta^0$, Corollary 6.1 ensures that

$$\|U^0\|_{L_\infty(0,T;\dot B^s_{p,1}(\mathbb{R}^n))}+\|U^0_t,\nabla^2U^0\|_{L_1(0,T;\dot B^s_{p,1}(\mathbb{R}^n))}$$
$$\lesssim\|\eta^0u_0\|_{\dot B^s_{p,1}(\mathbb{R}^n))}+\|\eta^0 f\|_{L_1(0,T;\dot B^s_{p,1}(\mathbb{R}^n))}+\|\widetilde\eta^0 u\|_{L_1(0,T;B^{s+1}_{p,1}(\mathbb{R}^n))}. \tag{6.20}$$

Now, by interpolation,

$$\|\widetilde\eta^0u\|_{B^{1+s}_{p,1}(\Omega)}\le C\|\widetilde\eta^0u\|^{\frac12}_{B^{2+s}_{p,1}(\Omega)}\|\widetilde\eta^0u\|^{\frac12}_{\dot B^s_{p,1}(\Omega)}. \tag{6.21}$$

As Supp $\widetilde\eta^0\subset K$ and as homogeneous and nonhomogeneous norms are equivalent on K, one may thus conclude that

$$\|U^0\|_{L_\infty(0,T;\dot B^s_{p,1}(\mathbb{R}^n))}+\|U^0_t,\nabla^2U^0\|_{L_1(0,T;\dot B^s_{p,1}(\mathbb{R}^n))}\lesssim\|f\|_{L_1(0,T;\dot B^s_{p,1}(\Omega))}$$
$$+T^{1/2}\|u\|_{L_1(0,T;\dot B^{2+s}_{p,1}(K))\cap L_\infty(0,T;\dot B^s_{p,1}(K))}+\|u_0\|_{\dot B^s_{p,1}(\Omega)}. \tag{6.22}$$

Note that starting from (6.21) and using Young's inequality also yield, for all $\varepsilon>0$,

$$\|U^0\|_{L_\infty(0,T;\dot B^s_{p,1}(\mathbb{R}^n))}+\|U^0_t,\nabla^2U^0\|_{L_1(0,T;\dot B^s_{p,1}(\mathbb{R}^n))}\le C\big(\|u_0\|_{\dot B^s_{p,1}(\Omega)}$$
$$+\|f\|_{L_1(0,T;\dot B^s_{p,1}(\Omega))}\big)+\varepsilon\|u\|_{L_1(0,T;\dot B^{2+s}_{p,1}(K))}+c(\varepsilon)\|u\|_{L_1(0,T;\dot B^s_{p,1}(K))}. \tag{6.23}$$

The terms U^ℓ with $1\le\ell\le M$ may be bounded exactly along the same lines because their support do not meet $\partial\Omega$; hence, their extension by 0 over $\mathbb{R}^n$ satisfies

$$\begin{aligned} U^\ell_t-\Delta U^\ell&=f^\ell\ \text{ in }\ (0,T)\times\mathbb{R}^n,\\ U^\ell|_{t=0}&=u_0\eta^\ell\ \text{ on }\ \mathbb{R}^n\end{aligned}$$

with

$$f^\ell:=-2\nabla\eta^\ell\cdot\nabla u-u\Delta\eta^\ell+\eta^\ell f. \tag{6.24}$$

Arguing as above and taking advantage of the fact that the functions η^ℓ are localized in balls of radius λ (that is, we use (6.18)), we now get

$$\|f^\ell\|_{L_1(0,T;\dot B^s_{p,1}(\Omega))}\lesssim\|\eta^\ell f\|_{L_1(0,T;\dot B^s_{p,1}(\Omega))}$$
$$+\lambda^{-2}\|\widetilde\eta^\ell u\|_{L_1(0,T;\dot B^s_{p,1}(\Omega))}+\lambda^{-1}\|\widetilde\eta^\ell\nabla u\|_{L_1(0,T;\dot B^s_{p,1}(\Omega))}. \tag{6.25}$$

Using again (6.21) (with $\widetilde{\eta}^\ell$ instead of $\widetilde{\eta}^0$), we end up with

$$\|U^\ell\|_{L_\infty(0,T;\dot{B}^s_{p,1}(\mathbb{R}^n))} + \|U^\ell_t,\nabla^2 U^\ell\|_{L_1(0,T;\dot{B}^s_{p,1}(\mathbb{R}^n))} \lesssim \|\eta^\ell f\|_{L_1(0,T;\dot{B}^s_{p,1}(\Omega))}$$
$$+ (\lambda^{-1}T^{1/2}+\lambda^{-2}T)\|u\|_{L_1(0,T;\dot{B}^{2+s}_{p,1}(K))\cap L_\infty(0,T;\dot{B}^s_{p,1}(K))} + \|u_0\eta^\ell\|_{\dot{B}^s_{p,1}(\Omega)}, \quad (6.26)$$

$$\|U^\ell\|_{L_\infty(0,T;\dot{B}^s_{p,1}(\mathbb{R}^n))} + \|U^\ell_t,\nabla^2 U^\ell\|_{L_1(0,T;\dot{B}^s_{p,1}(\mathbb{R}^n))} \le C\big(\|u_0\eta^\ell\|_{\dot{B}^s_{p,1}(\Omega)}$$
$$+ \|\eta^\ell f\|_{L_1(0,T;\dot{B}^s_{p,1}(\Omega))}\big) + \lambda^{-1}\|u\|^{1/2}_{L_1(0,T;\dot{B}^{2+s}_{p,1}(K))}\|u\|^{1/2}_{L_1(0,T;\dot{B}^s_{p,1}(K))}. \quad (6.27)$$

Second Step: The Boundary Estimate

We now consider an index $\ell \in \{L+1,\cdots,M\}$ so that $B(x^\ell,\lambda)$ is centered at a point of $\partial\Omega$. The localization leads to the following system for $U^\ell := u\eta^\ell$:

$$\begin{array}{lll} U^\ell_t - \Delta U^\ell = f^\ell & \text{in} & (0,T)\times\Omega, \\ U^\ell = 0 & \text{on} & (0,T)\times\partial\Omega, \\ U^\ell_t|_{t=0} = u_0\eta^\ell & \text{on} & \Omega, \end{array} \quad (6.28)$$

with f^ℓ defined by (6.24), hence satisfying (6.25).

Let us make a change of variables so as to recast (6.28) in the half-space. As $\partial\Omega$ is C^∞, if λ has been chosen small enough then for fixed ℓ, we are able to find a map Z_ℓ so that:

i) Z_ℓ is a C^∞ diffeomorphism from $B(x^\ell,\lambda)$ to $Z_\ell(B(x^\ell,\lambda))$.
ii) $Z_\ell(x^\ell)=0$ and $D_x Z(x^\ell)=\mathrm{Id}$.
iii) $Z_\ell(\Omega\cap B(x^\ell,\lambda))\subset \mathbb{R}^n_+$.
iv) $Z_\ell(\partial\Omega\cap B(x^\ell,\lambda)) = \partial\mathbb{R}^n_+ \cap Z_\ell(B(x^\ell,\lambda))$.

Setting $\nabla_x Z_\ell = \mathrm{Id}+A_\ell$, then one may assume in addition that there exist constants C_j depending only on Ω and on $j\in\mathbb{N}$ such that

$$\|D^j A_\ell\|_{L_\infty(B(x^\ell,\lambda))} \le C_j, \quad (6.29)$$

a property which implies (by the mean value formula) that

$$\|A_\ell\|_{L_\infty(B(x^\ell,\lambda))} \le C_1\lambda, \quad (6.30)$$

hence by interpolation between the spaces $L_q(B(x^\ell,\lambda))$ and $W^j_q(B(x^\ell,\lambda))$ (with j being any positive integer),

$$\|A_\ell\|_{B^{\frac{n}{q}}_{q,1}(B(x^\ell,\lambda))} \le C\lambda \quad \text{for all } 1\le q<\infty \text{ such that } n/q<j. \quad (6.31)$$

Let $V^\ell := Z^*_\ell U^\ell := U^\ell \circ Z_\ell^{-1}$. The system satisfied by V^ℓ reads

$$\begin{aligned} &V^\ell_t - \Delta_z V^\ell = F^\ell && \text{in } (0,T)\times\mathbb{R}^n_+,\\ &V^\ell|_{z_n=0} = 0 && \text{on } (0,T)\times\partial\mathbb{R}^n_+,\\ &V^\ell|_{t=0} = Z^*_\ell(U^\ell|_{t=0}) && \text{on } \partial\mathbb{R}^n_+, \end{aligned} \tag{6.32}$$

with

$$F^\ell := Z^*_\ell f^\ell + (\Delta_x - \Delta_z)V^\ell.$$

According to Theorem 6.2, we thus get

$$\begin{aligned} &\|V^\ell\|_{L_\infty(0,T;\dot B^s_{p,1}(\mathbb{R}^n_+))} + \|V^\ell_t, \nabla^2_z V^\ell\|_{L_1(0,T;\dot B^s_{p,1}(\mathbb{R}^n_+))}\\ &\quad\lesssim \|Z^*_\ell f^\ell\|_{L_1(0,T;\dot B^s_{p,1}(\mathbb{R}^n_+))} + \|(\Delta_x-\Delta_z)V^\ell)\|_{L_1(0,T;\dot B^s_{p,1}(\mathbb{R}^n_+))} + \|Z^*_\ell(U^\ell|_{t=0})\|_{\dot B^s_{p,1}(\mathbb{R}^n_+)}. \end{aligned}$$

Note that the first and last terms in the right-hand side may be dealt with thanks to Lemma 6.1: we have

$$\begin{aligned} \|Z^*_\ell f^\ell\|_{L_1(0,T;\dot B^s_{p,1}(\mathbb{R}^n_+))} &\lesssim \|f^\ell\|_{L_1(0,T;\dot B^s_{p,1}(\Omega))}\\ \|Z^*_\ell(U^\ell|_{t=0})\|_{\dot B^s_{p,1}(\mathbb{R}^n_+)} &\lesssim \|U^\ell|_{t=0}\|_{\dot B^s_{p,1}(\Omega)}. \end{aligned}$$

Compared to the first step, the only definitely new term is $(\Delta_x - \Delta_z)V^\ell$. Explicit computations (see, e.g., [5]) show that $(\Delta_z - \Delta_x)V^\ell$ is a linear combination of components of $\nabla^2_z A_\ell \otimes V^\ell$ and $\nabla_z A_\ell \otimes \nabla_z V^\ell$. Therefore

$$\begin{aligned} \|(\Delta_x-\Delta_z)V^\ell)\|_{L_1(0,T;\dot B^s_{p,1}(\mathbb{R}^n_+))} \lesssim &\|A_\ell\otimes\nabla^2_z V^\ell\|_{L_1(0,T;\dot B^s_{p,1}(\mathbb{R}^n_+))}\\ &+\|\nabla_z A_\ell\otimes\nabla_z V^\ell\|_{L_1(0,T;\dot B^s_{p,1}(\mathbb{R}^n_+))}. \end{aligned}$$

Now, according to Proposition 6.4 and owing to the support properties of the terms involved in the inequalities, we have

$$\|A_\ell\otimes\nabla^2_z V^\ell\|_{L_1(0,T;\dot B^s_{p,1}(\mathbb{R}^n_+))} \lesssim \|A_\ell\|_{\dot B^{\frac nq}_{q,1}(B(x^\ell,\lambda))}\|\nabla^2_z V^\ell\|_{\dot B^s_{p,1}(\mathbb{R}^n_+)} \quad\text{with } q=\min(p,p').$$

Therefore we have, thanks to (6.30) and to (6.31),

$$\|A_\ell\otimes\nabla^2_z V^\ell\|_{L_1(0,T;\dot B^s_{p,1}(\mathbb{R}^n_+))} \lesssim \lambda\|\nabla^2_z V^\ell\|_{\dot B^s_{p,1}(\mathbb{R}^n_+)}.$$

Similarly, we have

$$\|\nabla_z A_\ell\otimes\nabla_z V^\ell\|_{L_1(0,T;\dot B^s_{p,1}(\mathbb{R}^n_+))} \lesssim \|\nabla_z V^\ell\|_{L_1(0,T;\dot B^s_{p,1}(\mathbb{R}^n_+))}.$$

Therefore

$$\|(\Delta_x-\Delta_z)V^\ell\|_{L_1(0,T;\dot B^s_{p,1}(\mathbb{R}^n_+))} \lesssim \lambda\|\nabla^2_z V^\ell\|_{L_1(0,T;\dot B^s_{p,1}(\mathbb{R}^n_+))} + \|\nabla_z V^\ell\|_{L_1(0,T;\dot B^s_{p,1}(\mathbb{R}^n_+))}.$$

Putting together the above inequalities and remembering (6.25) and Lemma 6.1, we finally get, taking λ small enough,

$$
\begin{aligned}
\|V^\ell\|_{L_\infty(0,T;\dot B^s_{p,1}(\mathbb{R}^n_+))} &+ \|V^\ell_t, \nabla^2_z V^\ell\|_{L_1(0,T;\dot B^s_{p,1}(\mathbb{R}^n_+))} \\
&\lesssim \|\eta^\ell u_0\|_{\dot B^s_{p,1}(\Omega)} + \|\eta^\ell f\|_{L_1(0,T;\dot B^s_{p,1}(\Omega))} \\
+\lambda^{-2}\|\widetilde\eta^\ell u\|_{L_1(0,T;\dot B^s_{p,1}(\Omega))} &+ \lambda^{-1}\|\widetilde\eta^\ell \nabla u\|_{L_1(0,T;\dot B^s_{p,1}(\Omega))} + \|\nabla V^\ell\|_{L_1(0,T;\dot B^s_{p,1}(\mathbb{R}^n_+))}.
\end{aligned}
$$

By interpolation, we have

$$
\|\nabla V^\ell\|_{L_1(0,T;\dot B^s_{p,1}(\mathbb{R}^n_+))} \lesssim \|\nabla^2 V^\ell\|^{1/2}_{L_1(0,T;\dot B^s_{p,1}(\mathbb{R}^n_+))}\|V^\ell\|^{1/2}_{L_1(0,T;\dot B^s_{p,1}(\mathbb{R}^n_+))}.
$$

Therefore using Young's inequality enables us to reduce the above inequality to

$$
\begin{aligned}
\|V^\ell\|_{L_\infty(0,T;\dot B^s_{p,1}(\mathbb{R}^n_+))} &+ \|V^\ell_t, \nabla^2 V^\ell\|_{L_1(0,T;\dot B^s_{p,1}(\mathbb{R}^n_+))} \\
&\lesssim \|\eta^\ell u_0\|_{\dot B^s_{p,1}(\Omega)} + \|\eta^\ell f\|_{L_1(0,T;\dot B^s_{p,1}(\Omega))} + \|V^\ell\|_{L_1(0,T;\dot B^s_{p,1}(\mathbb{R}^n_+))} \\
&\qquad +\lambda^{-2}\|\widetilde\eta^\ell u\|_{L_1(0,T;\dot B^s_{p,1}(\Omega))} + \lambda^{-1}\|\widetilde\eta^\ell \nabla u\|_{L_1(0,T;\dot B^s_{p,1}(\Omega))}.
\end{aligned}
$$

In order to handle the last term, there are two ways of proceeding depending on whether we want a time-dependent constant or not. The first possibility is to write that, by interpolation and Hölder's inequality,

$$
\|\widetilde\eta^\ell \nabla u\|_{L_1(0,T;\dot B^s_{p,1}(\Omega))} \le T^{1/2}\|u\|_{L_1(0,T;\dot B^{s+2}_{p,1}(K))\cap L_\infty(0,T;\dot B^s_{p,1}(K))}.
$$

This yields

$$
\begin{aligned}
\|V^\ell\|_{L_\infty(0,T;\dot B^s_{p,1}(\mathbb{R}^n_+))} &+ \|V^\ell_t, \nabla^2 V^\ell\|_{L_1(0,T;\dot B^s_{p,1}(\mathbb{R}^n_+))} \\
&\lesssim \|\eta^\ell u_0\|_{\dot B^s_{p,1}(\Omega)} + \|\eta^\ell f\|_{L_1(0,T;\dot B^s_{p,1}(\Omega))} + T\|V^\ell\|_{L_\infty(0,T;\dot B^s_{p,1}(\mathbb{R}^n_+))} \\
&\qquad + (\lambda^{-1}T^{1/2} + \lambda^{-2}T)\|u\|_{L_1(0,T;\dot B^{s+2}_{p,1}(K))\cap L_\infty(0,T;\dot B^s_{p,1}(K))}. \qquad (6.33)
\end{aligned}
$$

The second possibility is to write that

$$
\|\widetilde\eta^\ell \nabla u\|_{L_1(0,T;\dot B^s_{p,1}(\Omega))} \le \|u\|^{\frac12}_{L_1(0,T;\dot B^{s+2}_{p,1}(K))}\|u\|^{\frac12}_{L_1(0,T;\dot B^s_{p,1}(K))}.
$$

We eventually get

$$
\begin{aligned}
\|V^\ell\|_{L_\infty(0,T;\dot B^s_{p,1}(\mathbb{R}^n_+))} &+ \|V^\ell_t, \nabla^2 V^\ell\|_{L_1(0,T;\dot B^s_{p,1}(\mathbb{R}^n_+))} \lesssim \|\eta^\ell u_0\|_{\dot B^s_{p,1}(\Omega)} \\
&+ \|\eta^\ell f\|_{L_1(0,T;\dot B^s_{p,1}(\Omega))} + \lambda^{-1}\|u\|^{1/2}_{L_1(0,T;\dot B^{2+s}_{p,1}(K))}\|u\|^{1/2}_{L_1(0,T;\dot B^s_{p,1}(K))} \\
&\qquad + \lambda^{-2}\|u\|_{L_1(0,T;\dot B^s_{p,1}(K))} + \|\nabla V^\ell\|_{L_1(0,T;\dot B^s_{p,1}(\mathbb{R}^n_+))}. \qquad (6.34)
\end{aligned}
$$

Third Step: Global a priori Estimates

Now, in view of Lemma 6.1, we may write

$$\|u\|_{L_\infty(0,T;\dot{B}^s_{p,1}(\Omega))} \leq \sum_\ell \|U^\ell\|_{L_\infty(0,T;\dot{B}^s_{p,1}(\Omega))}$$
$$\lesssim \sum_{0\leq\ell\leq M} \|U^\ell\|_{L_\infty(0,T;\dot{B}^s_{p,1}(\mathbb{R}^n))} + \sum_{M<\ell\leq L} \|V^\ell\|_{L_\infty(0,T;\dot{B}^s_{p,1}(\mathbb{R}^n_+))},$$

and similar inequalities for the other terms of the l.h.s of (6.33). Of course, Proposition 6.1 ensures that

$$\|\eta^\ell u_0\|_{\dot{B}^s_{p,1}(\Omega)} \lesssim \|u_0\|_{\dot{B}^s_{p,1}(\Omega)} \quad \text{and} \quad \|\widetilde{\eta}^\ell f\|_{L_1(0,T;\dot{B}^s_{p,1}(\Omega))} \lesssim \|f\|_{L_1(0,T;\dot{B}^s_{p,1}(\Omega))}.$$

So using also (6.22) and (6.26) and assuming that T is small enough, we end up with

$$\|u\|_{L_\infty(0,T;\dot{B}^s_{p,1}(\Omega))} + \|(u_t,\nabla^2 u)\|_{L_1(0,T;\dot{B}^s_{p,1}(\Omega))} \lesssim \|u_0\|_{\dot{B}^s_{p,1}(\Omega)}$$
$$+\|f\|_{L_1(0,T;\dot{B}^s_{p,1}(\Omega))} + (\lambda^{-1}T^{1/2}+\lambda^{-2}T)\|u\|_{L_1(0,T;\dot{B}^{s+2}_{p,1}(K))\cap L_\infty(0,T;\dot{B}^s_{p,1}(K))}.$$

Hence if in addition $\lambda^{-2}T$ is small enough,

$$\|u\|_{L_\infty(0,T;\dot{B}^s_{p,1}(\Omega))} + \|u_t,\nabla^2 u\|_{L_1(0,T;\dot{B}^s_{p,1}(\Omega))} \leq C\Big(\|u_0\|_{B^s_{p,1}(\mathbb{R}^n)} + \|f\|_{L_1(0,T;\dot{B}^s_{p,1}(\Omega))}\Big).$$

Repeating the argument over the interval $[T,2T]$ and so on, we get exactly Inequality (6.16).

If we want to remove the time dependency, then it is just a matter of starting from (6.34) to (6.27) instead of (6.33) to (6.22). After a few computation and thanks to Young's inequality, we get for some constant C depending on λ :

$$\|u_t,\nabla^2 u\|_{L_1(0,T;\dot{B}^s_{p,1}(\Omega))} \leq C(\|u_0\|_{\dot{B}^s_{p,1}(\Omega)} + \|f\|_{L_1(0,T;\dot{B}^s_{p,1}(\mathbb{R}^n_+))} + \|u\|_{L_1(0,T;\dot{B}^s_{p,1}(K))}).$$

For completeness, let us say a few words about the existence, which is a rather standard issue (see, e.g., [11]). As the domain is smooth, the easiest approach is via the L_2 framework and Galerkin method. We may consider smooth approximations of data f and u_0. Then the energy method provides us with approximate solutions in Sobolev spaces H^m with large m. In particular, the above a priori estimates (6.16) may be derived for such solutions. It is then easy to pass to the limit. □

Remark 6.4. Let us emphasize that the term $\|u\|_{L_1(0,T;\dot{B}^s_{p,1}(K))}$ may be replaced by other lower-order norms such as $\|u\|_{L_1(0,T;\dot{B}^{s'}_{p,1}(K))}$ with $s'\neq s$ close to 0. In particular, s' may be put to zero, and one may use $\|u\|_{L_1(0,T;L_p(K))}$.

In order to complete the proof of Theorem 6.3, we now have to bound the last term of (6.17), namely, $\|u\|_{L_1(0,T;\dot{B}^s_{p,1}(K))}$, *independently of* T. This is the goal of the next lemma (where we keep the assumption that $\nu = 1$). We here adapt to the heat equation an approach that has been proposed for the Stokes system in [13].

Lemma 6.2. *Assume that $n \geq 3$ and that $1 < p < n/2$. Then for any $s \in (-1 + 1/p, 1/p - 2/n)$, sufficiently smooth solutions to* (6.1) *fulfill*

$$\|u\|_{L_1(0,T;\dot{B}^s_{p,1}(K))} \leq C(\|f\|_{L_1(0,T;\dot{B}^s_{p,1}(\Omega))} + \|u_0\|_{\dot{B}^s_{p,1}(\Omega)}),$$

where C is independent of T.

Proof. Thanks to the linearity of the system, one may split the solution u into two parts: the first one u_1 being the solution of the system with zero initial data and source term f and the second one u_2 the solution of the system with no source term and initial data u_0. In other words, $u = u_1 + u_2$ with u_1 and u_2 satisfying

$$\begin{array}{llll} u_{1,t} - \Delta u_1 = f \text{ in } & (0,T) \times \Omega, & u_{2,t} - \Delta u_2 = 0 \text{ in } & (0,T) \times \Omega, \\ u_1 = 0 & \text{on } (0,T) \times \partial\Omega, & u_2 = 0 & \text{on } (0,T) \times \partial\Omega, \\ u_1|_{t=0} = 0 & \text{on } \quad \Omega, & u_2|_{t=0} = u_0 & \text{on } \quad \Omega. \end{array} \tag{6.35}$$

Let us first focus on u_1. Recall that up to a constant we have (see Proposition 6.3):

$$\|u_1(t)\|_{\dot{B}^s_{p,1}(K)} = \sup \int_K u_1(t,x)\eta_0(x)\,dx, \tag{6.36}$$

where the supremum is taken over all $\eta_0 \in \dot{B}^{-s}_{p',\infty}(K)$ such that $\|\eta_0\|_{\dot{B}^{-s}_{p',\infty}(K)} = 1$. Of course, by virtue of Remark 6.2, any such function η_0 may be extended by 0 over $\mathbb{R}^n$, and its extension still has a norm of order 1. So we may assume that the supremum is taken over all

$$\eta_0 \in \dot{B}^{-s}_{p',\infty}(\mathbb{R}^n) \quad \text{with } \|\eta_0\|_{\dot{B}^{-s}_{p',\infty}(\mathbb{R}^n)} = 1 \quad \text{and} \quad \operatorname{Supp} \eta_0 \subset K. \tag{6.37}$$

Consider the solution η to the following problem:

$$\begin{array}{ll} \eta_t - \Delta\eta = 0 & \text{in } \quad (0,T) \times \Omega, \\ \eta = 0 & \text{on } (0,T) \times \partial\Omega, \\ \eta|_{t=0} = \eta_0 & \text{on } \quad \Omega. \end{array} \tag{6.38}$$

Testing the equation for u_1 by $\eta(t - \cdot)$ we discover that

$$\int_\Omega u_1(t,x)\eta_0(x)\,dx = \int_0^t \int_\Omega f(\tau,x)\eta(t-\tau,x)\,dx\,d\tau. \tag{6.39}$$

The general theory for the heat operator in exterior domains implies the following estimates:

$$\|\eta(t)\|_{L_a(\Omega)} \leq C\|\eta_0\|_{L_b(\Omega)} t^{-\frac{n}{2}(\frac{1}{b}-\frac{1}{a})} \quad \text{for} \quad 1 < b \leq a < \infty, \tag{6.40}$$

as well as

$$\|\Delta\eta(t)\|_{L_a(\Omega)} \leq C\|\Delta\eta_0\|_{L_b(\Omega)} t^{-\frac{n}{2}(\frac{1}{b}-\frac{1}{a})} \quad \text{for} \quad 1<b\leq a<\infty. \tag{6.41}$$

In the case $\Omega=\mathbb{R}^n$, those two inequalities may be derived easily from the (explicit) heat kernel. To prove (6.40) in the case of an exterior domain, it is enough to look at solutions to (6.38) as subsolutions to the problem in the whole space. More precisely, if we assume that $\eta_0 \geq 0$ (this is not restrictive for one may consider the positive and negative part of the initial data separately), we get a solution to (6.38) defined over $(0,\infty)\times\Omega$ such that $\eta \geq 0$. Then we consider an extension $E\eta : \mathbb{R}^n \to \mathbb{R}$ of η, such that $E\eta=\eta$ for $x\in\Omega$ and $E\eta=0$ for $x\notin\Omega$. We claim that $E\eta$ is a subsolution to the Cauchy problem

$$\bar\eta_t - \Delta\bar\eta = 0 \text{ in } (0,T)\times\mathbb{R}^n \quad \text{with} \quad \bar\eta|_{t=0} = E\eta_0. \tag{6.42}$$

It is sufficient to show that $\eta \leq \bar\eta$, since $\bar\eta$ is always nonnegative. It is clear that

$$(\eta-\bar\eta)_t - \Delta(\eta-\bar\eta) = 0 \quad \text{in} \quad (0,T)\times\Omega. \tag{6.43}$$

Consider $(\eta-\bar\eta)_+ := \max\{\eta-\bar\eta,0\}$. It is obvious that $(\eta-\bar\eta)_+$ vanishes at the boundary, because η is zero and $\bar\eta$ is nonnegative there. Hence we conclude

$$\frac{1}{2}\frac{d}{dt}\int_\Omega (\eta-\bar\eta)_+^2\,dx + \int_\Omega |\nabla(\eta-\bar\eta)_+|^2 dx = 0. \tag{6.44}$$

Thus, $(\eta-\bar\eta)_+ \equiv 0$, since $(\eta-\bar\eta)_+|_{t=0}=0$. So η is bounded by $\bar\eta$.

To prove (6.41) we observe that for the smooth solutions the equation implies that $\Delta\eta|_{\partial\Omega}=0$, so we can consider the problem on $\Delta\eta$ instead of η. Now, as η vanishes at the boundary, we have (see, e.g., [8])

$$\|\nabla^2\eta\|_{L_c(\Omega)} \leq \|\Delta\eta\|_{L_c(\Omega)} \quad \text{for all} \quad 1<c<\infty. \tag{6.45}$$

Hence, interpolating between (6.40) and (6.41) yields for $0<s<1/a$:

$$\|\eta(t)\|_{\dot B^s_{a,r}(\Omega)} \leq C\|\eta_0\|_{\dot B^s_{b,r}(\Omega)} t^{-\frac{n}{2}(\frac{1}{b}-\frac{1}{a})} \quad \text{for} \quad 1<b\leq a<\infty \text{ and } 1\leq r\leq\infty. \tag{6.46}$$

In order to extend this inequality to negative indices s, we consider the following dual problem:

$$\begin{array}{ll} \zeta_t - \Delta\zeta = 0 & \text{in} \quad (0,T)\times\Omega, \\ \zeta = 0 & \text{on} \; (0,T)\times\partial\Omega, \\ \zeta|_{t=0} = \zeta_0 & \text{on} \qquad \Omega, \end{array} \tag{6.47}$$

where $\zeta_0 \in B^{-s}_{b',r'}(\Omega)$.

Now, testing (6.47) by $\eta(t-\cdot)$ yields

$$\int_\Omega \eta(t,x)\zeta_0(x)\,dx = \int_\Omega \eta_0(x)\zeta(t,x)\,dx. \tag{6.48}$$

Let us observe that

$$\|\eta(t)\|_{\dot{B}^s_{a,r}(\Omega)} = \sup_{\zeta_0} \int_\Omega \eta(t,x)\zeta_0(x)\,dx, \tag{6.49}$$

where the supremum is taken over all $\zeta_0 \in \dot{B}^{-s}_{a',r'}(\Omega)$ such that $\|\zeta_0\|_{\dot{B}^{-s}_{a',r'}(\Omega)} = 1$. Thus by virtue of (6.48), we get

$$\|\eta(t)\|_{\dot{B}^s_{a,r}(\Omega)} = \sup_{\zeta_0} \int_\Omega \eta_0(x)\zeta(t,x)\,dx \le \sup_{\zeta_0}\Big(\|\eta_0\|_{\dot{B}^s_{b,r}(\Omega)}\|\zeta(t)\|_{\dot{B}^{-s}_{b',r'}(\Omega)}\Big). \tag{6.50}$$

Since $-s$ is positive and $1 < a' \le b' < \infty$, we can apply (6.46) and get, if $0 < -s < 1/b'$,

$$\|\eta(t)\|_{\dot{B}^s_{a,r}(\Omega)} \le C\|\eta_0\|_{\dot{B}^s_{b,r}(\Omega)} t^{-\frac{n}{2}(\frac{1}{a'}-\frac{1}{b'})} \sup_{\zeta_0}\|\zeta_0\|_{\dot{B}^{-s}_{a',r'}(\Omega)}.$$

Since $\frac{1}{a'} - \frac{1}{b'} = \frac{1}{b} - \frac{1}{a}$, we conclude that

$$\|\eta(t)\|_{\dot{B}^s_{a,r}(\Omega)} \le C\|\eta_0\|_{\dot{B}^s_{b,r}(\Omega)} t^{-\frac{n}{2}(\frac{1}{b}-\frac{1}{a})} \quad \text{if } s > -1+1/b. \tag{6.51}$$

In order to get the remaining case $s = 0$, it suffices to argue by interpolation between (6.46) and (6.51). One can thus conclude that for all $1 < b \le a < \infty$, $q \in [1,\infty]$, and $-1+1/b < s < 1/a$, we have

$$\|\eta(t)\|_{\dot{B}^s_{a,r}(\Omega)} \le C\|\eta_0\|_{\dot{B}^s_{b,r}(\Omega)} t^{-\frac{n}{2}(\frac{1}{b}-\frac{1}{a})}. \tag{6.52}$$

Now we return to the initial problem of bounding u_1. Starting from (6.39) and using duality, one may write

$$\left|\int_\Omega u_1(t,x)\eta_0(x)\,dx\right| \lesssim \int_0^t \|f(\tau)\|_{\dot{B}^s_{p,1}(\Omega)}\|\eta(t-\tau)\|_{\dot{B}^{-s}_{p',\infty}(\Omega)}\,d\tau.$$

Hence splitting the interval $(0,t)$ into $(0,\max(0,t-1))$ and $(\max(0,t-1),t)$ and applying (6.52) yield, for any $\varepsilon \in (0,1+s)$,

$$\begin{aligned}\left|\int_\Omega u_1(t,x)\eta_0(x)\,dx\right| \lesssim &\int_{\max(0,t-1)}^t \|f(\tau)\|_{\dot{B}^s_{p,1}(\Omega)}\|\eta_0\|_{\dot{B}^{-s}_{p',\infty}(\Omega)}\,d\tau \\ &+ \int_0^{\max(0,t-1)} \|f(\tau)\|_{\dot{B}^s_{p,1}(\Omega)}\|\eta_0\|_{\dot{B}^{-s}_{\frac{1}{1-\varepsilon},\infty}(\Omega)}(t-\tau)^{-\frac{n}{2}(\frac{1}{p}-\varepsilon)}\,d\tau.\end{aligned}$$

Now, as η_0 is supported in K, one has $\|\eta_0\|_{\dot{B}^{-s}_{a,\infty}(\Omega)} \le C|K|^{\frac{1}{p}+\frac{1}{a}-1}\|\eta_0\|_{\dot{B}^{-s}_{p',\infty}(\Omega)}$. This may easily proved by introducing a suitable smooth cutoff function with value 1 over K and taking advantage of Proposition 6.1. A scaling argument yields the dependency of the norm of the embedding with respect to $|K|$. Hence we have for some constant C depending on K:

$$\|\eta_0\|_{\dot B^{-s}_{\frac{1}{1-\varepsilon},\infty}} \le C\|\eta_0\|_{\dot B^{-s}_{p',\infty}(\Omega)}.$$

So, keeping in mind (6.39) and the fact that the supremum is taken over all the functions η_0 satisfying (6.37), we deduce that

$$\|u_1(t)\|_{\dot B^s_{p,1}(K)} \le C\bigg(\int_{\max(0,t-1)}^t \|f(\tau)\|_{\dot B^s_{p,1}(\Omega)}\,d\tau + \int_0^{\max(0,t-1)} (t-\tau)^{-\frac{n}{2}(\frac{1}{p}-\varepsilon)}\|f(\tau)\|_{\dot B^s_{p,1}(\Omega)}\,d\tau\bigg).$$

Therefore,

$$\int_1^T \|u_1\|_{\dot B^s_{p,1}(K)}\,dt \le C\bigg(1+\int_1^T \tau^{-\frac{n}{2}(\frac{1}{p}-\varepsilon)}\,d\tau\bigg)\int_0^T \|f\|_{\dot B^s_{p,1}(\Omega)}\,dt. \tag{6.53}$$

For the time interval $[0,1]$, we merely have

$$\int_0^1 \|u_1\|_{\dot B^s_{p,1}(K)}\,dt \le C\int_0^1 \|f\|_{\dot B^s_{p,1}(\Omega)}\,dt.$$

Now, provided that one may find some $\varepsilon > 0$ such that

$$\frac{n}{2}\Big(\frac{1}{p}-\varepsilon\Big) > 1, \tag{6.54}$$

a condition which is equivalent to $p < n/2$, the constant in (6.53) may be made independent of T. Hence we conclude that

$$\int_0^T \|u_1\|_{\dot B^s_{p,1}(K)}\,dt \le C\int_0^T \|f\|_{\dot B^s_{p,1}(\Omega)}\,dt \tag{6.55}$$

with C independent of T.

Let us now bound u_2. We first write that

$$\|u_2(t)\|_{\dot B^s_{p,1}(K)} \le C\|u_0\|_{\dot B^s_{p,1}(\Omega)} \tag{6.56}$$

and, if $-1+\varepsilon < s < 1/p$,

$$\|u_2(t)\|_{\dot B^s_{p,1}(K)} \le C|K|^{\frac{1}{p}-\varepsilon}\|u_2(t)\|_{\dot B^s_{\frac{1}{\varepsilon},1}(K)} \le C|K|^{\frac{1}{p}-\varepsilon}\|u_0\|_{\dot B^s_{p,1}(\Omega)}t^{-\frac{n}{2}(\frac{1}{p}-\varepsilon)}.$$

Then decomposing the integral over $[0,T]$ into an integral over $[0,\min(1,T)]$ and $[\min(1,T),T]$, we easily get

$$\int_0^T \|u_2(t)\|_{\dot B^s_{p,1}(K)}\,dt \le C\bigg(1+\int_{\min(1,T)}^T t^{-\frac{n}{2}(\frac{1}{p}-\varepsilon)}\,dt\bigg)\|u_0\|_{\dot B^s_{p,1}(\Omega)}. \tag{6.57}$$

The integrant in the r.h.s. of (6.57) is finite whenever (6.54) holds. Hence,

$$\int_0^T \|u_2(t)\|_{\dot{B}^s_{p,1}(K)}\,dt \le C\|u_0\|_{\dot{B}^s_{p,1}(\Omega)}. \tag{6.58}$$

Putting this together with (6.53) and (6.3.2) completes the proof of the lemma.□

We are now ready to prove Theorem 6.3. Granted with Theorem 6.4, it is enough to show that $\|u\|_{L_1(0,T;\dot{B}^s_{p,1}(K)\cap\dot{B}^{s'}_{q,1}(K))}$ may be bounded by the right-hand side of (6.15).

As a matter of fact $\|u\|_{L_1(0,T;\dot{B}^{s'}_{q,1}(K))}$ may be directly bounded from Lemma 6.2, and the same holds for $\|u\|_{L_1(0,T;\dot{B}^s_{p,1}(K))}$ if $p < n/2$.

If $p \ge n/2$, then we use the fact that

$$\dot{B}^{s+2}_{q,1}(\Omega) \subset \dot{B}^s_{q^*,1}(\Omega) \quad \text{with} \quad \frac{1}{q^*} = \frac{1}{q} - \frac{2}{n}.$$

Therefore, if $q < n/2 \le p < q^*$, then one may combine interpolation and Lemma 6.2 so as to absorb $\|u\|_{L_1(0,T;\dot{B}^s_{p,1}(K))}$ by the left-hand side of (6.15), changing the constant C if necessary.

If $p \ge q^*$, then one may repeat the argument again and again until all possible values of p in $(n/2,\infty)$ are exhausted. Theorem 6.3 is proved.

6.3.3 The Bounded Domain Case

We end this section with a few remarks concerning the case where Ω is a bounded domain of $\mathbb{R}^n$ with $n \ge 2$. Then the proof of Theorem 6.4 is similar: we still have to introduce some suitable resolution of unity $(\eta^\ell)_{0\le\ell\le L}$. The only difference is that, now, η^0 has compact support. Hence Theorem 6.4 holds true with $K = \overline{\Omega}$.

In order to remove the time dependency in the estimates, we use the fact (see, e.g., [7]) that the solution η to (6.38) satisfies for some $c > 0$:

$$\|\eta(t)\|_{L_p(\Omega)} \le Ce^{-ct}\|\eta_0\|_{L_p(\Omega)},$$

which also implies that

$$\|\nabla^2\eta(t)\|_{L_p(\Omega)} \le Ce^{-ct}\|\nabla^2\eta_0\|_{L_p(\Omega)}.$$

Hence we have for any $1 < p < \infty$ and $-1+1/p < s < 1/p$,

$$\|\eta(t)\|_{\dot{B}^s_{p,1}(\Omega)} \le Ce^{-ct}\|\eta_0\|_{\dot{B}^s_{p,1}(\Omega)}. \tag{6.59}$$

Defining u_1 and u_2 as in (6.35), one may thus write

$$\left|\int_\Omega u_1(t,x)\eta_0(x)\,dx\right| \lesssim \|\eta_0\|_{\dot B^{-s}_{p',\infty}(\Omega)} \int_0^t \|f(\tau)\|_{\dot B^s_{p,1}(\Omega)} e^{-c(t-\tau)}\,d\tau,$$

thus giving

$$\|u_1\|_{L_1(0,T;\dot B^s_{p,1}(K))} \lesssim \|f\|_{L_1(0,T;\dot B^s_{p,1}(\Omega))}.$$

Of course, we also have

$$\|u_2\|_{L_1(0,T;\dot B^s_{p,1}(K))} \lesssim \|u_0\|_{\dot B^s_{p,1}(\Omega)}.$$

So one may conclude that Lemma 6.2 holds true for *any* $1<p<\infty$ and $-1+1/p< s<1/p$. Consequently, we get:

Theorem 6.5. *If* $1<p<\infty$ *and* $-1+1/p<s<1/p$, *then the statement of Theorem 6.2 remains true in any smooth bounded domain.*

6.4 Applications

In this last section, we give some application of the maximal regularity estimates that have been proved hitherto. As an example, we prove global stability results (in a critical functional framework) for trivial/constant solutions to the following system:

$$\begin{aligned} &\mathbf{u}_t-\nu\Delta\mathbf{u}+P\cdot\nabla^2\mathbf{u}=f_0(\mathbf{u})+f_1(\mathbf{u})\cdot\nabla\mathbf{u} && \text{in} \quad (0,T)\times\Omega,\\ &\mathbf{u}=0 && \text{at} \quad (0,T)\times\partial\Omega,\\ &\mathbf{u}|_{t=0}=u_0 && \text{on} \quad \Omega. \end{aligned} \tag{6.60}$$

Above, ν is a positive parameter, $\mathbf{u}$ stands for a r-dimensional vector, and $P=(P_1,\cdots,P_r)$ where the P_k's are $n\times n$ matrices with suitably smooth coefficients. The nonlinearities $f_0:\mathbb{R}^r\to\mathbb{R}^r$ and $f_1:\mathbb{R}^r\to\mathcal{M}_{r,n}(\mathbb{R})$ are C^1 and satisfy

$$f_0(0)=0, \quad df_0(0)=0 \text{ and } f_1(0)=0, \tag{6.61}$$

together with some growths conditions that will be detailed below.

As we have in mind applications to Theorem 6.3, we focus on the case where Ω is a smooth exterior domain of $\mathbb{R}^n$ with $n\geq 3$. Of course, based on our other maximal regularity results, similar (and somewhat easier) statements may be proved for bounded domains, $\mathbb{R}^n_+$ or $\mathbb{R}^n$.

Here are two important examples entering in the class of (6.60). The first one is *the nonlinear heat transfer equation* (see [15] and the references therein)

$$u_t-\nu\Delta u=f(u). \tag{6.62}$$

A classical form of nonlinearity is $f(u) = Ku^2(u-u^*)$. However one may consider more complex models describing a flame propagation like in [12].

The second example is the *viscous Burgers equation* [9, 10]

$$u_t + u\partial_{x_1} u - \nu\Delta u = 0. \tag{6.63}$$

which enters in the class of models like

$$\mathbf{u}_t - \nu\Delta\mathbf{u} = B(\mathbf{u}, \nabla\mathbf{u}). \tag{6.64}$$

In the case where $B(\mathbf{u},\nabla\mathbf{u}) = -\mathbf{u}\cdot\nabla\mathbf{u}$, this is just the equation for pressureless viscous gases with constant density.

Below, based on Theorem 6.3, we shall prove two global-in-time results concerning the stability of the trivial solution of System (6.60). In the first statement, to simplify the presentation, we only consider the case where the data belong to spaces with regularity index equals to 0. To simplify the notation, we omit the dependency with respect to the domain Ω in all that follows.

Theorem 6.6. *Let $1 < q < n/2$ and Ω be an exterior domain of $\mathbb{R}^n$ $(n \geq 3)$. There exist two positive constants η and c_ν such that for all $P : [0,\infty)\times\Omega \to \mathbb{R}^n\times\mathbb{R}^n\times\mathbb{R}^k$ satisfying*[4]

$$\|P\|_{L_\infty(0,\infty;\mathscr{M}(\dot{B}^0_{n,1}\cap\dot{B}^0_{q,1}))} \leq \eta\nu, \tag{6.65}$$

for all nonlinearities f_0 and f_1 fulfilling (6.61) *and*

$$|df_0(\mathbf{w})| \leq C|\mathbf{w}|, \qquad |df_1(\mathbf{w})| \leq C, \tag{6.66}$$

and for all $\mathbf{u}_0 \in \dot{B}^0_{n,1}\cap\dot{B}^0_{q,1}$ such that

$$\|\mathbf{u}_0\|_{\dot{B}^0_{n,1}\cap\dot{B}^0_{q,1}} \leq c_\nu, \tag{6.67}$$

System (6.60) *admits a unique global solution $\mathbf{u}$ in the space*

$$\mathscr{C}_b(0,\infty;\dot{B}^0_{n,1}\cap\dot{B}^0_{q,1})\cap L_1(0,\infty;\dot{B}^2_{n,1}\cap\dot{B}^2_{q,1}). \tag{6.68}$$

Proof. Granted with Theorem 6.3, the result mainly relies on embedding, composition, and product estimates in Besov spaces. We focus on the proof of a priori estimates for a global solution $\mathbf{u}$ to (6.60). First, applying Theorem 6.3 yields

$$\begin{aligned}\|\mathbf{u}\|_{L_\infty(0,\infty;\dot{B}^0_{n,1}\cap\dot{B}^0_{q,1})\cap L_1(0,\infty;\dot{B}^2_{n,1}\cap\dot{B}^2_{q,1})} &\lesssim \|P\|_{L_\infty(0,\infty;\mathscr{M}(\dot{B}^0_{n,1}\cap\dot{B}^0_{q,1}))}\|\mathbf{u}\|_{L_1(0,\infty;\dot{B}^2_{n,1}\cap\dot{B}^2_{q,1})}\\ &+ \|\mathbf{u}_0\|_{\dot{B}^0_{n,1}\cap\dot{B}^0_{q,1}} + \|f_0(\mathbf{u})\|_{L_1(0,\infty;\dot{B}^0_{n,1}\cap\dot{B}^0_{q,1})} + \|f_1(\mathbf{u})\cdot\nabla\mathbf{u}\|_{L_1(0,\infty;\dot{B}^0_{n,1}\cap\dot{B}^0_{q,1})}.\end{aligned} \tag{6.69}$$

[4] Below $\mathscr{M}(X)$ denotes the *multiplier space* associated to the Banach space X that is the set of those functions f such that $fg \in X$, whenever g is in X, endowed with the norm $\|f\|_{\mathscr{M}(X)} := \inf_g \|fg\|_X$ where the infimum is taken over all $g \in X$ with norm 1.

Bounding the last two terms follows from Propositions 6.1 and 6.6. More precisely, for $p = q, n$, we have

$$\begin{aligned}\|f_1(\mathbf{u})\cdot\nabla\mathbf{u}\|_{\dot{B}^0_{p,1}} &\lesssim \|f_1(\mathbf{u})\|_{\dot{B}^{1/2}_{n,1}}\|\nabla\mathbf{u}\|_{\dot{B}^{1/2}_{p,1}}\\ &\lesssim \|\mathbf{u}\|_{\dot{B}^{1/2}_{n,1}}\|\nabla\mathbf{u}\|_{\dot{B}^{1/2}_{p,1}}.\end{aligned}$$

Therefore, applying Hölder inequality,

$$\|f_1(\mathbf{u})\cdot\nabla\mathbf{u}\|_{L_1(0,\infty;\dot{B}^0_{p,1})} \lesssim \|\mathbf{u}\|_{L_4(0,\infty;\dot{B}^{1/2}_{n,1})}\|\nabla\mathbf{u}\|_{L_{4/3}(0,\infty;\dot{B}^{1/2}_{p,1})},$$

whence, using elementary interpolation,

$$\|f_1(\mathbf{u})\cdot\nabla\mathbf{u}\|_{L_1(0,\infty;\dot{B}^0_{n,1}\cap\dot{B}^0_{q,1})} \lesssim \|\mathbf{u}\|^2_{L_\infty(0,\infty;\dot{B}^0_{n,1}\cap\dot{B}^0_{q,1})\cap L_1(0,\infty;\dot{B}^2_{n,1}\cap\dot{B}^2_{q,1})}. \tag{6.70}$$

Bounding $f_0(\mathbf{u})$ is slightly more involved. To handle the norm in $L_1(0,\infty;\dot{B}^0_{n,1}(\Omega))$, we use the following critical embedding:

$$\dot{B}^1_{n/2,1} \hookrightarrow \dot{B}^{0+}_{n^-,1} \hookrightarrow \dot{B}^0_{n,1}.$$

Hence Proposition 6.1 enables us to write that

$$\begin{aligned}\|f_0(\mathbf{u})\|_{\dot{B}^0_{n,1}} &\lesssim \|f_0(\mathbf{u})\|_{\dot{B}^{0+}_{n^-,1}},\\ &\lesssim \|\mathbf{u}\|_{L_\infty}\|\mathbf{u}\|_{\dot{B}^{0+}_{n^-,1}},\\ &\lesssim \|\mathbf{u}\|_{L_\infty}\|\mathbf{u}\|_{\dot{B}^1_{n/2,1}},\\ &\lesssim \|\mathbf{u}\|_{\dot{B}^1_{n,1}}\|\mathbf{u}\|_{\dot{B}^1_{q,1}\cap\dot{B}^1_{n,1}}.\end{aligned}$$

The last inequality stems from the embedding $\dot{B}^1_{n,1} \hookrightarrow L_\infty$ and from the fact that $q < n/2 < n$, whence

$$\dot{B}^1_{q,1}\cap\dot{B}^1_{n,1} \hookrightarrow \dot{B}^1_{n/2,1}.$$

Therefore, using Hölder inequality and elementary interpolation, we deduce that

$$\|f_0(\mathbf{u})\|_{L_1(0,\infty;\dot{B}^0_{n,1})} \lesssim \|\mathbf{u}\|^2_{L_\infty(0,\infty;\dot{B}^0_{n,1}\cap\dot{B}^0_{q,1})\cap L_1(0,\infty;\dot{B}^2_{n,1}\cap\dot{B}^2_{q,1})}. \tag{6.71}$$

Finally we have to bound $f_0(\mathbf{u})$ in $L_1(0,\infty;\dot{B}^0_{q,1})$. For that, it suffices to estimate it in $L_1(0,\infty;\dot{B}^{0+}_{q,1})$ and in $L_1(0,\infty;L_{q^-})$. Indeed we observe that $L_{q^-} \hookrightarrow \dot{B}^{0-}_{q,\infty}$, and thus,

$$L_1(0,\infty;\dot{B}^{0+}_{q,1})\cap L_1(0,\infty;L_{q^-}) \hookrightarrow L_1(0,\infty;\dot{B}^0_{q,1}). \tag{6.72}$$

Now, on the one hand, according to Proposition 6.6 and Hölder inequality, we have

$$\|f_0(\mathbf{u})\|_{L_1(0,\infty;\dot{B}^{0+}_{q,1})} \lesssim \|\mathbf{u}\|_{L_{1^+}(0,\infty;L_\infty)}\|\mathbf{u}\|_{L_{\infty^-}(0,\infty;\dot{B}^{0+}_{q,1})}.$$

By interpolation, we easily get

$$\|\mathbf{u}\|_{L_{\infty-}(0,\infty;\dot{B}^{0+}_{q,1})} \lesssim \|\mathbf{u}\|_{L_\infty(0,\infty;\dot{B}^0_{q,1})\cap L_1(0,\infty;\dot{B}^2_{q,1})}$$

and because $q < n/2$,

$$\begin{aligned}\|\mathbf{u}\|_{L_{1+}(0,\infty;L_\infty)} &\lesssim \|\mathbf{u}\|_{L_1(0,\infty;\dot{B}^0_{\infty,1})\cap L_2(0,\infty;\dot{B}^0_{\infty,1})},\\ &\lesssim \|\mathbf{u}\|_{L_1(0,\infty;\dot{B}^2_{n/2,1})\cap L_2(0,\infty;\dot{B}^1_{n,1})},\\ &\lesssim \|\mathbf{u}\|_{L_1(0,\infty;\dot{B}^2_{q,1}\cap\dot{B}^2_{n,1})\cap L_1(0,\infty;\dot{B}^2_{n,1})\cap L_\infty(0,\infty;\dot{B}^0_{n,1})}.\end{aligned}$$

Therefore we have

$$\|f_0(\mathbf{u})\|_{L_1(0,\infty;\dot{B}^{0+}_{q,1})} \lesssim \|\mathbf{u}\|^2_{L_1(0,\infty;\dot{B}^2_{q,1}\cap\dot{B}^2_{n,1})\cap L_\infty(0,\infty;\dot{B}^0_{q,1}\cap\dot{B}^0_{n,1})}. \tag{6.73}$$

On the other hand, using the fact that $|f_0(\mathbf{u})| \le C|\mathbf{u}|^2$ and Hölder inequality, we may write

$$\|f_0(\mathbf{u})\|_{L_1(0,\infty;L_{q-})} \le \|\mathbf{u}\|_{L_\infty(0,\infty;L_q)}\|\mathbf{u}\|_{L_1(0,\infty;L_{\infty-})}.$$

We obviously have $\dot{B}^0_{q,1} \hookrightarrow L_q$ and, because $2 - n/q < 0$,

$$\dot{B}^2_{q,1} \cap \dot{B}^2_{n,1} \hookrightarrow L_{\infty-}.$$

Therefore

$$\|f_0(\mathbf{u})\|_{L_1(0,\infty;L_{q-})} \lesssim \|\mathbf{u}\|_{L_\infty(0,\infty;\dot{B}^0_{q,1})}\|\mathbf{u}\|_{L_1(\dot{B}^2_{q,1}\cap\dot{B}^2_{n,1})}. \tag{6.74}$$

So putting (6.73) and (6.74) together and taking advantage of (6.72), we end up with

$$\|f_0(\mathbf{u})\|_{L_1(0,\infty;\dot{B}^0_{q,1})} \lesssim \|\mathbf{u}\|^2_{L_\infty(0,\infty;\dot{B}^0_{n,1}\cap\dot{B}^0_{q,1})\cap L_1(0,\infty;\dot{B}^2_{n,1}\cap\dot{B}^2_{q,1})}. \tag{6.75}$$

It is now time to plug (6.70), (6.71), and (6.75) in (6.69). We get

$$\begin{aligned}&\|\mathbf{u}\|_{L_\infty(0,\infty;\dot{B}^0_{n,1}\cap\dot{B}^0_{q,1})\cap L_1(0,\infty;\dot{B}^2_{n,1}\cap\dot{B}^2_{q,1})}\\ &\qquad\le C\big(\|P\|_{L_\infty(0,\infty;\mathscr{M}(\dot{B}^0_{n,1}\cap\dot{B}^0_{q,1}))}\|\mathbf{u}\|_{L_1(0,\infty;\dot{B}^2_{n,1}\cap\dot{B}^2_{q,1})}\\ &\qquad\quad+\|\mathbf{u}\|^2_{L_\infty(0,\infty;\dot{B}^0_{n,1}\cap\dot{B}^0_{q,1})\cap L_1(0,\infty;\dot{B}^2_{n,1}\cap\dot{B}^2_{q,1})}+\|\mathbf{u}_0\|_{\dot{B}^0_{n,1}\cap\dot{B}^0_{q,1}}\big).\end{aligned} \tag{6.76}$$

Obviously, the above estimate enables us to get a *global-in-time* control of the solution in the desired functional space whenever (6.65) and (6.67) are satisfied. Starting from this observation and using the existence part of Theorem 6.3, it is easy to prove Theorem 6.6 by means of Banach fixed-point theorem as in [3] for instance. The details are left to the reader. □

Theorem 6.7. *Assume that $P \equiv 0$ and that $f_1 \equiv 0$. Suppose that f_0 satisfies* (6.61) *and*

$$|df_0(\mathbf{w})| \leq C(|\mathbf{w}|^{m-1} + |\mathbf{w}|) \text{ for some } m \geq 2.$$

Let $1 < q < \frac{n}{2}$ and $q \leq p < \infty$. Assume that

$$s_p := \frac{n}{p} - \frac{2}{m-1} \in \left(0, \frac{1}{p}\right) \quad \text{and} \quad 0 < s_q < \frac{1}{q} - \frac{2}{n}.$$

Then there exists a constant c_v such that if

$$\|\mathbf{u}_0\|_{\dot{B}^{s_p}_{p,1} \cap \dot{B}^{s_q}_{q,1}} \leq c_v \tag{6.77}$$

then System (6.60) *admits a unique global-in-time solution $\mathbf{u}$ such that*

$$\mathbf{u} \in \mathscr{C}_b(0,\infty; \dot{B}^{s_p}_{p,1} \cap \dot{B}^{s_q}_{q,1}) \cap L_1(0,\infty; \dot{B}^{2+s_p}_{p,1} \cap \dot{B}^{2+s_q}_{q,1}). \tag{6.78}$$

Proof. Once again, we start from Theorem 6.3 which implies the following inequality:

$$\|\mathbf{u}\|_{L_\infty(0,\infty;\dot{B}^{s_p}_{p,1}\cap\dot{B}^{s_q}_{q,1})\cap L_1(0,\infty;\dot{B}^{2+s_p}_{p,1}\cap\dot{B}^{2+s_q}_{q,1})} \lesssim \|f_0(\mathbf{u})\|_{L_1(0,\infty;\dot{B}^{s_p}_{p,1}\cap\dot{B}^{s_q}_{q,1})} + \|\mathbf{u}_0\|_{\dot{B}^{s_p}_{p,1}\cap\dot{B}^{s_q}_{q,1}}. \tag{6.79}$$

Now (a slight generalization of) Proposition 6.6 ensures that for $s = s_p, s_q$ and for $r = p, q$,

$$\|f_0(\mathbf{u})\|_{\dot{B}^s_{r,1}} \lesssim \left(\|\mathbf{u}\|_{L_\infty} + \|\mathbf{u}\|^{m-1}_{L_\infty}\right)\|\mathbf{u}\|_{\dot{B}^s_{r,1}}.$$

Therefore,

$$\begin{aligned}\|f_0(\mathbf{u})\|_{L_1(0,\infty;\dot{B}^{s_p}_{p,1}\cap\dot{B}^{s_q}_{q,1}} \leq C\big(&\|\mathbf{u}\|_{L_1(0,\infty;L_\infty)} \\ &+ \|\mathbf{u}\|^{m-1}_{L_{m-1}(0,\infty;L_\infty)}\big)\|\mathbf{u}\|_{L_\infty(0,\infty;\dot{B}^{s_p}_{p,1}\cap\dot{B}^{s_q}_{q,1})}.\end{aligned} \tag{6.80}$$

Hence it is only a matter of proving that the norm of $\mathbf{u}$ in $L_1(0,\infty;L_\infty)$ and in $L_{m-1}(0,\infty;L_\infty)$ may be bounded by means of the norm in $L_1(0,\infty;\dot{B}^2_{q,1} \cap \dot{B}^2_{n,1}) \cap L_\infty(0,\infty;\dot{B}^0_{q,1} \cap \dot{B}^0_{n,1})$. Now, we notice that $\dot{B}^{s_p+2/(m-1)}_{p,1}$ embeds continuously in L_∞ and that, by interpolation,

$$\|\mathbf{u}\|_{L_{m-1}(0,\infty;\dot{B}^{s_p+2/(m-1)}_{p,1})} \leq \|\mathbf{u}\|_{L_\infty(0,\infty;\dot{B}^{s_p}_{p,1})\cap L_1(0,\infty;\dot{B}^{s_p+2}_{p,1})}.$$

Hence we do have

$$\|\mathbf{u}\|_{L_{m-1}(0,\infty;L_\infty)} \lesssim \|\mathbf{u}\|_{L_\infty(0,\infty;\dot{B}^{s_p}_{p,1})\cap L_1(0,\infty;\dot{B}^{s_p+2}_{p,1})}. \tag{6.81}$$

Finally, we notice that $\dot{B}^2_{s_q+2,1} \hookrightarrow \dot{B}^{2+s_q-n/q}_{\infty,1}$ and that $2+s_q-n/q<0$. At the same time $\dot{B}^{s_p+2}_{p,1} \hookrightarrow \dot{B}^1_{\infty,1}$; therefore,

$$\dot{B}^{s_q+2}_{q,1} \cap \dot{B}^{s_p+2}_{p,1} \hookrightarrow L_\infty.$$

Hence we have

$$\|\mathbf{u}\|_{L_1(0,\infty;L_\infty)} \lesssim \|\mathbf{u}\|_{L_1(0,\infty;\dot{B}^{s_p+2}_{p,1}) \cap L_1(0,\infty;\dot{B}^{s_q+2}_{q,1})}. \tag{6.82}$$

Putting (6.81) and (6.82) into (6.80) and then into (6.79) we get

$$\|\mathbf{u}\|_{L_\infty(0,\infty;\dot{B}^{s_p}_{p,1}\cap\dot{B}^{s_q}_{q,1})\cap L_1(0,\infty;\dot{B}^{2+s_p}_{p,1}\cap\dot{B}^{2+s_q}_{q,1})} \lesssim \|\mathbf{u}_0\|_{\dot{B}^{s_p}_{p,1}\cap\dot{B}^{s_q}_{q,1}}$$
$$+(\|\mathbf{u}\|^{m-2}_{L_\infty(0,\infty;\dot{B}^{s_p}_{p,1}\cap\dot{B}^{s_q}_{q,1})\cap L_1(0,\infty;\dot{B}^{2+s_p}_{p,1}\cap\dot{B}^{2+s_q}_{q,1})}+1)\|\mathbf{u}\|^2_{L_\infty(0,\infty;\dot{B}^{s_p}_{p,1}\cap\dot{B}^{s_q}_{q,1})\cap L_1(0,\infty;\dot{B}^{2+s_p}_{p,1}\cap\dot{B}^{2+s_q}_{q,1})}.$$

The smallness of the initial data in (6.69) enables to close the estimate for the left-hand side of the above inequality. The existence issue is just a consequence of Banach fixed-point theorem. This completes the proof of the theorem. □

Remark 6.5. Even though System (6.60) does not have any scaling invariance in general, our two statements are somewhat critical from the regularity point of view. Indeed, in the functional framework used in Theorem 6.6 and under the growth condition (6.66), the nonlinearity $f_0(\mathbf{u})$ is lower order compared to $f_1(\mathbf{u})\cdot\nabla\mathbf{u}$. Now, we notice that if $f_0\equiv 0$ and $P\equiv 0$, then the initial value problem for System (6.60) (in the $\mathbb{R}^n$ case) is invariant for all $\lambda>0$ under the transform:

$$(u(t,x),u_0(x)) \longrightarrow \lambda(u(\lambda^2 t,\lambda x),u_0(\lambda x)).$$

At the same time, the norm $\|\cdot\|_{\dot{B}^0_{n,1}(\mathbb{R}^n)}$ is invariant by the above rescaling for u_0.

As regards Theorem 6.7, the nonlinearity $f_0(\mathbf{w})$ is at most of order m. Now, if (the coefficients of) $f_0(\mathbf{w})$ are homogeneous polynomials of degree m, then the system is invariant by

$$(u(t,x),u_0(x)) \longrightarrow \lambda^{\frac{2}{m-1}}(u(\lambda^2 t,\lambda x),u_0(\lambda x)).$$

Hence the regularity $\dot{B}^{s_p}_{p,1}$ is critical.

Acknowledgements The second author has been supported by the MN grant IdP2011 000661.

References

1. Bahouri, H., Chemin, J.-Y., Danchin, R.: In: Fourier Analysis and Nonlinear Partial Differential Equations. Grundlehren der mathematischen Wissenschaften, vol. 343. Springer, Berlin (2011)
2. Bennett, C., Sharpley, R.: Interpolation of operators. Pure and Applied Mathematics, vol. 129. Academic, Boston (1988)

3. Danchin, R., Mucha, P.: A critical functional framework for the inhomogeneous Navier–Stokes equations in the half-space. J. Funct. Anal. **256**(3), 881–927 (2009)
4. Danchin, R., Mucha, P.B.: A Lagrangian approach for solving the incompressible Navier–Stokes equations with variable density. Comm. Pure Appl. Math. **65**, 1458–1480 (2012)
5. Danchin, R., Mucha, P.: Critical functional framework and maximal regularity in action on systems of incompressible flows. in progress.
6. Denk, R., Hieber, M., Prüss, J.: *R*-boundedness, Fourier multipliers and problems of elliptic and parabolic type. Mem. Am. Math. Soc. **166** (2003)
7. Friedman, A.: Partial Differential Equations. Holt, Rinehart and Winston, Inc., New York-Montreal, Que.-London (1969)
8. Gilbarg, D., Trudinger, N.: Elliptic Partial Differential Equations of Second Order, 2nd edn. Grundlehren der Mathematischen Wissenschaften, vol. 224. Springer, Berlin (1983)
9. Hill, A.T., Süli, E.: Dynamics of a nonlinear convection-diffusion equation in multidimensional bounded domains. Proc. Roy. Soc. Edinburgh Sect. A **125**(2), 439–448 (1995)
10. Horsin, Th.: Local exact Lagrangian controllability of the Burgers viscous equation. Ann. Inst. H. Poincaré Anal. Non Linéaire **25**(2), 219–230 (2008)
11. Ladyzhenskaja, O., Solonnikov, V., Uraltseva, N.: In: Linear and quasilinear equations of parabolic type. Translations of Mathematical Monographs, vol. 23. American Mathematical Society, Providence (1967)
12. Lewicka, M., Mucha, P.B.: On the existence of traveling waves in the 3D Boussinesq system. Comm. Math. Phys. **292**(2), 417–429 (2009)
13. Maremonti, P., Solonnikov, V.A.: On nonstationary Stokes problem in exterior domains. Ann. Scuola Norm. Sup. Pisa Cl. Sci. (4) **24**(3), 395–449 (1997)
14. Triebel, H.: Interpolation theory, function spaces, differential operators. North-Holland Mathematical Library, vol. 18. North-Holland Publishing Co., Amsterdam (1978)
15. Xin, J.: Front propagation in heterogeneous media. SIAM Rev. **42**(2), 161–230 (2000)

Chapter 7
Cauchy Problem for Some 2×2 Hyperbolic Systems of Pseudo-differential Equations with Nondiagonalisable Principal Part

Todor Gramchev and Michael Ruzhansky

Abstract We investigate the well-posedness of the Cauchy problem for some first-order linear hyperbolic systems having nondiagonalisable principal parts.

Key words: Cauchy problem, Hyperbolic systems, Pseudo-differential equations

2010 Mathematics Subject Classification: Primary: 35L50; Secondary: 47G30, 35S30.

7.1 Introduction

We study the regularity and representation of solutions of first-order weakly hyperbolic pseudo-differential systems

$$D_t u = A(t,x,D_x)u + B(t,x,D_x)u + f(t,x), \quad u(0,x) = u^0(x), \tag{7.1}$$

$t \in [0,T]$, $x \in \mathbb{R}^n$, where $A(t,x,D_x) = \{A_{jk}(t,x,\xi)\}_{j,k=1}^2$ (respectively, $B(t,x,D_x) = \{b_{jk}(t,x,D_x)\}_{j,k=1}^2$) is a 2×2 matrix of first- (respectively, zero)order classical pseudo-differential operators in x depending smoothly on $t \in \mathbb{R}$. Here we use the traditional notation $D_x = i^{-1}\partial_x$.

The main novelty of the present work is the fact that we consider weakly hyperbolic systems which are not diagonalisable.

T. Gramchev (✉)
Dipartimento di Matematica e Informatica, Università di Cagliari,
via Ospedale 72, I-09124 Cagliari, Italy
e-mail: todor@unica.it

M. Ruzhansky
Mathematics Department, Imperial College London, Huxley Building,
180 Queen's Gate, London SW7 2AZ, UK
e-mail: m.ruzhansky@imperial.ac.uk

M. Cicognani et al. (eds.), *Studies in Phase Space Analysis with Applications to PDEs*, Progress in Nonlinear Differential Equations and Their Applications 84,
DOI 10.1007/978-1-4614-6348-1_7, © Springer Science+Business Media New York 2013

Our main goal is to apply techniques involving Fourier integral operators as in Kamotski and Ruzhansky [11, 12] combined with a generalisation of the reduction to normal forms of systems of pseudo-differential operators due to Taylor [20, 21], for nondiagonalisable hyperbolic systems of pseudo-differential operators. We note that relatively few results are available in the literature for hyperbolic systems in the presence of nontrivial Jordan blocks. Petkov [15] has studied the propagation of microlocal singularities of solutions of constant multiplicity hyperbolic systems with Jordan blocks, using the results of Arnold [1] on normal forms of matrices depending on parameters. As it concerns the well-posedness of Cauchy problem for hyperbolic systems admitting nontrivial Jordan blocks, we refer to Vaillant [22] for the complete description of constant coefficients hyperbolic systems of differential equations with constant multiplicity by means of subtle linear algebra arguments; see also the works [10, 13, 14, 19, 23]. We note that [22] also treats nonconstant coefficients provided the multiplicties are ≤ 5. Jordan blocks of dimension 2 appear in important models in mathematical physics (cf. Craig [4], see also the survey of Gramchev [8] on normal forms and dynamical systems problems in the presence of Jordan blocks and references therein). Here we can give an example of the system

$$D_t \begin{pmatrix} u_1 \\ u_2 \end{pmatrix} = \begin{pmatrix} 0 & 1 \\ 0 & 0 \end{pmatrix} D_x \begin{pmatrix} u_1 \\ u_2 \end{pmatrix} + \begin{pmatrix} 0 & 0 \\ 1 & 0 \end{pmatrix} D_x \begin{pmatrix} u_1 \\ u_2 \end{pmatrix},$$

which is equivalent to the ill-posed equation

$$D_t^2 u = D_x u.$$

Recently, Gramchev and Orrú [9] have studied $1D$ hyperbolic systems of differential equations with time-dependent coefficients admitting Jordan block structures. We also point out to a series of interesting results on well-posedness of Cauchy problems for weakly hyperbolic systems of differential equations with diagonalisable principal parts or under hypotheses of suitable behaviour of the characteristic roots near the points of multiplicity; see, e.g. [2, 5, 6] and the references therein.

In the case of pseudo-differential operators with variable coefficients, the new phenomena may appear. In particular, certain geometry of the set of multiplicities may enter into consideration. Systems with diagonal principal parts have been analysed by Kumano-go [13] who proved the well-posedness in L^2 and in C^∞ and studied the propagation of singularities using the representation of solution in terms of the Fourier integral operators with multiphases. If the (smooth) characteristics intersect transversally, such analysis can be refined, and it was used by Rozenblum [16] to derive the Weyl asymptotics for the corresponding elliptic systems. In [11, 12], Kamotski and the second author removed the transversality condition, obtaining also the L^p estimates for the systems with microlocally diagonalisable principal part.

The main novelties of our work can be summarised as follows:

- We reduce the problem to a system with the principal symbol in an upper triangular normal form.
- We show the local existence and uniqueness theorem in anisotropic Sobolev spaces for the normal form system provided the order of $b_{21}(t,x,D)$ is ≤ -1.

- We derive sufficient conditions for a local existence–uniqueness result for the Cauchy problem allowing symbols depending on t and x.

The paper is organised as follows: Sect. 7.2 deals with the reduction to an upper triangular normal form and the well-posedness in the C^∞ category and in anisotropic vector-valued Sobolev spaces provided the order of $\tilde{b}_{21}$ is not greater than -1. We derive the complete diagonalisation and the corresponding regularity estimates in Sect. 7.3. In Sect. 7.4 we investigate the Cauchy problem of the normal form system in the general case of symbols depending on t and x.

Throughout this paper, when talking about mth order symbols, we mean classical symbols from $S^m_{1,0}$.

The authors thank the anonymous referee for useful remarks leading to improvements of the paper.

7.2 Reduction and Well-Posedness in Anisotropic Sobolev Spaces

The first result is the reduction of the principal symbol to an upper triangular form in the case of the existence of one smooth eigenvector.

Theorem 7.1. *Suppose that the matrix symbol $A(t,x,\xi)$ admits smooth real eigenvalues $\lambda_j(t,x,\xi)$, $j=1,2$ and a smooth eigenvector $h(t,x,\xi) \in (S^0([0,T] \times \Omega \times \mathbb{R}^n))^2$ satisfying*

$$h_j(t,x,\xi) \neq 0, \qquad (t,x) \in [0,T] \times \Omega, |\xi| \gg 1, \tag{7.2}$$

for $j=1$ or $j=2$. Then we can find a 2×2 matrix-valued symbol $T(t,x,\xi)$, invertible for $|\xi| \gg 1$, such that the transformation

$$u = T(t,x,D_x)v \tag{7.3}$$

transforms the original system (7.1) *into*

$$D_t v = \tilde{A}(t,x,D_x)v + \tilde{B}(t,x,D_x)v + \tilde{f}, \qquad v(0,x) = v^0(x), \tag{7.4}$$

where $\tilde{A}(t,x,\xi) \in S^1$ is in the upper triangular form

$$\tilde{A}(t,x,\xi) = \begin{pmatrix} \lambda_1(t,x,\xi) & \tilde{A}_{12}(t,x,\xi) \\ 0 & \lambda_2(t,x,\xi) \end{pmatrix}, \tag{7.5}$$

and $\tilde{B}(t,x,\xi)$ is a zero-order symbol.

Proof. Without loss of generality we suppose that (7.2) holds for $j=1$. Hence $\begin{pmatrix} 1 \\ \omega(t,x,\xi) \end{pmatrix}$, $\omega = h_2/h_1$, is an eigenvector of A associated to the smooth eigenvalue $\lambda_1(t,x,\xi)$. We set

$$T(t,x,\xi) = \begin{pmatrix} 1 & 0 \\ \omega(t,x,\xi) & 1 \end{pmatrix}. \tag{7.6}$$

Straightforward calculations yield that the matrix T reduces A to an upper triangular form

$$(T(t,x,\xi))^{-1}A(t,x,\xi)T(t,x,\xi) = \begin{pmatrix} \lambda_1(t,x,\xi) & \tilde{A}_{12}(t,x,\xi) \\ 0 & \lambda_2(t,x,\xi) \end{pmatrix}, \tag{7.7}$$

for $(t,x) \in [0,T] \times \Omega$, $|\xi| \gg 1$, where the second diagonal entry must be λ_2 as an eigenvalue of A. Next, plugging (7.3) into (7.1) and multiplying (7.1) with the zero-order matrix-valued pseudo-differential operator $T^{-1}(t,x,D_x)$, we obtain, following the approach in [20, 21] and using the calculus, the desired form (7.4)–(7.5).

Next, we propose an assertion which illustrates for hyperbolic pseudo-differential systems that the obstacle for the C^∞ well-posedness in the presence of nilpotent parts is the zero-order part of $\tilde{b}_{21}(t,x,D_x)$, where we write $\tilde{B}(t,x,D_x) = \{\tilde{b}_{jk}(t,x,D_x)\}_{j,k=1}^2$ for $\tilde{B}$ in (7.4). We give conditions on $\tilde{b}_{21}(t,x,D_x)$ assuring the well-posedness.

Theorem 7.2. *Suppose that*

$$\text{the p.d.o. } \tilde{b}_{21} \text{ is of order not greater than } -1. \tag{7.8}$$

Then the Cauchy problem (7.4) *is well posed in* C^∞. *Moreover, it is well posed in the anisotropic Sobolev space* $\begin{pmatrix} H^{s_1}(\mathbb{R}^n) \\ H^{s_2}(\mathbb{R}^n) \end{pmatrix}$ *provided* $s_2 - s_1 \geq 1$. *In that case the solution satisfies the following estimates:*

$$\|v_1(t,\cdot)\|_{H^s} + \|v_2(t,\cdot)\|_{H^{s+1}} \leq c\, e^{ct} \left(\|v_1^0\|_{H^s} + \|v_2^0\|_{H^{s+1}}\right), \quad 0 \leq t \leq T,$$

for $v_j^0 \in H_{comp}^{s+j-1}(\mathbb{R}^n)$, $j = 1,2$ *with* $c > 0$ *depending on* s, T *and the support of the initial data.*

Proof. Let $G_j^0\theta$ (respectively, G_jg) be the solution of

$$D_t w = \lambda_j(t,x,D_x)w + \tilde{b}_{jj}(t,x,D_x)w, \quad w(0,x) = \theta(x) \tag{7.9}$$

(respectively,

$$D_t w = \lambda_j(t,x,D_x)w + \tilde{b}_{jj}(t,x,D_x)w + g, \quad w(0,x) = 0), \tag{7.10}$$

for $j = 1,2$. We note that $G_j^0\theta$ (respectively, G_jg) is represented by the Fourier integral operator

$$G_j^0\theta(t,x) = \int_{\mathbb{R}^n} e^{i\varphi_j(t,x,\xi)} a_j(t,x,\xi)\widehat{\theta}(\xi)d\xi \tag{7.11}$$

(respectively,

$$G_jg(t,x) = \int_0^t \int_{\mathbb{R}^n} e^{i\varphi_j(t,t_1,x,\xi)} A_j(t,t_1,x,\xi)\widehat{g}(t_1,\xi)d\xi dt_1), \tag{7.12}$$

with $\varphi_j(t,t_1,x,\xi)$ satisfying the eikonal equation

$$\partial_t \varphi_j = \lambda_j(t,x,\nabla_x \varphi_j), \qquad \varphi_j|_{t=t_1} = x\cdot\xi, \tag{7.13}$$

and with

$$\varphi_j(t,x,\xi) = \varphi_j(t,0,x,\xi), \tag{7.14}$$

whereas the amplitudes $A_{j,-k}(t,x,\xi)$ of order $-k$, $k\in\mathbb{N}$, with $A_j \sim \sum_{k=0}^{\infty} A_{j,-k}$, satisfy the transport equations with initial data at $t=t_1$, and we have $a_j(t,x,\xi) = A_j(t,0,x,\xi)$.

We rewrite the Cauchy problem for the normal form system (7.4) as follows:

$$v_1(t,x) = V_1^0(t,x) + G_1(\tilde{A}_{12}(t,x,D)v_2) + G_1(\tilde{b}_{12}(t,x,D)v_2), \tag{7.15}$$
$$v_2(t,x) = V_2^0(t,x) + G_2(\tilde{b}_{21}(t,x,D)v_1), \tag{7.16}$$

where

$$V_j^0 = G_j^0 v_j^0 + G_j(\tilde{f}_j), \quad j=1,2. \tag{7.17}$$

As in [9], we substitute v_2 from (7.16) in the right-hand side of (7.15). The result is

$$\begin{aligned} v_1(t,x) &= V^0(t,x) + G_1(\tilde{A}_{12}(t,x,D)G_2(\tilde{b}_{21}(t,x,D)v_1)) \\ &\quad + G_1(\tilde{b}_{12}(t,x,D)G_2(\tilde{b}_{21}(t,x,D)v_1)), \end{aligned} \tag{7.18}$$

with

$$V^0(t,x) = V_1^0 + G_1(\tilde{A}_{12}(t,x,D)V_2^0) + G_1(\tilde{b}_{12}(t,x,D)V_2^0). \tag{7.19}$$

The essential novelty in comparison to the case of diagonalisable weakly hyperbolic systems (e.g. see Kamotski, Ruzhansky [11, 12]) is the presence of the nilpotent part term $\tilde{A}_{12}(t,x,D)$. Using the composition rules for Fourier integral operators and the fact that $\tilde{A}_{12}(t,x,D)$ is of order 1, under the assumption (7.8) on the order of $\tilde{b}_{21}$, we obtain that

$$G_1(\tilde{A}_{12}(t,x,D)G_2(\tilde{b}_{21}(t,x,D)v_1))$$

acts continuously in L^2 and H^s and the arguments used, e.g. in [11, 12] hold.

7.3 Cauchy Problem for 2 × 2 Hyperbolic Pseudo-differential Systems

We consider

$$D_t u = A(t,x,D_x)u + B(t,x,D_x)u + f(t,x), \quad u(0,x) = u^0(x), \tag{7.20}$$

$t \in \mathbb{R}, x \in \mathbb{R}^n$, where

$$A(t,x,D_x) = \begin{pmatrix} \lambda_1(t,x,D_x) & \varkappa(t,x,D_x) \\ 0 & \lambda_2(t,x,D_x) \end{pmatrix}, \tag{7.21}$$

with $\lambda_j(t,x,\xi)$, $j = 1,2$, being real-valued symbols belonging to $S^1_{1,0}$, smoothly depending on t, whereas $\varkappa(t,x,\xi)$ is a complex-valued first-order symbol smoothly depending on t. The entries $B_{jk}(t,x,\xi)$, $j,k = 1,2$, of $B(t,x,\xi)$ are zero-order symbols depending smoothly on t.

Proposition 7.1. *In each sufficiently narrow frequency cone, there exists an invertible Fourier integral operator $J(t)$, smoothly depending on t, such that setting $u_j(t) = J(t)v_j(t)$, $j = 1,2$, the Cauchy problem* (7.20) *is microlocally reduced to*

$$D_t v = \tilde{A}(t,x,D_x)v + \tilde{B}(t,x,D_x)v + Rv + \tilde{f}(t,x), \quad v(0,x) = v^0(x), \tag{7.22}$$

where

$$\tilde{A}(t,x,D_x) = \begin{pmatrix} \lambda(t,x,D_x) & \mathcal{N}(t,x,D_x) \\ 0 & 0 \end{pmatrix}, \tag{7.23}$$

$$\tilde{B}(t,x,D_x) = \begin{pmatrix} b_{11}(t,x,D_x) & b_{12}(t,x,D_x) \\ b_{21}(t,x,D) & 0 \end{pmatrix}, \tag{7.24}$$

$\tilde{A}(t,x,D_x)$ and $\tilde{B}(t,x,D_x)$ are the first- and the zero-, resp., order matrix pseudo-differential operators and R is a smoothing pseudo-differential operator, smoothly depending on t. More precisely, we have

$$\begin{aligned} \lambda(t,x,D_x) &= J(t)^{-1}(\lambda_1(t,x,D_x) - \lambda_2(t,x,D_x))J(t), \\ \mathcal{N}(t,x,D_x) &= J(t)^{-1}\varkappa(t,x,D_x)J(t), \\ b_{11}(t,x,D_x) &= J(t)^{-1}(B_{11}(t,x,D_x) - B_{22}(t,x,D_x))J(t), \\ b_{12}(t,x,D_x) &= J(t)^{-1}B_{12}(t,x,D_x)J(t), \\ b_{21}(t,x,D_x) &= J(t)^{-1}B_{21}(t,x,D_x)J(t). \end{aligned}$$

We can also achieve $b_{11} \equiv 0$ in (7.23)–(7.24) *if we define*

$$\lambda(t,x,D_x) = J(t)^{-1}(\lambda_1(t,x,D_x) - \lambda_2(t,x,D_x) + B_{11}(t,x,D_x) - B_{22}(t,x,D_x))J(t).$$

We omit the proof as rather standard.

In view of Proposition 7.1 we will be now interested in the problem

$$D_t u = \tilde{A}(t,x,D_x)u + \tilde{B}(t,x,D_x)u + f(t,x), \quad u(0,x) = u^0(x), \tag{7.25}$$

with

$$\tilde{A}(t,x,D_x) = \begin{pmatrix} \lambda(t,x,D_x) & \mathcal{N}(t,x,D_x) \\ 0 & 0 \end{pmatrix}, \tag{7.26}$$

$$\tilde{B}(t,x,D_x) = \begin{pmatrix} b_{11}(t,x,D_x) & b_{12}(t,x,D_x) \\ b_{21}(t,x,D) & 0 \end{pmatrix}. \tag{7.27}$$

By Proposition 7.1 the results will cover the Cauchy problem (7.20) and (7.21) if we assume that the symbols of A and B are supported in a sufficiently narrow cones in the ξ-variable. In general, by Proposition 7.1 we have

$$\lambda(t,x,D_x) = \left(J(t)^{-1} \circ (\lambda_1 - \lambda_2) \circ J(t)\right)(t,x,D_x). \tag{7.28}$$

We set

$$L = D_t - \lambda(t,x,D_x) - b_{11}(t,x,D_x). \tag{7.29}$$

Following the well-known constructions of Fourier integral operators as parametrices (see, e.g. [7]), we define the Fourier integral operator

$$G_0 w_0(t,x) = \int_{\mathbb{R}^n} e^{i\varphi(t,x,\xi)} a(t,x,\xi) \widehat{w^0}(\xi) d\xi \tag{7.30}$$

as the solution (modulo a smoothing operator) of the Cauchy problem

$$Lw = 0, \qquad w(0,x) = w^0(x). \tag{7.31}$$

Recall that $\varphi(t,x,\xi)$ satisfies the eikonal equation

$$\varphi_t = \lambda(t,x,\varphi_x), \qquad \varphi|_{t=0} = x \cdot \xi, \tag{7.32}$$

whereas the amplitudes $a_{-j}(t,x,\xi)$ are of order $-j$, $j \in \mathbb{N}$, with $a \sim \sum_{j=0}^{\infty} a_{-j}$ satisfying the so-called transport equations.

We denote by

$$Gf(t,x) = \int_0^t \int_{\mathbb{R}^n} e^{i\varphi(t,t_1,x,\xi)} A(t,t_1,x,\xi) \widehat{f}(t_1,\xi) d\xi dt_1 \tag{7.33}$$

the solution (modulo a smoothing operator) of the Cauchy problem

$$Lw = f, \qquad w(0,x) = 0, \tag{7.34}$$

with $\varphi(t,t_1,x,\xi)$ satisfying the eikonal equation

$$\varphi_t = \lambda(t,x,\varphi_x), \qquad \varphi|_{t=t_1} = x \cdot \xi, \tag{7.35}$$

whereas the amplitudes $A_{-j}(t,x,\xi)$ are of order $-j$, $j \in \mathbb{N}$, with $A \sim \sum_{j=0}^{\infty} A_{-j}$ satisfying the transport equations with initial data at $t = t_1$.

We introduce the fundamental hypotheses for $0 \le \tau \le t$. But before it we introduce an important "singular" symbol which will play a crucial role in guaranteeing the compensation for the loss of 1 derivative due to the presence of first-order symbols in the nilpotent part. Let us denote

$$\mathscr{B}(t,t_1,\tau,x,\xi) := \frac{\mathscr{N}(t_1,\varphi_\xi(t,t_1,x,\xi),\xi) b_{21}(\tau,\varphi_\xi(t,t_1,x,\xi),\xi)}{\partial_{t_1}\varphi(t,t_1,x,\xi)}. \tag{7.36}$$

We observe that the eikonal equation (7.35) for φ yields

$$\partial_{t_1}\varphi(t,t_1,x,\xi) = \\ -\lambda(t_1,x,\xi) + \int_{t_1}^{t} \nabla_\xi \lambda(s,x,\varphi_x(s,t_1,x,\xi))\partial_{t_1}\varphi_x(s,t_1,x,\xi)ds. \quad (7.37)$$

In particular, if $\lambda = \lambda(t,\xi)$ is only time-dependent symbol, we obtain

$$\partial_{t_1}\varphi(t,t_1,x,\xi) = -\lambda(t_1,\xi). \quad (7.38)$$

Remark 7.1. If the system is strictly hyperbolic, namely, if

$$|\lambda(t,x,\xi)| \geq c|\xi| \qquad (c>0)$$

for large $|\xi|$, then we also have

$$|\partial_{t_1}\varphi(t,t_1,x,\xi)| \geq \tilde{c}|\xi| \qquad (\tilde{c}>0)$$

for large $|\xi|$ and sufficiently small $|t|,|t_1| \ll 1$. Hence the symbol $\mathscr{B}(t,t_1,\tau,x,\xi)$ has no "singularities" and becomes a standard zero-order symbol.

We also define

$$\mathscr{B}_0(t,\tau,x,\xi) := \mathscr{B}(t,t,\tau,x,\xi) = -\frac{\mathscr{N}(t,x,\xi)b_{21}(\tau,x,\xi)}{\lambda(t,x,\xi)}, \quad (7.39)$$

$$\mathscr{B}_1(t,\tau,x,\xi) := \mathscr{B}(t,\tau,\tau,x,\xi) \\ = \frac{\mathscr{N}(\tau,\varphi_\xi(t,\tau,x,\xi),\xi)b_{21}(\tau,\varphi_\xi(t,\tau,x,\xi),\xi)}{\partial_\tau\varphi(t,\tau,x,\xi)}. \quad (7.40)$$

We now introduce the quantities which will be responsible for the boundedness of certain operators appearing in the representation of solutions in the proof. We define

$$K_0(t,\tau) := \sup_{|\alpha|,|\beta|\leq[n/2]+1} \sup_{x,\xi} |\partial_x^\alpha\partial_\xi^\beta \mathscr{B}_0(t,\tau,x,\xi)|, \quad (7.41)$$

$$K_1(t,\tau) := \sup_{|\alpha|,|\beta|\leq 2n+1} \sup_{x,\xi} |\partial_x^\alpha\partial_\xi^\beta \mathscr{B}_1(t,\tau,x,\xi)|, \quad (7.42)$$

$$\widetilde{K}(t,t_1,\tau) := \sup_{|\alpha|,|\beta|\leq 2n+1} \sup_{x,\xi} |\partial_x^\alpha\partial_\xi^\beta \partial_{t_1}\mathscr{B}(t,t_1,\tau,x,\xi)|. \quad (7.43)$$

We observe that the numbers of derivatives needed in assumptions above are different. Indeed, the amplitude (7.41) appears in the pseudo-differential operator, while amplitudes (7.42) and (7.43) appear as amplitudes of Fourier integral operators, so different regularity suffices; see [17].

Since we are interested in the local well-posedness of (7.25)–(7.27), we may assume that the symbols of operators $\tilde{A}$ and $\tilde{B}$ are compactly supported in x. This means that the suprema in x in (7.41)–(7.43) is taken over the support of $\mathscr{B}_0, \mathscr{B}_1$

and $\mathscr{B}$. If the symbols do not depend on x, then this difference of local and global disappears. Moreover, since the appearing operators for small ξ are locally smoothing in x, the suprema in (7.41)–(7.43) with respect to ξ can be taken over $|\xi| \geq \delta > 0$ for some $\delta > 0$. We also set

$$K_2(t,\tau) := \int_\tau^t \widetilde{K}(t,t_1,\tau)dt_1. \tag{7.44}$$

We will make different types of assumptions in Theorem 7.3. On one hand, we may assume that

$$K_0(t,\tau), K_1(t,\tau), K_2(t,\tau) \in C([0,T]; L^1[0,T]) \tag{7.45}$$

and

$$\limsup_{t\to 0} \int_0^t |K_j(t,\tau)| d\tau \ll 1, \quad j = 0,1,2. \tag{7.46}$$

On the other hand, we may assume that

$$\sup_{0\leq\tau\leq t\leq T} |K_j(t,\tau)| \leq C, \qquad j = 0,1,2. \tag{7.47}$$

We now formulate the local well-posedness result. However, we note that the numbers and particular types of derivatives that we take in (7.41)–(7.43) may be taken differently depending on what boundedness result we will be using in the proof. We refer to Remark 7.3 for further details.

Theorem 7.3. *Under either of the hypotheses* (7.45), (7.46) *or* (7.47), *the Cauchy problem* (7.25)–(7.27) *is locally well posed in* L^2.

Proof. The second equation for v_2 in (7.25) is equivalent (modulo a smoothing operator) to

$$v_2(t,x) = v_2^0 + D_t^{-1}(f_2(t,x) + b_{21}(t,x,D_x)v_1), \qquad D_t^{-1} := \int_0^t. \tag{7.48}$$

We plug (7.48) into the first equation for v_1 in (7.25) and obtain

$$v_1(t,x) = F_1(t,x) + Mv_1 + Qv_1, \tag{7.49}$$

where

$$\begin{aligned} F_1(t,x) &= G^0 v_1^0 + Gf_1 + G\circ\mathscr{N}(t,x,D_x)v_2^0 + G\circ\mathscr{N}(t,x,D_x)\circ D_t^{-1}f_2 \\ &\quad + G\circ b_{12}v_2^0 + G\circ b_{12}\circ D_t^{-1}f_2, & (7.50) \\ Mv_1 &= G\circ\mathscr{N}(t,x,D_x)\circ D_t^{-1}b_{21}v_1, & (7.51) \\ Qv_1 &= G\circ b_{12}\circ D_t^{-1}\circ b_{21}v_1. & (7.52) \end{aligned}$$

The operator Q is of zero order and is essentially as in Kamotski–Ruzhansky [12], hence L^2 bounded for sufficiently small $0 \leq t \ll 1$. The main novelty and difficulty appears in the study of the operator M.

Remark 7.2. We observe that if b_{21} is a pseudo-differential operator of order -1, then the order of the amplitude of the (Fourier integral) operator M is zero. Consequently, it can be treated in the same way as Q, completing the whole argument. This is similar to the situation in Theorem 7.2.

Since b_{21} is, in general, a pseudo-differential operator of zero order, we analyse the operator M in detail. We have

$$\begin{aligned} Mv_1(t,x) &= \int_0^t \int_{\mathbb{R}^n_\xi} \int_{\mathbb{R}^n_y} e^{i(\varphi(t,t_1,x,\xi)-y\cdot\xi)} A(t,t_1,x,\xi) \\ &\quad \times \mathcal{N}(t_1,y,D_y) \circ D_{t_1}^{-1} \circ b_{21} v_1(t_1,y) dt_1 dy d\xi \\ &= \int_0^t \int_{\mathbb{R}^n_\xi} \int_{\mathbb{R}^n_y} e^{i(\varphi(t,t_1,x,\xi)-y\cdot\xi)} A(t,t_1,x,\xi) \\ &\quad \times \int_0^{t_1} \mathcal{N}(t_1,y,D_y) \circ b_{21}(\tau,y,D_y) v_1(\tau,y) dy d\tau d\xi dt_1. \end{aligned} \tag{7.53}$$

In view of the theorems on compositions of pseudo-differential operators and Fourier integral operators (e.g. see [7]), we obtain that, modulo a smoothing operator, Mv_1 can be represented as

$$Mv_1(t,x) = M_1 v_1 + M_0 v_1, \tag{7.54}$$

where

$$M_j v_1 = \int_0^t \int_0^{t_1} \int_{\mathbb{R}^n_\xi} \mathrm{e}^{i\varphi(t,t_1,x,\xi)} \mathcal{M}_j(t,t_1,\tau,x,\xi) \widehat{v_1}(\tau,\xi) d\xi dt_1 d\tau, \; j=0,1, \tag{7.55}$$

with $\mathcal{M}_1$ being the first-order symbol

$$\mathcal{M}_1(t,t_1,\tau,x,\xi) = \mathcal{N}(t_1,\varphi_\xi(t,t_1,x,\xi),\xi) b_{21}(\tau,\varphi_\xi(t,t_1,x,\xi),\xi), \tag{7.56}$$

while $\mathcal{M}_0(t,t_1,\tau,x,\xi)$ is of order zero. Here, the first-order symbol $\mathcal{M}_1$ is defined as the principal part coming from the wave front set in the composition of a Fourier integral operators with a pseudo-differential operator, where we use that the principal part of $A(t,t_1,x,\xi)$ as the solution of the transport equations can be taken to be identically equal to one, while $\mathcal{M}_0$ is the remaining part of the right-hand side in (7.54). Indeed, the wave front set of M is given by the equality $y = \varphi_\xi$, and away from it, we can reduce the order of the amplitude by the integration by parts in ξ. We observe that it follows from (7.36) that

$$\mathcal{M}_1(t,t_1,\tau,x,\xi) = \mathcal{B}(t,t_1,\tau,x,\xi) \partial_{t_1} \varphi(t,t_1,x,\xi). \tag{7.57}$$

The zero-order operator satisfies estimates of the type

$$\|M_0 v_1(t,\cdot)\|_{L^2} \leq C \int_0^t (t-\tau) \, \|v_1(\tau,\cdot)\|_{L^2} d\tau, \tag{7.58}$$

hence no loss of derivatives occurs. We rewrite $M_1 v_1$ in the following way:

$$M_1 v_1(t,x) = \int_0^t \int_{\mathbb{R}^n_\xi} \left(\int_\tau^t e^{i\varphi(t,t_1,x,\xi)} \mathscr{M}_1(t,t_1,\tau,x,\xi) dt_1 \right) \widehat{v}_1(\tau,\xi) d\xi d\tau. \quad (7.59)$$

The most difficult part, due to the appearance of a nonzero first-order pseudo-differential operator in the nilpotent part, is to derive optimal conditions on λ, $\mathscr{N}$ and b_{21} in order to "reduce" the order by integration by parts with respect to t_1 in (7.59).

Set

$$\Theta_1(t,\tau,x,\xi) = \int_\tau^t e^{i\varphi(t,t_1,x,\xi)} \mathscr{M}_1(t,t_1,\tau,x,\xi) dt_1. \quad (7.60)$$

The (formal) identity

$$e^{i\varphi(t,t_1,x,\xi)} = -i \frac{1}{\partial_{t_1} \varphi(t,t_1,x,\xi)} \partial_{t_1} e^{i\varphi(t,t_1,x,\xi)},$$

definitions (7.36), (7.39), (7.40) and the definition of $\mathscr{M}_1(t,t_1,\tau,x,\xi)$ allow us to integrate by parts in (7.60) to use the assumptions of theorem. Indeed, the definition of φ and the identity (7.37), together with definitions (7.36), (7.39), (7.40) and the identity (7.57), lead to

$$\begin{aligned} \Theta_1(t,\tau,x,\xi) &= i e^{ix\xi} \frac{\mathscr{N}(t,x,\xi) b_{21}(\tau,x,\xi)}{\lambda(t,x,\xi)} \\ &+ i e^{i\varphi(t,\tau,x,\xi)} \frac{\mathscr{M}_1(t,\tau,\tau,x,\xi)}{\partial_\tau \varphi(t,\tau,x,\xi)} \\ &+ i \int_\tau^t e^{i\varphi(t,t_1,x,\xi)} \partial_{t_1} \left(\frac{\mathscr{M}_1(t,t_1,\tau,x,\xi)}{\partial_{t_1} \varphi(t,t_1,x,\xi)} \right) dt_1 \\ &= -i e^{ix\xi} \mathscr{B}_0(t,\tau,x,\xi) + i e^{i\varphi(t,\tau,x,\xi)} \mathscr{B}_1(t,\tau,x,\xi) \\ &+ i \int_\tau^t e^{i\varphi(t,t_1,x,\xi)} \partial_{t_1} \mathscr{B}(t,t_1,\tau,x,\xi) dt_1. \end{aligned} \quad (7.61)$$

We observe that the operators corresponding to the amplitudes in (7.61) are bounded on $L^2(\mathbb{R}^n)$. In particular, the pseudo-differential operator with symbol $\mathscr{B}_0(t,\tau,x,\xi)$ is bounded by the criterion in Cordes [3], with the operator norm given by $K_0(t,\tau)$. The L^2 boundedness of Fourier integral operators with amplitudes $\mathscr{B}_1(t,\tau,x,\xi)$ and $\partial_{t_1} \mathscr{B}(t,t_1,\tau,x,\xi)$ follows from Ruzhansky–Sugimoto [17], with the operator norms given by $K_0(t,\tau)$ and $\widetilde{K}(t,t_1,\tau)$. Here we observe that the conditions required in [17] for the phase function $\varphi(t,t_1,x,\xi)$ are automatically satisfied for small t, t_1 since φ solves the eikonal equation (7.35) with the first-order symbol λ.

Consequently, under the assumptions (7.45), (7.46), we conclude by using the smallness argument. Under the assumption (7.47), we observe the convergence of the Picard series for the solution, similarly to Kamotski–Ruzhansky [11].

Remark 7.3. The numbers and particular types of derivatives that we take in (7.41)–(7.43) may be taken differently depending on what boundedness result we will be using in the proof. For example, since the conditions on K_0 guarantee the boundedness of the pseudo-differential term in (7.61), we may use any criteria guaranteeing the boundedness of pseudo-differential operators. Thus, if we use the Calderon–Vaillancourt criterion (or rather its refined version, see, e.g. [17]) and, instead of (7.41), define

$$K_0(t,\tau) := \sup_{\alpha,\beta\in\{0,1\}^n} \sup_{x,\xi} |\partial_x^\alpha \partial_\xi^\beta \mathscr{B}_0(t,\tau,x,\xi)|,$$

the statement of Theorem 7.3 is also valid.

Moreover, although local in time, the well-posedness can be proved globally in x provided that the suprema in (7.41)–(7.43) are taken globally over $(x,\xi)\in\mathbb{R}^n\times\mathbb{R}^n$.

Remark 7.4. In the case of only time-depending coefficients, the hypotheses (7.45), (7.46) of Theorem 7.3 recapture, in particular, the conditions in [9].

We note that the proof of Theorem 7.3 also yields the well-posedness in Sobolev spaces provided we assume more conditions on the derivatives in the definitions of K_0, K_1 and K_2. Indeed, in this case, instead of using the L^2-boundedness theorems for Fourier integral operators, we would use the Sobolev boundedness results (see [17] and also [18]), while the Sobolev boundedness of pseudo-differential operators follows from the calculus. Thus, we obtain

Corollary 7.1. *Let us define*

$$\begin{aligned}
K_0(t,\tau) &:= \sup_{\alpha} \sup_{|\beta|\le[n/2]+1} \sup_{x,\xi} |\partial_x^\alpha \partial_\xi^\beta \mathscr{B}_0(t,\tau,x,\xi)|,\\
K_1(t,\tau) &:= \sup_{\alpha} \sup_{|\beta|\le 2n+1} \sup_{x,\xi} |\partial_x^\alpha \partial_\xi^\beta \mathscr{B}_1(t,\tau,x,\xi)|,\\
\widetilde{K}(t,t_1,\tau) &:= \sup_{\alpha} \sup_{|\beta|\le 2n+1} \sup_{x,\xi} |\partial_x^\alpha \partial_\xi^\beta \partial_{t_1} \mathscr{B}(t,t_1,\tau,x,\xi)|,
\end{aligned}$$

with the supremum over α taken over all multi-indices $\alpha\ge 0$. Then under either of the hypotheses (7.45), (7.46) *or* (7.47), *the Cauchy problem* (7.25)–(7.27) *is locally well posed in H^s for all $s\in\mathbb{R}$.*

We start by illustrating Theorem 7.3 in the x-independent case, obtaining generalisations for the time-dependent pseudo-differential case of some results of Gramchev and Orrú [9].

Example 7.1. Assume that $\lambda(t,x,\xi)=\lambda(t,\xi)$ is independent of x. Then the solution φ of the eikonal equation (7.35) is given by

$$\varphi(t,t_1,x,\xi) = x\cdot\xi + \int_{t_1}^{t} \lambda(s,\xi)ds. \tag{7.62}$$

Consequently, we clearly have $\varphi_\xi(t,t_1,x,\xi)=x+\int_{t_1}^t \lambda_\xi(s,\xi)ds$ and $\partial_{t_1}\varphi(t,t_1,x,\xi) = -\lambda(t_1,\xi)$, so that the definition (7.36) becomes

$$\mathscr{B}(t,t_1,\tau,x,\xi) = -\frac{\mathscr{N}(t_1,x+\int_{t_1}^t \lambda_\xi(s,\xi)ds,\xi)b_{21}(\tau,x+\int_{t_1}^t \lambda_\xi(s,\xi)ds,\xi)}{\lambda(t_1,\xi)}.$$

Thus, if (7.25)–(7.27) contain only time-dependent symbols, also by (7.39) and (7.40), we obtain

$$\mathscr{B}(t,t_1,\tau,x,\xi) = -\frac{\mathscr{N}(t_1,\xi)b_{21}(\tau,\xi)}{\lambda(t_1,\xi)}, \tag{7.63}$$

$$\mathscr{B}_0(t,\tau,x,\xi) = -\frac{\mathscr{N}(t,\xi)b_{21}(\tau,\xi)}{\lambda(t,\xi)}, \tag{7.64}$$

$$\mathscr{B}_1(t,\tau,x,\xi) = -\frac{\mathscr{N}(\tau,\xi)b_{21}(\tau,\xi)}{\lambda(\tau,\xi)}. \tag{7.65}$$

The proof of Theorem 7.3 works if we assume that $\mathscr{N}$ is dominated by λ and we obtain

Corollary 7.2. *Assume that the symbols of $\tilde{A}(t,x,D_x)$ and $\tilde{B}(t,x,D_x)$ in* (7.25)–(7.27) *are independent of x. Assume that there is $C>0$ such that*

$$|N(t,\xi)| \le C|\lambda(t,\xi)| \tag{7.66}$$

for all $t\in[0,T]$ and $\xi\in\mathbb{R}^n$. Then the Cauchy problem (7.25)–(7.27) *is locally well posed in L^2, in H^s for any $s\in\mathbb{R}$ and in C^∞.*

We point out that in view of (7.40) the presence of x in the symbols, the estimate (7.66) is not sufficient for getting the L^2 estimates as above. More subtle estimates on the phase are required in this case.

Proof. We observe that for the L^2 boundedness of pseudo-differential operators or of Fourier integral operators with the phase function (7.62), with amplitudes in (7.61), it is enough to assume the boundedness of the amplitude. But this follows from (7.63)–(7.65) and the assumption (7.66).

We now give an example of equations with variable x-dependent multiplicities.

Example 7.2. Let

$$\lambda(t,x,\xi) = r(t)\sum_{j=1}^n \omega_j x_j \xi_j, \quad r\in C(\mathbb{R};\mathbb{R}),\ \omega_j\in\mathbb{R},\ j=1,\dots,n. \tag{7.67}$$

Then the phase function $\varphi(t,t_1,x,\xi)$ is defined by

$$\begin{aligned}\varphi(t,t_1,x,\xi) &= \langle \operatorname{diag}\{e^{\omega_1(R(t)-R(t_1))},\dots,e^{\omega_n(R(t)-R(t_1))}\}x,\xi\rangle \\ &= \sum_{j=1}^n e^{\omega_j(R(t)-R(t_1))}x_j\xi_j,\end{aligned} \tag{7.68}$$

where $R(t) = \int_0^t r(s)ds$. Therefore, we have

$$\begin{aligned}\partial_{t_1}\varphi(t,t_1,x,\xi) &= -r(t_1)\langle \operatorname{diag}\{\omega_1 e^{\omega_1(R(t)-R(t_1))},\dots,\omega_n e^{\omega_n(R(t)-R(t_1))}\}x,\xi\rangle \\ &= -r(t_1)\sum_{j=1}^n \omega_j e^{\omega_j(R(t)-R(t_1))}x_j\xi_j, \qquad (7.69)\\ \varphi_\xi(t,t_1,x,\xi) &= \operatorname{diag}\{e^{\omega_1(R(t)-R(t_1))},\dots,e^{\omega_n(R(t)-R(t_1))}\}x, \qquad (7.70)\\ \varphi_x(t,t_1,x,\xi) &= \operatorname{diag}\{e^{\omega_1(R(t)-R(t_1))},\dots,e^{\omega_n(R(t)-R(t_1))}\}\xi. \qquad (7.71)\end{aligned}$$

Suppose that for some $\mu,\nu\in\{0,1\}$, $\mu+\nu=1$, we have

$$\mathscr{N}(t,x,\xi) = \mathscr{N}_{1-\mu}(t,x,\xi)(\sum_{j=1}^n \omega_j x_j\xi_j)^\mu, \qquad (7.72)$$

$$b_{21}(\tau,x,\xi) = \tilde{b}_{-\nu}(t,x,\xi)(\sum_{j=1}^n \omega_j x_j\xi_j)^\nu, \qquad (7.73)$$

where the symbol $\mathscr{N}_{1-\mu}$ (respectively, $\tilde{b}_{-\nu}(t,x,\xi)$) is of order $1-\mu$ (respectively, $-\nu$).

We plug (7.69)–(7.73) into (7.36), (7.39) and (7.40) and observe that

$$\sum_{j=1}^n \omega_j\varphi_{\xi_j}(t,t_1,x,\xi)\xi_j = \sum_{j=1}^n \omega_j e^{\omega_j(R(t)-R(t_1))}x_j\xi_j.$$

Then, after straightforward calculations we obtain that the quantities (7.36), (7.39) and (7.40) become zero-order symbols

$$\mathscr{B}(t,t_1,\tau,x,\xi) = -\frac{\mathscr{N}_{1-\mu}(t,\varphi_\xi(t,t_1,x,\xi),\xi)\tilde{b}_{-\nu}(\tau,\varphi_\xi(t,t_1,x,\xi),\xi)}{r(t_1)}, \qquad (7.74)$$

$$\mathscr{B}_0(t,\tau,x,\xi) = -\frac{\mathscr{N}_{1-\mu}(t,x,\xi)\tilde{b}_{-\nu}(\tau,x,\xi)}{r(t)}, \qquad (7.75)$$

$$\mathscr{B}_1(t,\tau,x,\xi) = -\frac{\mathscr{N}_{1-\mu}(\tau,\varphi_\xi(t,\tau,x,\xi),\xi)\tilde{b}_{-\nu}(\tau,\varphi_\xi(t,\tau,x,\xi),\xi)}{r(\tau)}. \qquad (7.76)$$

One can now observe that in the case $r(t)\neq 0$, $|t|\ll 1$, the hypotheses of Theorem 7.3 hold. In the case of $r(t)$ admitting zeros, we need to impose the same zero set and essentially the same order conditions as in the case of time-dependent coefficients. However, in this case, we also need conditions on the ∂_x and ∂_ξ derivatives, as in (7.41)–(7.43).

In conclusion, we observe that the example above also covers the case when the difference of the characteristics conjugated via the Fourier integral operator $J(t)$ in Proposition 7.1 equals the linear symbol (7.77) below. Indeed, we can generalise the example above further, to consider

$$\lambda(t,x,\xi) = r(t)\langle Ax,\xi\rangle, \quad A\in M_n(\mathbb{R}). \qquad (7.77)$$

Then the phase function $\varphi(t,t_1,x,\xi)$ is defined by

$$\varphi(t,t_1,x,\xi) = \langle e^{(R(t)-R(t_1))A}x,\xi\rangle. \tag{7.78}$$

In this case, we note that in (7.29), we can work with the operator

$$L = D_t - r(t)\langle Ax, D_x\rangle, \quad A \in M_n(\mathbb{R}) \tag{7.79}$$

with

$$\begin{aligned} L^{-1}f(t,x) &= \int_0^t f(t_1, e^{(R(t)-R(t_1))A}x)dt_1 \\ &= \int_0^t \int_{\mathbb{R}^n} e^{i\exp((R(t)-R(t_1))A)x\cdot\xi}\widehat{f}(t_1,\xi)\,\bar{d}\xi dt_1. \end{aligned} \tag{7.80}$$

7.4 Final Remarks

We outline some open problems and plans for investigations of hyperbolic systems of pseudo-differential equations with principal parts admitting Jordan block structures.

First, we plan to study the class of first-order hyperbolic equations $D_t - \lambda(t,x,D_x)$ such that the corresponding phase function in the Fourier integral operator J (parametrix for the Cauchy problem) satisfies (7.40).

Next, we point out that we are able to generalise some of the results for $m \times m$ systems, $m \geq 3$, provided the principal part is in triangular form. It is difficult to find a complete description in the general case, but we are able to derive sufficient conditions involving lower-order terms and the phase functions associated to the characteristic roots.

Another interesting and challenging issue is whether one can find some nondegeneracy conditions on lower-order term guaranteeing the reduction of the system to a diagonal normal form with nonclassical symbols of the type

$$D_t + \lambda_j(t,x,D_x) + \kappa_j(t,x,D_x) + r_j(t,,D_x), \quad j = 1,\dots,m,$$

where κ_j are pseudo-differential operators of order $\mu \in]0,1[$ with real symbols and r_j are of order 0. Such result is impossible for differential operators. We are able to extend and/or generalise assertions of Taylor [20] on complete decoupling of microlocal strictly hyperbolic systems and the smoothing effect type estimates for diagonalisable systems in Kamotski and Ruzhansky [12] for hyperbolic systems ad mitting Jordan block structures under suitable nondegereavy conditions on the terms of the nilpotent part of the principal symbols and lower-order terms. We recapture as a particular case the linear part of some particular cases of second-order weakly hyperbolic equation in one space dimension studied in [4].

It will be done in another work.

Acknowledgements This work was completed with the support of the research impulse grant of Imperial College London 2011. The first author was also supported by a research grant of the Università di Cagliari (regional funds 2010). The second author was also supported by the EPSRC Leadership Fellowship.

References

1. Arnold, V.: Matrices depending on parameters. Uspekhi Nauk U.S.S.R. **26**, 101–114 (1971)
2. Colombini, F., Nishitani, T.: Two by two strongly hyperbolic systems and Gevrey classes. In: Workshop on Partial Differential Equations (Ferrara, 1999). Ann. Univ. Ferrara Sez. VII (N.S.), vol. 45 (1999), suppl., pp. 79–108 (2000)
3. Cordes, H.O.: On compactness of commutators of multiplications and convolutions, and boundedness of pseudodifferential operators. J. Funct. Anal. **18**, 115–131 (1975)
4. Craig, W.: Nonstrictly hyperbolic nonlinear systems. Math. Ann. **277**, 213–232 (1987)
5. D'Abbicco, M., Taglialatela, G.: Some results on the well-posedness for systems with time dependent coefficients. Ann. Fac. Sci. Toulouse Math., (6) **18**, 247–284 (2009)
6. D'Ancona, P., Kinoshita, T., Spagnolo, S.: On the 2 by 2 weakly hyperbolic systems. Osaka J. Math. **45**, 921–939 (2008)
7. Duistermaat, J.J.: Fourier integral operators. Progress in Mathematics, vol. 130. Birkhäuser Boston, Inc., Boston (1996)
8. Gramchev, T.: Perturbation of systems with Nilpotent real part. In: Gaeta, G. (ed.) Editor–in–Chief R. A. Meyers, Section Perturbation Theory. Encyclopedia of Complexity and Systems Science, pp. 6649–6667. Springer, Berlin (2009)
9. Gramchev, T., Orrù, N.: Cauchy problem for a class of nondiagonalisable hyperbolic systems. Discrete and Continuous Dynamical Systems. In: Dynamical Systems, Differential Equations and Applications Proceedings of the 8th AIMS Conference at Dresden, Germany, May 25–28, 2010, vol. Supplement 2011, pp. 533–542. American Institute of Mathematical Sciences, Springfield (MO)
10. Kajitani, K.: Cauchy problem for nonstrictly hyperbolic systems in Gevrey classes. J. Math. Kyoto Univ. **23**, 599–616 (1983)
11. Kamotski, I., Ruzhansky, M.: Representation of solutions and regularity properties for weakly hyperbolic systems. In: Pseudo-Differential Operators and Related Topics. Operator Theory: Advances and Applications, vol. 164, pp. 53–63. Birkhäuser, Basel (2006)
12. Kamotski, I., Ruzhansky, M.: Regularity properties, representation of solutions, and spectral asymptotics of systems with multiplicities. Comm. Part. Differ. Equat. **32**, 1–35 (2007)
13. Kumano-go, H.: Pseudodifferential Operators. MIT, Cambridge (1981)
14. Mizohata, S.: On the Cauchy problem. Notes and Reports in Mathematics in Science and Engineering, 3. Academic Press, Inc., Orlando, FL; vi+177 Science Press, Beijing, (1985)
15. Petkov, V.: Microlocal forms for hyperbolic systems. Math. Nachr. **93**, 117–131 (1979)
16. Rozenblum, G.: Spectral asymptotic behavior of elliptic systems. (Russian) Zap. LOMI **96**, 255–271 (1980)
17. Ruzhansky, M., Sugimoto, M.: Global L^2-boundedness theorems for a class of Fourier integral operators. Comm. Part. Differ. Equat. **31**, 547–569 (2006)
18. Ruzhansky, M., Sugimoto, M.: Weighted Sobolev L2 estimates for a class of Fourier integral operators. Math. Nachr. **284**, 1715–1738 (2011)
19. Taglialatela, G., Vaillant, J.: Remarks on the Levi conditions for differential systems. Hokkaido Math. J. **37**, 463–492 (2008)
20. Taylor, M.: Reflection of singularities of solutions to systems of differential equations. Comm. Pure Appl. Math. **28**, 457–478 (1975)
21. Taylor, M.: In: Pseudodifferential Operators. Princeton Mathematical Series, vol. 34. Princeton University Press, Princeton (1981)

22. Vaillant, J.: Invariants des systèmes d'opérateurs différentiels et sommes formelles asymptotiques. Jpn. J. Math. New Ser. **25**, 1–153 (1999)
23. Yamahara, H.: Cauchy problem for hyperbolic systems in Gevrey class. A note on Gevrey indices. Ann. Fac. Sci. Toulouse Math., (6) **9**, 147–160 (2000)

Chapter 8
Scattering Problem for Quadratic Nonlinear Klein–Gordon Equation in 2D

Nakao Hayashi and Pavel I. Naumkin

Abstract We consider the scattering problem for the two-dimensional nonlinear Klein–Gordon equation with a quadratic nonlinearity. In this paper we find more natural conditions for the initial data than those of previous works to ensure the existence of scattering states.

Key words: Nonlinear Klein–Gordon equation, Quadratic nonlinearity, Scattering operator, Two space dimensions

2010 Mathematics Subject Classification: 35Q55, 35B40, 35P25, 81Q05.

8.1 Introduction

We study the scattering problem for the two-dimensional nonlinear Klein–Gordon equation with a quadratic nonlinearity:

$$\partial_t^2 v - \Delta v + v = \lambda v^2,\ (t,x) \in \mathbf{R} \times \mathbf{R}^2, \tag{8.1}$$

where $\lambda \in \mathbf{R}$. By changing $u = \left(v + i \langle i\nabla \rangle^{-1} v_t\right)/2$, we find that u satisfies the following Cauchy problem:

$$\mathscr{L}u = i\lambda \langle i\nabla \rangle^{-1} (u + \overline{u})^2 \tag{8.2}$$

N. Hayashi (✉)
Department of Mathematics, Graduate School of Science, Osaka University, Osaka, Toyonaka, 560-0043, Japan
e-mail: nhayashi@math.sci.osaka-u.ac.jp

P.I. Naumkin
Centro de Ciencias Matemáticas UNAM, Campus Morelia, AP 61-3 (Xangari), Morelia CP 58089, Michoacán, México
e-mail: pavelni@matmor.unam.mx

M. Cicognani et al. (eds.), *Studies in Phase Space Analysis with Applications to PDEs*, Progress in Nonlinear Differential Equations and Their Applications 84,
DOI 10.1007/978-1-4614-6348-1_8,

if we consider a real-valued solution v, where $\mathscr{L} = \partial_t + i\langle i\nabla\rangle$, $\langle i\nabla\rangle = \sqrt{1-\Delta}$. In what follows we study (8.2). Our aim is to find more natural requirements on the initial data

$$u_0 = \frac{1}{2}\left(v(0) + i\langle i\nabla\rangle^{-1}\partial_t v(0)\right),$$

comparing with the previous papers [4, 6, 7], where the assumption on the initial data was $u_0 \in \mathbf{H}^{m,m-1}$, $m \geq 13$ and works [2, 3], where the assumption on the initial data was $u_0 \in \mathbf{H}^{\alpha,1}$, $\alpha > 1$.

We denote the Lebesgue space by $\mathbf{L}^p$ with $1 \leq p \leq \infty$. The weighted Sobolev space is $\mathbf{H}^{m,s}_p = \{\varphi \in \mathbf{L}^p; \|\langle x\rangle^s \langle i\nabla\rangle^m \varphi\|_{\mathbf{L}^p} < \infty\}$, for $m, s \in \mathbf{R}$, $1 \leq p \leq \infty$, where $\langle x\rangle = \sqrt{1+|x|^2}$. For simplicity we write $\mathbf{H}^{m,s} = \mathbf{H}^{m,s}_2$ and $\mathbf{H}^m = \mathbf{H}^{m,0}$. We denote the Fourier transform of the function φ by

$$\mathscr{F}\varphi \equiv \hat{\varphi} = \frac{1}{2\pi}\int_{\mathbf{R}^2} e^{-ix\cdot\xi}\varphi(x)\,dx.$$

Define the function space $\mathbf{X}_\infty = \left\{\varphi \in \mathbf{C}\left([0,\infty);\mathbf{L}^2\right); \|\varphi\|_{\mathbf{X}_\infty} < \infty\right\}$, where the norm

$$\|\varphi\|_{\mathbf{X}_\infty} = \sup_{t\in[0,\infty)}\left(\sup_{2\leq q\leq\frac{4}{2-\alpha}} \langle t\rangle^{1-\frac{2}{q}}\|\varphi(t)\|_{\mathbf{H}_q^{\mu-2\left(1-\frac{2}{q}\right)}}\right)$$

with $1 < \alpha < 2, \mu \geq \alpha$.

We first state the existence result for the Cauchy problem (8.2) with the initial data u_0. Define the free Klein–Gordon evolution group $\mathscr{U}(t) = e^{-it\langle i\nabla\rangle}$.

Theorem 8.1. *Let $\alpha \in \left(1, \frac{8}{7}\right], \mu \geq \alpha$. Assume that $u_0 \in \mathbf{H}^{\mu}_{\frac{4}{2+\alpha}} \cap \mathbf{H}^\mu$, with a norm $\|u_0\|_{\mathbf{H}^{\mu}_{\frac{4}{2+\alpha}} \cap \mathbf{H}^\mu} \leq \varepsilon$. Then there exists $\varepsilon > 0$ such that the Cauchy problem (8.2) with the initial data u_0 has a unique global solution $u \in \mathbf{C}([0,\infty);\mathbf{H}^\mu)$ satisfying the estimate $\|u\|_{\mathbf{X}_\infty} \leq C\varepsilon^{\frac{2}{3}}$. Furthermore for any small $u_0 \in \mathbf{H}^{\mu}_{\frac{4}{2+\alpha}} \cap \mathbf{H}^\mu$ there exists a unique scattering state $u_+ \in \mathbf{H}^\mu$ such that*

$$\lim_{t\to\infty}\|u(t) - \mathscr{U}(t)u_+\|_{\mathbf{H}^\mu} = 0.$$

Remark 8.1. In our previous paper [3], we proved the global existence of solutions to the Cauchy problem (8.2) with the initial data $u_0 \in \mathbf{H}^{\alpha,1}$ with $\alpha > 1$.

Remark 8.2. Note that

$$\sup_{t\in[0,\infty)}\left(\|\varphi(t)\|_{\mathbf{H}^\mu} + \langle t\rangle^{\frac{\alpha}{2}}\|\varphi(t)\|_{\mathbf{L}^{\frac{4}{2-\alpha}}}\right) \leq C\|\varphi\|_{\mathbf{X}_\infty}.$$

On the other hand by the Sobolev embedding theorem $\|\varphi\|_{\mathbf{H}^\beta_q} \leq C\|\varphi\|^\theta_{\mathbf{H}^\alpha}\|\varphi\|^{1-\theta}_{\mathbf{L}^p}$ with $\frac{1}{q} = \beta + \theta\left(\frac{1}{2} - \alpha\right) + \frac{1-\theta}{p}$, $2 \leq q \leq p$, $0 \leq \beta \leq \alpha$, $\frac{\beta}{\alpha} \leq \theta \leq 1$, we have

$$\|\varphi\|_{\mathbf{H}_q^{\alpha-2\left(1-\frac{2}{q}\right)}} \leq C\|\varphi\|_{\mathbf{H}^\alpha}^{1-\frac{2}{\alpha}\left(1-\frac{2}{q}\right)} \|\varphi\|_{\mathbf{L}^{\frac{4}{2-\alpha}}}^{\frac{2}{\alpha}\left(1-\frac{2}{q}\right)}$$

for $2 \leq q \leq \frac{4}{2-\alpha}$, $1 < \alpha < 2$. From which it follows that

$$\langle t\rangle^{1-\frac{2}{q}} \|\varphi(t)\|_{\mathbf{H}_q^{\alpha-2\left(1-\frac{2}{q}\right)}} \leq C\|\varphi(t)\|_{\mathbf{H}^\alpha}^{1-\frac{2}{\alpha}\left(1-\frac{2}{q}\right)} \left(\langle t\rangle^{\frac{\alpha}{2}} \|\varphi(t)\|_{\mathbf{L}^{\frac{4}{2-\alpha}}}\right)^{\frac{2}{\alpha}\left(1-\frac{2}{q}\right)}.$$

Therefore

$$\|\varphi\|_{\mathbf{X}_\infty} \leq C \sup_{t\in[0,\infty)} \left(\|\varphi(t)\|_{\mathbf{H}^\mu} + \langle t\rangle^{\frac{\alpha}{2}} \|\varphi(t)\|_{\mathbf{L}^{\frac{4}{2-\alpha}}}\right),$$

which means that the norm

$$\sup_{t\in[0,\infty)} \left(\|\varphi(t)\|_{\mathbf{H}^\mu} + \langle t\rangle^{\frac{\alpha}{2}} \|\varphi(t)\|_{\mathbf{L}^{\frac{4}{2-\alpha}}}\right)$$

is equivalent to the norm $\|\varphi\|_{\mathbf{X}_\infty}$.

We next consider the final value problem. We introduce the function space

$$\mathbf{Z}_\infty = \left\{\varphi \in \mathbf{C}\left([0,\infty);\mathbf{L}^2\right); \|\varphi\|_{\mathbf{Z}_\infty} = \|\varphi - \mathscr{U}(\cdot)u_+\|_{\mathbf{Y}_\infty} < \infty\right\},$$

where the norm

$$\|\varphi\|_{\mathbf{Y}_\infty} = \sup_{t\in[0,\infty)} \left(\sup_{2\leq q\leq\frac{4}{2-\alpha}} \langle t\rangle^{1-\frac{2}{q}+\gamma} \|\varphi(t)\|_{\mathbf{H}_q^{\mu-2\left(1-\frac{2}{q}\right)}}\right)$$

with $\gamma = \nu^3$, $\nu = \alpha - 1 > 0, \mu \geq \alpha$.

Theorem 8.2. *Let $\alpha \in \left(1, \frac{8}{7}\right], \mu \geq \alpha$. Assume that the final value $u_+ \in \mathbf{H}^\mu_{\frac{4}{2+\alpha}} \cap \mathbf{H}^\mu$, with the norm $\|u_+\|_{\mathbf{H}^\mu_{\frac{4}{2+\alpha}} \cap \mathbf{H}^\mu} \leq \varepsilon$. Then there exists $\varepsilon > 0$ such that (8.2) has a unique global solution $u \in \mathbf{C}([0,\infty);\mathbf{H}^\mu)$ satisfying the estimate $\|u\|_{\mathbf{Z}_\infty} \leq C\varepsilon^{\frac{2}{3}}$.*

Remark 8.3. The estimate $\|u\|_{\mathbf{Z}_\infty} \leq C\varepsilon^{\frac{2}{3}}$ means that the solution of (8.2) lies in the neighborhood of a free solution in the sense

$$\|u(t) - \mathscr{U}(\cdot)u_+\|_{\mathbf{H}^\mu} + \langle t\rangle^{\frac{\alpha}{2}} \|u(t) - \mathscr{U}(\cdot)u_+\|_{\mathbf{L}^{\frac{4}{2-\alpha}}} \leq C\langle t\rangle^{-\gamma}.$$

Remark 8.4. Theorem 8.2 gives an improvement of the previous work [2].

As in the previous papers [2, 3] we split the nonlinearity in (8.2)

$$\begin{aligned}\mathscr{L}u = {} & i\lambda\left\langle i\langle t\rangle^{\nu-1}\nabla\right\rangle^{-2} \langle i\nabla\rangle^{-1}(u+\overline{u})^2 \\ & -i\lambda\langle t\rangle^{2\nu-2}\Delta\langle i\nabla\rangle^{-1}\left\langle i\langle t\rangle^{\nu-1}\nabla\right\rangle^{-2}(u+\overline{u})^2,\end{aligned} \tag{8.3}$$

where the first term has a gain of regularity and the second one has a better time decay. We apply the method of normal forms by Shatah [8] to remove the quadratic nonlinearity $i\lambda \left\langle i\langle t\rangle^{\nu-1}\nabla\right\rangle^{-2} \langle i\nabla\rangle^{-1}(u+\overline{u})^2$ from the right-hand side of (8.3). In order to do it, we define the bilinear operators

$$\mathscr{T}_j(\varphi,\psi)(x) = \int_{\mathbf{R}^2}\int_{\mathbf{R}^2} e^{ix\cdot(\xi+\eta)}\widehat{\varphi}(\xi)\,\widehat{\psi}(\eta)\frac{d\xi d\eta}{4\pi^2 S_j(\xi,\eta)}$$

for $j=1,2,3$ with the symbols

$$S_1(\xi,\eta) = \langle\xi+\eta\rangle + \langle\xi\rangle + \langle\eta\rangle, S_2(\xi,\eta) = \langle\xi+\eta\rangle - \langle\xi\rangle - \langle\eta\rangle,$$
$$S_3(\xi,\eta) = \langle\xi+\eta\rangle + \langle\xi\rangle - \langle\eta\rangle.$$

The bilinear operators $\mathscr{T}_1,\mathscr{T}_2,\mathscr{T}_3$ correspond to the nonlinear terms $\overline{u}^2, u^2, |u|^2$, respectively. Then we find from (8.3)

$$\mathscr{L}(u+\mathscr{N}_1(t,u)) = \mathscr{N}_2(t,u) + \mathscr{N}_3(t,u) + \mathscr{N}_4(t,u), \tag{8.4}$$

where

$$\begin{aligned}
\mathscr{N}_1(t,u) &= -\lambda\left\langle i\langle t\rangle^{\nu-1}\nabla\right\rangle^{-2}\langle i\nabla\rangle^{-1}(\mathscr{T}_1(\overline{u},\overline{u}) + \mathscr{T}_2(u,u) + 2\mathscr{T}_3(u,\overline{u})),\\
\mathscr{N}_2(t,u) &= -i\lambda\langle t\rangle^{2\nu-2}\Delta\langle i\nabla\rangle^{-1}\left\langle i\langle t\rangle^{\nu-1}\nabla\right\rangle^{-2}(u+\overline{u})^2,\\
\mathscr{N}_3(t,u) &= 2(\nu-1)t\langle t\rangle^{2\nu-4}\Delta\left\langle i\langle t\rangle^{\nu-1}\nabla\right\rangle^{-2}\mathscr{N}_1(u),\\
\mathscr{N}_4(t,u) &= -2\lambda\left\langle i\langle t\rangle^{\nu-1}\nabla\right\rangle^{-2}\langle i\nabla\rangle^{-1}\left(\mathscr{T}_1\left(\overline{u},\overline{\mathscr{L}u}\right)\right.\\
&\quad \left.+\mathscr{T}_2(u,\mathscr{L}u) + \mathscr{T}_3\left(u,\overline{\mathscr{L}u}\right) + \mathscr{T}_3(\mathscr{L}u,\overline{u})\right).
\end{aligned}$$

The first and second terms in the right-hand side of (8.4) are the quadratic nonlinearities with an explicit additional time decay, whereas the third term is a cubic nonlocal nonlinearity since in view of (8.2) $\mathscr{T}_1\left(\overline{u},\overline{\mathscr{L}u}\right) = -i\lambda\mathscr{T}_1\left(\overline{u},\langle i\nabla\rangle^{-1}(u+\overline{u})^2\right)$ and so on. If we could apply the Hölder inequality to the bilinear operators, $\mathscr{T}_j$, we would get the desired result easily. Unfortunately it is impossible, since we encounter the derivative loss difficulty applying Proposition 8.1 below. In order to compensate a derivative loss through the bilinear operators, we use the splitting argument which implies that we divide the identity operator into two terms such that

$$\left\langle i\langle t\rangle^{\nu-1}\nabla\right\rangle^{-2} - \langle t\rangle^{2\nu-2}\Delta = 1.$$

8.2 Preliminary Estimates

We first state a time decay estimate from paper [5].

Lemma 8.1. *The estimate is valid*

$$\left\| e^{it\langle i\nabla\rangle}\varphi \right\|_{\mathbf{L}^{\frac{4}{2-\beta}}} \le Ct^{-\frac{\beta}{2}} \|\varphi\|_{\mathbf{H}^{\beta}_{\frac{4}{2+\beta}}}$$

for all $t > 0$, *where* $0 \le \beta \le 2$, *provided that the right-hand side is finite.*

The following estimate for the bilinear operators $\mathscr{T}_j$ was proved in paper [2].

Proposition 8.1. *The bilinear operators* $\mathscr{T}_j$, $j = 1,2,3$, *are bounded from* $\mathbf{H}^{\beta}_{s}\left(\mathbf{R}^2\right) \times \mathbf{H}^{\mu}_{r}\left(\mathbf{R}^2\right)$ *to* $\mathbf{H}^{-\sigma}_{p}\left(\mathbf{R}^2\right)$, *i.e.*

$$\left\| \mathscr{T}_j(f,g) \right\|_{\mathbf{H}^{-\sigma}_{p}} \le C \|f\|_{\mathbf{H}^{\beta}_{s}} \|g\|_{\mathbf{H}^{\mu}_{r}}$$

where $1 \le p \le r \le \infty$, $\frac{1}{s} + \frac{1}{r} = 1 + \frac{1}{p} - \frac{1}{l}$, $1 \le l \le 2$, $\sigma, \beta, \mu \ge 0$ *are such that* $\sigma + \beta > 1, \mu > 1$, *or* $\beta > 1, \sigma + \mu > 1$.

Remark 8.5. If we do not use the splitting argument, and consider the problem in $\mathbf{H}^{\alpha}$ space, then we have to consider the norm $\left\| (-\Delta)^{\frac{\alpha}{2}} \mathscr{T}_2(u,u) \right\|_{\mathbf{H}^{-1}}$, for example, which requires us to the estimates of solutions such that $\left\| (-\Delta)^{\frac{\alpha}{2}} u \right\|_{\mathbf{H}^{\varepsilon}} \|u\|_{\mathbf{H}^{1+\varepsilon}}$ by Proposition 8.1 with $\sigma = 1, \beta = \varepsilon, s = r = l = 2$. Thus we encounter the derivative loss.

8.3 Proof of Theorem 8.1

We only consider the case $\mu = \alpha$, since the case $\mu > \alpha$ can be treated similarly. We start with the linearized version of (8.4) written as

$$\mathscr{L}\left(u + \mathscr{N}_1(t,v)\right) = \sum_{j=2}^{4} \mathscr{N}_j(t,v), \tag{8.5}$$

with the initial data $u(0) = u_0$, and a given function v, such that $v(0) = u_0$ and $v \in \mathbf{X}_{\infty,\rho} = \left\{\varphi \in \mathbf{X}_{\infty}; \|\varphi\|_{\mathbf{X}_{\infty}} \le \rho\right\}$, where $\rho = \varepsilon^{\frac{2}{3}}$, $\varepsilon > 0$. Solving (8.5) we define the mapping $\mathscr{M}$ by $u = \mathscr{M}v$. By the integral equation associated with (8.5) we get

$$\begin{aligned} &\|u(t) + \mathscr{N}_1(t,v(t))\|_{\mathbf{H}^{\alpha}} \\ &\le \|u_0 + \mathscr{N}_1(0,u_0)\|_{\mathbf{H}^{\alpha}} + C\sum_{j=2}^{4} \int_0^t \left\| \mathscr{N}_j(\tau, v(\tau)) \right\|_{\mathbf{H}^{\alpha}} d\tau. \end{aligned} \tag{8.6}$$

Denote $\nu = \alpha - 1 \in \left(0, \frac{1}{7}\right]$, $\gamma = \nu^3$ and $\widetilde{\mu} = 1 + \gamma$. Note that $1 < \widetilde{\mu} < \alpha$. For simplicity, we denote $\widetilde{\mu}$ by μ (note that μ is different to one given in theorems). By the definition of the norm $\mathbf{X}_{\infty}$ we have the estimate

$$\|v\|_{\mathbf{H}_r^\mu} \le C\langle t\rangle^{-\frac{\alpha-\mu}{2}}\|v\|_{\mathbf{X}_\infty} \le C\rho\langle t\rangle^{-\frac{\alpha-\mu}{2}}. \tag{8.7}$$

for $r=\frac{4}{2+\mu-\alpha}$. Then applying Proposition 8.1 with $s=p=2$, $r=\frac{4}{2+\mu-\alpha}$, $\sigma=0$, $\beta=\mu$ and estimate (8.7) we obtain

$$\begin{aligned}
&\|\mathscr{N}_1(t,v(t))\|_{\mathbf{H}^\alpha} \\
&\le C\left\|\langle i\nabla\rangle^\nu\left\langle i\langle t\rangle^{\nu-1}\nabla\right\rangle^{-2}(\mathscr{T}_1(\overline{v},\overline{v})+\mathscr{T}_2(v,v)+2\mathscr{T}_3(v,\overline{v}))\right\|_{\mathbf{L}^2} \\
&\le C\langle t\rangle^{\nu(1-\nu)}\|\mathscr{T}_1(\overline{v},\overline{v})+\mathscr{T}_2(v,v)+2\mathscr{T}_3(v,\overline{v})\|_{\mathbf{L}^2} \\
&\le C\langle t\rangle^{\nu(1-\nu)}\|v\|_{\mathbf{H}_r^\mu}^2 \le C\rho^2\langle t\rangle^{-\alpha+\mu+\nu(1-\nu)} \le C\rho^2\langle t\rangle^{-\gamma}.
\end{aligned} \tag{8.8}$$

By the definition of the norm $\mathbf{X}_\infty$ we have the estimate

$$\|v\|_{\mathbf{H}^{\alpha-4\nu}_{\frac{2}{1-2\nu}}} \le C\langle t\rangle^{-2\nu}\|v\|_{\mathbf{X}_\infty} \le C\rho\langle t\rangle^{-2\nu}. \tag{8.9}$$

Then applying the Hölder inequality, the Sobolev embedding theorem $\|\varphi\|_{\mathbf{L}^{\frac{2}{1-\nu}}} \le C\|\varphi\|_{\mathbf{H}^\nu}$ and $\|v\|_{\mathbf{L}^{\frac{2}{\nu}}} \le C\|v\|_{\mathbf{H}^{\alpha-4\nu}_{\frac{2}{1-2\nu}}}$, and (8.9), we get

$$\begin{aligned}
&\|\mathscr{N}_2(t,v(t))\|_{\mathbf{H}^\alpha} \\
&\le C\langle t\rangle^{-2(1-\nu)}\left\|\Delta\langle i\nabla\rangle^{\alpha-1}\left\langle i\langle t\rangle^{\nu-1}\nabla\right\rangle^{-2}(v+\overline{v})^2\right\|_{\mathbf{L}^2} \\
&\le C\langle t\rangle^{(\alpha-2)(1-\nu)}\left\|\nabla(v+\overline{v})^2\right\|_{\mathbf{L}^2} \le C\langle t\rangle^{-(1-\nu)^2}\|v\|_{\mathbf{L}^{\frac{2}{\nu}}}\|\nabla v\|_{\mathbf{L}^{\frac{2}{1-\nu}}} \\
&\le C\langle t\rangle^{-(1-\nu)^2}\|v\|_{\mathbf{L}^{\frac{2}{\nu}}}\|v\|_{\mathbf{H}^\alpha} \le C\langle t\rangle^{-(1-\nu)^2}\|v\|_{\mathbf{H}^{\alpha-4\nu}_{\frac{2}{1-2\nu}}}\|v\|_{\mathbf{H}^\alpha} \\
&\le C\rho^2\langle t\rangle^{-(1-\nu)^2-2\nu} \le C\rho^2\langle t\rangle^{-\mu}.
\end{aligned} \tag{8.10}$$

The nonlinear term $\mathscr{N}_3$ is estimated similarly to $\mathscr{N}_1$

$$\begin{aligned}
\|\mathscr{N}_3(t,v(t))\|_{\mathbf{H}^\alpha} &\le C\langle t\rangle^{2\nu-3}\left\|\Delta\left\langle i\langle t\rangle^{\nu-1}\nabla\right\rangle^{-2}\mathscr{N}_1(v)\right\|_{\mathbf{H}^\alpha} \\
&\le C\langle t\rangle^{-1}\|\mathscr{N}_1(v)\|_{\mathbf{H}^\alpha} \le C\rho^2\langle t\rangle^{-\mu}.
\end{aligned} \tag{8.11}$$

We next estimate the nonlinearity $\mathscr{N}_4$

$$\begin{aligned}
&\|\mathscr{N}_4(t,v(t))\|_{\mathbf{H}^\alpha} \le C\left\|\left\langle i\langle t\rangle^{\nu-1}\nabla\right\rangle^{-2}\langle i\nabla\rangle^{\alpha+\gamma-1}\left(\mathscr{T}_1\left(\overline{v},\overline{\mathscr{L}v}\right)\right.\right. \\
&\left.\left.+\mathscr{T}_2(v,\mathscr{L}v)+\mathscr{T}_3\left(v,\overline{\mathscr{L}v}\right)+\mathscr{T}_3(\mathscr{L}v,\overline{v})\right)\right\|_{\mathbf{H}^{-\gamma}} \\
&\le C\langle t\rangle^{(\nu+\gamma)(1-\nu)}\left\|\mathscr{T}_1\left(\overline{v},\overline{\mathscr{L}v}\right)+\mathscr{T}_2(v,\mathscr{L}v)+\mathscr{T}_3\left(v,\overline{\mathscr{L}v}\right)+\mathscr{T}_3(\mathscr{L}v,\overline{v})\right\|_{\mathbf{H}^{-\gamma}}.
\end{aligned}$$

Since $\mathscr{T}_1\left(\overline{u},\overline{\mathscr{L}u}\right)=-i\lambda\,\mathscr{T}_1\left(\overline{u},\langle i\nabla\rangle^{-1}(u+\overline{u})^2\right)$ and so on, applying Proposition 8.1 with $s=\frac{2}{2-\alpha}$, $r=\frac{4}{2+\mu-\alpha}$ and estimate (8.7), we find

$$\begin{aligned}&\|\mathscr{N}_4(t,v(t))\|_{\mathbf{H}^\alpha}\le C\langle t\rangle^{(\nu+\gamma)(1-\nu)}\|v\|_{\mathbf{H}^\mu_r}\|v\|^2_{\mathbf{L}^{2s}}\\&\le C\rho^3\langle t\rangle^{(\nu+\gamma)(1-\nu)-\frac{\alpha-\mu}{2}-\alpha}\le C\rho^3\langle t\rangle^{-\mu}.\end{aligned}\tag{8.12}$$

Now from inequalities (8.6), (8.8), and (8.10)–(8.12), it follows that

$$\|u(t)\|_{\mathbf{H}^\alpha}\le C\varepsilon+C\rho^2\le C\varepsilon.\tag{8.13}$$

We next estimate the norm $\|u(t)\|_{\mathbf{L}^{\frac{4}{2-\alpha}}}$. By Lemma 8.1, we have

$$\begin{aligned}&\|u(t)+\mathscr{N}_1(t,v(t))\|_{\mathbf{L}^{\frac{4}{2-\alpha}}}\le Ct^{-\frac{\alpha}{2}}\|u_0+\mathscr{N}_1(0,u_0)\|_{\mathbf{H}^\alpha_{\frac{4}{2+\alpha}}}\\&+C\sum_{j=2}^{4}\int_0^t(t-\tau)^{-\frac{\alpha}{2}}\|\mathscr{N}_j(\tau,v(\tau))\|_{\mathbf{H}^\alpha_{\frac{4}{2+\alpha}}}\,d\tau.\end{aligned}\tag{8.14}$$

The nonlinear term $\mathscr{N}_1(t,v(t))$ is estimated by Proposition 8.1 with $s=\frac{4}{2-\alpha}$, $r=\frac{4}{2+\mu-\alpha}$ and estimate (8.7)

$$\begin{aligned}&\|\mathscr{N}_1(t,v(t))\|_{\mathbf{L}^{\frac{4}{2-\alpha}}}\\&\le C\left\|\left\langle i\langle t\rangle^{\nu-1}\nabla\right\rangle^{-2}\langle i\nabla\rangle^\gamma\left(\mathscr{T}_1(\overline{u},\overline{u})+\mathscr{T}_2(u,u)+2\mathscr{T}_3(u,\overline{u})\right)\right\|_{\mathbf{H}^{-\mu}_{\frac{4}{2-\alpha}}}\\&\le C\langle t\rangle^{\gamma(1-\nu)}\|\mathscr{T}_1(\overline{u},\overline{u})+\mathscr{T}_2(u,u)+2\mathscr{T}_3(u,\overline{u})\|_{\mathbf{H}^{-\mu}_{\frac{4}{2-\alpha}}}\\&\le C\langle t\rangle^{\gamma(1-\nu)}\|v\|_{\mathbf{H}^\mu_r}\|v\|_{\mathbf{L}^{\frac{4}{2-\alpha}}}\\&\le C\rho^2\langle t\rangle^{\gamma(1-\nu)-\frac{\alpha-\mu}{2}-\frac{\alpha}{2}}\le C\rho^2\langle t\rangle^{-\frac{\alpha}{2}-\gamma}.\end{aligned}\tag{8.15}$$

The nonlinear term $\mathscr{N}_1(0,u_0)$ is estimated by Proposition 8.1 with $s=r=2$

$$\begin{aligned}&\|\mathscr{N}_1(0,u_0)\|_{\mathbf{H}^\alpha_{\frac{4}{2+\alpha}}}\\&\le C\|\mathscr{T}_1(\overline{u_0},\overline{u_0})+\mathscr{T}_2(u_0,u_0)+2\mathscr{T}_3(u_0,\overline{u_0})\|_{\mathbf{H}^{\alpha-3}_{\frac{4}{2+\alpha}}}\\&\le C\|u_0\|^2_{\mathbf{H}^\alpha}\le C\varepsilon^2.\end{aligned}$$

Next applying the Hölder inequality, the Sobolev embedding theorem, and (8.7), we get

$$\begin{aligned}&\|\mathcal{N}_2(t,v(t))\|_{\mathbf{H}^{\alpha}_{\frac{4}{2+\alpha}}} \leq C\langle t\rangle^{-2(1-\nu)}\left\|\Delta\langle i\nabla\rangle^{\alpha-1}\left\langle i\langle t\rangle^{\nu-1}\nabla\right\rangle^{-2}(v+\overline{v})^2\right\|_{\mathbf{L}^{\frac{4}{2+\alpha}}}\\
&\leq C\langle t\rangle^{(\alpha-2)(1-\nu)}\left\|\nabla(v+\overline{v})^2\right\|_{\mathbf{L}^{\frac{4}{2+\alpha}}} \leq C\langle t\rangle^{(\alpha-2)(1-\nu)}\|v\|_{\mathbf{L}^{\frac{4}{3\alpha-2}}}\|\nabla v\|_{\mathbf{L}^{\frac{2}{2-\alpha}}}\\
&\leq C\langle t\rangle^{(\alpha-2)(1-\nu)}\|v\|_{\mathbf{L}^{\frac{4}{3\alpha-2}}}\|v\|_{\mathbf{H}^{\alpha}}\\
&\leq C\langle t\rangle^{(\alpha-2)(1-\nu)-\left(1-\frac{3\alpha-2}{2}\right)}\rho^2 \leq C\langle t\rangle^{-\mu}\rho^2, \qquad (8.16)\end{aligned}$$

since $\alpha\in\left[1,\frac{8}{7}\right)$. In view of Proposition 8.1 with $s=r=\frac{4}{2+\mu-\alpha}$ and estimate (8.7),

$$\begin{aligned}&\|\mathcal{N}_3(t,v(t))\|_{\mathbf{H}^{\alpha}_{\frac{4}{2+\alpha}}}\\
&\leq C\langle t\rangle^{2\nu-3}\left\|\Delta\langle i\nabla\rangle^{\alpha-1}\left\langle i\langle t\rangle^{\nu-1}\nabla\right\rangle^{-4}(\mathcal{T}_1(\overline{v},\overline{v})\right.\\
&\quad\left.+\mathcal{T}_2(v,v)+2\mathcal{T}_3(v,\overline{v}))\right\|_{\mathbf{L}^{\frac{4}{2+\alpha}}}\\
&\leq C\langle t\rangle^{\nu(1-\nu)-1}\|\mathcal{T}_1(\overline{v},\overline{v})+\mathcal{T}_2(v,v)+2\mathcal{T}_3(v,\overline{v})\|_{\mathbf{L}^{\frac{4}{2+\alpha}}}\\
&\leq C\langle t\rangle^{\nu(1-\nu)-1}\|v\|^2_{\mathbf{H}^{\mu}_r} \leq C\rho^2\langle t\rangle^{\nu(1-\nu)-1-\alpha+\mu} = C\rho^2\langle t\rangle^{-\mu}. \qquad (8.17)\end{aligned}$$

Finally by Proposition 8.1 with $s=\frac{2}{2-\alpha}$, $r=\frac{4}{2+\mu-\alpha}$ and (8.7), we get

$$\begin{aligned}&\|\mathcal{N}_4(t,v(t))\|_{\mathbf{H}^{\alpha}_{\frac{4}{2+\alpha}}}\\
&\leq C\langle t\rangle^{(1-\nu)(\nu+\gamma)}\left\|\mathcal{T}_1\left(\overline{v},\langle i\nabla\rangle^{-1}(v+\overline{v})^2\right)\right\|_{\mathbf{H}^{-\gamma}_{\frac{4}{2+\alpha}}}\\
&\quad+C\langle t\rangle^{(1-\nu)(\nu+\gamma)}\left\|\mathcal{T}_2\left(v,\langle i\nabla\rangle^{-1}(v+\overline{v})^2\right)\right\|_{\mathbf{H}^{-\gamma}_{\frac{4}{2+\alpha}}}\\
&\quad+C\langle t\rangle^{(1-\nu)(\nu+\gamma)}\left\|\mathcal{T}_3\left(v,\langle i\nabla\rangle^{-1}(v+\overline{v})^2\right)\right\|_{\mathbf{H}^{-\gamma}_{\frac{4}{2+\alpha}}}\\
&\quad+C\langle t\rangle^{(1-\nu)(\nu+\gamma)}\left\|\mathcal{T}_3\left(\langle i\nabla\rangle^{-1}(v+\overline{v})^2,\overline{v}\right)\right\|_{\mathbf{H}^{-\gamma}_{\frac{4}{2+\alpha}}}\\
&\leq C\langle t\rangle^{(1-\nu)(\nu+\gamma)}\|v\|_{\mathbf{H}^{\mu}_r}\|v\|^2_{\mathbf{L}^{2s}}\\
&\leq C\rho^3\langle t\rangle^{(1-\nu)(\nu+\gamma)-\frac{\alpha-\mu}{2}-\alpha} \leq C\rho^3\langle t\rangle^{-\mu}. \qquad (8.18)\end{aligned}$$

Therefore via (8.14)–(8.18) we obtain

$$\begin{aligned}&\|u(t)\|_{\mathbf{L}^{\frac{4}{2-\alpha}}} \leq C\varepsilon t^{-\frac{\alpha}{2}}\\
&+C\rho^2\int_0^t(t-\tau)^{-\frac{\alpha}{2}}\langle\tau\rangle^{-\mu}d\tau \leq C\varepsilon t^{-\frac{\alpha}{2}}+C\rho^2\langle t\rangle^{-\frac{\alpha}{2}} \leq C\varepsilon\langle t\rangle^{-\frac{\alpha}{2}}.\end{aligned}$$

for $t \geq 1$. In view of (8.13), by the Sobolev embedding theorem we have $\|u\|_{\mathbf{L}^{\frac{4}{2-\alpha}}} \leq C\|u\|_{\mathbf{H}^\alpha} \leq C\varepsilon$ for $t \leq 1$. Therefore

$$\|u(t)\|_{\mathbf{L}^{\frac{4}{2-\alpha}}} \leq C\varepsilon \langle t \rangle^{-\frac{\alpha}{2}} \tag{8.19}$$

for all $t \geq 0$. Thus from (8.13) and (8.19) we have

$$\|\mathscr{M}v\|_{\mathbf{X}_\infty} \leq C\varepsilon \leq \rho. \tag{8.20}$$

In the same way as in the proof of (8.20) we obtain

$$\|\mathscr{M}v_1 - \mathscr{M}v_2\|_{\mathbf{X}_\infty} \leq \frac{1}{2}\|v_1 - v_2\|_{\mathbf{X}_\infty}.$$

Hence there exists a unique solution u of (8.4) such that $\|u\|_{\mathbf{X}_\infty} \leq \rho$.

The integral equation associated with (8.4) is written by

$$w(t) = e^{-i\langle i\nabla\rangle t}u_0 + \int_0^t e^{-i\langle i\nabla\rangle(t-\tau)}F(u)\,d\tau,$$

where $w = u + \mathscr{N}_1$, $F(u) = \mathscr{N}_2 + \mathscr{N}_3 + \mathscr{N}_4$. Then we obtain

$$e^{i\langle i\nabla\rangle t}w(t) - e^{i\langle i\nabla\rangle s}w(s) = \int_s^t e^{i\langle i\nabla\rangle \tau}F(u)\,d\tau.$$

As above we get the estimate

$$\left\|e^{i\langle i\nabla\rangle t}w(t) - e^{i\langle i\nabla\rangle s}w(s)\right\|_{\mathbf{H}^\alpha} \leq C\langle s\rangle^{-\gamma}$$

for all $t > s > 0$. By the relation $w = u + \mathscr{N}_1$ and estimate (8.8): $\|\mathscr{N}_1\|_{\mathbf{H}^\alpha} \leq C\langle t\rangle^{-\gamma}$, we find

$$\left\|e^{i\langle i\nabla\rangle t}u(t) - e^{i\langle i\nabla\rangle s}u(s)\right\|_{\mathbf{H}^\alpha} \leq C\langle s\rangle^{-\gamma}.$$

Hence there exists a unique scattering state $u_+ \in \mathbf{H}^\alpha$ such that

$$\|u(t) - \mathscr{U}(t)u_+\|_{\mathbf{H}^\alpha} \to 0$$

as $t \to \infty$. This completes the proof of Theorem 8.1.

8.4 Proof of Theorem 8.2

We only consider the case $\mu = \alpha$, since the case $\mu > \alpha$ can be treated similarly. The final value problem is formulated as follows: we pose a final value $u_+ \in \mathbf{H}^\alpha_{\frac{4}{2+\alpha}} \cap \mathbf{H}^\alpha$ and solve (8.2) under the final state condition

$$\|u(t)-\mathscr{U}(t)u_+\|_{\mathbf{H}^\alpha}+\langle t\rangle^{\frac{\alpha}{2}}\|u(t)-\mathscr{U}(t)u_+\|_{\mathbf{L}^{\frac{4}{2-\alpha}}}\to 0 \tag{8.21}$$

as $t\to\infty$. We make a transformation to (8.4)

$$\partial_t\mathscr{U}(-t)(u(t)+\mathscr{N}_1(t,u(t)))=\sum_{j=2}^{4}\mathscr{U}(-t)\mathscr{N}_j(t,u(t)).$$

By the condition (8.21) it follows that

$$\langle t\rangle^{\frac{\alpha-\widetilde{\mu}}{2}}\|u(t)-\mathscr{U}(t)u_+\|_{\mathbf{H}_r^{\widetilde{\mu}}}\to 0, \tag{8.22}$$

where $r=\frac{4}{2+\widetilde{\mu}-\alpha}$, $\widetilde{\mu}=1+\gamma$, $\gamma=\nu^3$, $\nu=\alpha-1>0$. In what follows we denote $\widetilde{\mu}$ by μ for simplicity as in the proof of Theorem 8.1. Then in view of Proposition 8.1 with $s=p=2$, $r=\frac{4}{2+\mu-\alpha}$, $\sigma=0$, $\beta=\mu$, and estimate (8.22) we obtain

$$\begin{aligned}
&\|\mathscr{N}_1(t,u(t))\|_{\mathbf{H}^\alpha}\\
&\le C\left\|\langle i\nabla\rangle^{\nu}\left\langle i\langle t\rangle^{\nu-1}\nabla\right\rangle^{-2}(\mathscr{T}_1(\overline{u},\overline{u})+\mathscr{T}_2(u,u)+2\mathscr{T}_3(u,\overline{u}))\right\|_{\mathbf{L}^2}\\
&\le C\langle t\rangle^{\nu(1-\nu)}\|\mathscr{T}_1(\overline{u},\overline{u})+\mathscr{T}_2(u,u)+2\mathscr{T}_3(u,\overline{u})\|_{\mathbf{L}^2}\\
&\le C\langle t\rangle^{\nu(1-\nu)}\|u\|^2_{\mathbf{H}_r^{\mu}}\le C\langle t\rangle^{-\gamma}.
\end{aligned}$$

Hence we have

$$\|\mathscr{N}_1(t,u(t))\|_{\mathbf{H}^\alpha}\to 0 \tag{8.23}$$

as $t\to\infty$. In the same manner by Proposition 8.1 with $s=\frac{4}{2-\alpha}$, $r=\frac{4}{2+\mu-\alpha}$ and estimate (8.22) we find

$$\begin{aligned}
&\|\mathscr{N}_1(t,u(t))\|_{\mathbf{L}^{\frac{4}{2-\alpha}}}\\
&\le C\langle t\rangle^{\gamma(1-\nu)}\|\mathscr{T}_1(\overline{u},\overline{u})+\mathscr{T}_2(u,u)+2\mathscr{T}_3(u,\overline{u})\|_{\mathbf{H}^{-\mu}_{\frac{4}{2-\alpha}}}\\
&\le C\langle t\rangle^{\gamma(1-\nu)}\|u\|_{\mathbf{H}_r^{\mu}}\|u\|_{\mathbf{L}^{\frac{4}{2-\alpha}}}\le C\langle t\rangle^{-\frac{\alpha}{2}-\gamma}.
\end{aligned}$$

Thus

$$\langle t\rangle^{\frac{\alpha}{2}}\|\mathscr{N}_1(t,u(t))\|_{\mathbf{L}^{\frac{4}{2-\alpha}}}\to 0 \tag{8.24}$$

as $t\to\infty$. Hence the integral equation associated with the final state problem for (8.4) can be written as

$$u(t)+\mathscr{N}_1(t,u(t))=\mathscr{U}(t)u_+-\sum_{j=2}^{4}\int_t^\infty\mathscr{U}(t-\tau)\mathscr{N}_j(\tau,u(\tau))\,d\tau. \tag{8.25}$$

We next assume that $v\in\mathbf{Z}_{\infty,\rho}=\left\{\varphi\in\mathbf{Z}_\infty;\|\varphi\|_{\mathbf{Z}_\infty}\le\rho\right\}$ and consider the linearized version of (8.25)

$$\begin{aligned} &u(t) + \mathcal{N}_1(t, v(t)) \\ &= \mathcal{U}(t) u_+ - \sum_{j=2}^{4} \int_t^\infty \mathcal{U}(t-\tau) \mathcal{N}_j(\tau, v(\tau)) \, d\tau. \end{aligned} \tag{8.26}$$

Note that the condition $v \in \mathbf{Z}_{\infty,\rho}$ implies that

$$\begin{aligned} \|v\|_{\mathbf{X}_\infty} &\le \|v - \mathcal{U}(\cdot) u_+\|_{\mathbf{X}_\infty} + \|\mathcal{U}(\cdot) u_+\|_{\mathbf{X}_\infty} \\ &\le \|v - \mathcal{U}(\cdot) u_+\|_{\mathbf{Y}_\infty} + \|\mathcal{U}(\cdot) u_+\|_{\mathbf{X}_\infty} \le \rho + C\varepsilon. \end{aligned} \tag{8.27}$$

Hence we can apply estimates (8.10)–(8.12) and (8.15)–(8.18) to get

$$\|\mathcal{N}_1(t, v(t))\|_{\mathbf{H}^\alpha} + \langle t \rangle^{\frac{\alpha}{2}} \|\mathcal{N}_1(t, v(t))\|_{\mathbf{L}^{\frac{4}{2-\alpha}}} \le C\rho^2 \langle t \rangle^{-\gamma} \tag{8.28}$$

and

$$\sum_{j=2}^{4} \left(\|\mathcal{N}_j(t, v(t))\|_{\mathbf{H}^\alpha} + \|\mathcal{N}_j(t, v(t))\|_{\mathbf{H}^\alpha_{\frac{4}{2+\alpha}}} \right) \le C\rho^2 \langle t \rangle^{-\mu}. \tag{8.29}$$

By (8.26), (8.28), and (8.29) we find

$$\begin{aligned} &\|u(t) - \mathcal{U}(t) u_+\|_{\mathbf{H}^\alpha} \\ &\le C\rho^2 \langle t \rangle^{-\gamma} + \sum_{j=2}^{4} \int_t^\infty \|\mathcal{N}_j(\tau, v(\tau))\|_{\mathbf{H}^\alpha} \, d\tau \le C\rho^2 \langle t \rangle^{-\gamma} \end{aligned}$$

and

$$\begin{aligned} &\|u(t) - \mathcal{U}(t) u_+\|_{\mathbf{L}^{\frac{4}{2-\alpha}}} \\ &\le C\rho^2 \langle t \rangle^{-\frac{\alpha}{2}-\gamma} + \sum_{j=2}^{4} \int_t^\infty (t-\tau)^{-\frac{\alpha}{2}} \|\mathcal{N}_j(\tau, v(\tau))\|_{\mathbf{H}^\alpha_{\frac{4}{2+\alpha}}} \, d\tau \\ &\le C\rho^2 \langle t \rangle^{-\frac{\alpha}{2}-\gamma} + \sum_{j=2}^{4} \int_t^\infty (t-\tau)^{-\frac{\alpha}{2}} \langle \tau \rangle^{-\mu} \, d\tau \le C\rho^2 \langle t \rangle^{-\frac{\alpha}{2}-\gamma}. \end{aligned}$$

In the same manner we can estimate the differences

$$\|u_1 - u_2\|_{\mathbf{Z}_\infty} \le \frac{1}{2} \|v_1 - v_2\|_{\mathbf{Z}_\infty}$$

to show that there exists a unique global solution $u \in \mathbf{C}([0,\infty); \mathbf{H}^\alpha)$ of (8.4) satisfying the estimate

$$\begin{aligned} &\|u\|_{\mathbf{Z}_\infty} = \|u - \mathcal{U}(\cdot) u_+\|_{\mathbf{Y}_\infty} \\ &\le \sup_{t \in [0,\infty)} \langle t \rangle^{\gamma} \left(\|u(t) - \mathcal{U}(t) u_+\|_{\mathbf{H}^\alpha} + \langle t \rangle^{\frac{\alpha}{2}} \|u(t) - \mathcal{U}(t) u_+\|_{\mathbf{L}^{\frac{4}{2-\alpha}}} \right) \le C\rho^2. \end{aligned}$$

This completes the proof of Theorem 8.2.

Acknowledgements The authors would like to thank the referee for letting us know a preprint [1] by Georgiev and Stefanov in which global existence of small solutions is shown in almost energy space. The work of P.I.N. is partially supported by CONACYT and PAPIIT.

References

1. Georgiev, V., Stefanov, A.: Global regularity for the quadratic Klein–Gordon equation in R^{1+2} (2011, preprint)
2. Hayashi, N., Naumkin, P.I.: A bilinear estimate and its application to a quadratic nonlinear Klein–Gordon equation in two space dimensions (2011, preprint)
3. Hayashi, N., Naumkin, P.I.: Quadratic nonlinear Klein–Gordon equation in 2d, Cauchy problem, to appear in Proceedings of 4thMSJ-SI Nonlinear Dynamics in Partial Differential Equations (xxxx)
4. Kosecki, R.: The unit condition and global existence for a class of nonlinear Klein–Gordon equations. J. Differ. Equat. **100**, 257–268 (1992)
5. Marshall, B., Strauss, W., Wainger, S.: $\mathbf{L}^p - \mathbf{L}^q$ estimates for the Klein–Gordon equation. J. Math. Pures Appl. **59**, 417–440 (1980)
6. Ozawa, T., Tustaya, K., Tsutsumi, Y.: Global existence and asymptotic behavior of solutions for the Klein–Gordon equations with quadratic nonlinearity in two space dimensions. Math. Z. **222**, 341–362 (1996)
7. Ozawa, T., Tustaya, K., Tsutsumi, Y.: Remarks on the Klein–Gordon equation with quadratic nonlinearity in two space dimensions. In: Nonlinear Waves. Gakuto International Series, Mathematical Sciences and Applications, vol. 10, pp. 383–392. Gakkotosha, Tokyo, Japan (1997)
8. Shatah, J.: Normal forms and quadratic nonlinear Klein–Gordon equations. Comm. Pure Appl. Math. **38**, 685–696 (1985)

Chapter 9
Global Solutions to the 3-D Incompressible Inhomogeneous Navier–Stokes System with Rough Density

Jingchi Huang, Marius Paicu, and Ping Zhang

Abstract In this paper, we prove the global well-posedness of the 3-D incompressible inhomogeneous Navier–Stokes equations with initial data $a_0 \in \mathscr{M}(B_{p,1}^{-1+\frac{3}{p}}(\mathbb{R}^3))$, $u_0 = (u_0^h, u_0^3) \in B_{p,1}^{-1+\frac{3}{p}}(\mathbb{R}^3)$, which satisfies

$$\left(\mu \|a_0\|_{\mathscr{M}(B_{p,1}^{-1+\frac{3}{p}})} + \|u_0^h\|_{B_{p,1}^{-1+\frac{3}{p}}}\right) \exp\left(C_0 \|u_0^3\|^2_{B_{p,1}^{-1+\frac{3}{p}}} \Big/ \mu^2\right) \le c_0 \mu$$

for some positive constants c_0, C_0 and $\frac{3}{2} < p < 6$. The novelty of this paper is to replace $\|a_0\|_{B_{q,1}^{\frac{3}{q}}}$ in the smallness condition of [20] by the rough norm in the multiplier space $\|a_0\|_{\mathscr{M}(B_{p,1}^{-1+\frac{3}{p}})}$ here.

Key words: Inhomogeneous Navier–Stokes equations, Littlewood–Paley theory, Wellposedness.

2010 Mathematics Subject Classification: 35Q30, 76D03.

J. Huang (✉)
Academy of Mathematics & Systems Science, Chinese Academy of Sciences, Beijing 100190, P. R. China
e-mail: jchuang@amss.ac.cn

M. Paicu
Institut de Mathématiques de Bordeaux, Université Bordeaux 1, 351, cours de la Libération, F-33405 Talence Cedex, France
e-mail: marius.paicu@math.u-bordeaux1.fr

P. Zhang
Academy of Mathematics & Systems Science and Hua Loo-Keng Key Laboratory of Mathematics, The Chinese Academy of Sciences, Beijing 100190, P. R. China
e-mail: zp@amss.ac.cn

M. Cicognani et al. (eds.), *Studies in Phase Space Analysis with Applications to PDEs*, Progress in Nonlinear Differential Equations and Their Applications 84,
DOI 10.1007/978-1-4614-6348-1_9,

9.1 Introduction

In this paper, we consider the global well-posedness of the following 3-D incompressible inhomogeneous Navier–Stokes equations with initial data in the critical spaces

$$\begin{cases} \partial_t \rho + \operatorname{div}(\rho u) = 0, \qquad (t,x) \in \mathbb{R}^+ \times \mathbb{R}^3, \\ \partial_t(\rho u) + \operatorname{div}(\rho u \otimes u) - \operatorname{div}(\mu \mathscr{M}) + \nabla \Pi = 0, \\ \operatorname{div} u = 0, \\ \rho|_{t=0} = \rho_0, \quad \rho u|_{t=0} = m_0, \end{cases} \tag{9.1}$$

where $\rho, u = (u_1, u_2, u_3)$ stand for the density and velocity of the fluid, respectively, $\mathscr{M} = \frac{1}{2}(\partial_i u_j + \partial_j u_i)$, Π is a scalar pressure function, and in general, the viscosity coefficient $\mu(\rho)$ is a smooth, positive function on $[0, \infty)$. Such system describes a fluid which is obtained by mixing two miscible fluids that are incompressible and that have different densities. It may also describe a fluid containing a melted substance. One may check [17] for the detailed derivation of this system.

When $\mu(\rho)$ is independent of ρ, i. e., μ is a positive constant and ρ_0 is bounded away from 0, Kazhikov [15] proved that the inhomogeneous Navier–Stokes equations (9.1) have at least one global weak solutions in the energy space. In addition, he also proved the global existence of strong solutions to this system for small data in three space dimensions and all data in two dimensions. However, the uniqueness of both types of weak solutions has not been solved. Ladyženskaja and Solonnikov [16] first addressed the question of unique resolvability of (9.1). More precisely, they considered the system (9.1) in bounded domain Ω with homogeneous Dirichlet boundary condition for u. Under the assumption that $u_0 \in W^{2-\frac{2}{p},p}(\Omega)$ $(p > N)$ is divergence-free and vanishes on $\partial\Omega$ and that $\rho_0 \in C^1(\Omega)$ is bounded away from zero, then they [16] proved:

- Global well-posedness in dimension $N = 2$.
- Local well-posedness in dimension $N{=}3$. If in addition u_0 is small in $W^{2-\frac{2}{p},p}(\Omega)$, then global well-posedness holds true.

Similar results were obtained by Danchin [10] in $\mathbb{R}^N$ with initial data in the almost critical Sobolev spaces.

In the general case when $\mu(\rho)$ depends on ρ, Lions [17] proved the global existence of weak solutions to (9.1) in any space dimensions. Yet the uniqueness and regularities of such weak solutions are big open questions even in two space dimension, as was mentioned by Lions in [17]. On the other hand, Abidi, Gui, and Zhang [4] investigated the large time decay and stability to any given global smooth solutions of (9.1), which in particular implies the global well-posedness of 3-D inhomogeneous Navier–Stokes equations with axisymmetric initial data and without swirl for the initial velocity field provided that the initial density is close enough to a positive constant.

When the density ρ is away from zero, we denote by $a \stackrel{\text{def}}{=} \frac{1}{\rho} - 1$ and $\tilde{\mu}(a) \stackrel{\text{def}}{=} \mu(\rho)$, then the system (9.1) can be equivalently reformulated as

$$\begin{cases} \partial_t a + u\cdot\nabla a = 0, \qquad (t,x)\in\mathbb{R}^+\times\mathbb{R}^N,\\ \partial_t u + u\cdot\nabla u + (1+a)(\nabla\Pi - \operatorname{div}(\tilde{\mu}(a)\mathscr{M})) = 0,\\ \operatorname{div} u = 0,\\ (a,u)|_{t=0} = (a_0,u_0). \end{cases} \tag{9.2}$$

Notice that just as the classical Navier–Stokes system, the inhomogeneous Navier–Stokes system (9.2) also has a scaling. More precisely, if (a,u) solves (INS) with initial data (a_0,u_0), then for $\forall\ell>0$,

$$(a,u)_\ell \stackrel{\text{def}}{=} (a(\ell^2\cdot,\ell\cdot),\ell u(\ell^2\cdot,\ell\cdot)) \quad\text{and}\quad (a_0,u_0)_\ell \stackrel{\text{def}}{=} (a_0(\ell\cdot),\ell u_0(\ell\cdot)) \tag{9.3}$$

$(a,u)_\ell$ is also a solution of (9.2) with initial data $(a_0,u_0)_\ell$.

In [9], Danchin studied in general space dimension N the unique solvability of the system (9.2) with constant viscosity coefficient and in scaling invariant (or critical) homogeneous Besov spaces, which generalized the celebrated results by Fujita and Kato [12] devoted to the classical Navier–Stokes system. In particular, the norm of $(a,u)\in B^{\frac{N}{2}}_{2,\infty}(\mathbb{R}^N)\cap L^\infty(\mathbb{R}^N)\times B^{\frac{N}{2}-1}_{2,1}(\mathbb{R}^N)$ is scaling invariant under the change of scale of (9.3). In this case, Danchin proved that if the initial data $(a_0,u_0)\in B^{\frac{N}{2}}_{2,\infty}(\mathbb{R}^N)\cap L^\infty(\mathbb{R}^N)\times B^{\frac{N}{2}-1}_{2,1}(\mathbb{R}^N)$ with a_0 sufficiently small in $B^{\frac{N}{2}}_{2,\infty}(\mathbb{R}^N)\cap L^\infty(\mathbb{R}^N)$, then the system (9.2) has a unique local-in-time solution. In [1], Abidi proved that if $1<p<2N$, $0<\underline{\mu}<\tilde{\mu}(a)$, $u_0\in B^{\frac{N}{p}-1}_{p,1}(\mathbb{R}^N)$ and $a_0\in B^{\frac{N}{p}}_{p,1}(\mathbb{R}^N)$, then (9.2) has a global solution provided that $\|a_0\|_{B^{\frac{N}{p}}_{p,1}}+\|u_0\|_{B^{\frac{N}{p}-1}_{p,1}}\le c_0$ for some c_0 sufficiently small. Furthermore, this obtained solution is unique if $1<p\le N$. This result generalized the corresponding results in [9, 10] and was improved by Abidi and Paicu in [2] when $\tilde{\mu}(a)$ is a positive constant, by using different Lebesgue exponents for the density a_0 and for the velocity u_0 (see also [3] for an application of this method to a more complex system). Very recently, Danchin and Mucha [11] improved the uniqueness result for $p\in(N,2N)$ through Lagrange approach, and Abidi, Gui, and Zhang relaxed the smallness condition for a_0 in [5, 6].

For simplicity, in what follows we just take $\mu(\rho)=\mu$ and the space dimension $N=3$. In this case, (9.2) becomes

$$\begin{cases} \partial_t a + u\cdot\nabla a = 0, \qquad (t,x)\in\mathbb{R}^+\times\mathbb{R}^3,\\ \partial_t u + u\cdot\nabla u + (1+a)(\nabla\Pi - \mu\Delta u) = 0,\\ \operatorname{div} u = 0,\\ (a,u)|_{t=0} = (a_0,u_0). \end{cases} \tag{9.4}$$

Motivated by [13, 19, 22] concerning the global Well-posedness of 3-D incompressible anisotropic Navier–Stokes system with the third component of the initial velocity field being large, we [20] relaxed the smallness condition in [2] so that (9.4) still has a unique global solution. We emphasize that our proof uses in a fundamental way the algebraical structure of (9.4). The first step is to obtain energy estimates on the horizontal components of the velocity field on the one hand and then on the vertical component on the other hand. Compared with [13, 19, 22], the additional difficulties with this strategy are that there appears a hyperbolic-type equation in (9.4) and due to the appearance of a in the momentum equation of (9.4), the pressure term is more difficult to be estimated.

Before going further, we first recall the functional space framework from [7]:

$$\Delta_j a = \mathscr{F}^{-1}(\varphi(2^{-j}|\xi|)\widehat{a}), \qquad S_j a = \sum_{j'\le j-1} \Delta_{j'} a, \tag{9.5}$$

where $\mathscr{F}a$ and $\widehat{a}$ denote the Fourier transform of the distribution a, and $\varphi(\tau)$ is a smooth function such that

$$\operatorname{Supp}\varphi \subset \left\{\tau \in \mathbb{R} \,/\, \frac{3}{4} \le |\tau| \le \frac{8}{3}\right\} \quad \text{and} \quad \forall \tau > 0,\ \sum_{j\in\mathbb{Z}} \varphi(2^{-j}\tau) = 1.$$

Definition 9.1. *Let $(p,r) \in [1,+\infty]^2$, $s \in \mathbb{R}$ and $u \in \mathscr{S}_h'(\mathbb{R}^3)$, we set*

$$\|u\|_{B^s_{p,r}} \stackrel{\text{def}}{=} \left(2^{qs}\|\Delta_q u\|_{L^p}\right)_{\ell^r}.$$

- *For $s < \frac{3}{p}$ (or $s = \frac{3}{p}$ if $r = 1$), we define $B^s_{p,r}(\mathbb{R}^3) \stackrel{\text{def}}{=} \left\{u \in \mathscr{S}_h'(\mathbb{R}^3) \mid \|u\|_{B^s_{p,r}} < \infty\right\}$.*
- *If $k \in \mathbb{N}$ and $\frac{3}{p} + k \le s < \frac{3}{p} + k + 1$ (or $s = \frac{3}{p} + k + 1$ if $r = 1$), then $B^s_{p,r}(\mathbb{R}^3)$ is defined as the subset of distributions $u \in \mathscr{S}_h'(\mathbb{R}^3)$ such that $\partial^\beta u \in B^{s-k}_{p,r}(\mathbb{R}^3)$ whenever $|\beta| = k$.*

In order to obtain a better description of the regularizing effect of the transport-diffusion equation, we will use Chemin-Lerner-type spaces $\widetilde{L}^\lambda_T(B^s_{p,r}(\mathbb{R}^n))$.

Definition 9.2. Let $(r,\lambda,p) \in [1,+\infty]^3$ and $T \in]0,+\infty]$. We define $\widetilde{L}^\lambda_T(B^s_{pr}(\mathbb{R}^n))$ as the completion of $C([0,T];\mathscr{S}(\mathbb{R}^n))$ by the norm

$$\|f\|_{\widetilde{L}^\lambda_T(B^s_{p,r})} \stackrel{\text{def}}{=} \Big(\sum_{q\in\mathbb{Z}} 2^{qrs}\Big(\int_0^T \|\Delta_q f(t)\|^\lambda_{L^p}\,dt\Big)^{\frac{r}{\lambda}}\Big)^{\frac{1}{r}} < \infty,$$

with the usual change if $r = \infty$. For short, we just denote this space by $\widetilde{L}^\lambda_T(B^s_{p,r})$.

We also need the following form of functional framework, which is a sort of generalization to the weighted Chemin-Lerner-type norm from [19, 20]:

Definition 9.3. Let $f(t) \in L^1_{loc}(\mathbb{R}^+)$, $f(t) \ge 0$ and X be a Banach space. We define

$$\|u\|_{L^1_{T,f}(X)} \stackrel{\text{def}}{=} \int_0^T f(t)\|u(t)\|_X\,dt.$$

The following theorem was proved by the authors in [20]:

Theorem 9.1. *Let $1 < q \le p < 6$ with $\frac{1}{q} - \frac{1}{p} \le \frac{1}{3}$. Let $a_0 \in B^{\frac{3}{q}}_{q,1}(\mathbb{R}^3)$ and $u_0 = (u_0^h, u_0^3) \in B^{-1+\frac{3}{p}}_{p,1}(\mathbb{R}^3)$. Then there exist positive constants c_0 and C_0 such that if*

$$\eta \stackrel{\text{def}}{=} (\mu\|a_0\|_{B^{\frac{3}{q}}_{q,1}} + \|u_0^h\|_{B^{-1+\frac{3}{p}}_{p,1}}) \exp\Big\{C_0\|u_0^3\|^2_{B^{-1+\frac{3}{p}}_{p,1}} / \mu^2\Big\} \le c_0\mu, \tag{9.6}$$

(9.4) has a unique global solution $a \in \mathscr{C}([0,\infty); B^{\frac{3}{q}}_{q,1}(\mathbb{R}^3)) \cap \widetilde{L}^\infty(\mathbb{R}^+; B^{\frac{3}{q}}_{q,1}(\mathbb{R}^3))$ *and* $u \in \mathscr{C}([0,\infty); B^{-1+\frac{3}{p}}_{p,1}(\mathbb{R}^3)) \cap \widetilde{L}^\infty(\mathbb{R}^+; B^{-1+\frac{3}{p}}_{p,1}(\mathbb{R}^3)) \cap L^1(\mathbb{R}^+, B^{1+\frac{3}{p}}_{p,1}(\mathbb{R}^3))$. *Moreover, there holds*

$$\|u - e^{\mu t\Delta}u_0\|_{\widetilde{L}^\infty(\mathbb{R}^+; B^{-1+\frac{3}{p}}_{p,1})} + \mu\|u - e^{\mu t\Delta}u_0\|_{L^1(\mathbb{R}^+; B^{1+\frac{3}{p}}_{p,1})} \le C(\|u_0^3\|_{B^{-1+\frac{3}{p}}_{p,1}} + \mu + \eta)\frac{\eta}{\mu}.$$

We emphasize that the main feature of the density space used in this theorem is to be a multiplier on the velocity space. This allow to define the nonlinear terms containing products between the density and the velocity, in the system (9.4). Very recently, Danchin and Mucha obtained in [11] a more general result by considering very rough densities in some multiplier spaces on the Besov spaces $B^{-1+\frac{d}{p}}_{p,1}(\mathbb{R}^d)$. In particular, they are able to consider the physical case of mixture of fluids with piecewise constant density. For the convenience of the readers, we also recall the definition of multiplier spaces to Besov spaces from [18].

Definition 9.4. We call f belonging to the multiplier space, $\mathscr{M}(B^s_{p,1}(\mathbb{R}^n))$, of $B^s_{p,1}(\mathbb{R}^n)$ if the distributions f satisfies $\psi f \in B^s_{p,1}(\mathbb{R}^n)$ whenever $\psi \in B^s_{p,1}(\mathbb{R}^n)$. We endow this space with the norm

$$\|f\|_{\mathscr{M}(B^s_{p,1})} \stackrel{\text{def}}{=} \sup_{\|\psi\|_{B^s_{p,1}}=1} \|\psi f\|_{B^s_{p,1}} \quad \text{for} \quad f \in \mathscr{M}(B^s_{p,1}(\mathbb{R}^n)).$$

The following interesting result was proved by Danchin and Mucha in [11]:

Theorem 9.2 (Theorems 1 and 2 of [11]). *Let* $p \in [1, 2n)$ *and* u_0 *be a divergence-free vector field in* $B^{\frac{n}{p}-1}_{p,1}(\mathbb{R}^n)$. *Assume that the initial density* ρ_0 *belongs to the multiplier space* $\mathscr{M}(B^{\frac{n}{p}-1}_{p,1}(\mathbb{R}^n))$. *There exists a constant c depending only on p and n such that if*

$$\|\rho_0 - 1\|_{\mathscr{M}(B^{\frac{n}{p}-1}_{p,1}(\mathbb{R}^n))} \le c$$

then there exists some $T > 0$ *such that system (9.1) with* $\mu(\rho) = \mu$, *namely,*

$$\begin{cases} \partial_t\rho + u\cdot\nabla\rho = 0, & (t,x) \in \mathrm{R}^+ \times \mathrm{R}^n, \\ \partial_t(\rho u) + \mathrm{div}(\rho u \otimes u) - \mu\Delta u + \nabla\Pi = 0, \\ \mathrm{div} u = 0, \end{cases} \tag{9.7}$$

has a unique local solution (ρ, u) *with* $\rho \in L^\infty([0,T); \mathscr{M}(B^{\frac{n}{p}-1}_{p,1}(\mathbb{R}^n)))$ *and* $u \in C_b([0,T); B^{\frac{n}{p}-1}_{p,1}(\mathbb{R}^n)) \cap L^1([0,T); B^{\frac{n}{p}+1}_{p,1}(\mathbb{R}^n))$. *Moreover, if* $\|u_0\|_{B^{\frac{n}{p}-1}_{p,1}} \le c\mu$, *thus obtained solution is global.*

Motivated by [11], the object of this paper is to replace the $\|a_0\|_{B^{\frac{3}{q}}_{q,1}}$ in the smallness condition (9.6) by $\|a_0\|_{\mathscr{M}(B^{-1+\frac{3}{p}}_{p,1})}$ and to prove a similar version of Theorem 9.1. More precisely,

Theorem 9.3. *Let* $\frac{3}{2}<p<6$*. Let* $a_0 \in \mathscr{M}(B^{-1+\frac{3}{p}}_{p,1}(\mathbb{R}^3))$ *and* $u_0=(u_0^h,u_0^3)\in B^{-1+\frac{3}{p}}_{p,1}(\mathbb{R}^3)$*. Then there exist positive constants* c_0 *and* C_0 *such that if*

$$\delta \stackrel{\text{def}}{=} \big(\mu\|a_0\|_{\mathscr{M}(B^{-1+\frac{3}{p}}_{p,1})}+\|u_0^h\|_{B^{-1+\frac{3}{p}}_{p,1}}\big)\exp\Big\{C_0\|u_0^3\|^2_{B^{-1+\frac{3}{p}}_{p,1}}\Big/\mu^2\Big\}\leq c_0\mu, \tag{9.8}$$

(9.4) has a unique global solution $a\in L^\infty(\mathbb{R}^+;\mathscr{M}(B^{-1+\frac{3}{p}}_{p,1}(\mathbb{R}^3))$ *and* $u\in\mathscr{C}([0,\infty);B^{-1+\frac{3}{p}}_{p,1}(\mathbb{R}^3))\cap\widetilde{L}^\infty(\mathbb{R}^+;B^{-1+\frac{3}{p}}_{p,1}(\mathbb{R}^3))\cap L^1(\mathbb{R}^+,B^{1+\frac{3}{p}}_{p,1}(\mathbb{R}^3))$*. Moreover, there holds*

$$\|u-e^{\mu t\Delta}u_0\|_{\widetilde{L}^\infty(\mathbb{R}^+;B^{-1+\frac{3}{p}}_{p,1})}+\mu\|u-e^{\mu t\Delta}u_0\|_{L^1(\mathbb{R}^+;B^{1+\frac{3}{p}}_{p,1})}\leq C(\|u_0^3\|_{B^{-1+\frac{3}{p}}_{p,1}}+\mu+\delta)\frac{\delta}{\mu}. \tag{9.9}$$

Remark 9.1. We claim that similar version of Theorem 9.3 for $p\in(3,6)$ remains to be true for the system (9.2) with the variable viscous coefficient $\widetilde{\mu}(a)$ which satisfies $\widetilde{\mu}(a)\geq\underline{\mu}>0$ and $\widetilde{\mu}(a)-\widetilde{\mu}(0)$ being small in $\mathscr{M}(B^{\frac{3}{p}}_{p,1}(\mathbb{R}^3))$. For simplicity, we shall not pursue this point here (for more details, see [14]).

Remark 9.2. In particular, as in [11], we can consider the case of a mixture of fluids with initial velocity with large vertical component, that is, $\rho_0=1+c\chi_\Omega$ and $u_0=(u_0^h,u_0^3)\in B^{-1+\frac{3}{p}}_{p,1}$, $2<p<6$, with c and u_0^h small enough (compared with u_0^3), and Ω a bounded or exterior C^1 domain.

Scheme of the proof and organization of the paper. In the second section, we shall apply the Littlewood–Paley theory to study the free transport equation with initial data in the multiplier spaces of Besov spaces. In Sect. 9.3, we shall present the estimate to the pressure function. Finally in the last section, we shall complete the proof of Theorem 9.3.

Let us complete this section by the notations we shall use in this context.

Notation. Let A,B be two operators, we denote $[A;B]=AB-BA$, the commutator between A and B. For $a\lesssim b$, we mean that there is a uniform constant C, which may be different on different lines, such that $a\leq Cb$. We shall denote by $(a|b)$ the $L^2(\mathbb{R}^3)$ inner product of a and b. $(d_j)_{j\in\mathbb{Z}}$ will be a generic element of $\ell^1(\mathbb{Z})$ so that $d_j\geq 0$ and $\sum_{j\in\mathbb{Z}}d_j=1$.

For X a Banach space and I an interval of $\mathbb{R}$, we denote by $\mathscr{C}(I;X)$ the set of continuous functions on I with values in X, and by $L^q(I;X)$ stands for the set of measurable functions on I with values in X, such that $t\longmapsto\|f(t)\|_X$ belongs to $L^q(I)$.

9.2 The Estimate of the Transport Equation

For the convenience of the reader, we recall some basic facts on Littlewood–Paley theory from [7].

Lemma 9.1. *Let $\mathscr{B}$ be a ball of $\mathbb{R}^3$, and $\mathscr{C}$ be a ring of $\mathbb{R}^3$; let $1 \leq p_2 \leq p_1 \leq \infty$. Then there hold:*

If the support of $\hat{a}$ is included in $2^k\mathscr{B}$, then

$$\|\partial_x^\alpha a\|_{L^{p_1}} \lesssim 2^{k(|\alpha|+3(\frac{1}{p_2}-\frac{1}{p_1}))}\|a\|_{L^{p_2}}.$$

If the support of $\hat{a}$ is included in $2^k\mathscr{C}$, then

$$\|a\|_{L^{p_1}} \lesssim 2^{-kN} \sup_{|\alpha|=N} \|\partial^\alpha a\|_{L^{p_1}}.$$

Lemma 9.2. *Let θ be a smooth function supported in an annulus $\mathscr{C}$ of $\mathbb{R}^n$. There exists a constant C such that for any $C^{0,1}$ measure-preserving global diffeomorphism ψ over $\mathbb{R}^n$ with inverse ϕ, any tempered distribution u with $\hat{u}$ supported in $\lambda\mathscr{C}$, any $p \in [1,\infty]$, and any $(\lambda,\mu) \in (0,\infty)^2$, we have*

$$\|\theta(\mu^{-1}D)(u\circ\psi)\|_{L^p} \leq C\|u\|_{L^p} \min\Big(\frac{\mu}{\lambda}\|D\phi\|_{L^\infty}, \frac{\lambda}{\mu}\|D\psi\|_{L^\infty}\Big).$$

Lemma 9.3. *If the support of $\hat{u}$ is included in $\lambda\mathscr{C}$, then there exists a positive constant c, such that*

$$\|e^{t\triangle}u\|_{L^p} \lesssim e^{-c\lambda^2 t}\|u\|_{L^p} \quad \text{for} \quad p \in [1,\infty].$$

Lemma 9.4. *Let $p_2 \geq p_1 \geq 1$, and $s_1 \leq \frac{3}{p_1}, s_2 \leq \frac{3}{p_2}$ with $s_1+s_2 > 3\max(0, \frac{1}{p_1}+\frac{1}{p_2}-1)$. Let $a \in B^{s_1}_{p_1,1}(\mathbb{R}^3), b \in B^{s_2}_{p_2,1}(\mathbb{R}^3)$. Then $ab \in B^{s_1+s_2-\frac{3}{p_1}}_{p_2,1}(\mathbb{R}^3)$, and*

$$\|ab\|_{B^{s_1+s_2-\frac{3}{p_1}}_{p_2,1}} \lesssim \|a\|_{B^{s_1}_{p_1,1}}\|b\|_{B^{s_2}_{p_2,1}}.$$

This section basically follows from Sect. 2 of [14]. Indeed as we shall not use Lagrange approach as that in [11], we need first to investigate the following transport equation:

$$\partial_t a + u\cdot\nabla a = 0, \qquad a|_{t=0} = a_0, \tag{9.10}$$

with initial data a_0 in the multiplier space of $B^s_{p,1}(\mathbb{R}^n)$. We denote $X_u(t,y)$ to be the flow map determined by u, namely,

$$X_u(t,y) = y + \int_0^t u(\tau, X_u(\tau,y))\,d\tau. \tag{9.11}$$

Lemma 9.5. *Let $s \in (-1,1)$ and $p \geq 1$. Let $a \in B^s_{p,1}(\mathbb{R}^n)$, $u \in L^1((0,T);Lip(\mathbb{R}^n))$, and X_u the flow map determined by (9.11). Then $a \circ X_u \in L^\infty((0,T);B^s_{p,1}(\mathbb{R}^n))$, and there holds*

$$\|a \circ X_u\|_{L^\infty_t(B^s_{p,1})} \leq C\|a\|_{B^s_{p,1}} \exp\Big\{C\int_0^t \|\nabla u(\tau)\|_{L^\infty}\,d\tau\Big\}. \tag{9.12}$$

Proof. Indeed let $a_\ell \stackrel{\text{def}}{=} \Delta_\ell a$; we deduce from Lemma 9.2 that

$$\|\Delta_j(a_\ell \circ X_u)\|_{L^\infty_t(L^p)} \leq Cd_\ell 2^{-\ell s}\|a\|_{B^s_{p,1}} \min\big(2^{j-\ell}, 2^{\ell-j}\big)\exp\Big\{C\int_0^t \|\nabla u(\tau)\|_{L^\infty} d\tau\Big\},$$

from which and $-1 < s < 1$, we infer for any $j \in \mathbb{Z}$

$$\begin{aligned}
\|\Delta_j(a \circ X_u)\|_{L^\infty_t(L^p)} \leq& \big(\sum_{\ell<j} + \sum_{\ell\geq j}\big)\|\Delta_j(a_\ell \circ X_u)\|_{L^\infty_t(L^p)}\\
\leq& C\|a\|_{B^s_{p,1}}\big(\sum_{\ell<j} d_\ell 2^{-\ell s}2^{\ell-j} + \sum_{\ell\geq j} d_\ell 2^{-\ell s}2^{j-\ell}\big)\\
&\times \exp\Big\{C\int_0^t \|\nabla u(\tau)\|_{L^\infty}\,d\tau\Big\}\\
\leq& Cd_j 2^{-js}\|a\|_{B^s_{p,1}} \exp\Big\{C\int_0^t \|\nabla u(\tau)\|_{L^\infty}\,d\tau\Big\},
\end{aligned} \tag{9.13}$$

This completes the proof of the lemma.

Proposition 9.1. *Let $s \in (-1,1)$ and $p \geq 1$. Let $u \in L^1_T(Lip(\mathbb{R}^3))$ and $a_0 \in \mathcal{M}(B^s_{p,1}(\mathbb{R}^3))$. Then (9.10) has a unique solution $a \in L^\infty([0,T];\mathcal{M}(B^s_{p,1}(\mathbb{R}^3)))$ so that*

$$\|a\|_{L^\infty_t(\mathcal{M}(B^s_{p,1}))} \leq \|a_0\|_{\mathcal{M}(B^s_{p,1})} \exp\Big\{C\|\nabla u\|_{L^1_t(L^\infty)}\Big\} \quad \text{for any} \quad t \in [0,T]. \tag{9.14}$$

Proof. Thanks to (9.11), we deduce from (9.10) that $a(t,x) = a_0(X_u^{-1}(t,x))$. Then thanks to Definition 9.4 and Lemma 9.5, we obtain

$$\begin{aligned}
\|a(t)\|_{\mathcal{M}(B^s_{p,1})} =& \sup_{\|\psi\|_{B^s_{p,1}}=1} \|\psi a(t)\|_{B^s_{p,1}}\\
=& \sup_{\|\psi\|_{B^s_{p,1}}=1} \|(\psi \circ X_u(t)a_0) \circ X_u^{-1}(t)\|_{B^s_{p,1}}\\
\leq& C \sup_{\|\psi\|_{B^s_{p,1}}=1} \|(\psi \circ X_u(t))a_0\|_{B^s_{p,1}} \exp\Big\{C\int_0^t \|\nabla u(\tau)\|_{L^\infty}\,d\tau\Big\},
\end{aligned}$$

applying Lemma 9.5 once again leads to

$$\begin{aligned}\|a(t)\|_{\mathcal{M}(B^s_{p,1})} \leq & C\|a_0\|_{\mathcal{M}(B^s_{p,1})}\exp\Big\{C\int_0^t \|\nabla u(\tau)\|_{L^\infty}\,d\tau\Big\}\sup_{\|\psi\|_{B^s_{p,1}}=1}\|\psi\circ X_u(t)\|_{B^s_{p,1}}\\ \leq & C\|a_0\|_{\mathcal{M}(B^s_{p,1})}\exp\Big\{C\int_0^t \|\nabla u(\tau)\|_{L^\infty}\,d\tau\Big\}\sup_{\|\psi\|_{B^s_{p,1}}=1}\|\psi\|_{B^s_{p,1}}\\ \leq & C\|a_0\|_{\mathcal{M}(B^s_{p,1})}\exp\Big\{C\int_0^t \|\nabla u(\tau)\|_{L^\infty}\,d\tau\Big\}.\end{aligned}$$

This completes the proof of Proposition 9.1.

9.3 The Estimate of the Pressure

As is well known, the main difficulty in the study of the Well-posedness of incompressible inhomogeneous Navier–Stokes equations is to derive the estimate for the pressure term. The goal of this section is to provide the pressure estimates in the framework of weighted Chemin-Lerner-type norms. We first get by taking div to the momentum equation of (9.4) that

$$-\Delta\Pi = \mathrm{div}(a\nabla\Pi) + \mathrm{div}_h\mathrm{div}_h(u^h\otimes u^h) + 2\partial_3\,\mathrm{div}_h(u^3u^h) + \partial_3^2(u^3)^2 - \mu\,\mathrm{div}(a\Delta u). \tag{9.15}$$

The following proposition concerning the estimate of the pressure will be the main ingredient used in the estimate of u^h.

Proposition 9.2. *Let $1\leq p<6$ and $a\in L^\infty_T(\mathcal{M}(B_{p,1}^{-1+\frac{3}{p}}))$, $u\in \widetilde{L}^\infty_T(B_{p,1}^{-1+\frac{3}{p}})\cap L^1_T(B_{p,1}^{1+\frac{3}{p}})$. We denote*

$$\begin{aligned} & f_1(t)\stackrel{\text{def}}{=}\|u^3(t)\|_{B_{p,1}^{1+\frac{3}{p}}},\quad f_2(t)\stackrel{\text{def}}{=}\|u^3(t)\|^2_{B_{p,1}^{\frac{3}{p}}}\quad\text{and}\\ & \Pi_\lambda\stackrel{\text{def}}{=}\Pi\exp\Big(-\lambda_1\int_0^t f_1(t')\,dt'-\lambda_2\int_0^t f_2(t')\,dt'\Big)\quad\text{for}\quad \lambda_1,\lambda_2>0,\end{aligned}\tag{9.16}$$

and similar notations for u_λ. Then (9.15) has a unique solution $\nabla\Pi\in L^1_T(B_{p,1}^{-1+\frac{3}{p}})$ which decays to zero when $|x|\to\infty$ so that for all $t\in[0,T]$, there holds

$$\begin{aligned}\|\nabla\Pi_\lambda\|_{L^1_t(B_{p,1}^{-1+\frac{3}{p}})}\leq & \frac{C}{1-C\|a\|_{L^\infty_t(\mathcal{M}(B_{p,1}^{-1+\frac{3}{p}}))}}\Big\{\big(\mu\|a\|_{L^\infty_t(\mathcal{M}(B_{p,1}^{-1+\frac{3}{p}}))}\\ & +\|u^h\|_{L^\infty_t(B_{p,1}^{-1+\frac{3}{p}})}\big)\|u^h_\lambda\|_{L^1_t(B_{p,1}^{1+\frac{3}{p}})}+\|u^h_\lambda\|^{\frac12}_{L^1_t(B_{p,1}^{1+\frac{3}{p}})}\|u^h_\lambda\|^{\frac12}_{L^1_{t,f_2}(B_{p,1}^{-1+\frac{3}{p}})}\\ & +\|u^h_\lambda\|_{L^1_{t,f_1}(B_{p,1}^{-1+\frac{3}{p}})}+\mu\|a\|_{L^\infty_t(\mathcal{M}(B_{p,1}^{-1+\frac{3}{p}}))}\|u^3\|_{L^1_t(B_{p,1}^{1+\frac{3}{p}})}\Big\}\end{aligned}\tag{9.17}$$

provided that $C\|a\|_{L^\infty_t(\mathcal{M}(B^{-1+\frac{3}{p}}_{p,1}))} \leq \frac{1}{2}$.

The proof of this proposition basically follows from that of Proposition 4.1 in [20]. We first recall the following two lemmas from [20]:

Lemma 9.6 (Lemma 4.1 of [20]). *Under the assumptions of Proposition 9.2 and* f_1, f_2 *being given by* (9.16)*, one has*

$$\|\Delta_j(u^3u^h)\|_{L^1_t(L^p)} \lesssim d_j 2^{-\frac{3j}{p}} \big(\|u^h\|^{\frac{1}{2}}_{L^1_t(B^{1+\frac{3}{p}}_{p,1})} \|u^h\|^{\frac{1}{2}}_{L^1_{t,f_2}(B^{-1+\frac{3}{p}}_{p,1})} + \|u^h\|_{L^1_{t,f_1}(B^{-1+\frac{3}{p}}_{p,1})}\big)$$

and

$$\|\Delta_j(u^3u^h)\|_{L^1_t(L^p)} \lesssim d_j 2^{-\frac{3j}{p}} \big(\|u^h\|_{\widetilde{L}^\infty_t(B^{-1+\frac{3}{p}}_{p,1})} \|u^3\|_{L^1_t(B^{1+\frac{3}{p}}_{p,1})} + \|u^h\|_{L^1_t(B^{1+\frac{3}{p}}_{p,1})} \|u^3\|_{\widetilde{L}^\infty_t(B^{-1+\frac{3}{p}}_{p,1})}\big).$$

Proof. For completeness, we outline its proof here. Indeed thanks to Bony's decomposition, we have

$$\Delta_j(u^3u^h) = \sum_{j' \geq j-N_0} \Delta_j\big(S_{j'}u^3\Delta_{j'}u^h + \Delta_{j'}u^3S_{j'+1}u^h\big). \tag{9.18}$$

Notice that on the one hand,

$$\begin{aligned}
\|\sum_{j' \geq j-N_0} \Delta_j\big(S_{j'}u^3\Delta_{j'}u^h\big)\|_{L^1_t(L^p)} &\lesssim \sum_{j' \geq j-N_0} \int_0^t \|u^3(t')\|_{B^{\frac{3}{p}}_{p,1}} \|\Delta_{j'}u^h(t')\|_{L^p}\, dt' \\
&\lesssim \sum_{j' \geq j-N_0} \Big\{\int_0^t \|u^3(t')\|^2_{B^{\frac{3}{p}}_{p,1}} \|\Delta_{j'}u^h(t')\|_{L^p}\, dt'\Big\}^{\frac{1}{2}} \|\Delta_{j'}u^h\|^{\frac{1}{2}}_{L^1_t(L^p)} \\
&\lesssim d_j 2^{-\frac{3j}{p}} \|u^h\|^{\frac{1}{2}}_{L^1_t(B^{1+\frac{3}{p}}_{p,1})} \|u^h\|^{\frac{1}{2}}_{L^1_{t,f_2}(B^{-1+\frac{3}{p}}_{p,1})}.
\end{aligned}$$

On the other hand, again thanks to Lemma 9.1, we obtain

$$\begin{aligned}
&\|\sum_{j' \geq j-N_0} \Delta_j\big(\Delta_{j'}u^3S_{j'+1}u^h\big)\|_{L^1_t(L^p)} \\
&\quad\lesssim \sum_{j' \geq j-N_0} 2^{-j'(1+\frac{3}{p})} \int_0^t \|u^3(t')\|_{B^{1+\frac{3}{p}}_{p,1}} \|S_{j'+1}u^h(t')\|_{L^\infty}\, dt' \\
&\quad\lesssim \sum_{j' \geq j-N_0} 2^{-j'(1+\frac{3}{p})} \sum_{\ell \leq j'} 2^{\frac{3\ell}{p}} \int_0^t \|u^3(t')\|_{B^{1+\frac{3}{p}}_{p,1}} \|\Delta_\ell u^h(t')\|_{L^p}\, dt' \\
&\quad\lesssim d_j 2^{-\frac{3j}{p}} \|u^h\|_{L^1_{t,f_1}(B^{-1+\frac{3}{p}}_{p,1})}.
\end{aligned}$$

This proves the first inequality of the lemma.

Whereas again thanks to (9.18), we get by applying Lemma 9.1 that

$$\begin{aligned}
&\|\Delta_j(u^3u^h)\|_{L^1_t(L^p)}\\
&\lesssim \sum_{j'\geq j-N_0}\big(\|S_{j'}u^3\|_{L^\infty_t(L^\infty)}\|\Delta_{j'}u^h\|_{L^1_t(L^p)}+\|\Delta_{j'}u^3\|_{L^1_t(L^p)}\|S_{j'+1}u^h\|_{L^\infty_t(L^\infty)}\big)\\
&\lesssim d_j2^{-\frac{3j}{p}}\big(\|u^h\|_{L^1_t(B^{1+\frac{3}{p}}_{p,1})}\|u^3\|_{\widetilde{L}^\infty_t(B^{-1+\frac{3}{p}}_{p,1})}+\|u^h\|_{\widetilde{L}^\infty_t(B^{-1+\frac{3}{p}}_{p,1})}\|u^3\|_{L^1_t(B^{1+\frac{3}{p}}_{p,1})}\big).
\end{aligned}$$

This completes the proof of the lemma.

Lemma 9.7 (Lemma 4.2 of [20]). *Let* $1\leq p<6$ *and* f_1,f_2 *be given by* (9.16)*. Then under the assumptions of Proposition 9.2, one has*

$$\begin{aligned}
&\|\Delta_j(u^3\operatorname{div}_h u^h)\|_{L^1_t(L^p)}\\
&\qquad\lesssim d_j2^{j(1-\frac{3}{p})}\big(\|u^h\|^{\frac12}_{L^1_t(B^{1+\frac{3}{p}}_{p,1})}\|u^h\|^{\frac12}_{L^1_{t,f_2}(B^{-1+\frac{3}{p}}_{p,1})}+\|u^h\|_{L^1_{t,f_1}(B^{-1+\frac{3}{p}}_{p,1})}\big),
\end{aligned}$$

and

$$\begin{aligned}
\|\Delta_j(u^3\operatorname{div}_h u^h)\|_{L^1_t(L^p)}\lesssim d_j2^{j(1-\frac{3}{p})}\big(&\|u^h\|_{\widetilde{L}^\infty_t(B^{-1+\frac{3}{p}}_{p,1})}\|u^3\|_{L^1_t(B^{1+\frac{3}{p}}_{p,1})}\\
&+\|u^h\|_{L^1_t(B^{1+\frac{3}{p}}_{p,1})}\|u^3\|_{\widetilde{L}^\infty_t(B^{-1+\frac{3}{p}}_{p,1})}\big).
\end{aligned}$$

Proof. Again we outline its proof here. We first get by applying Bony's decomposition that

$$u^3\operatorname{div}_h u^h=T_{u^3}\operatorname{div}_h u^h+T_{\operatorname{div}_h u^h}u^3+R(u^3,\operatorname{div}_h u^h). \tag{9.19}$$

Applying Lemma 9.1 gives

$$\begin{aligned}
\|\Delta_j\big(T_{u^3}\operatorname{div}_h u^h\big)\|_{L^1_t(L^p)}&\lesssim\sum_{|j'-j|\leq5}2^{j'}\int_0^t\|u^3(t')\|_{B^{\frac{3}{p}}_{p,1}}\|\Delta_{j'}u^h(t')\|_{L^p}\,dt'\\
&\lesssim d_j2^{j(1-\frac{3}{p})}\|u^h\|^{\frac12}_{L^1_t(B^{1+\frac{3}{p}}_{p,1})}\|u^h\|^{\frac12}_{L^1_{t,f_2}(B^{-1+\frac{3}{p}}_{p,1})}.
\end{aligned}$$

Similarly, we have

$$\begin{aligned}
\|\Delta_j\big(T_{\operatorname{div}_h u^h}u^3\big)\|_{L^1_t(L^p)}&\lesssim\sum_{|j'-j|\leq5}2^{-j'(1+\frac{3}{p})}\int_0^t\|u^3(t')\|_{B^{1+\frac{3}{p}}_{p,1}}\|S_{j'-1}\operatorname{div}_h u^h(t')\|_{L^\infty}\,dt'\\
&\lesssim d_j2^{j(1-\frac{3}{p})}\|u^h\|_{L^1_{t,f_1}(B^{-1+\frac{3}{p}}_{p,1})}.
\end{aligned}$$

Finally, for $2\leq p<6$, we get by applying Lemma 9.1 that

$$\|\Delta_j\big(R(u^3,\mathrm{div}_h u^h)\big)\|_{L^1_t(L^p)} \lesssim 2^{\frac{3j}{p}} \sum_{j'\geq j-N_0} 2^{j'} \int_0^t \|\widetilde{\Delta}_{j'}u^3(t')\|_{L^p}\|\Delta_{j'}u^h(t')\|_{L^p}\,dt'$$
$$\lesssim 2^{\frac{3j}{p}} \sum_{j'\geq j-N_0} 2^{-\frac{3j'}{p}} \int_0^t \|u^3(t')\|_{B^{1+\frac{3}{p}}_{p,1}}\|\Delta_{j'}u^h(t')\|_{L^p}\,dt'$$
$$\lesssim d_j 2^{j(1-\frac{3}{p})}\|u^h\|_{L^1_{t,f_1}(B^{-1+\frac{3}{p}}_{p,1})}.$$

For $1\leq p<2$, we get by applying Lemma 9.1 that

$$\|\Delta_j\big(R(u^3,\mathrm{div}_h u^h)\big)\|_{L^1_t(L^p)}$$
$$\lesssim 2^{3j(1-\frac{1}{p})} \sum_{j'\geq j-N_0} 2^{j'} \int_0^t \|\widetilde{\Delta}_{j'}u^3(t')\|_{L^{\frac{p}{p-1}}}\|\Delta_{j'}u^h(t')\|_{L^p}\,dt'$$
$$\lesssim 2^{3j(1-\frac{1}{p})} \sum_{j'\geq j-N_0} 2^{-\frac{3j'}{p}}2^{3j'(\frac{2}{p}-1)} \int_0^t \|u^3(t')\|_{B^{1+\frac{3}{p}}_{p,1}}\|\Delta_{j'}u^h(t')\|_{L^p}\,dt'$$
$$\lesssim d_j 2^{j(1-\frac{3}{p})}\|u^h\|_{L^1_{t,f_1}(B^{-1+\frac{3}{p}}_{p,1})}.$$

Along with (9.19), we prove the first inequality of Lemma 9.7.

On the other hand, it is easy to observe that

$$\|\Delta_j\big(T_{u^3}\mathrm{div}_h u^h + T_{\mathrm{div}_h u^h}u^3\big)\|_{L^1_t(L^p)}$$
$$\lesssim \sum_{|j'-j|\leq 5}\big(\|S_{j'-1}u^3\|_{L^\infty_t(L^\infty)}\|\Delta_{j'}\mathrm{div}_h u^h\|_{L^1_t(L^p)}$$
$$+\|S_{j'-1}\mathrm{div}_h u^h\|_{L^\infty_t(L^\infty)}\|\Delta_{j'}u^3\|_{L^1(L^p)}\big)$$
$$\lesssim d_j 2^{j(1-\frac{3}{p})}\big(\|u^3\|_{\widetilde{L}^\infty_t(B^{-1+\frac{3}{p}}_{p,1})}\|u^h\|_{L^1_t(B^{1+\frac{3}{p}}_{p,1})}+\|u^h\|_{\widetilde{L}^\infty_t(B^{-1+\frac{3}{p}}_{p,1})}\|u^3\|_{L^1_t(B^{1+\frac{3}{p}}_{p,1})}\big).$$

Whereas again for $2\leq p<6$, we get by applying Lemma 9.1 that

$$\|\Delta_j(R(u^3,\mathrm{div}_h u^h))\|_{L^1_t(L^p)} \lesssim 2^{\frac{3j}{p}} \sum_{j'\geq j-N_0} 2^{j'}\|\Delta_{j'}u^3\|_{L^1_t(L^p)}\|\widetilde{\Delta}_{j'}u^h\|_{L^\infty_t(L^p)}$$
$$\lesssim d_j 2^{j(1-\frac{3}{p})}\|u^h\|_{\widetilde{L}^\infty_t(B^{-1+\frac{3}{p}}_{p,1})}\|u^3\|_{L^1_t(B^{1+\frac{3}{p}}_{p,1})}.$$

The case $1\leq p<2$ follows the same lines as above. Whence thanks to (9.19), we obtain the second inequality of the lemma.

Now we present the proof of Proposition 9.2.

Proof (Proof of Proposition 9.2). Again as both the proof of the existence and uniqueness of solutions to (9.15) are essentially followed by the estimates (9.17) for some appropriate approximate solutions of (9.15). For simplicity, we just prove (9.17) for smooth enough solutions of (9.15). Indeed thanks to (9.15) and $\mathrm{div}\,u=0$, we have

$$\nabla\Pi_\lambda = \nabla(-\Delta)^{-1}\big[\mathrm{div}(a\nabla\Pi_\lambda) + \mathrm{div}_h\,\mathrm{div}_h(u^h\otimes u_\lambda^h) + 2\partial_3\,\mathrm{div}_h(u^3u_\lambda^h) \\ - 2\partial_3(u^3\,\mathrm{div}_h\,u_\lambda^h) - \mu\,\mathrm{div}_h(a\Delta u_\lambda^h) - \mu\partial_3(a\Delta u_\lambda^3)\big]. \tag{9.20}$$

Applying Δ_j to the above equation and using Lemma 9.1 leads to

$$\|\Delta_j(\nabla\Pi_\lambda)\|_{L_t^1(L^p)} \lesssim \|\Delta_j(a\nabla\Pi_\lambda)\|_{L_t^1(L^p)} + 2^j\big(\|\Delta_j(u^h\otimes u_\lambda^h)\|_{L_t^1(L^p)} \\ + \|\Delta_j(u^3u_\lambda^h)\|_{L_t^1(L^p)}\big) + \|\Delta_j(u^3\,\mathrm{div}_h\,u_\lambda^h)\|_{L_t^1(L^p)} \\ + \mu\|\Delta_j(a\Delta u_\lambda^h)\|_{L_t^1(L^p)} + \mu\|\Delta_j(a\Delta u_\lambda^3)\|_{L_t^1(L^p)}. \tag{9.21}$$

However as $1\le p<6$, applying Definition 9.4, Lemma 9.4 and standard product laws in Besov space gives rise to

$$\|\Delta_j(a\nabla\Pi_\lambda)\|_{L_t^1(L^p)} \lesssim d_j 2^{j(1-\frac{3}{p})}\|a\|_{L_t^\infty(\mathcal{M}(B_{p,1}^{-1+\frac{3}{p}}))}\|\nabla\Pi_\lambda\|_{L_t^1(B_{p,1}^{-1+\frac{3}{p}})} \quad \text{and}$$

$$\|\Delta_j(u^h\otimes u_\lambda^h)\|_{L_t^1(L^p)} \lesssim d_j 2^{-\frac{3j}{p}}\|u^h\|_{L_t^\infty(B_{p,1}^{-1+\frac{3}{p}})}\|u_\lambda^h\|_{L_t^1(B_{p,1}^{1+\frac{3}{p}})}, \tag{9.22}$$

$$\|\Delta_j(a\Delta u_\lambda)\|_{L_t^1(L^p)} \lesssim d_j 2^{j(1-\frac{3}{p})}\|a\|_{L_t^\infty(\mathcal{M}(B_{p,1}^{-1+\frac{3}{p}}))}\|u_\lambda\|_{L_t^1(B_{p,1}^{1+\frac{3}{p}})}.$$

While thanks to Definition 9.3 and (9.16), one has

$$\|u_\lambda^3\|_{L_t^1(B_{p,1}^{1+\frac{3}{p}})} \le \|u^3\|_{L_t^1(B_{p,1}^{1+\frac{3}{p}})},$$

which along with Lemma 9.6 to Lemma 9.7 and (9.21) implies that

$$\|\Delta_j(\nabla\Pi_\lambda)\|_{L_t^1(L^p)} \\ \lesssim d_j 2^{j(1-\frac{3}{p})}\Big\{\|a\|_{L_t^\infty(\mathcal{M}(B_{p,1}^{-1+\frac{3}{p}}))}\|\nabla\Pi_\lambda\|_{L_t^1(B_{p,1}^{-1+\frac{3}{p}})} + \big(\mu\|a\|_{L_t^\infty(\mathcal{M}(B_{p,1}^{-1+\frac{3}{p}}))} \\ + \|u^h\|_{L_t^\infty(B_{p,1}^{-1+\frac{3}{p}})}\big)\|u_\lambda^h\|_{L_t^1(B_{p,1}^{1+\frac{3}{p}})} + \|u_\lambda^h\|_{L_t^1(B_{p,1}^{1+\frac{3}{p}})}^{\frac{1}{2}}\|u_\lambda^h\|_{L_{t,f_2}^1(B_{p,1}^{-1+\frac{3}{p}})}^{\frac{1}{2}} \\ + \|u_\lambda^h\|_{L_{t,f_1}^1(B_{p,1}^{-1+\frac{3}{p}})} + \mu\|a\|_{L_t^\infty(\mathcal{M}(B_{p,1}^{-1+\frac{3}{p}}))}\|u^3\|_{L_t^1(B_{p,1}^{1+\frac{3}{p}})}\Big\},$$

from which, we deduce that

$$
\begin{aligned}
&\|\nabla\Pi_\lambda\|_{L^1_t(B_{p,1}^{-1+\frac{3}{p}})}\\
&\leq C\Big\{\|a\|_{L^\infty_t(\mathcal{M}(B_{p,1}^{-1+\frac{3}{p}}))}\|\nabla\Pi_\lambda\|_{L^1_t(B_{p,1}^{-1+\frac{3}{p}})}+\big(\mu\|a\|_{L^\infty_t(\mathcal{M}(B_{p,1}^{-1+\frac{3}{p}}))}\\
&\quad+\|u^h\|_{L^\infty_t(B_{p,1}^{-1+\frac{3}{p}})}\big)\|u^h_\lambda\|_{L^1_t(B_{p,1}^{1+\frac{3}{p}})}+\|u^h_\lambda\|^{\frac{1}{2}}_{L^1_t(B_{p,1}^{1+\frac{3}{p}})}\|u^h_\lambda\|^{\frac{1}{2}}_{L^1_{t,f_2}(B_{p,1}^{-1+\frac{3}{p}})}\\
&\quad+\|u^h_\lambda\|_{L^1_{t,f_1}(B_{p,1}^{-1+\frac{3}{p}})}+\mu\|a\|_{L^\infty_t(\mathcal{M}(B_{p,1}^{-1+\frac{3}{p}}))}\|u^3\|_{L^1_t(B_{p,1}^{1+\frac{3}{p}})}\Big\}\quad\text{for}\quad t\leq T.
\end{aligned}
$$

This in particular implies (9.17) if $C\|a\|_{L^\infty_t(\mathcal{M}(B_{p,1}^{-1+\frac{3}{p}}))}\leq\frac{1}{2}$. This finishes the proof of Proposition 9.2.

To deal with the estimate of u^3, we also need the following proposition:

Proposition 9.3. *Under the assumptions of Proposition 9.2, one has*

$$
\begin{aligned}
&\|\nabla\Pi\|_{L^1_t(B_{p,1}^{-1+\frac{3}{p}})}\\
&\leq\frac{C}{1-C\|a\|_{L^\infty_t(\mathcal{M}(B_{p,1}^{-1+\frac{3}{p}}))}}\Big\{\|u^h\|_{L^1_t(B_{p,1}^{1+\frac{3}{p}})}\|u^3\|_{\widetilde{L}^\infty_t(B_{p,1}^{-1+\frac{3}{p}})}\\
&\quad+\big(\|u^h\|_{L^1_t(B_{p,1}^{1+\frac{3}{p}})}+\|u^3\|_{L^1_t(B_{p,1}^{1+\frac{3}{p}})}\big)\big(\mu\|a\|_{L^\infty_t(\mathcal{M}(B_{p,1}^{-1+\frac{3}{p}}))}+\|u^h\|_{\widetilde{L}^\infty_t(B_{p,1}^{-1+\frac{3}{p}})}\big)\Big\}
\end{aligned}
\tag{9.23}
$$

for $t\leq T$ provided that $C\|a\|_{L^\infty_t(\mathcal{M}(B_{p,1}^{-1+\frac{3}{p}}))}\leq\frac{1}{2}$.

Proof. The proof of this proposition exactly follows the same line as that of Proposition 9.2. Indeed taking $\lambda_1=\lambda_2=0$ in (9.21) and (9.22) and applying Lemma 9.6 to Lemma 9.7, we arrive at

$$
\begin{aligned}
&\|\nabla\Pi\|_{L^1_t(B_{p,1}^{-1+\frac{3}{p}})}\\
&\leq C\Big\{\|a\|_{L^\infty_t(\mathcal{M}(B_{p,1}^{-1+\frac{3}{p}}))}\|\nabla\Pi\|_{L^1_t(B_{p,1}^{-1+\frac{3}{p}})}+\|u^h\|_{L^1_t(B_{p,1}^{1+\frac{3}{p}})}\|u^3\|_{\widetilde{L}^\infty_t(B_{p,1}^{-1+\frac{3}{p}})}\\
&\quad+\big(\|u^h\|_{L^1_t(B_{p,1}^{1+\frac{3}{p}})}+\|u^3\|_{L^1_t(B_{p,1}^{1+\frac{3}{p}})}\big)\big(\mu\|a\|_{L^\infty_t(\mathcal{M}(B_{p,1}^{-1+\frac{3}{p}}))}+\|u^h\|_{\widetilde{L}^\infty_t(B_{p,1}^{-1+\frac{3}{p}})}\big)\Big\}
\end{aligned}
$$

for $t\leq T$, from which and the fact that $C\|a\|_{L^\infty_t(\mathcal{M}(B_{p,1}^{-1+\frac{3}{p}}))}\leq\frac{1}{2}$, we conclude the proof of (9.23).

9.4 The Proof of Theorem 9.3

The goal of this section is to present the proof of Theorem 9.3. Indeed given $a_0 \in \mathscr{M}(B_{p,1}^{-1+\frac{3}{p}}(\mathbb{R}^3)), u_0 \in B_{p,1}^{-1+\frac{3}{p}}(\mathbb{R}^3)$ with $\|a_0\|_{L_t^\infty(\mathscr{M}(B_{p,1}^{-1+\frac{3}{p}}))}$ sufficiently small and p satisfying the conditions listed in Theorem 9.3, repeating the proof of Theorem 1 of [11] ensures that there exists a positive time T so that (9.4) has a uniqueness of solution (a,u,Π) with

$$a \in L^\infty([0,T];\mathscr{M}(B_{p,1}^{-1+\frac{3}{p}}(\mathbb{R}^3))),\ u \in \mathscr{C}([0,T];B_{p,1}^{-1+\frac{3}{p}}(\mathbb{R}^3)) \cap L^1((0,T);B_{p,1}^{1+\frac{3}{p}}(\mathbb{R}^3))$$
$$\text{and}\quad \nabla\Pi \in L^1((0,T);B_{p,1}^{-1+\frac{3}{p}}(\mathbb{R}^3)). \tag{9.24}$$

We denote T^* to be the largest time so that there holds (9.24). Hence to prove Theorem 9.3, we only need to prove that $T^* = \infty$ and there holds (9.9). Toward this and motivated by [13, 19, 20], we shall deal with the L^p-type energy estimate for u^h and u^3 separately.

9.4.1 *The Estimate of u^h*

As in Proposition 9.2, let $f_1(t)$, $f_2(t)$, u_λ, Π_λ be given by (9.16). Then thanks to (9.4), we have

$$\partial_t u_\lambda^h + (\lambda_1 f_1(t) + \lambda_2 f_2(t))u_\lambda - \mu\Delta u_\lambda^h = -u\cdot\nabla u_\lambda^h - (1+a)\nabla_h\Pi_\lambda + \mu a\Delta u_\lambda^h.$$

Applying the operator Δ_j to the above equation and taking the L^2 inner product of the resulting equation with $|\Delta_j u_\lambda^h|^{p-2}\Delta_j u_\lambda^h$ (again in the case when $p \in (1,2)$, we need to make some modification as that [8]), we obtain

$$\begin{aligned}\frac{1}{p}\frac{d}{dt}\|\Delta_j u_\lambda^h(t)\|_{L^p}^p + (\lambda_1 f_1(t) + \lambda_2 f_2(t))\|\Delta_j u_\lambda^h(t)\|_{L^p}^p&\\ -\mu\int_{\mathbb{R}^3}\Delta\Delta_j u_\lambda^h \mid |\Delta_j u_\lambda^h|^{p-2}\Delta_j u_\lambda^h\,dx&\\ = -\int_{\mathbb{R}^3}(\Delta_j(u\cdot\nabla u_\lambda^h) + \Delta_j((1+a)\nabla_h\Pi_\lambda) - \mu\Delta_j(a\Delta u_\lambda^h)) \mid |\Delta_j u_\lambda^h|^{p-2}\Delta_j u_\lambda^h\,dx.&\end{aligned}$$

However thanks to [8, 21], there exists a positive constant $\bar{c}$ so that

$$-\int_{\mathbb{R}^3}\Delta\Delta_j u_\lambda^h \mid |\Delta_j u_\lambda^h|^{p-2}\Delta_j u_\lambda^h\,dx \geq \bar{c}2^{2j}\|\Delta_j u_\lambda^h\|_{L^p}^p,$$

whence a similar argument as that in [8] gives rise to

$$\frac{d}{dt}\|\Delta_j u^h_\lambda(t)\|_{L^p} + (\lambda_1 f_1(t) + \lambda_2 f_2(t))\|\Delta_j u^h_\lambda(t)\|_{L^p} + \bar{c}\mu 2^{2j}\|\Delta_j u^h_\lambda(t)\|_{L^p}$$
$$\leq \|\Delta_j(u\cdot\nabla u^h_\lambda)\|_{L^p} + \|\Delta_j((1+a)\nabla_h\Pi_\lambda)\|_{L^p} + \mu\|\Delta_j(a\Delta u^h_\lambda)\|_{L^p}. \tag{9.25}$$

Applying Lemmas 9.4 and 9.6, we obtain

$$\begin{aligned}\|\Delta_j(u\cdot\nabla u^h_\lambda)\|_{L^1_t(L^p)} \leq & 2^j\big(\|\Delta_j(u^h\otimes u^h_\lambda)\|_{L^1_t(L^p)} + \|\Delta_j(u^3 u^h_\lambda)\|_{L^1_t(L^p)}\big)\\ \leq & Cd_j 2^{j(1-\frac{3}{p})}\big(\|u^h\|_{\widetilde{L}^\infty_t(B^{-1+\frac{3}{p}}_{p,1})}\|u^h_\lambda\|_{L^1_t(B^{1+\frac{3}{p}}_{p,1})}\\ & + \|u^h_\lambda\|^{\frac{1}{2}}_{L^1_t(B^{1+\frac{3}{p}}_{p,1})}\|u^h_\lambda\|^{\frac{1}{2}}_{L^1_{t,f_2}(B^{-1+\frac{3}{p}}_{p,1})} + \|u^h_\lambda\|_{L^1_{t,f_1}(B^{-1+\frac{3}{p}}_{p,1})}\big).\end{aligned}$$

While applying Definition 9.4 leads to

$$\|\Delta_j(a\Delta u^h_\lambda)\|_{L^1_t(L^p)} \leq Cd_j 2^{j(1-\frac{3}{p})}\|a\|_{L^\infty_t(\mathcal{M}(B^{-1+\frac{3}{p}}_{p,1}))}\|u^h_\lambda\|_{L^1_t(B^{1+\frac{3}{p}}_{p,1})},$$

and

$$\|\Delta_j((1+a)\nabla_h\Pi_\lambda)\|_{L^1_t(L^p)} \leq Cd_j 2^{j(1-\frac{3}{p})}(1+\|a\|_{L^\infty_t(\mathcal{M}(B^{-1+\frac{3}{p}}_{p,1}))})\|\nabla_h\Pi_\lambda\|_{L^1_t(B^{-1+\frac{3}{p}}_{p,1})},$$

Moreover under the assumption that

$$\|a\|_{L^\infty_t(\mathcal{M}(B^{-1+\frac{3}{p}}_{p,1}))} \leq c_1 \tag{9.26}$$

for some c_1 sufficiently small, (9.17) ensures that

$$\begin{aligned}&\|\Delta_j((1+a)\nabla_h\Pi_\lambda)\|_{L^1_t(L^p)}\\ &\leq Cd_j 2^{j(1-\frac{3}{p})}\Big\{\big(\mu\|a\|_{L^\infty_t(\mathcal{M}(B^{-1+\frac{3}{p}}_{p,1}))} + \|u^h\|_{\widetilde{L}^\infty_t(B^{-1+\frac{3}{p}}_{p,1})}\big)\|u^h_\lambda\|_{L^1_t(B^{1+\frac{3}{p}}_{p,1})}\\ &\qquad + \|u^h_\lambda\|^{\frac{1}{2}}_{L^1_t(B^{1+\frac{3}{p}}_{p,1})}\|u^h_\lambda\|^{\frac{1}{2}}_{L^1_{t,f_2}(B^{-1+\frac{3}{p}}_{p,1})} + \|u^h_\lambda\|_{L^1_{t,f_1}(B^{-1+\frac{3}{p}}_{p,1})}\\ &\qquad + \mu\|a\|_{L^\infty_t(\mathcal{M}(B^{-1+\frac{3}{p}}_{p,1}))}\|u^3\|_{L^1_t(B^{1+\frac{3}{p}}_{p,1})}\Big\}.\end{aligned}$$

Thanks to (9.26), integrating (9.25) over $[0,t]$ and substituting the above estimates into the resulting inequality, we obtain

$$\begin{aligned}
&\|u_\lambda^h\|_{\widetilde{L}_t^\infty(B_{p,1}^{-1+\frac{3}{p}})}+\lambda_1\|u_\lambda^h\|_{L^1_{t,f_1}(B_{p,1}^{-1+\frac{3}{p}})}+\lambda_2\|u_\lambda^h\|_{L^1_{t,f_2}(B_{p,1}^{-1+\frac{3}{p}})}+\bar{c}\mu\|u_\lambda^h\|_{L^1_t(B_{p,1}^{1+\frac{3}{p}})}\\
&\leq\|u_0^h\|_{B_{p,1}^{-1+\frac{3}{p}}}+\frac{\bar{c}\mu}{2}\|u_\lambda^h\|_{L^1_t(B_{p,1}^{1+\frac{3}{p}})}+C\Big\{\big(\mu\|a\|_{L_t^\infty(\mathcal{M}(B_{p,1}^{-1+\frac{3}{p}}))}\\
&\quad+\|u^h\|_{\widetilde{L}_t^\infty(B_{p,1}^{-1+\frac{3}{p}})}\big)\|u_\lambda^h\|_{L^1_t(B_{p,1}^{1+\frac{3}{p}})}+\frac{1}{\mu}\|u_\lambda^h\|_{L^1_{t,f_2}(B_{p,1}^{-1+\frac{3}{p}})}\\
&\quad+\|u_\lambda^h\|_{L^1_{t,f_1}(B_{p,1}^{-1+\frac{3}{p}})}+\mu\|a\|_{L_t^\infty(\mathcal{M}(B_{p,1}^{-1+\frac{3}{p}}))}\|u^3\|_{L^1_t(B_{p,1}^{1+\frac{3}{p}})}\Big\}\quad\text{for}\quad t\leq T.
\end{aligned}\tag{9.27}$$

Taking $\lambda_1=4C$ and $\lambda_2=\frac{2C}{\mu}$ in (9.27) resulting in

$$\begin{aligned}
&\|u_\lambda^h\|_{\widetilde{L}_t^\infty(B_{p,1}^{-1+\frac{3}{p}})}+\frac{\lambda_1}{2}\|u_\lambda^h\|_{L^1_{t,f_1}(B_{p,1}^{-1+\frac{3}{p}})}+\frac{\lambda_2}{2}\|u_\lambda^h\|_{L^1_{t,f_2}(B_{p,1}^{-1+\frac{3}{p}})}+\frac{\bar{c}\mu}{2}\|u_\lambda^h\|_{L^1_t(B_{p,1}^{1+\frac{3}{p}})}\\
&\leq\|u_0^h\|_{B_{p,1}^{-1+\frac{3}{p}}}+C_1\Big\{\big(\mu\|a\|_{L_t^\infty(\mathcal{M}(B_{p,1}^{-1+\frac{3}{p}}))}+\|u^h\|_{\widetilde{L}_t^\infty(B_{p,1}^{-1+\frac{3}{p}})}\big)\|u_\lambda^h\|_{L^1_t(B_{p,1}^{1+\frac{3}{p}})}\\
&\quad+\mu\|a\|_{L_t^\infty(\mathcal{M}(B_{p,1}^{-1+\frac{3}{p}}))}\|u^3\|_{L^1_t(B_{p,1}^{1+\frac{3}{p}})}\Big\}.
\end{aligned}\tag{9.28}$$

Now let c_2 be a small enough positive constant, which will be determined later on, we define $\mathfrak{T}$ by

$$\mathfrak{T}\overset{\text{def}}{=}\max\Big\{t\in[0,T^*):\|u^h\|_{\widetilde{L}_t^\infty(B_{p,1}^{-1+\frac{3}{p}})}+\mu\big(\|a\|_{L_t^\infty(\mathcal{M}(B_{p,1}^{-1+\frac{3}{p}}))}+\|u^h\|_{L^1_t(B_{p,1}^{1+\frac{3}{p}})}\big)\leq c_2\mu\Big\}.\tag{9.29}$$

(9.29) implies that $\|a\|_{L_{\mathfrak{T}}^\infty(\mathcal{M}(B_{p,1}^{-1+\frac{3}{p}}))}\leq c_2$, in particular, if we take $c_2\leq c_1$ in (9.26), then there automatically holds (9.26) for $t\leq\mathfrak{T}$. In what follows, we shall prove that $\mathfrak{T}=\infty$ under the assumption of (9.8). Otherwise, taking $c_2\leq\bar{c}_2\overset{\text{def}}{=}\min\big(c_1,\frac{\bar{c}}{4C_1}\big)$, we deduce from (9.28) that

$$\|u_\lambda^h\|_{\widetilde{L}_t^\infty(B_{p,1}^{-1+\frac{3}{p}})}+\frac{\bar{c}\mu}{4}\|u_\lambda^h\|_{L^1_t(B_{p,1}^{1+\frac{3}{p}})}\leq\|u_0^h\|_{B_{p,1}^{-1+\frac{3}{p}}}+C_1\mu\|a\|_{L_t^\infty(\mathcal{M}(B_{p,1}^{-1+\frac{3}{p}}))}\|u^3\|_{L^1_t(B_{p,1}^{1+\frac{3}{p}})},\tag{9.30}$$

for $t\leq\mathfrak{T}$.

On the other hand, it is easy to observe from (9.16) that

$$\begin{aligned}
&\big(\|u^h\|_{\widetilde{L}_t^\infty(B_{p,1}^{-1+\frac{3}{p}})}+\frac{\bar{c}\mu}{4}\|u^h\|_{L^1_t(B_{p,1}^{1+\frac{3}{p}})}\big)\exp\Big\{-\int_0^t(\lambda_1f_1+\lambda_2f_2)(t')\,dt'\Big\}\\
&\leq\|u_\lambda^h\|_{\widetilde{L}_t^\infty(B_{p,1}^{-1+\frac{3}{p}})}+\frac{\bar{c}\mu}{4}\|u_\lambda^h\|_{L^1_t(B_{p,1}^{1+\frac{3}{p}})}\big),
\end{aligned}$$

As a consequence, we deduce from (9.29) to (9.30) that

$$\|u^h\|_{\widetilde{L}^\infty_t(B_{p,1}^{-1+\frac{3}{p}})}+\mu\|u^h\|_{L^1_t(B_{p,1}^{1+\frac{3}{p}})}\leq\left(\|u_0^h\|_{B_{p,1}^{-1+\frac{3}{p}}}+\mu\|a\|_{L^\infty_t(\mathcal{M}(B_{p,1}^{-1+\frac{3}{p}}))}\right)$$
$$\times\exp\Big\{C\int_0^t\big(\|u^3(t')\|_{B_{p,1}^{1+\frac{3}{p}}}+\frac{1}{\mu}\|u^3(t')\|^2_{B_{p,1}^{\frac{3}{p}}}\big)\,dt'\Big\}\quad\text{for}\quad t\leq\mathfrak{T}. \tag{9.31}$$

9.4.2 The Estimate of u^3

We use the fact that the equation on the vertical component of the velocity field is a linear equation with coefficients depending on its horizontal components and a. Thanks to the u^3 equation of (9.4)

$$\partial_t u^3+u^h\nabla_h u^3-\operatorname{div}_h u^h u^3-\mu\Delta u^3+\partial_3\Pi=a(\mu\Delta u^3-\partial_3\Pi)$$

with $\partial_3\Pi$ the linear expression in u^3 given by (9.20) (with $\lambda=0$), we get by a similar derivation of (9.25) that

$$\|\Delta_j u^3\|_{L^\infty_t(L^p)}+\bar{c}\mu 2^{2j}\|\Delta_j u^3\|_{L^1_t(L^p)}\leq\|\Delta_j u_0^3\|_{L^p}$$
$$+C\Big(\|\Delta_j(u\cdot\nabla u^3)\|_{L^1_t(L^p)}+\|\Delta_j((1+a)\partial_3\Pi)\|_{L^1_t(L^p)}+\mu\|\Delta_j(a\Delta u^3)\|_{L^1_t(L^p)}\Big). \tag{9.32}$$

Applying LemmaS 9.6 and 9.7 ensures that

$$\|\Delta_j(u\cdot\nabla u^3)\|_{L^1_t(L^p)}\lesssim 2^j\|\Delta_j(u^hu^3)\|_{L^1_t(L^p)}+\|\Delta_j(u^3\operatorname{div}_h u^h)\|_{L^1_t(L^p)}$$
$$\lesssim d_j 2^{j(1-\frac{3}{p})}\big(\|u^h\|_{\widetilde{L}^\infty_t(B_{p,1}^{-1+\frac{3}{p}})}\|u^3\|_{L^1_t(B_{p,1}^{1+\frac{3}{p}})}+\|u^h\|_{L^1_t(B_{p,1}^{1+\frac{3}{p}})}\|u^3\|_{\widetilde{L}^\infty_t(B_{p,1}^{-1+\frac{3}{p}})}\big).$$

Whereas under the assumption of (9.26), we get by applying Lemma 9.4 and Proposition 9.3 that

$$\|\Delta_j((1+a)\partial_3\Pi)\|_{L^1_t(L^p)}\leq Cd_j 2^{j(1-\frac{3}{p})}(1+\|a\|_{L^\infty_t(\mathcal{M}(B_{p,1}^{-1+\frac{3}{p}}))})\|\partial_3\Pi\|_{L^1_t(B_{p,1}^{-1+\frac{3}{p}})}$$
$$\leq Cd_j 2^{j(1-\frac{3}{p})}\Big\{\|u^h\|_{L^1_t(B_{p,1}^{1+\frac{3}{p}})}\|u^3\|_{\widetilde{L}^\infty_t(B_{p,1}^{-1+\frac{3}{p}})}$$
$$+\big(\|u^h\|_{L^1_t(B_{p,1}^{1+\frac{3}{p}})}+\|u^3\|_{L^1_t(B_{p,1}^{1+\frac{3}{p}})}\big)\big(\mu\|a\|_{L^\infty_t(\mathcal{M}(B_{p,1}^{-1+\frac{3}{p}}))}+\|u^h\|_{\widetilde{L}^\infty_t(B_{p,1}^{-1+\frac{3}{p}})}\big)\Big\}.$$

Then we get by substituting the above estimates into (9.32) that

$$
\begin{aligned}
&\|u^3\|_{\widetilde{L}^\infty_t(B_{p,1}^{-1+\frac{3}{p}})} + \bar{c}\mu\|u^3\|_{L^1_t(B_{p,1}^{1+\frac{3}{p}})} \\
&\leq \|u_0^3\|_{B_{p,1}^{-1+\frac{3}{p}}} + C\Big\{\|u^h\|_{L^1_t(B_{p,1}^{1+\frac{3}{p}})}\|u^3\|_{\widetilde{L}^\infty_t(B_{p,1}^{-1+\frac{3}{p}})} \\
&\quad + \big(\|u^h\|_{L^1_t(B_{p,1}^{1+\frac{3}{p}})} + \|u^3\|_{L^1_t(B_{p,1}^{1+\frac{3}{p}})}\big)\big(\mu\|a\|_{L^\infty_t(\mathcal{M}(B_{p,1}^{-1+\frac{3}{p}}))} + \|u^h\|_{\widetilde{L}^\infty_t(B_{p,1}^{-1+\frac{3}{p}})}\big)\Big\}.
\end{aligned}
\tag{9.33}
$$

9.4.3 The Proof of Theorem 9.3

With (9.31) and (9.33), we can prove that $T^* = \infty$ provided that there holds (9.8). In fact, notice that $-1+\frac{3}{p} \in (-1,1)$ if $p > \frac{3}{2}$, we get by taking $s = -1+\frac{3}{p}$ in Proposition 9.1 that

$$
\|a\|_{L^\infty_t(\mathcal{M}(B_{p,1}^{-1+\frac{3}{p}}))} \leq \|a_0\|_{\mathcal{M}(B_{p,1}^{-1+\frac{3}{p}})} \exp\Big\{C(\|u^h\|_{L^1_t(B_{p,1}^{1+\frac{3}{p}})} + \|u^3\|_{L^1_t(B_{p,1}^{1+\frac{3}{p}})})\Big\}. \tag{9.34}
$$

Substituting (9.34) into (9.31) gives rise to

$$
\begin{aligned}
&\|u^h\|_{\widetilde{L}^\infty_t(B_{p,1}^{-1+\frac{3}{p}})} + \mu\big(\|a\|_{L^\infty_t(\mathcal{M}(B_{p,1}^{-1+\frac{3}{p}}))} + \|u^h\|_{L^1_t(B_{p,1}^{1+\frac{3}{p}})}\big) \\
&\leq \big(\mu\|a_0\|_{\mathcal{M}(B_{p,1}^{-1+\frac{3}{p}})} + \|u_0^h\|_{B_{p,1}^{-1+\frac{3}{p}}}\big)\exp\Big\{C\int_0^t \big(\|u^3(t')\|_{B_{p,1}^{1+\frac{3}{p}}} + \frac{1}{\mu}\|u^3(t')\|^2_{B_{p,1}^{\frac{3}{p}}}\big)\,dt'\Big\}
\end{aligned}
\tag{9.35}
$$

for $t \leq \mathfrak{T}$.

While thanks to (9.33), we get by taking $c_2 \leq \min\{\bar{c}_2, \frac{1}{2C}, \frac{\bar{c}}{2C}\}$ in (9.29) that

$$
\|u^3\|_{\widetilde{L}^\infty_t(B_{p,1}^{-1+\frac{3}{p}})} + \mu\bar{c}\|u^3\|_{L^1_t(B_{p,1}^{1+\frac{3}{p}})} \leq 2\|u_0^3\|_{B_{p,1}^{-1+\frac{3}{p}}} + c_2\mu \quad \text{for} \quad t \leq \mathfrak{T}. \tag{9.36}
$$

Combining (9.35) with (9.36), we reach

$$
\begin{aligned}
&\|u^h\|_{\widetilde{L}^\infty_t(B_{p,1}^{-1+\frac{3}{p}})} + \mu\big(\|a\|_{L^\infty_t(\mathcal{M}(B_{p,1}^{-1+\frac{3}{p}}))} + \|u^h\|_{L^1_t(B_{p,1}^{1+\frac{3}{p}})}\big) \\
&\leq \mathfrak{C}_1\big(\mu\|a_0\|_{\mathcal{M}(B_{p,1}^{-1+\frac{3}{p}})} + \|u_0^h\|_{B_{p,1}^{-1+\frac{3}{p}}}\big)\exp\{\frac{\mathfrak{C}_2}{\mu^2}\|u_0^3\|^2_{B_{p,1}^{-1+\frac{3}{p}}}\}
\end{aligned}
\tag{9.37}
$$

for $t \leq \mathfrak{T}$ and some positive constants $\mathfrak{C}_1, \mathfrak{C}_2$ which depends on $\bar{c}$ and c_2. In particular, (9.37) implies that if we take C_0 large enough and c_0 sufficiently small in (9.8), there holds

$$
\|u^h\|_{\widetilde{L}^\infty_t(B_{p,1}^{-1+\frac{3}{p}})} + \mu\big(\|a\|_{L^\infty_t(\mathcal{M}(B_{p,1}^{-1+\frac{3}{p}}))} + \|u^h\|_{L^1_t(B_{p,1}^{1+\frac{3}{p}})}\big) \leq C\delta \leq \frac{c_2}{2}\mu \quad \text{for} \quad t \leq \mathfrak{T},
$$

which contradicts with (9.29). Whence we conclude that $\mathfrak{T}=\infty$, and there holds

$$\begin{aligned}
&\|u^h\|_{\widetilde{L}^\infty(\mathbb{R}^+;B_{p,1}^{-1+\frac{3}{p}})}+\mu\big(\|a\|_{L_t^\infty(\mathcal{M}(B_{p,1}^{-1+\frac{3}{p}}))}+\|u^h\|_{L^1(\mathbb{R}^+;B_{p,1}^{1+\frac{3}{p}})}\big)\le C\delta,\\
&\|u^3\|_{\widetilde{L}^\infty(\mathbb{R}^+;B_{p,1}^{-1+\frac{3}{p}})}+\mu\|u^3\|_{L^1(\mathbb{R}^+;B_{p,1}^{1+\frac{3}{p}})}\le 2\|u_0^3\|_{B_{p,1}^{-1+\frac{3}{p}}}+c_2\mu.
\end{aligned}\tag{9.38}$$

Finally let us turn to the proof of (9.9). Indeed thanks to (9.4), we have

$$u=e^{\mu t\Delta}u_0+\int_0^t e^{\mu(t-s)\Delta}\mathbb{P}\{-\mathrm{div}_h(u^h\otimes u)-\partial_3(u^3u)+\mu a\Delta u-a\nabla\Pi\}(s)\,ds,$$

where $\mathbb{P}$ denotes the Leray projection operator to the divergence-free vector field space. Then we get by applying Lemma 9.3 that

$$\begin{aligned}
\|\Delta_j(u-e^{\mu t\Delta}u_0)(t)\|_{L^p}\le C\int_0^t e^{-\mu(t-s)2^{2j}}&\big(2^j\|\Delta_j(u^h\otimes u)(s)\|_{L^p}+\|\Delta_j\partial_3(u^3u)(s)\|_{L^p}\\
&+\mu\|\Delta_j(a\Delta u)(s)\|_{L^p}+\|\Delta_j(a\nabla\Pi)(s)\|_{L^p}\big)\,ds.
\end{aligned}\tag{9.39}$$

Applying Lemmas 9.6 and 9.7 and (9.38) that

$$\begin{aligned}
&\|\Delta_j\partial_3(u^3u)\|_{L^1(L^p)}\\
&\quad\lesssim d_j2^{j(1-\frac{3}{p})}\big(\|u^h\|_{\widetilde{L}_t^\infty(B_{p,1}^{-1+\frac{3}{p}})}\|u^3\|_{L_t^1(B_{p,1}^{1+\frac{3}{p}})}+\|u^h\|_{L_t^1(B_{p,1}^{1+\frac{3}{p}})}\|u^3\|_{\widetilde{L}_t^\infty(B_{p,1}^{-1+\frac{3}{p}})}\big)\\
&\quad\lesssim d_j2^{j(1-\frac{3}{p})}(2\|u_0^3\|_{B_{p,1}^{-1+\frac{3}{p}}}+c_2\mu)\frac{\delta}{\mu}.
\end{aligned}$$

While taking $\lambda_1=\lambda_2=0$ in (9.22) leads to

$$\begin{aligned}
&2^j\|\Delta_j(u^h\otimes u)\|_{L_t^1(L^p)}+\mu\|\Delta_j(a\Delta u)\|_{L_t^1(L^p)}+\|\Delta_j(a\nabla\Pi)\|_{L_t^1(L^p)}\\
&\quad\lesssim d_j2^{j(1-\frac{3}{p})}\Big\{\big(\|u^h\|_{\widetilde{L}_t^\infty(B_{p,1}^{-1+\frac{3}{p}})}+\mu\|a\|_{L_t^\infty(\mathcal{M}(B_{p,1}^{-1+\frac{3}{p}}))}\big)\|u\|_{L_t^1(B_{p,1}^{1+\frac{3}{p}})}\\
&\qquad+\|u^h\|_{L_t^1(B_{p,1}^{1+\frac{3}{p}})}\|u\|_{\widetilde{L}_t^\infty(B_{p,1}^{-1+\frac{3}{p}})}+\|a\|_{L_t^\infty(\mathcal{M}(B_{p,1}^{-1+\frac{3}{p}}))}\|\nabla\Pi\|_{L_t^1(B_{p,1}^{-1+\frac{3}{p}})}\Big\}\\
&\quad\lesssim d_j2^{j(1-\frac{3}{p})}(2\|u_0^3\|_{B_{p,1}^{-1+\frac{3}{p}}}+c_2\mu+\delta)\frac{\delta}{\mu},
\end{aligned}$$

where we used (9.23) and (9.38) in the last step to obtain the estimate of $\|\nabla\Pi\|_{L_t^1\left(B_{p,1}^{-1+\frac{3}{p}}\right)}$.

Whence thanks to (9.39), we arrive at

$$\begin{aligned}
\|\Delta_j(u-e^{\mu t\Delta}u_0)\|_{L_t^\infty(L^p)}&+\mu 2^{2j}\|\Delta_j(u-e^{\mu t\Delta}u_0)\|_{L_t^1(L^p)}\\
&\le Cd_j2^{j(1-\frac{3}{p})}(2\|u_0^3\|_{B_{p,1}^{-1+\frac{3}{p}}}+c_2\mu+\delta)\frac{\delta}{\mu},
\end{aligned}$$

which implies (9.9), and we complete the proof of Theorem 9.3.

Acknowledgements Part of this work was done when Marius Paicu was visiting Morningside Center of the Chinese Academy of Sciences in the Spring of 2012. We appreciate the hospitality of MCM and the financial support from the Chinese Academy of Sciences. P. Zhang is partially supported by NSF of China under Grant 10421101 and 10931007, the one hundred talents' plan from Chinese Academy of Sciences under Grant GJHZ200829 and innovation grant from National Center for Mathematics and Interdisciplinary Sciences.

References

1. Abidi, H.: Équation de Navier–Stokes avec densité et viscosité variables dans l'espace critique. Rev. Mat. Iberoam. **23**(2), 537–586 (2007)
2. Abidi, H., Paicu, M.: Existence globale pour un fluide inhomogéne. Ann. Inst. Fourier (Grenoble) **57**, 883–917 (2007)
3. Abidi, H., Paicu, M.: Global existence for the magnetohydrodynamic system in critical spaces. In: Proceedings of the Royal Society of Edinburgh, Section A, Mathematics, **138**, 447–476 (2008)
4. Abidi, H., Gui, G., Zhang, P.: On the decay and stability to global solutions of the 3−D inhomogeneous Navier–Stokes equations. Comm. Pure. Appl. Math. **64**, 832–881 (2011)
5. Abidi, H., Gui, G., Zhang, P.: On the wellposedness of 3−D inhomogeneous Navier–Stokes equations in the critical spaces. Arch. Ration. Mech. Anal. **204**(1), 189–230 (2012)
6. Abidi, H., Gui, G., Zhang, P.: Comm. Pure. Appl. Math. **65**(10), 1458–1480 (2012)
7. Bahouri, H., Chemin, J.Y., Danchin, R.: Fourier analysis and nonlinear partial differential equations. Grundlehren der mathematischen Wissenschaften, vol. 343. Springer, Berlin (2011)
8. Danchin, R.: Local theory in critical spaces for compressible viscous and heat-conducting gases. Comm. P. D. E. **26**, 1183–1233 (2001)
9. Danchin, R.: Density-dependent incompressible viscous fluids in critical spaces. Proc. Roy. Soc. Edinburgh Sect. A **133**, 1311–1334 (2003)
10. Danchin, R.: Local and global well-posedness results for flows of inhomogeneous viscous fluids. Adv. Differ. Equat. **9**, 353–386 (2004)
11. Danchin, R., Mucha, P.B.: A Lagrangian approach for the incompressible Navier–Stokes equations with variable density. Comm. Pure. Appl. Math. **65**(10), 1458–1480 (2012)
12. Fujita, H., Kato, T.: On the Navier–Stokes initial value problem I. Arch. Ration. Mech. Anal. **16**, 269–315 (1964)
13. Gui, G., Zhang, P.: Stability to the global solutions of 3-D Navier–Stokes equations. Adv. Math. **225**, 1248–1284 (2010)
14. Huang, J., Paicu, M., Zhang, P.: Global solutions to 2-D inhomogeneous Navier–Stokes system with general velocity (2012, preprint) arXiv:1212.3916
15. Kazhikov, A.V.: Solvability of the initial-boundary value problem for the equations of the motion of an inhomogeneous viscous incompressible fluid (Russian). Dokl. Akad. Nauk SSSR **216**, 1008–1010 (1974)
16. Ladyženskaja, O.A., Solonnikov, V.A.: The unique solvability of an initial-boundary value problem for viscous incompressible inhomogeneous fluids. (Russian) Boundary value problems of mathematical physics, and related questions of the theory of functions, 8. Zap. Naučn. Sem. Leningrad. Otdel. Mat. Inst. Steklov. (LOMI) **52**, 52–109, 218–219 (1975)
17. Lions, P.-L.: Mathematical topics in fluid mechanics. Vol. 1. Incompressible models. Oxford Lecture Series in Mathematics and its Applications, 3. Oxford Science Publications. The Clarendon Press, Oxford University Press, New York, (1996)
18. Maz'ya, V., Shaposhnikova, T.: Theory of Sobolev multipliers. With applications to differential and integral operators. Grundlehren der mathematischen Wissenschaften, vol. 337. Springer, Berlin (2009)
19. Paicu, M., Zhang, P.: Global solutions to the 3-D incompressible anisotropic Navier–Stokes system in the critical spaces. Comm. Math. Phys. **307**, 713–759 (2011)

20. Paicu, M., Zhang, P.: Global solutions to the 3-D incompressible inhomogeneous Navier–Stokes system. J. Funct. Anal. **262**(8), 3556–3584 (2012)
21. Planchon, F.: Sur un inégalité de type Poincaré. C. R. Acad. Sci. Paris Sér. I Math. **330**, 21–23 (2000)
22. Zhang, T.: Global wellposedness problem for the 3-D incompressible anisotropic Navier–Stokes equations in an anisotropic space. Comm. Math. Phys. **287**, 211–224 (2009)

Chapter 10
The Cauchy Problem for the Euler–Poisson System and Derivation of the Zakharov–Kuznetsov Equation

David Lannes, Felipe Linares, and Jean-Claude Saut

Abstract We consider in this paper the rigorous justification of the Zakharov–Kuznetsov equation from the Euler–Poisson system for uniformly magnetized plasmas. We first provide a proof of the local well-posedness of the Cauchy problem for the aforementioned system in dimensions two and three. Then we prove that the long-wave small-amplitude limit is described by the Zakharov–Kuznetsov equation. This is done first in the case of cold plasma; we then show how to extend this result in presence of the isothermal pressure term with uniform estimates when this latter goes to zero.

Key words: Zakharov–Kuznetsov, Euler–Poisson

2010 Mathematics Subject Classification: 35B40, 35Q53.

D. Lannes (✉)
Département de Mathématiques et Applications, École Normale Supérieure,
45, rue d'Ulm, F-75230 Paris Cedex 05, France
e-mail: lannes@ens.fr

F. Linares
IMPA, Estrada Dona Castorina 110, Rio de Janeiro 22460-320, RJ Brasil
e-mail: linares@impa.br

J.-C. Saut
Laboratoire de Mathématiques, UMR 8628, Université Paris-Sud 11 et CNRS,
F-91405 Orsay Cedex, France
e-mail: jean-claude.saut@math.u-psud.fr

M. Cicognani et al. (eds.), *Studies in Phase Space Analysis with Applications to PDEs*, Progress in Nonlinear Differential Equations and Their Applications 84,
DOI 10.1007/978-1-4614-6348-1_10,

10.1 Introduction

10.1.1 General Setting

The Zakharov–Kuznetsov equation

$$u_t + u\partial_x u + \partial_x \Delta u = 0, \qquad u = u(x,y,z,t), \qquad (x,y,z) \in \mathbb{R}^d, t \in \mathbb{R}, d = 2,3 \quad (10.1)$$

was introduced as an asymptotic model in [36] (see also [7, 17, 19], and [35] for some generalizations) to describe the propagation of nonlinear ionic-sonic waves in a magnetized plasma.

The Zakharov–Kuznetsov is a natural multidimensional extension of the Korteweg-de Vries equation, quite different from the well-known Kadomtsev–Petviashvili (KP) equation which is obtained as an asymptotic model of various nonlinear dispersive systems under a different scaling.

Contrary to the Korteweg-de Vries or the Kadomtsev–Petviashvili equations, the Zakharov–Kuznetsov equation is not completely integrable, but it has a Hamiltonian structure and possesses two invariants, namely, (for $u_0 = u(\cdot,0)$) :

$$M(t) = \int_{R^d} u^2(x,t) = \int_{R^d} u_0^2(x) = M(0) \quad (10.2)$$

and the Hamiltonian

$$H(t) = \frac{1}{2}\int_{R^d} [|\nabla u|^2 - \frac{u^3}{3}] = \frac{1}{2}\int_{R^d} [|\nabla u_0|^2 - \frac{u_0^3}{3}] = H(0). \quad (10.3)$$

The Cauchy problem for the Zakharov–Kuznetsov equation has been proven to be globally well posed in the two-dimensional case for data in $H^1(\mathbb{R}^2)$ [8] and locally well posed in the three-dimensional case for data in $H^s(\mathbb{R}^3)$, $s > \frac{3}{2}$ [25] and recently in $H^s(\mathbb{R}^3)$, $s > 1$ [30]. We also refer to [26] for solutions on a nontrivial background and to [9, 23, 24, 31] for well-posedness results of the Cauchy problem for *generalized* Zakharov–Kuznetsov equations in $\mathbb{R}^2$. Unique continuation properties for the Zakharov–Kuznetsov equation were established in [3, 27]. Stability properties of the ground state solutions of the Zakharov–Kuznetsov equation are studied in [2].

The Zakharov–Kuznetsov posed in a strip or in a bounded domain (with ad hoc boundary conditions) has global dissipative properties leading in particular to well-posedness results with L^2 initial data (see [32, 33]). Applications to the decay rates of small solutions are given in [21] and to control theory in [28].

The Zakharov–Kuznetsov equation was formally derived in [36] as a long-wave small-amplitude limit of the following Euler–Poisson system in the "cold plasma" approximation:

$$\begin{cases} \partial_t n + \nabla\cdot\mathbf{v} + \nabla\cdot(n\mathbf{v}) = 0, \\ \partial_t \mathbf{v} + (\mathbf{v}\cdot\nabla)\mathbf{v} + \nabla\varphi + a\mathbf{e}\wedge\mathbf{v} = 0, \\ \Delta\varphi - e^{\varphi} + 1 + n = 0. \end{cases} \quad (10.4)$$

Here n is the deviation of the ion density from 1, $\mathbf{v}$ is the ion velocity, φ is the electric potential, and a is a measure of the uniform magnetic field, applied along the vector $\mathbf{e} = (1,0,0)^T$ so that if $\mathbf{v} = (v_1,v_2,v_3)^T$, $\mathbf{e} \wedge \mathbf{v} = (0,-v_3,v_2)^T$.

Note that this skew-adjoint term is similar to a Coriolis term in the Euler equations for inviscid incompressible fluids.

The main goal of this paper is to justify rigorously this formal long-wave limit. The one-dimensional case (leading to the Korteweg-de Vries equation) has been partially justified in [6], and Guo-Pu gave in a recent preprint a full justification of this limit [13].

Setting $\rho = (1+n)$, (10.4) writes

$$\begin{cases} \partial_t \rho + \nabla \cdot (\rho \mathbf{v}) = 0, \\ \partial_t \mathbf{v} + (\mathbf{v} \cdot \nabla)\mathbf{v} + \nabla \varphi + a\mathbf{e} \wedge \mathbf{v} = 0, \\ \Delta \varphi - e^{\varphi} + \rho = 0. \end{cases} \tag{10.5}$$

In this formulation, the Euler–Poisson system possesses a (formally) Hamiltonian conserved energy

$$\begin{aligned} H(\rho,\mathbf{v},\varphi) &= \int_{\mathbb{R}^d} [\frac{1}{2}\rho|\mathbf{v}|^2 + \rho\varphi - \frac{1}{2}|\nabla\varphi|^2 - e^{\varphi} + 1]dx \\ &= \int_{\mathbb{R}^d} [\frac{1}{2}\rho|\mathbf{v}|^2 + \frac{1}{2}|\nabla\varphi|^2 + e^{\varphi}(\varphi - 1) + 1]dx, \end{aligned} \tag{10.6}$$

as it is easily seen by multiplying the first (resp. second) equation in (10.5) by $\frac{1}{2}|\mathbf{v}|^2 + \varphi$ (resp. $\rho\mathbf{v}$), then adding and integrating over $\mathbb{R}^d$.

Of course one has to find a correct functional setting in order to justify the definition of H and its conservation (see Remark 10.1 in Sect. 10.2.1 below).

Note that the Euler–Poisson system with $a = 0$ has another formally conserved quantity, namely, the impulse

$$\frac{d}{dt}\int_{\mathbb{R}^d} \rho\mathbf{v}dx = 0. \tag{10.7}$$

This conservation law is easily derived (formally) from the equation for $\rho\mathbf{v}$:

$$\partial_t(\rho\mathbf{v}) + \nabla \cdot (\rho\mathbf{v} \otimes \mathbf{v}) + \rho\nabla\varphi = 0. \tag{10.8}$$

Linearizing (10.4) around the constant solution $n = 1$, $\mathbf{v} = 0$, $\varphi = 0$, one finds the dispersion relation for a plane wave $e^{i(\omega t - \mathbf{k}.x)}$, $\mathbf{k} = (k_1,k_2,k_3)$:

$$\omega^4(\mathbf{k}) + \omega^2(\mathbf{k})(a^2 - \frac{|\mathbf{k}|^2}{1+|\mathbf{k}|^2}) - a^2\frac{k_1^2}{1+|\mathbf{k}|^2} = 0 \tag{10.9}$$

or

$$(1+|\mathbf{k}|^2) - \frac{k_1^2}{\omega^2(\mathbf{k})} - \frac{|\mathbf{k}_\perp|^2}{\omega^2(\mathbf{k}) + a^2} = 0 \tag{10.10}$$

where $|\mathbf{k}_\perp|^2 = k_2^2 + k_3^2$.

In the absence of applied magnetic field ($a = 0$), this relation reduces to

$$\omega^2(\mathbf{k}) = \frac{|\mathbf{k}|^2}{1+|\mathbf{k}|^2}. \tag{10.11}$$

Those relations display the weakly dispersive character of the Euler–Poisson system.

Contrary to the KP case (see [1] for the justification of various asymptotic models of surface waves), the rigorous justification of the long-wave limit of the Euler–Poisson system has not been carried out so far (see, however, [6, 13] in the one-dimensional case). This justification is the main goal of this paper.

To start with, we investigate the local well-posedness of the Euler–Poisson system (10.4) which does not raise any particular difficulty, but for which, to our knowledge, no result seems to be explicitly available in the literature in dimensions 2 and 3. This paper [34] concerns a "linearized" version of (10.4), namely, the term $\Delta\varphi + 1 - e^{\varphi}$ is replaced by its linearization at $\varphi = 0$, that is, $(\Delta - 1)\varphi$ (see also [16] where a unique continuation property is established for the one-dimensional version of this "modified" Euler–Poisson system). Note that this "linearized" version of (10.4) is somewhat reminiscent of the pressureless Euler–Poisson system which has been intensively studied (see, for instance, [4, 5] and the references therein).

In the one-dimensional case (10.4) has the very simple form

$$\begin{cases} \partial_t n + [(1+n)\mathbf{v}]_x = 0, \\ \partial_t \mathbf{v} + \mathbf{v}\mathbf{v}_x + \varphi_x = 0, \\ \varphi_{xx} - e^{\varphi} + 1 + n = 0. \end{cases} \tag{10.12}$$

The existence of supersonic solitary waves for (10.12) has been proven in [22]. The linear stability of those solitary waves was investigated in [15], and their interactions were studied in [14], in particular in comparison with their approximations in the long-wave limit by KdV solitary waves.

Though our analysis is mainly concerned with the higher dimensional case $d = 2,3$, our results apply as well to the system (10.12) and to provide an alternative proof to [13] of the justification of the KdV approximation.

Let us mention, finally, that (10.4) is valid for cold plasmas only; in the general case, an isothermal pressure term must be added,

$$\begin{cases} \partial_t n + \nabla\cdot\mathbf{v} + \nabla\cdot(n\mathbf{v}) = 0, \\ \partial_t \mathbf{v} + (\mathbf{v}\cdot\nabla)\mathbf{v} + \nabla\varphi + \alpha\dfrac{\nabla n}{1+n} + a\mathbf{e}\wedge\mathbf{v} = 0, \\ \Delta\varphi - e^{\varphi} + 1 + n = 0, \end{cases} \tag{10.13}$$

with $\alpha \geq 0$. For this system (with $\alpha > 0$), global existence for small data has been proved [12] in dimension $d = 3$ in absence of magnetic field ($a = 0$) and in the irrotational case. Still for $\alpha > 0$, in [10], the authors provide for the full equations uniform energy estimate in the quasineutral limit (i.e., $e^{\varphi} = 1 + n$) for well-prepared initial data (note that they also handle the initial boundary value problem). The derivation and justification of the KdV approximation are generalized in [13] using a different proof as in the case $\alpha = 0$. We show how to extend our results on the ZK approximation to (10.13) and provide uniform estimates with respect to α that allow one to handle the convergence of solutions of (10.13) to solutions of (10.4) when $\alpha \to 0$.

10.1.2 Organization of the Paper

This paper is organized as follows.

In Sect. 10.2, we prove that the Cauchy problem for the Euler–Poisson system (10.4) is locally well posed. The main step is to express φ as a function of n by solving the elliptic equation in (10.4) by the super- and sub-solution methods. This step is of course trivial when one considers the "linearized" Euler–Poisson system. Then we establish the local well-posedness for data $(n_0, \mathbf{v}_0)$ in $H^{s-1}(\mathbb{R}^d) \times H^s(\mathbb{R}^d)^d$, $s > \frac{d}{2} + 1$ such that $|n_0|_\infty < 1$, thus generalizing the result of [34]. We use in a crucial way the smoothing property of the map $n \mapsto \varphi$.

In Sect. 10.3 we derive rigorously the Zakharov–Kuznetsov equation as a long-wave limit of the Euler–Poisson system. In order to do this, we need to introduce a small parameter ε and to establish for a scaled version of the Euler–Poisson system existence and bounds on the correct time scale. However the elliptic equation for φ provides a smoothing effect which is not uniform with respect to ε, and this makes the Cauchy problem more delicate (we cannot apply the previous strategy which would give an existence time shrinking to zero with ε). We are thus led to view the Euler–Poisson system as a semilinear perturbation of a symmetrizable quasilinear system, and we have to find the correct symmetrizer. We obtain in this framework well-posedness (with uniform bounds on the correct time scale) for data in $H^s(\mathbb{R}^d) \times H_\varepsilon^{s+1}(\mathbb{R}^d)^d$, $s > \frac{d}{2} + 1$ (we refer to (10.34) for the definition of $H_\varepsilon^{s+1}(\mathbb{R}^d)$).

We then prove that this solution is well approximated by the solution of the ZK equation on the relevant time scales.

We finally show in Sect. 10.4 how to modify the results of Sect. 10.3 when the isothermal pressure is not neglected, i.e. when one works with (10.13) instead of (10.4). The presence of a nonzero coefficient $\alpha > 0$ in front of the isothermal pressure term induces a smoothing effect on the variable n; however, this smoothing effect vanishes as $\alpha \to 0$, and in order to obtain an existence time uniform with respect to ε and α, it is necessary to work in a Banach scale indexed by these parameters and that is adapted to measure this smoothing effect.

10.1.3 Notations

Partial differentiation are denoted by subscripts, ∂_x, ∂_t, $\partial_j = \partial_{x_j}$, etc.

- We denote by $|\cdot|_p$ $(1 \le p \le \infty)$ the standard norm of the Lebesgue spaces $L^p(\mathbb{R}^d)$ $(d = 2, 3)$.

- We use the Fourier multiplier notation: $f(D)u$ is defined as $\mathscr{F}(f(D)u)(\xi) = f(\xi)\widehat{u}(\xi)$, where $\mathscr{F}$ and $\widehat{\cdot}$ stand for the Fourier transform.
- The operator $\Lambda = (1-\Delta)^{1/2}$ is equivalently defined using the Fourier multiplier notation to be $\Lambda = (1+|D|^2)^{1/2}$.
- The standard notation $H^s(\mathbb{R}^d)$, or simply H^s if the underlying domain is clear from the context, is used for the L^2-based Sobolev spaces; their norm is written as $|\cdot|_{H^s}$.
- For a given Banach space X we will denote $|.|_{X,T}$ the norm in $C([0,T];X)$. When $X = L^p$, the corresponding norm will be denoted $|.|_{p,T}$.
- We will denote by C various absolute constants.
- The notation $A + \langle B\rangle_{s>t_0}$ refers to A if $s \le t_0$ and $A+B$ if $s > t_0$.

10.2 The Cauchy Problem for the Euler–Poisson System

The aim of this section is to prove the local well-posedness of the Cauchy problem associated to the Euler–Poisson system.

$$\begin{cases} \partial_t n + \nabla\cdot\mathbf{v} + \nabla\cdot(n\mathbf{v}) = 0, \\ \partial_t \mathbf{v} + (\mathbf{v}\cdot\nabla)\mathbf{v} + \nabla\varphi + a\mathbf{e}\wedge\mathbf{v} = 0, \\ \Delta\varphi - e^{\varphi} + 1 + n = 0 \\ n(\cdot,0) = n_0, \\ \mathbf{v}(\cdot,0) = \mathbf{v}_0. \end{cases} \tag{10.14}$$

10.2.1 *Solving the Elliptic Part*

We consider here, for a given n, the elliptic equation

$$L(\varphi) = -\Delta\varphi + e^{\varphi} - 1 = n. \tag{10.15}$$

Proposition 10.1. *Let $n \in L^\infty(\mathbb{R}^d)\cap H^1(\mathbb{R}^d)$, $d = 1,2,3$, such that $\inf_{\mathbb{R}^d} 1+n > 0$. Then there exists a unique solution $\varphi \in H^3(\mathbb{R}^d)$ of (10.15) such that:*

(i) The following estimate holds:

$$-K_- = \ln(1-|n|_\infty) \le \varphi \le K_+ = \ln(1+|n|_\infty).$$

(ii) Defining $c_\infty(n) = |(1+n)^{-1}|_\infty$ and $I_1(n) = \frac{1}{c_\infty(n)}|n|_2^2 + |n|_{H^1}^2$, one has

$$\int_{\mathbb{R}^d}\Big[\frac{c_\infty(n)}{2}|\varphi|^2 + \frac{1}{2}|\nabla\varphi|^2 + |\Delta\varphi|^2 + 2e^{\varphi}|\nabla\varphi|^2 + \frac{1}{2}(e^{\varphi}-1)^2\Big] \le \frac{1}{2}I_1(n).$$

(iii) If furthermore, $n \in H^s(\mathbb{R}^d)$, $s \geq 0$, then $\varphi \in H^{s+2}(\mathbb{R}^d)$ and

$$|\Lambda^{s+1}\nabla\varphi|_2 \leq F_s(I_1(n), |n|_\infty)(1 + |n|_{H^s}), \tag{10.16}$$

where $F_s(.,.)$ is an increasing function of its arguments.

Remark 10.1. Writing $\xi_i\xi_j\hat{\varphi} = -\frac{\xi_i\xi_j}{|\xi|^2}|\xi|^2\hat{\varphi}$ and using the L^2 continuity of the Riesz transforms, we see that we can replace $\Delta\varphi$ in the left-hand side of (ii) in Proposition 10.1 by any derivative $\partial^\alpha\varphi$, replacing possibly the right-hand side by $CI_1(n)$, where C is an absolute constant.

Remark 10.2. Notice that by (ii), one has $|\varphi|_2^2 \leq \frac{I_1(n)}{c_\infty(n)}$.

Proof. We use the method of sub- and super-solutions to construct the solution φ. As a super-solution, we take $\varphi_+ = K_+$ where K_+ is a positive constant satisfying

$$K_+ \geq \ln(1 + |n|_\infty),$$

so that

$$L(\varphi_+) \geq n.$$

As a sub-solution, we choose $\varphi_- = -K_- < 0$ where K_- is a positive constant satisfying

$$K_- \geq \ln\Big(\frac{1}{\inf_{\mathbb{R}^d}(1+n)}\Big) = \ln\Big(\frac{1}{c_\infty(n)}\Big),$$

so that

$$L(\varphi_-) \leq n.$$

The next elementary lemma will be useful.

Lemma 10.1. *Let Ω be an arbitrary open subset of $\mathbb{R}^n$ and $\varphi \in L^2(\Omega) \cap L^\infty(\Omega)$, $\varphi \neq 0$. Then*

$$|e^\varphi - 1|_2 \leq \frac{e^{|\varphi|_\infty} - 1}{|\varphi|_\infty}|\varphi|_2.$$

If moreover $\varphi \in H^1(\Omega)$, then $e^\varphi - 1 \in H^1(\Omega)$.

Proof. To prove the first point we expand

$$|e^\varphi - 1| = |\varphi \sum_{n\geq 1}\frac{\varphi^{n-1}}{n!}| \leq \frac{|\varphi|}{|\varphi|_\infty}\sum_{n\geq 1}\frac{|\varphi|_\infty^n}{n!}.$$

The last assertion results from the estimate

$$|\nabla e^\varphi|_2 \leq e^{|\varphi|_\infty}|\nabla\varphi|_2.$$

□

Let $F(\varphi) = e^\varphi - 1$. Then $F'(\varphi)$ is bounded by a positive constant K on the interval $[-K_-, K_+]$. Let $\lambda > K$ be so that the function $\lambda I - F$ is strictly increasing in $[-K_-, K_+]$.

Let $B_p = \{x \in \mathbb{R}^d, |x| < p\}$. We consider the auxiliary problem, for a given $\varphi \in H^2(B_p)$ satisfying $-K_- \le \varphi \le K_+$,

$$\begin{cases} -\Delta\psi + \lambda\psi = \lambda\varphi - F(\varphi) + n \\ \psi_{|\partial B_p} = 0, \end{cases} \tag{10.17}$$

and we write $S(\varphi) = \psi$. Since $F(\varphi) \in L^2(B_p)$, $\psi = S(\varphi) \in H_0^1 \cap H^2(B_p)$. Moreover, one checks easily by the maximum principle that, since $-K_- = \varphi_- \le \varphi \le K_+ = \varphi_+$, then also

$$-K_- = \varphi_- \le S(\varphi) \le K_+ = \varphi_+.$$

We now define inductively $\varphi_0 = \varphi_-$, $\varphi_{k+1} = S(\varphi_k)$. By the maximum principle, one checks that (φ_k) is increasing, $\varphi_{k+1} \ge \varphi_k$, $k \in \mathbb{N}$, and that

$$\varphi_- \le \varphi_k \le \varphi_+,\ k \in \mathbb{N}.$$

Moreover, $S : L^2(B_p) \to L^2(B_p)$ is continuous since $z \mapsto \lambda z - F(z)$ is Lipschitz. The sequence (φ_k) is increasing and bounded from above. It converges almost everywhere to some φ which belongs to $L^2(B_p)$ since B_p is bounded. By Lebesgue's theorem, the convergence holds also in $L^2(B_p)$. On the other hand, using that F is Lipschitz on $[-K_-, K_+]$, one checks that (φ_k) is Cauchy in $H_0^1(B_p)$, proving that $\varphi \in H_0^1(B_p)$, and is a solution of (10.15) in B_p. Moreover $e^\varphi - 1 \in L^2(B_p)$ and $\varphi \in H^2(B_p)$.

Assuming that φ_1 and φ_2 are two $H_0^1 \cap H^2(B_p)$ solutions, we deduce immediately that, setting $\varphi = \varphi_1 - \varphi_2$,

$$|\nabla\varphi|_2^2 + \int_{B_p} (e^{\varphi_1} - e^{\varphi_2})\varphi = 0,$$

and we conclude that $\varphi = 0$ by the monotonicity of the exponential.

To summarize, for any $p \in \mathbb{N}$, we have proven the existence of a unique solution in $H_0^1 \cap H^2(B_p)$ (which we will denote φ_p from now on) of the elliptic problem

$$\begin{cases} -\Delta\varphi + e^\varphi - 1 = n \\ \varphi_{|\partial B_p} = 0. \end{cases} \tag{10.18}$$

Moreover, φ_p satisfies the bounds

$$\varphi_- = -K_- \le \varphi_p \le \varphi_+ = K_+. \tag{10.19}$$

We derive now a series of (uniform in p) estimates on φ_p. We first notice that for $\varphi \ge -K_-$, one has with $\alpha_0 = \frac{1-e^{-K_-}}{K_-}$,

$$(e^\varphi - 1)\varphi \ge \alpha_0\varphi^2. \tag{10.20}$$

Multiplying (10.18) by φ_p and integrating over B_p, we thus deduce

$$\int_{B_p} [\frac{\alpha_0}{2}|\varphi_p|^2 + |\nabla \varphi_p|^2] \le \frac{1}{2\alpha_0}|n|_2^2. \tag{10.21}$$

Multiplying (10.18) by $-\Delta \varphi_p$ and integrating over B_p, we obtain

$$\int_{Bp} |\Delta \varphi_p|^2 + \int_{Bp} e^{\varphi_p}|\nabla \varphi_p|^2 = \int_{Bp} \nabla \varphi_p \cdot \nabla n. \tag{10.22}$$

Finally we integrate (10.18) against $(e^{\varphi_p} - 1)$ to get

$$\int_{B_p} e^{\varphi_p}|\nabla \varphi_p|^2 + \frac{1}{2}\int_{B_p} (e^{\varphi_p} - 1)^2 \le \frac{1}{2}\int_{B_p} n^2. \tag{10.23}$$

Adding (10.21)–(10.23), we obtain

$$\int_{B_p} [\frac{\alpha_0}{2}|\varphi_p|^2 + \frac{1}{2}|\nabla \varphi_p|^2 + |\Delta \varphi_p|^2 + 2e^{\varphi_p}|\nabla \varphi_p|^2 + \frac{1}{2}(e^{\varphi_p} - 1)^2]$$
$$\le \frac{1}{2\alpha_0}|n|_2^2 + \frac{1}{2}|\nabla n|_2^2 + \frac{1}{2}|n|_2^2. \tag{10.24}$$

Now we extend φ_p outside B_p by 0 to get a $H^1(\mathbb{R}^d)$ function $\tilde{\varphi}_p$. Obviously $\tilde{\varphi}_p$ satisfies the bound (10.21) and (10.23) and up to a subsequence, $\tilde{\varphi}_p$ converges weakly in $H^1(\mathbb{R}^d)$, strongly in $L^2_{loc}(\mathbb{R}^d)$, and almost everywhere to some function $\varphi \in H^1(\mathbb{R}^d)$ which satisfies the bound (10.19).

Let us prove that φ is solution of the elliptic equation $L\varphi = n$ (see (10.15)). Let $\chi \in \mathfrak{D}(\mathbb{R}^d)$. Then supp $\chi \subset B_p$ for some p. Since φ_p solves $L\varphi_p = n$ in B_p, one has

$$\int_{B_p} \nabla \varphi_p \cdot \nabla \chi + \int_{B_p} (e^{\varphi_p} - 1)\chi = \int_{B_p} n\chi,$$

and thus

$$\int_{\mathbb{R}^d} \nabla \tilde{\varphi}_p \cdot \nabla \chi + \int_{\mathbb{R}^d} (e^{\tilde{\varphi}_p} - 1)\chi = \int_{\mathbb{R}^d} n\chi.$$

One then infers by weak convergence and Lebesgue theorem (using Lemma 10.1) that

$$\int_{\mathbb{R}^d} \nabla \varphi \cdot \nabla \chi + \int_{\mathbb{R}^d} (e^{\varphi} - 1)\chi = \int_{\mathbb{R}^d} n\chi,$$

proving that φ solves

$$-\Delta \varphi + e^{\varphi} - 1 = n. \tag{10.25}$$

Uniqueness is derived by the same argument used for φ_p. By passing to the limit in (10.21) and (10.23), one obtains the corresponding estimates for φ. Since $e^{\varphi} - 1 - n \in L^2(\mathbb{R}^d)$, $\varphi \in H^2(\mathbb{R}^d)$, and (10.22) for φ follows. Finally the estimate (ii) in Proposition 10.1, that is, (10.24) for φ, follows by adding the previous estimates.

We now prove the higher regularity estimates (iii) assuming that $n \in H^s(\mathbb{R}^d)$. From the continuity of the Riesz transforms (see Remark 10.1), it is enough to control $|\Lambda^s \Delta \varphi|_2$. We get from (10.25) that

$$\begin{aligned} |\Lambda^s \Delta \varphi|_2 &\le |n|_{H^s} + |e^{\varphi} - 1|_{H^s} \\ &\le |n|_{H^s} + C(|\varphi|_\infty)|\varphi|_{H^s}, \end{aligned}$$

the second line being a consequence of Moser's inequality. We can therefore deduce the result by a simple induction using the first two points of the proposition. □

Remark 10.3. One easily checks that the energy (see (10.6)) makes sense for $(n, \mathbf{v}, \varphi) \in H^1(\mathbb{R}^d) \times H^1(\mathbb{R}^d)^d \times L^2(\mathbb{R}^d) \cap L^\infty(\mathbb{R}^d)$. In fact one has then (recalling that $\rho = 1 + n$),

$$\rho\varphi - (e^{\varphi} - 1) = \rho\varphi - (\varphi + g) = n\varphi - g, \qquad g \in L^1(\mathbb{R}^d),$$

and on the other hand,

$$\rho|\mathbf{v}|^2 = |\mathbf{v}|^2 + n|\mathbf{v}|^2 \in L^1(\mathbb{R}^d),$$

by Sobolev embedding.

10.2.2 Local Well-Posedness

We establish here the local well-posedness of the Cauchy problem for (10.14).

Theorem 10.1. *Let $s > \frac{d}{2} + 1$, $n_0 \in H^{s-1}(\mathbb{R}^d)$, $\mathbf{v}_0 \in H^s(\mathbb{R}^d)^d$ such that $\inf_{\mathbb{R}^d} 1 + n_0 > 0$.*

There exist $T > 0$ and a unique solution $(n, \mathbf{v}) \in C([0,T]; H^{s-1}(\mathbb{R}^d) \times H^s(\mathbb{R}^d)^d)$ of (10.14) such that $|(1+n)^{-1}|_{\infty,T} < 1$ and $\varphi \in C([0,T]; H^{s+1}(\mathbb{R}^d))$.

Moreover the energy (10.6) is conserved on $[0,T]$ and so is the impulse (if $a = 0$).

Proof. Solving (10.15) for φ, we set $\nabla\varphi = F(n) = \nabla L^{-1}(n)$ and rewrite (10.14) as

$$\begin{cases} \partial_t n + \nabla \cdot \mathbf{v} + \nabla \cdot (n\mathbf{v}) = 0, \\ \partial_t \mathbf{v} + (\mathbf{v} \cdot \nabla)\mathbf{v} + F(n) + a\mathbf{e} \wedge \mathbf{v} = 0, \\ n(\cdot, 0) = n_0, \\ \mathbf{v}(\cdot, 0) = \mathbf{v}_0. \end{cases} \tag{10.26}$$

We derive first energy estimates. We apply the operator Λ^{s-1} to the first equation in (10.26) and Λ^s to the second one and take the L^2 scalar product with $\Lambda^{s-1}n$ and $\Lambda^s \mathbf{v}$, respectively. Using the Kato–Ponce commutator estimates [18] and integration by parts in the first equation, we obtain (note that the skew-adjoint term does not play a role here)

$$\begin{cases} \frac{1}{2}\frac{d}{dt}|\Lambda^{s-1}n|_2^2 \le |\Lambda^s \mathbf{v}|_2(1 + |\Lambda^{s-1}n|_2) + C|\nabla\mathbf{v}|_\infty |\Lambda^{s-1}n|_2^2, \\ \frac{1}{2}\frac{d}{dt}|\Lambda^s \mathbf{v}|_2^2 \le C|\nabla\mathbf{v}|_\infty |\Lambda^s \mathbf{v}|_2^2 + C|\Lambda^s F(n)|_2 |\Lambda^s \mathbf{v}|_2. \end{cases} \tag{10.27}$$

As soon as $|(1+n)^{-1}|_{L^\infty_{(x,t)}} < \infty$, using Proposition 10.1 (for the existence of φ given n), we infer that

$$|\Lambda^s F(n)|_2 \le F_{s-1}(I_1(n), |n|_\infty)\big(1+|n|_{H^{s-1}}\big). \tag{10.28}$$

One has plainly that

$$I_1(n) \le C|n|^2_{H^1}\Big(\frac{1}{1-|n|_\infty}+1\Big) \le C|n|^2_{H^1}\Big(\frac{2}{c_0}+1\Big),$$

if $1+n \ge \frac{c_0}{2}$.

Gathering those inequalities, we obtain the system of differential inequalities for $|n(\cdot,t)|_{H^{s-1}}$ and $|\mathbf{v}(\cdot,t)|_{H^s}$ and provided that $1-|n|_\infty \ge \frac{c_0}{2}$,

$$\begin{cases} \dfrac{d}{dt}|n|_{H^{s-1}} \le C_1(|\mathbf{n}|_{H^{s-1}}, |\mathbf{v}|_{H^s}) \\ \dfrac{d}{dt}|\mathbf{v}|_{H^s} \le C_2(|\mathbf{n}|_{H^{s-1}}, |\mathbf{v}|_{H^s}) \end{cases} \tag{10.29}$$

where C_1 and C_2 are smooth functions. Let $\alpha(t) = |n(\cdot,t)|_{H^{s-1}}$ and $\beta(t) = |\mathbf{v}(\cdot,t)|_{H^s}$ and consider the differential system:

$$\begin{cases} \alpha' = C_1(\alpha,\beta)), \\ \beta' = C_2(\alpha,\beta). \end{cases} \tag{10.30}$$

Let (A,B) be the local solution of (10.30) with initial data $(A_0,B_0)=(|n_0|_{H^{s-1}}, |\mathbf{v}_0|_{H^s})$ satisfying $1+n_0 \ge c_0$. The solution to (10.30) exists on a time interval which length depends only on (A_0,B_0).

Coming back to (10.29), one deduces that $(|n(\cdot,t)|_{H^{s-1}}, |\mathbf{v}(\cdot,t)|_{H^s})$ is bounded from above by (A,B) on a time interval I which length depends only on $(|n_0|_{H^{s-1}}, |\mathbf{v}_0|_{H^s})$ (and possibly shortened to ensure that $1+n \ge \frac{c_0}{2}$ by continuity).

To complete the proof we need to smooth out (10.4). This can be done, for instance, by truncating the high frequencies, that is, using $\chi(jD)$ where χ is a cutoff function and $j = 1,2,\ldots$. We obtain an ODE system in $H^{s-1} \times H^s$. The energy estimates are derived as above, and one passes to the limit in a standard way.

Finally, the conservation of both the energy and the impulse (if $a = 0$) is obvious since the functional setting of Theorem 10.1 allows to justify their formal derivation. □

10.3 The Long-Wave Limit of the Euler–Poisson System

In order to justify the Zakharov–Kuznetsov equation as a long-wave limit of the Euler–Poisson system with an applied magnetic field, we have to introduce an appropriate scaling.

We set $\mathbf{v} = (v_x, v_y, v_z)$. Laedke and Spatschek [19] derived formally the Zakharov–Kuznetsov equation by looking for approximate solutions of (10.4) of the form

$$\begin{aligned}
n^\varepsilon &= \varepsilon n^{(1)}(\varepsilon^{1/2}(x-t), \varepsilon^{1/2}y, \varepsilon^{1/2}z, \varepsilon^{3/2}t) + \varepsilon^2 n^{(2)} + \varepsilon^3 n^{(3)} \\
\varphi^\varepsilon &= \varepsilon \varphi^{(1)}(\varepsilon^{1/2}(x-t), \varepsilon^{1/2}y, \varepsilon^{1/2}z, \varepsilon^{3/2}t) + \varepsilon^2 \varphi^{(2)} + \varepsilon^3 \varphi^{(3)} \\
v_x^\varepsilon &= \varepsilon v_x^{(1)}(\varepsilon^{1/2}(x-t), \varepsilon^{1/2}y, \varepsilon^{1/2}z, \varepsilon^{3/2}t) + \varepsilon^2 v_x^{(2)} + \varepsilon^3 v_x^{(3)} \\
v_y^\varepsilon &= \varepsilon^{3/2} v_y^{(1)}(\varepsilon^{1/2}(x-t), \varepsilon^{1/2}y, \varepsilon^{1/2}z, \varepsilon^{3/2}t) + \varepsilon^2 v_y^{(2)} + \varepsilon^{5/2} v_y^{(3)} \\
v_z^\varepsilon &= \varepsilon^{3/2} v_z^{(1)}(\varepsilon^{1/2}(x-t), \varepsilon^{1/2}y, \varepsilon^{1/2}z, \varepsilon^{3/2}t) + \varepsilon^2 v_z^{(2)} + \varepsilon^{5/2} v_z^{(3)}.
\end{aligned} \tag{10.31}$$

The asymptotic analysis of the Euler–Poisson system is easier to handle if we work with rescaled variables and unknowns adapted to this ansatz. More precisely, if we introduce

$$\tilde{x}=\varepsilon^{1/2}x, \quad \tilde{y}=\varepsilon^{1/2}y, \quad \tilde{z}=\varepsilon^{1/2}z, \quad \tilde{t}=\varepsilon^{1/2}t, \quad \tilde{n}=\varepsilon^{-1}n, \quad \tilde{\varphi}=\varepsilon^{-1}\varphi, \quad \tilde{\mathbf{v}}=\varepsilon^{-1}\mathbf{v},$$

the Euler–Poisson equation (10.4) becomes (dropping the tilde superscripts)

$$\begin{cases}
\partial_t n + \nabla\cdot((1+\varepsilon n)\mathbf{v}) = 0, \\
\partial_t \mathbf{v} + \varepsilon(\mathbf{v}\cdot\nabla)\mathbf{v} + \nabla\varphi + a\varepsilon^{-1/2}\mathbf{e}\wedge\mathbf{v} = 0, \\
-\varepsilon^2\Delta\varphi + e^{\varepsilon\varphi} - 1 = \varepsilon n,
\end{cases} \tag{10.32}$$

and the ZK equation is derived by looking for approximate solutions to this system under the form[1]

$$\begin{aligned}
n^\varepsilon &= n^{(1)}(x-t, y, z, \varepsilon t) + \varepsilon n^{(2)} + \varepsilon^2 n^{(3)} \\
\varphi^\varepsilon &= \varphi^{(1)}(x-t, y, z, \varepsilon t) + \varepsilon \varphi^{(2)} + \varepsilon^2 \varphi^{(3)} \\
v_x^\varepsilon &= v_x^{(1)}(x-t, y, z, \varepsilon t) + \varepsilon v_x^{(2)} + \varepsilon^2 v_x^{(3)} \\
v_y^\varepsilon &= \varepsilon^{1/2} v_y^{(1)}(x-t, y, z, \varepsilon t) + \varepsilon v_y^{(2)} + \varepsilon^{3/2} v_y^{(3)} \\
v_z^\varepsilon &= \varepsilon^{1/2} v_z^{(1)}(x-t, y, z, \varepsilon t) + \varepsilon v_z^{(2)} + \varepsilon^{3/2} v_z^{(3)}.
\end{aligned} \tag{10.33}$$

10.3.1 The Cauchy Problem Revisited

It is easily checked that when applied to (10.32), Theorem 10.1 provides an existence time which is of order $O(1)$ with respect to ε, while the time scale $O(1/\varepsilon)$ must be reached to observe the dynamics of the Zakharov–Kuznetsov equation that occur along the slow time scale εt.

As explained in the Introduction, we therefore need, in order to justify the Zakharov–Kuznetsov equation as a long-wave limit of the Euler–Poisson system, to

[1] Actually we will not use the third-order profiles.

solve the Cauchy problem associated to (10.32) on a time interval of order $O(1/\varepsilon)$. In order to do so, we will consider (10.32) as a perturbation of a hyperbolic quasilinear system and give a proof which does not use the smoothing effect of the φ equation for a fixed ε. This is the reason why more regularity is required on the initial data in the statement of the theorem below. Before stating it, let us introduce the space H^{s+1}_ε which is the standard Sobolev space $H^{s+1}(\mathbb{R}^d)$ endowed with the norm

$$\forall s \geq 0, \quad \forall f \in H^{s+1}, \quad |f|^2_{H^{s+1}_\varepsilon} = |f|^2_{H^s} + \varepsilon |\nabla f|^2_{H^s}; \tag{10.34}$$

the presence of the small parameter ε in front of the second term is here to make this norm adapted to measure the smoothing effects of the φ equation; the fact that these smoothing effects are small explains why Theorem 10.1, which relies on them, provides an existence time much too small to observe the dynamics of the Zakharov–Kuznetsov equation.

Theorem 10.2. *Let $s > \frac{d}{2} + 1$ and $n_0 \in H^s(\mathbb{R}^d)$, $\mathbf{v}_0 \in H^{s+1}(\mathbb{R}^d)^d$ such that $1 + n_0 \geq c_0$ for some $c_0 > 0$.*

Then there exists $\underline{T} > 0$ such that for all $\varepsilon \in (0,1)$, there is a unique solution $(n^\varepsilon, \mathbf{v}^\varepsilon, \varphi^\varepsilon) \in C([0, \frac{\underline{T}}{\varepsilon}]; H^s(\mathbb{R}^d) \times H^{s+1}_\varepsilon(\mathbb{R}^d)^d \times H^{s+1}(\mathbb{R}^d))$ of (10.32) such that $1 + \varepsilon n^\varepsilon > c_0/2$.

Moreover the family $(n^\varepsilon, \mathbf{v}^\varepsilon, \nabla\varphi^\varepsilon)_{\varepsilon \in (0,1)}$ is uniformly bounded in $H^s \times H^{s+1}_\varepsilon \times H^{s-1}$.

Proof. **Step 1. Preliminary results.**
If we differentiate the third equation of (10.32) with respect to ∂_j $(j = x, y, z, t)$, we get

$$M_\varepsilon(\phi)\partial_j\varphi = \partial_j n \quad \text{with} \quad M_\varepsilon(\phi) = -\varepsilon\Delta + e^{\varepsilon\varphi}; \tag{10.35}$$

the operator $M_\varepsilon(\phi)$ plays a central role in the energy estimates below, and we state here some basic estimates. The first is that $(u, M_\varepsilon(\phi)u)^{1/2}$ defines a norm which is uniformly equivalent to $|\cdot|_{H^1_\varepsilon}$; more precisely, for all $u \in H^1(\mathbb{R}^d)$,

$$(u, M_\varepsilon(\phi)u) \leq C(|\varphi|_\infty)|u|^2_{H^1_\varepsilon} \quad \text{and} \quad |u|^2_{H^1_\varepsilon} \leq C(|\varphi|_\infty)(u, M_\varepsilon(\phi)u). \tag{10.36}$$

We also have for all $u \in H^1(\mathbb{R}^d)$ and $f \in W^{1,\infty}(\mathbb{R}^d)$

$$\big(u, fM_\varepsilon(\phi)u\big) \leq C(|\varphi|_\infty, |f|_\infty, \sqrt{\varepsilon}|\nabla f|_\infty)\,|u|^2_{H^1_\varepsilon},$$

and, for all $u \in H^1(\mathbb{R}^d)$ and $f \in W^{2,\infty}(\mathbb{R}^d)$,

$$\big(u, [f\partial_j, M_\varepsilon(\phi)]u\big) \leq C(|\varphi|_{W^{1,\infty}}, |f|_{W^{1,\infty}}, \sqrt{\varepsilon}|\nabla\partial_j f|_\infty)\,|u|^2_{H^1_\varepsilon} \qquad (j - x, y, z);$$

these two estimates are readily obtained from the definition of $M_\varepsilon(\phi)$ and integration by parts.

Let us finally prove that $M_\varepsilon(\phi)$ is invertible and give estimates on its inverse.

Lemma 10.2. *Let $\varphi \in L^\infty(\mathbb{R}^d)$. Then $M_\varepsilon(\phi) : H^2(\mathbb{R}^d) \to L^2(\mathbb{R}^d)$ is an isomorphism and*

$$\forall v \in L^2(\mathbb{R}^d), \qquad e^{-\frac{\varepsilon}{2}|\varphi|_\infty}|M_\varepsilon(\phi)^{-1}v|_2 + \sqrt{\varepsilon}|\nabla M_\varepsilon(\phi)^{-1}v|_2 \le e^{\frac{\varepsilon}{2}|\varphi|_\infty}|v|_2.$$

If moreover $t_0 > d/2$, $s \ge 0$ and $\varphi \in H^{t_0+1} \cap H^s(\mathbb{R}^d)$, then we also have

$$\forall v \in H^s(\mathbb{R}^d), \qquad |M_\varepsilon(\phi)^{-1}v|_{H^s} + \sqrt{\varepsilon}|\nabla M_\varepsilon(\phi)^{-1}v|_{H^s} \le C(|\varphi|_{H^{t_0+1}\cap H^s})|v|_{H^s}.$$

Proof (Proof of the lemma). The invertibility property of $M_\varepsilon(\phi)$ follows classically from Lax–Milgram's theorem, and the first estimate of the lemma follows from the coercivity property of $M_\varepsilon(\phi)$, namely,

$$e^{-\varepsilon|\varphi|_\infty}|u|_2^2 + \varepsilon|\nabla u|_2^2 \le (M_\varepsilon(\phi)u, u). \tag{10.37}$$

In order to prove the higher-order estimates, let us write $u = M_\varepsilon(\phi)^{-1}v$. We have by definition $(-\varepsilon\Delta + e^{\varepsilon\varphi})u = v$, so that applying Λ^s on both sides, we get

$$(-\varepsilon\Delta + e^{\varepsilon\varphi})\Lambda^s u = \Lambda^s v - [\Lambda^s, e^{\varepsilon\varphi}]u. \tag{10.38}$$

Using the first estimate and recalling that $u = M_\varepsilon(\phi)^{-1}v$, we deduce that

$$|M_\varepsilon(\phi)^{-1}v|_{H^s} + \sqrt{\varepsilon}|\nabla M_\varepsilon(\phi)^{-1}v|_{H^s} \le e^{\varepsilon|\varphi|_\infty}\big(|v|_{H^s} + |[\Lambda^s, e^{\varepsilon\varphi}]M_\varepsilon(\phi)^{-1}v|_2\big).$$

Now, using Kato–Ponce and Coifman–Meyer commutator estimates and Moser's inequality shows that

$$\begin{aligned}|[\Lambda^s, e^{\varepsilon\varphi}]M_\varepsilon(\phi)^{-1}v|_2 &\le \varepsilon C(|\varphi|_\infty)\\ &\times \Big(|\varphi|_{H^{t_0+1}}|M_\varepsilon(\phi)^{-1}v|_{H^{s-1}} + \langle|\varphi|_{H^s}|M_\varepsilon(\phi)^{-1}v|_{H^{t_0}}\rangle_{s>t_0+1}\Big),\end{aligned}$$

and we get therefore

$$\begin{aligned}|M_\varepsilon(\phi)^{-1}v|_{H^s} + \sqrt{\varepsilon}|\nabla M_\varepsilon(\phi)^{-1}u|_{H^s} \le C(|\varphi|_\infty)\big(|v|_{H^s} + |\varphi|_{H^{t_0+1}}|M_\varepsilon(\phi)^{-1}v|_{H^{s-1}}\\ + \langle|\varphi|_{H^s}|M_\varepsilon(\phi)^{-1}v|_{H^{t_0}}\rangle_{s>t_0+1}\big),\end{aligned}$$

and the results follow by a continuous induction. □

Step 2. L^2 estimates for a linearized system.
Let $T > 0$ and $(\underline{n}, \underline{\mathbf{v}}) \in L^\infty([0,T]; W^{1,\infty}(\mathbb{R}^d)) \cap W^{1,\infty}([0,T]; L^\infty(\mathbb{R}^d))$ be such that $1 + \varepsilon\underline{n} \ge c_0 > 0$ uniformly on $[0,T]$ and $\underline{\varphi} \in W^{1,\infty}$ with $\partial_t\underline{\varphi} \in L^\infty$. Let us consider a classical solution $(n, \mathbf{v}, \varphi)$ of the linear system

$$\begin{cases}\partial_t n + (1 + \varepsilon\underline{n})\nabla\cdot\mathbf{v} + \varepsilon\underline{\mathbf{v}}\cdot\nabla n = \varepsilon f,\\ \partial_t\mathbf{v} + \varepsilon(\underline{\mathbf{v}}\cdot\nabla)\mathbf{v} + \nabla\varphi + a\varepsilon^{-1/2}\mathbf{e}\wedge\mathbf{v} = \varepsilon\mathbf{g},\\ M_\varepsilon(\underline{\phi})\nabla\varphi = \nabla n + \varepsilon\mathbf{h}.\end{cases} \tag{10.39}$$

We want to prove here that

$$\sup_{[0,T]}\big(|n|_2^2+|\mathbf{v}|_{H^1_\varepsilon}^2\big)\le \exp(\varepsilon C_0 T)\\ \times\big(|n_{|_{t=0}}|_2^2+|\mathbf{v}_{|_{t=0}}|_{H^1_\varepsilon}^2\big)+\varepsilon T\big(|f|_{L^2_T}^2+|\mathbf{g}|_{H^1_{\varepsilon,T}}^2+|\mathbf{h}|_{L^2_T}^2\big)\big), \tag{10.40}$$

with $C_0=C(\frac{1}{c_0},|(\underline{n},\underline{\mathbf{v}},\varphi)|_{W_T^{1,\infty}},|\partial_t(\underline{n},\underline{\mathbf{v}},\underline{\varphi})|_{L_T^\infty},\sqrt{\varepsilon}|\nabla\nabla\cdot\underline{\mathbf{v}}|_{L_T^\infty})$.

Taking the L^2 scalar product with $\frac{1}{1+\varepsilon\underline{n}}n$, we get

$$\big(\frac{1}{1+\varepsilon\underline{n}}\partial_t n,n\big)+(\nabla\cdot\mathbf{v},n)+\varepsilon\big(\frac{\underline{\mathbf{v}}}{1+\varepsilon\underline{n}}\cdot\nabla n,n\big)=\varepsilon\big(\frac{f}{1+\varepsilon\underline{n}},n\big),$$

which can be rewritten under the form

$$\frac{1}{2}\partial_t\big(\frac{1}{1+\varepsilon\underline{n}}n,n\big)+\varepsilon\frac{1}{2}\big(\frac{\partial_t\underline{n}}{(1+\varepsilon\underline{n})^2}n,n\big)+(\nabla\cdot\mathbf{v},n)\\ -\varepsilon\frac{1}{2}\big(n,\nabla\cdot\big(\frac{\underline{\mathbf{v}}}{1+\varepsilon\underline{n}}\big)n\big)=\varepsilon\big(\frac{f}{1+\varepsilon\underline{n}},n\big). \tag{10.41}$$

We now take the scalar product of the second equation with $M_\varepsilon(\underline{\phi})\mathbf{v}$ to obtain, after recalling that $M_\varepsilon(\underline{\phi})\nabla\varphi=\nabla n+\varepsilon\mathbf{h}$,

$$\big(M_\varepsilon(\underline{\phi})\partial_t\mathbf{v},\mathbf{v}\big)+\varepsilon\big(M_\varepsilon(\underline{\phi})\underline{\mathbf{v}}\cdot\nabla\mathbf{v},\mathbf{v}\big)+(\nabla n,\mathbf{v})\\ =-\varepsilon(\mathbf{h},\mathbf{v})+\varepsilon\big(M_\varepsilon(\underline{\phi})\mathbf{g},\mathbf{v}\big),$$

or equivalently

$$\frac{1}{2}\partial_t\big(M_\varepsilon(\underline{\phi})\mathbf{v},\mathbf{v}\big)-\frac{1}{2}\varepsilon\big(\partial_t\underline{\varphi}e^{\varepsilon\underline{\varphi}}\mathbf{v},\mathbf{v}\big)-\frac{1}{2}\varepsilon\big(\mathbf{v},(\nabla\cdot\underline{\mathbf{v}})M_\varepsilon(\underline{\phi})\mathbf{v}\big)\\ -\frac{1}{2}\varepsilon\big(\mathbf{v},[\underline{\mathbf{v}}\cdot\nabla,M_\varepsilon(\underline{\phi})]\mathbf{v}\big)+(\nabla n,\mathbf{v})=-\varepsilon(\mathbf{h},\mathbf{v})+\varepsilon\big(M_\varepsilon(\underline{\phi})\mathbf{g},\mathbf{v}\big). \tag{10.42}$$

Adding (10.42) to (10.41), we get therefore

$$\frac{1}{2}\partial_t\big(\frac{1}{1+\varepsilon\underline{n}}n,n\big)+\frac{1}{2}\partial_t\big(M_\varepsilon(\underline{\phi})\mathbf{v},\mathbf{v}\big)=\varepsilon\frac{1}{2}\big(\big[\frac{\partial_t\underline{n}}{(1+\varepsilon\underline{n})^2}+\nabla\cdot\big(\frac{\underline{\mathbf{v}}}{1+\varepsilon\underline{n}}\big)\big]n,n\big)\\ +\frac{1}{2}\varepsilon\big(\partial_t\underline{\varphi}e^{\varepsilon\underline{\varphi}}\mathbf{v},\mathbf{v}\big)+\frac{1}{2}\varepsilon\big(\mathbf{v},(\nabla\cdot\underline{\mathbf{v}})M_\varepsilon(\underline{\phi})\mathbf{v}\big)+\frac{1}{2}\varepsilon\big(\mathbf{v},[\underline{\mathbf{v}}\cdot\nabla,M_\varepsilon(\underline{\phi})]\mathbf{v}\big)\\ +\varepsilon\big(\frac{1}{1+\varepsilon\underline{n}}f,n\big)-\varepsilon(\mathbf{h},\mathbf{v})+\varepsilon\big(M_\varepsilon(\underline{\psi})\mathbf{g},\mathbf{v}\big).$$

All the terms on the right-hand side are easily controlled (with the help of step 1 for the third and fourth terms) to obtain

$$\partial_t\{(\frac{n}{1+\varepsilon\underline{n}},n)+(M_\varepsilon(\underline{\phi})\mathbf{v},\mathbf{v})\}\le \varepsilon C(\frac{1}{c_0},|(\underline{n},\underline{\mathbf{v}},\underline{\varphi})|_{W^{1,\infty}},|\partial_t(\underline{n},\underline{\mathbf{v}},\underline{\varphi})|_\infty,\varepsilon|\nabla\nabla\cdot\underline{\mathbf{v}}|_2)$$
$$\times((\frac{1}{1+\varepsilon\underline{n}}n,n)+|\mathbf{v}|^2_{H^1_\varepsilon}+|f|_2^2+|\mathbf{g}|^2_{H^1_\varepsilon}+|\mathbf{h}|_2^2).$$

Using (10.36) and a Gronwall inequality, we readily deduce (10.40).

Step 3. H^s estimates for a linearized system.
We want to prove here that the solution $(n,\mathbf{v},\varphi)$ to (10.39) satisfies, for all $s\ge t_0+1$,

$$\sup_{[0,T]}\big(|n|^2_{H^s}+|\mathbf{v}|^2_{H^{s+1}_\varepsilon}\big)\le \exp(\varepsilon C_s T)\times\Big(|n_{|_{t=0}}|^2_{H^s}+|\mathbf{v}_{|_{t=0}}|^2_{H^{s+1}_\varepsilon}$$
$$+\varepsilon T(|f|^2_{H^s_T}+|\mathbf{g}|^2_{H^{s+1}_{\varepsilon,T}}+|\mathbf{h}|^2_{H^s_T}+|n|^2_{H^s}+|\mathbf{v}|^2_{H^{s+1}_\varepsilon})\Big),\tag{10.43}$$

with $C_s=C(\frac{1}{c_0},|\underline{n}|_{H^s_T},|\underline{\mathbf{v}}|_{H^{s+1}_{\varepsilon,T}},|\underline{\varphi}|_{H^s_T},|\partial_t(\underline{n},\underline{\mathbf{v}},\underline{\varphi})|_{L^\infty_T})$.

Applying Λ^s to the three equations of (10.39) and writing $\tilde{n}=\Lambda^s n$, $\tilde{\mathbf{v}}=\Lambda^s\mathbf{v}$, and $\tilde{\varphi}=\Lambda^s\varphi$, we get

$$\begin{cases}\partial_t\tilde{n}+(1+\varepsilon\underline{n})\nabla\cdot\tilde{\mathbf{v}}+\varepsilon\underline{\mathbf{v}}\cdot\nabla\tilde{n}=\varepsilon\tilde{f},\\ \partial_t\tilde{\mathbf{v}}+\varepsilon(\underline{\mathbf{v}}\cdot\nabla)\tilde{\mathbf{v}}+\nabla\tilde{\varphi}+a\varepsilon^{-1/2}\mathbf{e}\wedge\tilde{\mathbf{v}}=\varepsilon\tilde{\mathbf{g}},\\ M_\varepsilon(\underline{\phi})\nabla\tilde{\varphi}=\nabla\tilde{n}+\varepsilon\tilde{\mathbf{h}}.\end{cases}\tag{10.44}$$

with

$$\begin{aligned}\tilde{f}&=\Lambda^s f-[\Lambda^s,\underline{n}]\nabla\cdot\mathbf{v}-[\Lambda^s,\underline{\mathbf{v}}]\cdot\nabla n,\\ \tilde{\mathbf{g}}&=\Lambda^s\mathbf{g}-[\Lambda^s,\underline{\mathbf{v}}]\cdot\nabla\mathbf{v},\\ \tilde{\mathbf{h}}&=\Lambda^s\mathbf{h}-\frac{1}{\varepsilon}[\Lambda^s,M_\varepsilon(\underline{\phi})]\nabla\varphi.\end{aligned}$$

From step 2, we get that

$$\sup_{[0,T]}\big(|\tilde{n}|_2^2+|\tilde{\mathbf{v}}|^2_{H^1_\varepsilon}\big)\le\exp(\varepsilon C_0 T)$$
$$\times\big(|\tilde{n}_{|_{t=0}}|_2^2+|\tilde{\mathbf{v}}_{|_{t=0}}|^2_{H^1_\varepsilon}\big)+\varepsilon T(|\tilde{f}|^2_{L^2_T}+|\tilde{\mathbf{g}}|^2_{H^1_{\varepsilon,T}}+|\tilde{\mathbf{h}}|^2_{L^2_T})).$$

Now, by standard commutator estimates, we have for all $s\ge t_0+1$

$$|\tilde{f}|_2+|\tilde{\mathbf{g}}|_{H^1_\varepsilon}\le|f|_{H^s}+|\mathbf{g}|_{H^{s+1}_\varepsilon}+\big(|\underline{n}|_{H^s}+|\underline{\mathbf{v}}|_{H^{s+1}_\varepsilon}\big)\times\big(|\mathbf{v}|_{H^{s+1}_\varepsilon}+|n|_{H^s}\big)$$

and

$$\begin{aligned}|\tilde{\mathbf{h}}|_2&\le|\mathbf{h}|_{H^s}+C(|\underline{\varphi}|_{H^s})|\nabla\varphi|_{H^{s-1}},\\ &\le C(|\underline{\varphi}|_{H^s})\big(|\mathbf{h}|_{H^s}+|n|_{H^s}\big),\end{aligned}$$

where we used Lemma 10.2 to get the last inequality. We can now directly deduce (10.43) from the Sobolev embedding $W^{1,\infty}(\mathbb{R}^d) \subset H^s(\mathbb{R}^d)$ ($s \geq t_0 + 1$).

Step 4. End of the proof.
The exact solution provided by Theorem 10.1 solves (10.39) with $(\underline{n}, \underline{\mathbf{v}}, \underline{\varphi}) = (n, \mathbf{v}, \varphi)$ and $f = 0$, $\mathbf{g} = 0$, and $\mathbf{h} = 0$; we deduce therefore from step 3 that it satisfies the estimate

$$\sup_{[0,T]} \big(|n|_{H^s}^2 + |\mathbf{v}|_{H_\varepsilon^{s+1}}^2\big) \leq \exp(\varepsilon C_s T) \times \Big(|n_{|_{t=0}}|_{H^s}^2 + |\mathbf{v}_{|_{t=0}}|_{H_\varepsilon^{s+1}}^2\big) + \varepsilon T\big(|n|_{H^s}^2 + |\mathbf{v}|_{H_\varepsilon^{s+1}}^2\big)\Big),$$

with

$$\begin{aligned} C_s &= C(\frac{1}{c_0}, |n|_{H_T^s}, |\mathbf{v}|_{H_{\varepsilon,T}^{s+1}}, |\varphi|_{H^s}, |\partial_t(n, \mathbf{v}, \varphi)|_{L_T^\infty}) \\ &= C(\frac{1}{c_0}, |n|_{H_T^s}, |\mathbf{v}|_{H_{\varepsilon,T}^{s+1}}), \end{aligned}$$

where we used (10.35) and the equation to control the time derivatives in terms of space derivatives and Proposition 10.1 to control the L^2 norm of φ (control on $\nabla\varphi$ being provided by Lemma 10.2). This provides us with a $\underline{T} > 0$ such that $|n|_{H^s}^2 + |\mathbf{v}|_{H_\varepsilon^{s+1}}^2$ remains uniformly bounded with respect to ε on $[0, \underline{T}/\varepsilon]$. □

10.3.2 The ZK Approximation to the Euler–Poisson System

We construct here an approximate solution to (10.32) based on the ZK equation. Following Laedke and Spatschek [22], but with the rescaled variables and unknowns introduced in (10.31), we look for approximate solutions of (10.32) under the form (10.33), namely,

$$\begin{aligned} n^\varepsilon &= n^{(1)}(x-t, y, z, \varepsilon t) + \varepsilon n^{(2)} \\ \varphi^\varepsilon &= \varphi^{(1)}(x-t, y, z, \varepsilon t) + \varepsilon \varphi^{(2)} \\ v_x^\varepsilon &= v_x^{(1)}(x-t, y, z, \varepsilon t) + \varepsilon v_x^{(2)} \\ v_y^\varepsilon &= \varepsilon^{1/2} v_y^{(1)}(x-t, y, z, \varepsilon t) + \varepsilon v_y^{(2)} \\ v_z^\varepsilon &= \varepsilon^{1/2} v_z^{(1)}(x-t, y, z, \varepsilon t) + \varepsilon v_z^{(2)}. \end{aligned} \tag{10.45}$$

Notation 1 *We will denote by X the variable $x - t$ and by T the slow time variable εt.*

Plugging this ansatz into the first equation of (10.32), we obtain

$$\partial_t n^\varepsilon + \nabla \cdot (1 + \varepsilon n^\varepsilon)\mathbf{v}^\varepsilon = \sum_{j=0}^{6} \varepsilon^{j/2} N^j, \tag{10.46}$$

where

$$\begin{aligned}
N^0 &= -\frac{\partial}{\partial X}n^{(1)} + \frac{\partial}{\partial X}v_x^{(1)} \\
N^1 &= \frac{\partial}{\partial y}v_y^{(1)} + \frac{\partial}{\partial z}v_z^{(1)} \\
N^2 &= \frac{\partial}{\partial T}n^{(1)} - \frac{\partial}{\partial X}n^{(2)} + \frac{\partial}{\partial X}(n^{(1)}v_x^{(1)}) + \frac{\partial}{\partial X}v_x^{(2)} + \frac{\partial}{\partial y}v_y^{(2)} + \frac{\partial}{\partial z}v_z^{(2)} \\
N^3 &= \frac{\partial}{\partial y}(n^{(1)}v_y^{(1)}) + \frac{\partial}{\partial z}(n^{(1)}v_z^{(1)}) \\
N^4 &= \frac{\partial}{\partial T}n^{(2)} + \frac{\partial}{\partial X}(n^{(1)}v_x^{(2)}) + \frac{\partial}{\partial y}(n^{(1)}v_y^{(2)}) \\
&\quad + \frac{\partial}{\partial z}(n^{(1)}v_z^{(2)}) + \frac{\partial}{\partial X}(n^{(2)}v_x^{(1)}), \\
N^5 &= \frac{\partial}{\partial y}(n^{(2)}v_y^{(1)}) + \frac{\partial}{\partial z}(n^{(2)}v_z^{(1)}), \\
N^6 &= \frac{\partial}{\partial X}(n^{(2)}v_x^{(2)}) + \frac{\partial}{\partial y}(n^{(2)}v_y^{(2)}) + \frac{\partial}{\partial z}(n^{(2)}v_z^{(2)}).
\end{aligned}$$

Similarly, one has

$$\frac{\partial}{\partial t}v_x^\varepsilon + \varepsilon v_x^\varepsilon\frac{\partial}{\partial x}v_x^\varepsilon + \varepsilon v_y^\varepsilon\frac{\partial}{\partial y}v_x^\varepsilon + \varepsilon v_z^\varepsilon\frac{\partial}{\partial z}v_x^\varepsilon + \frac{\partial}{\partial x}\varphi^\varepsilon = \sum_{j=0}^{6}\varepsilon^{j/2}R_j^1 \tag{10.47}$$

with $R_1^1 = 0$ and

$$\begin{aligned}
R_0^1 &= -\frac{\partial}{\partial X}v_x^{(1)} + \frac{\partial}{\partial X}\varphi^{(1)} \\
R_2^1 &= \frac{\partial}{\partial T}v_x^{(1)} - \frac{\partial}{\partial X}v_x^{(2)} + v_x^{(1)}\frac{\partial}{\partial X}v_x^{(1)} + \frac{\partial}{\partial X}\varphi^{(2)} \\
R_3^1 &= v_y^{(1)}\frac{\partial}{\partial y}v_x^{(1)} + v_z^{(1)}\frac{\partial}{\partial z}v_x^{(1)} \\
R_4^1 &= \frac{\partial}{\partial T}v_x^{(2)} + \frac{\partial}{\partial X}(v_x^{(1)}v_x^{(2)}) + v_y^{(2)}\frac{\partial}{\partial y}v_x^{(1)} + v_z^{(2)}\frac{\partial}{\partial z}v_x^{(1)} \\
R_5^1 &= v_y^{(1)}\frac{\partial}{\partial y}v_x^{(2)} + v_z^{(1)}\frac{\partial}{\partial z}v_x^{(2)} \\
R_6^1 &= v_y^{(2)}\frac{\partial}{\partial y}v_x^{(2)} + v_z^{(2)}\frac{\partial}{\partial z}v_x^{(2)} + v_x^{(2)}\frac{\partial}{\partial X}v_x^{(2)};
\end{aligned}$$

for the second component of the velocity equation, we get

$$\frac{\partial}{\partial t}v_y^\varepsilon + \varepsilon v_x^\varepsilon\frac{\partial}{\partial x}v_y^\varepsilon + \varepsilon v_y^\varepsilon\frac{\partial}{\partial y}v_y^\varepsilon + \varepsilon v_z^\varepsilon\frac{\partial}{\partial z}v_y^\varepsilon + \frac{\partial}{\partial y}\varphi^\varepsilon - a\varepsilon^{-1/2}v_z^\varepsilon = \sum_{j=0}^{6}\varepsilon^{j/2}R_j^2 \tag{10.48}$$

with

$$
\begin{aligned}
R_0^2 &= \frac{\partial}{\partial y}\varphi^{(1)} - a v_z^{(1)} \\
R_1^2 &= -\left(\frac{\partial}{\partial X} v_y^{(1)} + a v_z^{(2)}\right) \\
R_2^2 &= -\frac{\partial}{\partial X} v_y^{(2)} + \frac{\partial}{\partial y}\varphi^{(2)} \\
R_3^2 &= \frac{\partial}{\partial T} v_y^{(1)} + v_x^{(1)} \frac{\partial}{\partial X} v_y^{(1)} \\
R_4^2 &= \frac{\partial}{\partial T} v_y^{(2)} + v_x^{(1)} \frac{\partial}{\partial X} v_y^{(2)} + v_y^{(1)} \frac{\partial}{\partial y} v_y^{(1)} + v_z^{(1)} \frac{\partial}{\partial z} v_y^{(1)} \\
R_5^2 &= v_x^{(2)} \frac{\partial}{\partial X} v_y^{(1)} + \frac{\partial}{\partial y}(v_y^{(1)} v_y^{(2)}) + v_z^{(1)} \frac{\partial}{\partial z} v_y^{(2)} + v_z^{(2)} \frac{\partial}{\partial z} v_y^{(1)} \\
R_6^2 &= v_x^{(2)} \frac{\partial}{\partial X} v_y^{(2)} + v_y^{(2)} \frac{\partial}{\partial y} v_y^{(2)} + v_z^{(2)} \frac{\partial}{\partial z} v_y^{(2)},
\end{aligned}
$$

while for the third component, the equations are

$$
\frac{\partial}{\partial t} v_z^\varepsilon + \varepsilon v_x^\varepsilon \frac{\partial}{\partial x} v_z^\varepsilon + \varepsilon v_y^\varepsilon \frac{\partial}{\partial y} v_z^\varepsilon + \varepsilon v_z^\varepsilon \frac{\partial}{\partial z} v_z^\varepsilon + \frac{\partial}{\partial z}\varphi^\varepsilon + a\varepsilon^{-1/2} v_y^\varepsilon = \sum_{j=0}^{6} \varepsilon^{j/2} R_j^3 \quad (10.49)
$$

with

$$
\begin{aligned}
R_0^3 &= \frac{\partial}{\partial z}\varphi^{(1)} + a v_y^{(1)} \\
R_1^3 &= -\frac{\partial}{\partial X} v_z^{(1)} + a v_y^{(2)} \\
R_2^3 &= -\frac{\partial}{\partial X} v_z^{(2)} + \frac{\partial}{\partial z}\varphi^{(2)} \\
R_3^3 &= \frac{\partial}{\partial T} v_z^{(1)} + v_x^{(1)} \frac{\partial}{\partial X} v_z^{(1)} \\
R_4^3 &= \frac{\partial}{\partial T} v_z^{(2)} + v_x^{(1)} \frac{\partial}{\partial X} v_z^{(2)} + v_y^{(1)} \frac{\partial}{\partial y} v_z^{(1)} + v_z^{(1)} \frac{\partial}{\partial z} v_z^{(1)} \\
R_5^3 &= v_x^{(2)} \frac{\partial}{\partial X} v_z^{(1)} + \frac{\partial}{\partial z}(v_z^{(1)} v_z(2)) + v_y^{(1)} \frac{\partial}{\partial y} v_z^{(2)} + v_y^{(2)} \frac{\partial}{\partial y} v_z^{(1)} \\
R_6^3 &= v_x^{(2)} \frac{\partial}{\partial X} v_z^{(2)} + v_y^{(2)} \frac{\partial}{\partial y} v_z^{(2)} + v_z^{(2)} \frac{\partial}{\partial z} v_z^{(2)}.
\end{aligned}
$$

Finally, for the equation on the potential, we obtain

$$
-\varepsilon^2 \Delta\varphi^\varepsilon + e^{\varepsilon\varphi} - 1 - \varepsilon n = \varepsilon r^2 + \varepsilon^2 r^4 + \varepsilon^3 r^6 + O(\varepsilon^4) \quad (10.50)
$$

with

$$
\begin{aligned}
r^2 &= \varphi^{(1)} - n^{(1)} \\
r^4 &= -\Delta\varphi^{(1)} + \varphi^{(2)} + \frac{1}{2}(\varphi^{(1)})^2 - n^{(2)} \\
r^6 &= -\Delta\varphi^{(2)} + \varphi^{(1)}\varphi^{(2)} + \frac{1}{6}(\varphi^{(1)})^3.
\end{aligned}
$$

We first derive (following essentially [22]) the equations corresponding to the successive cancellation of the leading-order remainder terms; we then show that they imply that $n^{(1)}$ must solve the Zakharov–Kuznetsov equation and then turn to show in the spirit of Ben Youssef and Colin who considered the one-dimensional case in [6] (see also [13]) that it is indeed possible to construct an approximate solution (10.45) satisfying all the cancellation conditions previously derived. The consequence is the consistency property of (10.45) stated in Proposition 10.2.

10.3.2.1 Cancellation of Terms of Order Zero in ε

Cancelling the terms N^0 and R_0^j $(j = 1,2,3)$ is equivalent to the following conditions:

$$
v_x^{(1)} = \varphi^{(1)} = n^{(1)} \qquad \text{(assuming that } v_x^{(1)},\ \varphi^{(1)},\ n^{(1)} \text{ vanish as } |X| \to +\infty) \tag{10.51}
$$

$$
v_y^{(1)} = -\frac{1}{a}\partial_z n^{(1)} \tag{10.52}
$$

$$
v_z^{(1)} = \frac{1}{a}\partial_y n^{(1)}. \tag{10.53}
$$

10.3.2.2 Cancellation of Terms of Order $\varepsilon^{1/2}$

Using (10.52) and (10.53), the cancellation of the terms N^1 and R_1^j $(j = 1,2,3)$ is equivalent to

$$
v_y^{(2)} = \frac{1}{a^2}\partial_{Xy}^2 n^{(1)} \quad \text{and} \quad v_z^{(2)} = \frac{1}{a^2}\partial_{Xz}^2 n^{(1)}. \tag{10.54}
$$

10.3.2.3 Cancellation of Terms of Order ε

Using the conditions derived above, the cancellation of N^2 and R_2^j $(j = 1,2,3)$ is

$$\partial_T n^{(1)} + 2n^{(1)}\partial_X n^{(1)} + \frac{1}{a^2}\partial_X \Delta_\perp n^{(1)} = -\partial_X (v_x^{(2)} - n^{(2)}) \tag{10.55}$$

$$\partial_X (v_x^{(2)} - \varphi^{(2)}) = \partial_T n^{(1)} + n^{(1)}\partial_X n^{(1)} \tag{10.56}$$

$$\partial_y \varphi^{(2)} = \frac{1}{a^2}\partial^3_{XXy} n^{(1)} \tag{10.57}$$

$$\partial_z \varphi^{(2)} = \frac{1}{a^2}\partial^3_{XXz} n^{(1)} \tag{10.58}$$

while the cancellation of r_2 is equivalent to $\varphi^{(1)} = n^{(1)}$ which has already been imposed.

10.3.2.4 Cancellation of Terms of Order $\varepsilon^{3/2}$

It is possible to cancel the terms of order $\varepsilon^{3/2}$ for the equation on the density and on the first component of the velocity; the fact that $N^3 = R_3^1 = 0$ is actually a direct consequence of (10.51)–(10.53) and (10.54). Looking for the cancellation of the other components of the velocity equation, namely, setting $R_3^2 = R_3^3 = 0$ yields, respectively,

$$0 = -\frac{1}{a}[\partial^2_{zT} n^{(1)} + n^{(1)}\partial^2_{zX} n^{(1)}],$$

$$0 = \frac{1}{a}[\partial^2_{yT} n^{(1)} + n^{(1)}\partial^2_{yX} n^{(1)}] \,,$$

which are inconsistent with the other equations on $n^{(1)}$; consequently, *we cannot expect a better error than $O(\varepsilon^{3/2})$ on the equations for the transverse components of the velocity unless we add higher-order terms in the ansatz.*

10.3.2.5 Cancellation of Terms of Order ε^2

In order to justify the ZK approximation, we need to cancel the $O(\varepsilon^2)$ terms in the equation for φ^ε, that is, to impose $r^4 = 0$, leading to the equation

$$\varphi^{(2)} - n^{(2)} = \Delta n^{(1)} - \frac{1}{2}(n^{(1)})^2. \tag{10.59}$$

10.3.2.6 Derivation of the Zakharov–Kuznetsov Equation

Combining (10.55), (10.56), and (10.59), we find that $n^{(1)}$ must solve the Zakharov–Kuznetsov equation,

$$2\partial_T n^{(1)} + 2n^{(1)}\partial_X n^{(1)} + (\frac{1}{a^2}\Delta_\perp + \Delta)\partial_X n^{(1)} = 0. \tag{10.60}$$

10.3.2.7 Construction of the Profiles

The ZK equation being locally well posed on $H^s(\mathbb{R}^d)$, for all $s > d/2+1$,[2] we can consider a solution $n^{(1)} \in C([0,T_0];H^s(\mathbb{R}^d))$, $s \geq 5$ to (10.60) for some $T_0 > 0$. We show here how to construct all the quantities involved in (10.45) in terms of $n^{(1)}$.

- In agreement with (10.51), we set $\varphi^{(1)} = v_x^{(1)} = n^{(1)}$.
- Equations (10.52) and (10.53) then give $v_y^{(1)}, v_z^{(1)} \in C([0,T_0];H^{s-1}(\mathbb{R}^d))$.
- We then use (10.54) to obtain $v_y^{(2)}, v_z^{(2)} \in C([0,T_0];H^{s-2}(\mathbb{R}^d))$.
- Taking $\varphi^{(2)} = \frac{1}{a^2}\partial_{XX}^2 n^{(1)} \in C([0,T_0];H^{s-2}(\mathbb{R}^d))$ then ensures that (10.57) and (10.58) hold.
- We then get the density corrector $n^{(2)} \in C([0,T_0];H^{s-2}(\mathbb{R}^d))$ by (10.59).
- We recover $v_x^{(2)} \in C([0,T_0];H^{s-2}(\mathbb{R}^d))$ from (10.55) or (10.56)—the fact that $n^{(1)}$ solves the ZK equation ensures that we find the same expression with (10.55) or (10.56) .

The computations above and the explicit expression of the remaining residual terms in (10.46)–(10.50) imply the following consistency result.

Proposition 10.2. *Let $T_0 > 0$, $n_0 \in H^s(\mathbb{R}^d)$ $(s \geq 5)$ and $n^{(1)} \in C([0,T_0];H^s(\mathbb{R}^d))$, solving*

$$2\partial_T n^{(1)} + 2n^{(1)}\partial_X n^{(1)} + (\frac{1}{a^2}\Delta_\perp + \Delta)\partial_X n^{(1)} = 0, \qquad n^{(1)}_{|t=0} = n_0.$$

Constructing the other profiles as indicated above, the approximate solution $(n^\varepsilon, \mathbf{v}^\varepsilon, \varphi^\varepsilon)$ given by (10.45) solves (10.32) up to order ε^3 in φ^ε, ε^2 in n^ε, v_x^ε, and up to order $\varepsilon^{3/2}$ in v_y^ε, v_z^ε:

$$\begin{cases} \partial_t n^\varepsilon + \nabla\cdot((1+\varepsilon n^\varepsilon)\mathbf{v}^\varepsilon) = \varepsilon^2 N^\varepsilon, \\ \partial_t \mathbf{v}^\varepsilon + \varepsilon(\mathbf{v}^\varepsilon\cdot\nabla)\mathbf{v}^\varepsilon + \nabla\varphi^\varepsilon + a\varepsilon^{-1/2}\mathbf{e}\wedge\mathbf{v}^\varepsilon = \varepsilon^{3/2}R^\varepsilon, \\ -\varepsilon^2\Delta\varphi^\varepsilon + e^{\varepsilon\varphi^\varepsilon} - 1 = \varepsilon n^\varepsilon + \varepsilon^3 r^\varepsilon, \end{cases} \tag{10.61}$$

with $R^\varepsilon = (\varepsilon^{1/2}R_1^\varepsilon, R_2^\varepsilon, R_3^\varepsilon)$ and

$$|N^\varepsilon|_{L^\infty([0,\frac{T_0}{\varepsilon}];H^{s-5})} + \sum_{j=1}^{3}|R_j^\varepsilon|_{L^\infty([0,\frac{T_0}{\varepsilon}];H^{s-5})} + |r^\varepsilon|_{L^\infty([0,\frac{T_0}{\varepsilon}];H^{s-4})} \leq C(T_0, |n_0|_{H^s}).$$

10.3.3 Justification of the Zakharov–Kuznetsov Approximation

We are now set to justify the Zakharov–Kuznetsov approximation.

[2] See [8, 25, 30] for the Cauchy theory in larger spaces.

Theorem 10.3. *Let $n^0 \in H^{s+5}$, with $s > d/2+1$, such that $1+n^0 \geq c_0$ on $\mathbb{R}^d$ for some constant $c_0 > 0$. There exists $T_1 > 0$ such that for all $\varepsilon \in (0,1)$:*

i. *The Zakharov–Kuznetsov approximation $(n^\varepsilon, \mathbf{v}^\varepsilon, \varphi^\varepsilon)$ of Proposition 10.2 exists on the time interval $[0, T_1/\varepsilon]$.*
ii. *There exists a unique solution*

$$(\underline{n}, \underline{\mathbf{v}}, \underline{\varphi}) \in C([0, \frac{T_1}{\varepsilon}]; H^s(\mathbb{R}^d) \times H^{s+1}_\varepsilon(\mathbb{R}^d)^d \times H^{s+1}(\mathbb{R}^d))$$

provided by Theorem 10.2 to the Euler–Poisson equations (10.32) *with initial condition $(\underline{n}^0, \underline{\mathbf{v}}^0, \underline{\varphi}^0) = (n^\varepsilon, \mathbf{v}^\varepsilon, \varphi^\varepsilon)_{|t=0}$.*
Moreover, one has the error estimate

$$\forall 0 \leq t \leq T_1/\varepsilon, \quad |\underline{n}(t) - n^\varepsilon(t)|^2_{H^s} + |\underline{\mathbf{v}}(t) - \mathbf{v}^\varepsilon(t)|^2_{H^{s+1}_\varepsilon} \leq \varepsilon^{3/2} t C(\frac{1}{c_0}, T_1, |n^0|_{H^{s+5}}).$$

Remark 10.4. The error of the approximation is $O(\varepsilon^{3/2})$ for times of order $O(1)$ but of size $O(\varepsilon^{1/2})$ for times of order $O(1/\varepsilon)$. Looking at (10.45), this is a relative error of size $O(\varepsilon^{1/2})$ for n and v_x but of size $O(1)$ for v_y and v_z. Consequently, the Zakharov–Kuznetsov approximation provides a good approximation for the density and the longitudinal velocity, but not, for large times, for the transverse velocity (at least, we did not prove it in Theorem 10.3).

Remark 10.5. In the one-dimensional case (KdV approximation), all the terms of order $\varepsilon^{3/2}$ can be cancelled (see Sect. 10.3.2.4), and the residual in the second equation of (10.61) is of size $O(\varepsilon^2)$ instead of $O(\varepsilon^{3/2})$. The error in the theorem then becomes $O(\varepsilon^2 t)$, which gives a relative error of size $O(\varepsilon)$ for both the density and the velocity over large times $O(1/\varepsilon)$. In the one-dimensional case, it is also possible to construct higher-order approximations by including the order three and higher terms in the ansatz (10.33). This has been done in [13].

Remark 10.6. In [13], the authors justify the KdV approximation (corresponding to the one- dimensional ZK approximation) by looking at an exact solution as a perturbation of the approximate solution, $(n_{ex}, v_{ex}) = (n_{app}, v_{app}) + \varepsilon^k (n_R, v_R)$ (with $k > 0$ depending on the order of the approximation). They study the equations satisfied by (n_R, v_R), which requires subtle estimates. Our approach is much simpler: we prove uniform (with respect to ε) well-posedness of the Euler–Poisson equation, from which we deduce very easily that any consistent approximation remains close to the exact solution (but of course, in a lower norm).

Proof. Let us take $0 < T_1 \leq \min\{\underline{T}, T_0\}$, where $\underline{T}/\varepsilon$ is the existence time of the exact solution provided by Theorem 10.2 and T_0/ε the existence time of the approximate solution in Proposition 10.2. Denote by $(\underline{n}, \underline{\mathbf{v}}, \underline{\varphi}) \in C([0, \frac{T_1}{\varepsilon}]; H^s(\mathbb{R}^d) \times H^{s+1}_\varepsilon(\mathbb{R}^d)^d \times H^{s+1}(\mathbb{R}^d)$ the exact solution to (10.32) with the same initial conditions as $(n^\varepsilon, \mathbf{v}^\varepsilon, \varphi^\varepsilon)$ furnished by Theorem 10.2. We also write

$$(n, \mathbf{v}, \varphi) = (n^\varepsilon, \mathbf{v}^\varepsilon, \varphi^\varepsilon) - (\underline{n}, \underline{v}, \underline{\varphi}).$$

Taking the difference between (10.61) and (10.32), we get

$$\begin{cases} \partial_t n + (1+\varepsilon \underline{n})\nabla\cdot\mathbf{v} + \varepsilon\underline{\mathbf{v}}\cdot\nabla n + \varepsilon\nabla n^\varepsilon\cdot\mathbf{v} + \varepsilon(\nabla\cdot\mathbf{v}^\varepsilon)n = \varepsilon f, \\ \partial_t\mathbf{v} + \varepsilon(\underline{\mathbf{v}}\cdot\nabla)\mathbf{v} + \varepsilon\mathbf{v}\cdot\nabla\mathbf{v}^\varepsilon + \nabla\varphi + a\varepsilon^{-1/2}\mathbf{e}\wedge\mathbf{v} = \varepsilon\mathbf{g}, \\ M_\varepsilon(\underline{\phi})\nabla\varphi = \nabla n - \varepsilon\varphi e^{\varepsilon\varphi^\varepsilon}\nabla\varphi^\varepsilon + \varepsilon\mathbf{h}. \end{cases} \tag{10.62}$$

with $f = \varepsilon N^\varepsilon$, $\mathbf{g} = \varepsilon^{1/2}R^\varepsilon$ and $\mathbf{h} = \varepsilon\nabla r^\varepsilon + \big(\frac{e^{\varepsilon\underline{\varphi}} - e^{\varepsilon\varphi^\varepsilon}}{\varepsilon} - (\underline{\varphi} - \varphi^\varepsilon)e^{\varepsilon\varphi^\varepsilon}\big)\nabla\varphi^\varepsilon$. This system is of the form (10.39) with additional linear terms (namely, $\varepsilon\nabla n^\varepsilon\cdot\mathbf{v} + \varepsilon(\nabla\cdot\mathbf{v}^\varepsilon)n$ in the first equation, $\varepsilon\mathbf{v}\cdot\nabla\mathbf{v}^\varepsilon$ in the second one, and $-\varepsilon\varphi e^{\varepsilon\varphi^\varepsilon}\nabla\varphi^\varepsilon$ in the third one) that do not affect the derivation of the energy estimate (10.43). The only difference is that the constant C_s in (10.43) must also depend on $|(n^\varepsilon, \mathbf{v}^\varepsilon, \varphi^\varepsilon)|_{H_T^{s+1}}$. Since the initial conditions for (10.62) are identically zero, this yields, for all $0 \le T \le T_1/\varepsilon$,

$$\sup_{[0,T]} \big(|n|_{H^s}^2 + |\mathbf{v}|_{H_\varepsilon^{s+1}}^2\big) \le \exp(\varepsilon\widetilde{C}_s T) \times \varepsilon T\big(|f|_{H_T^s}^2 + |\mathbf{g}|_{H_{\varepsilon,T}^{s+1}}^2 + |\mathbf{h}|_{H_T^s}^2 + |n|_{H^s}^2 + |\mathbf{v}|_{H_\varepsilon^{s+1}}^2\big), \tag{10.63}$$

with

$$\widetilde{C}_s = C\big(\frac{1}{c_0}, |\underline{n}|_{H_T^s}, |\underline{\mathbf{v}}|_{H_{\varepsilon,T}^{s+1}}, |\underline{\varphi}|_{H_T^s}, |\partial_t(\underline{n}, \underline{\mathbf{v}}, \underline{\varphi})|_{L_T^\infty}, |(n^\varepsilon, \mathbf{v}^\varepsilon, \varphi^\varepsilon)|_{H_T^{s+1}}\big).$$

Taking if necessary a smaller $T_1 > 0$, this implies that for all $0 \le t \le T_1/\varepsilon$,

$$|n(t)|_{H^s}^2 + |\mathbf{v}(t)|_{H_\varepsilon^{s+1}}^2 \le \varepsilon t \exp(\widetilde{C}_s T_1) \times \big(|f|_{H_t^s}^2 + |\mathbf{g}|_{H_{\varepsilon,t}^{s+1}}^2 + |\mathbf{h}|_{H_t^s}^2\big). \tag{10.64}$$

Now, on the one hand, we have

$$\begin{aligned} \widetilde{C}_s &= C\big(\frac{1}{c_0}, |\underline{n}|_{H_T^s}, |\underline{\mathbf{v}}|_{H_{\varepsilon,T}^{s+1}}, |\underline{\varphi}|_{H_T^s}, |\partial_t(\underline{n}, \underline{\mathbf{v}}, \underline{\varphi})|_{L_T^\infty}, |(n^\varepsilon, \mathbf{v}^\varepsilon, \varphi^\varepsilon)|_{H_T^{s+1}}\big) \\ &= C\big(\frac{1}{n_0}, |n^0|_{H^{s+5}}\big), \end{aligned}$$

where we used Theorem 10.2 to control the norms of $(\underline{n}, \underline{\mathbf{v}}, \underline{\varphi})$ in terms of $(n^0, \mathbf{v}^0)$ (with $\mathbf{v}^0$ given in terms of n^0 by $\mathbf{v}^0 = \mathbf{v}^\varepsilon_{|t=0}$) and the expression of all the components of $(n^\varepsilon, \mathbf{v}^\varepsilon, \varphi^\varepsilon)$ in terms of $n^{(1)}$ to control the norms of the approximate solution. On the other hand, we get from the definition of f, $\mathbf{g}$, $\mathbf{h}$, and Proposition 10.2 that

$$|f|_{H_t^s}^2 + |\mathbf{g}|_{H_{\varepsilon,t}^{s+1}}^2 + |\mathbf{h}|_{H_t^s}^2 \le \varepsilon^{1/2} C(T_1, |n^0|_{H^{s+5}}).$$

We deduce therefore from (10.64) that

$$\forall 0 \le t \le T_1/\varepsilon, \quad |n(t)|_{H^s}^2 + |\mathbf{v}(t)|_{H_\varepsilon^{s+1}}^2 \le \varepsilon^{3/2} t C\big(\frac{1}{c_0}, T_1, |n^0|_{H^{s+5}}\big).$$

□

10.4 The Euler–Poisson System with Isothermal Pressure

In [13], the authors derived and justified a version of the KdV (and therefore $d=1$) equation in the case where the isothermal pressure is not neglected. In the general dimension of $d \geq 1$, (10.4) are then given by

$$\begin{cases} \partial_t n + \nabla\cdot\mathbf{v} + \nabla\cdot(n\mathbf{v}) = 0, \\ \partial_t \mathbf{v} + (\mathbf{v}\cdot\nabla)\mathbf{v} + \nabla\varphi + \alpha\dfrac{\nabla n}{1+n} + a\mathbf{e}\wedge\mathbf{v} = 0, \\ \Delta\varphi - e^{\varphi} + 1 + n = 0, \end{cases} \tag{10.65}$$

where α is a positive constant related to the ratio of the ion temperature over the ion mass. In the case of cold plasmas considered in the previous sections, one has $\alpha = 0$. In [13], the cases $\alpha = 0$ and $\alpha > 0$ are treated differently, and the limit $\alpha \to 0$ (or, for instance, $\alpha = O(\varepsilon)$) cannot be handled. We show here that the proof of Theorem 10.2 can easily be adapted to the general case $\alpha \geq 0$, hereby allowing the limit $\alpha \to 0$ and providing a generalization of the results of [13] to the case $d \geq 1$. We first extend Theorem 10.2 to the general Euler–Poisson system with isothermal pressure (10.65). We then indicate how to derive and justify a generalization of the Zakharov–Kuznetsov approximation taking into account this new term, in the same spirit as the KdV approximation derived in the one-dimensional case in [13].

10.4.1 The Cauchy Problem for the Euler–Poisson System with Isothermal Pressure

As in Sect. 10.3, we work with rescaled equations. More precisely, we perform the same rescaling as for (10.32); without the "cold plasma" assumption, this system must be replaced by

$$\begin{cases} \partial_t n + \nabla\cdot((1+\varepsilon n)\mathbf{v}) = 0, \\ \partial_t \mathbf{v} + \varepsilon(\mathbf{v}\cdot\nabla)\mathbf{v} + \nabla\varphi + \alpha\dfrac{\nabla n}{1+\varepsilon n} + a\varepsilon^{-1/2}\mathbf{e}\wedge\mathbf{v} = 0, \\ -\varepsilon^2\Delta\varphi + e^{\varepsilon\varphi} - 1 = \varepsilon n, \end{cases} \tag{10.66}$$

with $\alpha \geq 0$. The presence of the extra term $\alpha\frac{\nabla n}{1+\varepsilon n}$ in the second equation induces a smoothing effect that allows the authors of [13] to use the pseudo-differential estimates of Grenier [11]. However, these smoothing effects disappear when $\alpha \to 0$, and the existence time thus obtained is not uniform with respect to α. We provide here a generalization of Theorem 10.2 that gives a uniform existence time with respect to ε *and* α (so that solutions to (10.66) provided by Theorem 10.4 can be seen as limits when $\alpha \to 0$ of solutions to (10.66)). This evanescent smoothing effect is taken into account by working with $n \in H^{s+1}_{\varepsilon\alpha}(\mathbb{R}^d)$ rather than $n \in H^s(\mathbb{R}^d)$ as in Theorem 10.2 (from the definition (10.34) of $H^s_{\varepsilon\alpha}$, $H^{s+1}_{\varepsilon\alpha}(\mathbb{R}^d)$ coincides with $H^s(\mathbb{R}^d)$ when $\alpha = 0$).

Theorem 10.4. *Let $s > \frac{d}{2}+1$, $\alpha_0 > 0$ and $n_0 \in H^{s+1}_{\varepsilon\alpha_0}(\mathbb{R}^d)$, $\mathbf{v}_0 \in H^{s+1}(\mathbb{R}^d)^d$ such that $1-|n_0|_\infty \geq c_0$ for some $c_0 > 0$.*

Then there exist $\underline{T} > 0$ such that for all $\varepsilon \in (0,1)$ and $\alpha \in (0,\alpha_0)$, there is a unique solution $(n^{\varepsilon,\alpha}, \mathbf{v}^{\varepsilon,\alpha}) \in C([0,\frac{\underline{T}}{\varepsilon}]; H^{s+1}_{\varepsilon\alpha}(\mathbb{R}^d)\times H^{s+1}_{\varepsilon}(\mathbb{R}^d)^d)$ of (10.32) such that $1+\varepsilon n > c_0/2$ and $\varphi^{\varepsilon,\alpha} \in C([0,T]; H^{s+1}(\mathbb{R}^d))$.

Moreover the family $(n^{\varepsilon,\alpha}, \mathbf{v}^{\varepsilon,\alpha}, \nabla\varphi^{\varepsilon,\alpha})_{\varepsilon\in(0,1),\alpha\in(0,\alpha_0)}$ is uniformly bounded in $H^{s+1}_{\varepsilon\alpha}\times H^{s+1}_{\varepsilon}\times H^{s-1}$.

Proof. The proof follows the same steps as the proof of Theorem 10.2; in addition to the operator $M_\varepsilon(\phi)$ defined in (10.35), we also need to define another second-order self-adjoint operator $N_{\varepsilon,\alpha}(\phi,n)$ as

$$N_{\varepsilon,\alpha}(\phi,n) = \frac{1}{1+n} + \frac{1}{1+n} M_\varepsilon(\phi)\frac{1}{1+n}, \tag{10.67}$$

provided that $\inf_{\mathbb{R}^d}(1+n) > 0$.

Step 1. Preliminary results.
The operator $N_{\varepsilon,\alpha}(\phi,n)$ defined in (10.67) inherits from the properties of $M_\varepsilon(\phi)$ the following estimates that echo (10.36);

$$\begin{aligned}
&\big(u, N_{\varepsilon,\alpha}(\phi,n)v\big) \leq C\big(|\varphi|_\infty, |n|_\infty\big)\Big|\frac{u}{1+\varepsilon n}\Big|_{H^1_{\varepsilon\alpha}}\Big|\frac{v}{1+\varepsilon n}\Big|_{H^1_{\varepsilon\alpha}},\\
&\Big|\frac{u}{1+\varepsilon n}\Big|^2_{H^1_{\varepsilon\alpha}} \leq C\big(|\varphi|_\infty, |n|_\infty\big)\big(u, N_{\varepsilon,\alpha}(\phi,n)u\big).
\end{aligned}$$

We also have the following commutator estimates that are similar to those satisfied by $M_\varepsilon(\phi)$ (see step 1 in the proof of Theorem 10.2),

$$\begin{aligned}
\big(u, [\partial_t, N_{\varepsilon,\alpha}(\phi,n)]u\big) &\leq \varepsilon C\big(\frac{1}{c_0}, |\partial_t n|_\infty, |n|_{W^{1,\infty}}, |\partial_t\varphi|_\infty, |\varphi|_\infty\big)\Big|\frac{u}{1+\varepsilon n}\Big|^2_{H^1_{\varepsilon\alpha}}\\
\big(u, [f\partial_j, N_{\varepsilon,\alpha}(\phi,n)]u\big) &\leq C\big(\frac{1}{c_0}, |(n,\varphi,f)|_{W^{1,\infty}}, \sqrt{\varepsilon\alpha}|\nabla\partial_j n|_\infty, \sqrt{\varepsilon}|\nabla\partial_j f|_\infty\big)\\
&\quad\times\Big|\frac{u}{1+\varepsilon n}\Big|^2_{H^1_{\varepsilon\alpha}}.
\end{aligned}$$

Step 2.
L^2 estimates for a linearized system. Without the "cold plasma" approximation, one must replace (10.39) by

$$\begin{cases}
\partial_t n + (1+\varepsilon\underline{n})\nabla\cdot\mathbf{v} + \varepsilon\underline{\mathbf{v}}\cdot\nabla n = \varepsilon f,\\
\partial_t\mathbf{v} + \varepsilon(\underline{\mathbf{v}}\cdot\nabla)\mathbf{v} + \nabla\varphi + \alpha\dfrac{\nabla n}{1+\varepsilon\underline{n}} + a\varepsilon^{-1/2}\mathbf{e}\wedge\mathbf{v} = \varepsilon\mathbf{g},\\
M_\varepsilon(\underline{\phi})\nabla\varphi = \nabla n + \varepsilon\mathbf{h}.
\end{cases} \tag{10.68}$$

The presence of the new term in the second equation of (10.68) yields some smoothing effect on the estimate on n that are absent when $\alpha = 0$; this smoothing is measured with the $H^1_{\varepsilon\alpha}(\mathbb{R}^d)$ norm, which coincides with the L^2 norm used for (10.68) when $\alpha = 0$. More precisely, we want to prove here that

$$\sup_{[0,T]} \big(|n|^2_{H^1_{\varepsilon\alpha}} + |\mathbf{v}|^2_{H^1_\varepsilon}\big) \leq \exp(\varepsilon C_0 T)$$
$$\times \big(|n_{|_{t=0}}|^2_{H^1_{\varepsilon\alpha}} + |\mathbf{v}_{|_{t=0}}|^2_{H^1_\varepsilon}\big) + \varepsilon T (|f|^2_{H^1_{\varepsilon\alpha,T}} + |\mathbf{g}|^2_{H^1_{\varepsilon,T}} + |\mathbf{h}|^2_{L^2_T})\big), \tag{10.69}$$

with $C_0 = C(\frac{1}{c_0}, |(\underline{n}, \underline{\mathbf{v}}, \varphi)|_{W^{1,\infty}_T}, |\partial_t(\underline{n}, \underline{\mathbf{v}}, \underline{\varphi})|_{L^\infty_T}, \sqrt{\varepsilon}|\underline{\mathbf{v}}|_{W^{2,\infty}_T}, \sqrt{\varepsilon\alpha}|n|_{W^{2,\infty}_T})$.

Instead of multiplying the first equation of (10.68) by $(1+\varepsilon\underline{n})^{-1}$ as in the case $\alpha = 0$, we multiply it by $N_{\varepsilon,\alpha}(\underline{\phi}, \underline{n})$ to obtain

$$\big(N_{\varepsilon,\alpha}(\underline{\phi},\underline{n})\partial_t n, n\big) + \big([1+\alpha\frac{1}{1+\varepsilon\underline{n}} M_\varepsilon(\underline{\phi})]\nabla\cdot\mathbf{v}, n\big) + \varepsilon\big(N_{\varepsilon,\alpha}(\underline{\phi},\underline{n})\mathbf{v}\cdot\nabla n, n\big)$$
$$= \varepsilon\big(N_{\varepsilon,\alpha}(\underline{\phi},\underline{n}) f, n\big),$$

which can be rewritten under the form

$$\frac{1}{2}\partial_t\big(N_{\varepsilon,\alpha}(\underline{\phi},\underline{n})n, n\big) - \frac{1}{2}\big([\partial_t, N_{\varepsilon,\alpha}(\underline{\phi},\underline{n})]n, n\big) + \big([1+\alpha\frac{1}{1+\varepsilon\underline{n}} M_\varepsilon(\underline{\phi})]\nabla\cdot\mathbf{v}, n\big)$$
$$-\varepsilon\frac{1}{2}\big(n, [\underline{\mathbf{v}}\cdot\nabla, N_{\varepsilon,\alpha}(\underline{\phi},\underline{n})]n\big) = \varepsilon\big(N_{\varepsilon,\alpha}(\underline{\phi},\underline{n}) f, n\big). \tag{10.70}$$

As in the proof of Theorem 10.2, we take the L^2 scalar product of the second equation with $M_\varepsilon(\phi)\mathbf{v}$; after remarking that

$$M_\varepsilon(\underline{\phi})\big(\nabla\varphi + \alpha\frac{\nabla n}{1+\varepsilon\underline{n}}\big) = [1+\alpha M_\varepsilon(\underline{\phi})\frac{1}{1+\varepsilon\underline{n}}]\nabla n + \varepsilon\mathbf{h}$$

we obtain with the same computations the following generalization of (10.42):

$$\frac{1}{2}\partial_t\big(M_\varepsilon(\underline{\phi})\mathbf{v}, \mathbf{v}\big) - \frac{1}{2}\varepsilon\big(\partial_t\underline{\varphi} e^{\varepsilon\underline{\varphi}}\mathbf{v}, \mathbf{v}\big) - \frac{1}{2}\varepsilon\big(\mathbf{v}, (\nabla\cdot\underline{\mathbf{v}})M_\varepsilon(\underline{\phi})\mathbf{v}\big)$$
$$-\frac{1}{2}\varepsilon\big(\mathbf{v}, [\underline{\mathbf{v}}\cdot\nabla, M_\varepsilon(\underline{\phi})]\mathbf{v}\big) + \big([1+\alpha M_\varepsilon(\underline{\phi})\frac{1}{1+\varepsilon\underline{n}}]\nabla n, \mathbf{v}\big)$$
$$= -\varepsilon\big(\mathbf{h}, \mathbf{v}\big) + \varepsilon\big(M_\varepsilon(\underline{\phi})\mathbf{g}, \mathbf{v}\big). \tag{10.71}$$

Adding (10.71) to (10.70) and proceeding exactly as in the proof of Theorem 10.2, we get (10.69).

Step 3. H^s **estimates for a linearized system**

We want to prove here that for all $s \geq 0$, the solution $(n, \mathbf{v}, \varphi)$ to (10.68) satisfies, for all $s \geq t_0 + 1$,

$$\sup_{[0,T]} \big(|n|^2_{H^{s+1}_{\varepsilon\alpha}} + |\mathbf{v}|^2_{H^{s+1}_\varepsilon}\big) \leq \exp(\varepsilon C_s T) \times \Big(|n_{|_{t=0}}|^2_{H^{s+1}_{\varepsilon\alpha}} + |\mathbf{v}_{|_{t=0}}|^2_{H^{s+1}_\varepsilon}\big)$$
$$+\varepsilon T (|f|^2_{H^{s+1}_{\varepsilon\alpha,T}} + |\mathbf{g}|^2_{H^{s+1}_{\varepsilon,T}} + |\mathbf{h}|^2_{H^s_T} + |n|^2_{H^{s+1}_{\varepsilon\alpha}} + |\mathbf{v}|^2_{H^{s+1}_\varepsilon})\Big), \tag{10.72}$$

with $C_s = C(\frac{1}{c_0}, |\underline{n}|_{H^{s+1}_{\varepsilon\alpha,T}}, |\underline{\mathbf{v}}|_{H^{s+1}_{\varepsilon,T}}, |\underline{\varphi}|_{H^s_T}, |\partial_t(\underline{n}, \underline{\mathbf{v}}, \underline{\varphi})|_{H^s_T})$.

Applying Λ^s to the three equations of (10.39) and writing $\tilde{n} = \Lambda^s n$, $\tilde{\mathbf{v}} = \Lambda^s \mathbf{v}$, and $\tilde{\varphi} = \Lambda^s \varphi$, we get

$$\begin{cases} \partial_t \tilde{n} + (1+\varepsilon \underline{n})\nabla\cdot\tilde{\mathbf{v}} + \varepsilon\underline{\mathbf{v}}\cdot\nabla\tilde{n} = \varepsilon\tilde{f}, \\ \partial_t\tilde{\mathbf{v}} + \varepsilon(\underline{\mathbf{v}}\cdot\nabla)\tilde{\mathbf{v}} + \nabla\tilde{\varphi} + \alpha\dfrac{\nabla\tilde{n}}{1+\varepsilon\underline{n}} + a\varepsilon^{-1/2}\mathbf{e}\wedge\tilde{\mathbf{v}} = \varepsilon\tilde{\mathbf{g}}, \\ M_\varepsilon(\underline{\phi})\nabla\tilde{\varphi} = \nabla\tilde{n} + \varepsilon\tilde{\mathbf{h}}. \end{cases} \tag{10.73}$$

with $\tilde{f}$ and $\tilde{\mathbf{h}}$ as in the proof of Theorem 10.2, while $\tilde{\mathbf{g}}$ must be changed into $\tilde{\mathbf{g}} = \tilde{\mathbf{g}} + \alpha[\Lambda^s, \frac{n}{1+\varepsilon\underline{n}}]\nabla\tilde{n}$. Using step 2, one can mimic the proof of the Theorem 10.2. The only new ingredients needed are a control in $H^1_{\varepsilon\alpha}(\mathbb{R}^d)$ of $\alpha[\Lambda^s, \frac{n}{1+\varepsilon\underline{n}}]\nabla\tilde{n}$ (the new term in $\tilde{\mathbf{g}}$) and a control of $\tilde{f}$ in $H^1_{\varepsilon\alpha}(\mathbb{R}^d)$ instead of $L^2(\mathbb{R}^d)$. Classical commutator estimates (see, for instance, [20]) yield for $s > d/2+1$

$$\begin{aligned} &\big|\alpha[\Lambda^s, \frac{\underline{n}}{1+\varepsilon\underline{n}}]\nabla\tilde{n}\big|_{H^1_\varepsilon} \le C(\frac{1}{c_0}, |\underline{n}|_{H^{s+1}_{\varepsilon\alpha}})|\tilde{n}|_{H^{s+1}_{\varepsilon\alpha}}, \\ &|\tilde{f}|_{H^1_{\varepsilon\alpha}} \le |f|_{H^{s+1}_{\varepsilon\alpha}} + \big(|\underline{n}|_{H^{s+1}_{\varepsilon\alpha}} + |\underline{\mathbf{v}}|_{H^{s+1}_\varepsilon}\big)\times\big(|\mathbf{v}|_{H^{s+1}_\varepsilon} + |n|_{H^{s+1}_{\varepsilon\alpha}}\big), \end{aligned}$$

so that (10.72) follows exactly as (10.43).

Step 4. End of the proof.
The end of the proof is exactly similar the same as for Theorem 10.2 (the exact solution is no longer furnished by Theorem 10.1 but by a standard iterative scheme).□

Remark 10.7. The proof given in Theorem 10.1 for the case $\alpha = 0$ does not work when $\alpha > 0$.

10.4.2 Derivation of a Zakharov–Kuznetsov Equation in Presence of Isothermal Pressure

We proceed similarly to the cold plasma case, but we replace the ansatz (10.45) by

$$\begin{aligned} n^\varepsilon &= n^{(1)}(x-ct,y,z,\varepsilon t) + \varepsilon n^{(2)} \\ \varphi^\varepsilon &= \varphi^{(1)}(x-ct,y,z,\varepsilon t) + \varepsilon\varphi^{(2)} \\ v_x^\varepsilon &= v_x^{(1)}(x-ct,y,z,\varepsilon t) + \varepsilon v_x^{(2)} \\ v_y^\varepsilon &= \varepsilon^{1/2} v_y^{(1)}(x-ct,y,z,\varepsilon t) + \varepsilon v_y^{(2)} \\ v_z^\varepsilon &= \varepsilon^{1/2} v_z^{(1)}(x-ct,y,z,\varepsilon t) + \varepsilon v_z^{(2)}, \end{aligned} \tag{10.74}$$

where the velocity c has to be determined.

Following the strategy of Sect. 10.3.2, we plug this ansatz into (10.66) and choose the profiles in (10.74) in order to cancel the leading-order terms.

10.4.2.1 Cancellation of Terms of Order ε^0

Cancelling the leading-order $O(1)$ yields

$$\begin{aligned} -c\partial_X n^{(1)} + \partial_X v_x^{(1)} &= 0, \\ -c\partial_X v_x^{(1)} + \partial_X \varphi^{(1)} + \alpha\partial_X n^{(1)} &= 0, \\ (1+\alpha)\partial_y n^{(1)} - a v_z^{(1)} &= 0, \\ (1+\alpha)\partial_z n^{(1)} + a v_y^{(1)} &= 0. \end{aligned} \tag{10.75}$$

10.4.2.2 Cancellation of Terms of Order $\varepsilon^{1/2}$

We get at this step

$$v_y^{(2)} = \frac{(1+\alpha)^{3/2}}{a^2}\partial_{Xy}^2 n^{(1)}, \qquad v_z^{(2)} = \frac{(1+\alpha)^{3/2}}{a^2}\partial_{Xz}^2 n^{(1)}. \tag{10.76}$$

10.4.2.3 Cancellation of Terms of Order ε

Proceeding as in Sect. 10.3.2.3, we get

$$\partial_T n^{(1)} + 2n^{(1)}\partial_X n^{(1)} + \frac{1}{a^2}\partial_X \Delta_\perp n^{(1)} = -\partial_X (v_x^{(2)} - cn^{(2)}) \tag{10.77}$$

$$\partial_X (cv_x^{(2)} - \alpha n^{(2)} - \varphi^{(2)}) = c\partial_T n^{(1)} + n^{(1)}\partial_X n^{(1)} \tag{10.78}$$

$$\partial_y \varphi^{(2)} = \frac{(1+\alpha)^{3/2}}{a^2}\partial_{XXy}^3 n^{(1)} \tag{10.79}$$

$$\partial_z \varphi^{(2)} = \frac{(1+\alpha)^{3/2}}{a^2}\partial_{XXz}^3 n^{(1)} \tag{10.80}$$

$$\varphi^{(1)} = n^{(1)} \tag{10.81}$$

After replacing $\varphi^{(1)}$ by $n^{(1)}$ according to (10.81), one readily checks that the first two equations of (10.75) are consistent if and only if $c = \sqrt{1+\alpha}$.

10.4.2.4 Cancellation of Terms of Order $\varepsilon^{3/2}$

As in Sect. 10.3.2.4, the cancellation of the $O(\varepsilon^{3/2})$ terms for the density and the longitudinal velocity equations is automatic, but it is not possible for the equations on the transverse velocity.

10.4.2.5 Cancellation of Terms of Order ε^2

As in Sect. 10.3.2.5 we only need to cancel the $O(\varepsilon^2)$ terms in the equation for φ^ε, which yields here

$$\varphi^{(2)} - n^{(2)} = \Delta n^{(1)} - \frac{1}{2}(n^{(1)})^2. \tag{10.82}$$

10.4.2.6 Derivation of the Zakharov–Kuznetsov Equation

Combining (10.77), (10.78), and (10.82), we find that $n^{(1)}$ must solve the following Zakharov–Kuznetsov equation (which coincides with (10.60) when $\alpha = 0$):

$$2c\partial_T n^{(1)} + 2cn^{(1)}\partial_X n^{(1)} + (\frac{c}{a^2}\Delta_\perp + \Delta)\partial_X n^{(1)} = 0. \tag{10.83}$$

10.4.2.7 Construction of the Profiles

The profiles involved in (10.74) are constructed in terms of $n^{(1)}$ as follows:

- In agreement with (10.81) and the first two equations of (10.75), we set $\varphi^{(1)} = n^{(1)}$ and $v_x^{(1)} = cn^{(1)}$, with $c = \sqrt{1+\alpha}$.
- The last two equations of (10.75) then give $v_y^{(1)}, v_z^{(1)}$.
- We then use (10.76) to obtain $v_y^{(2)}, v_z^{(2)}$.
- We take $\varphi^{(2)} = \frac{(1+\alpha)^{3/2}}{a^2}\partial_{XX}^2 n^{(1)}$ to satisfy (10.79) and (10.80).
- We then get the density corrector $n^{(2)}$ by (10.82).
- We recover $v_x^{(2)}$ from (10.77) or (10.78)—this is equivalent since $n^{(1)}$ solves the ZK equation (10.83).

Finally we get the following consistency result that generalized Proposition 10.2 when isothermal pressure is taken into account.

Proposition 10.3. *Let $T_0 > 0$, $n_0 \in H^s(\mathbb{R}^d)$ $(s \geq 5)$ and $n^{(1)} \in C([0,T_0]; H^s(\mathbb{R}^d))$, solving (with $c = \sqrt{1+\alpha}$)*

$$2c\partial_T n^{(1)} + 2cn^{(1)}\partial_X n^{(1)} + (\frac{c}{a^2}\Delta_\perp + \Delta)\partial_X n^{(1)} = 0, \qquad n^{(1)}_{|t=0} = n_0.$$

Constructing the other profiles as indicated above, the approximate solution $(n^\varepsilon, \mathbf{v}^\varepsilon, \varphi^\varepsilon)$ given by (10.74) solves (10.66) up to order ε^3 in φ^ε, ε^2 in n^ε, v_x^ε, and up to order $\varepsilon^{3/2}$ in v_y^ε, v_z^ε:

$$\begin{cases} \partial_t n^\varepsilon + \nabla \cdot ((1+\varepsilon n^\varepsilon)\mathbf{v}^\varepsilon) = \varepsilon^2 N^\varepsilon, \\ \partial_t \mathbf{v}^\varepsilon + \varepsilon(\mathbf{v}^\varepsilon \cdot \nabla)\mathbf{v}^\varepsilon + \nabla\varphi^\varepsilon + \alpha\dfrac{\nabla n}{1+\varepsilon n} + a\varepsilon^{-1/2}\mathbf{e} \wedge \mathbf{v}^\varepsilon = \varepsilon^{3/2} R^\varepsilon, \\ -\varepsilon^2\Delta\varphi^\varepsilon + e^{\varepsilon\varphi^\varepsilon} - 1 = \varepsilon n^\varepsilon + \varepsilon^3 r^\varepsilon, \end{cases} \tag{10.84}$$

with $R^\varepsilon = (\varepsilon^{1/2}R_1^\varepsilon, R_2^\varepsilon, R_3^\varepsilon)$ *and*

$$|N^\varepsilon|_{L^\infty([0,\frac{T_0}{\varepsilon}];H^{s-5})} + \sum_{j=1}^{3} |R_j^\varepsilon|_{L^\infty([0,\frac{T_0}{\varepsilon}];H^{s-5})} + |r^\varepsilon|_{L^\infty([0,\frac{T_0}{\varepsilon}];H^{s-4})} \leq C(T_0, |n_0|_{H^s}).$$

10.4.3 Justification of the Zakharov–Kuznetsov Approximation

Proceeding exactly as for Theorem 10.3 but replacing Theorem 10.2 by Theorem 10.4, we get the following justification of the Zakharov–Kuznetsov approximation in presence of an isothermal pressure.

Theorem 10.5. *Let* $n^0 \in H^{s+5}$, *with* $s > d/2+1$, *such that* $1+\varepsilon n^0 \geq c_0$ *on* $\mathbb{R}^d$ *for some constant* $c_0 > 0$. *There exists* $T_1 > 0$ *such that:*

i. *The Zakharov–Kuznetsov approximation* $(n^\varepsilon, \mathbf{v}^\varepsilon, \varphi^\varepsilon)$ *of Proposition 10.3 exists on the time interval* $[0, T_1/\varepsilon]$.
ii. *There exists a unique solution* $(\underline{n}, \underline{\mathbf{v}}, \underline{\varphi}) \in C([0,\frac{T_1}{\varepsilon}]; H_{\varepsilon\alpha}^{s+1}(\mathbb{R}^d) \times H_\varepsilon^{s+1}(\mathbb{R}^d)^d \times H^{s+1}(\mathbb{R}^d))$ *provided by Theorem 10.4 to the Euler–Poisson equations with isothermal pressure* (10.66) *with initial condition* $(\underline{n}^0, \underline{\mathbf{v}}^0, \underline{\varphi}^0) = (n^\varepsilon, \mathbf{v}^\varepsilon, \varphi^\varepsilon)_{|_{t=0}}$.

Moreover, one has the error estimate

$$\forall 0 \leq t \leq T_1/\varepsilon, \quad |\underline{n}(t) - n^\varepsilon(t)|_{H_{\varepsilon\alpha}^{s+1}}^2 + |\underline{\mathbf{v}}(t) - \mathbf{v}^\varepsilon(t)|_{H_\varepsilon^{s+1}}^2 \leq \varepsilon^{3/2} t C(\frac{1}{c_0}, T_1, |n^0|_{H^{s+5}}).$$

Remark 10.8. The comments made in Remarks 10.4 and 10.5 on the precision of the Zakharov–Kuznetsov approximation for cold plasmas can be transposed to the more general case considered here.

Remark 10.9. As already said, the ZK equation (10.83) coincides in dimension $d = 1$ with the KdV equation derived in [13]. A consequence of the uniformity of the existence time with respect to α in Theorem 10.4 is that Theorem 10.5 provides a justification on a time scale of order $O(1/\varepsilon)$ which is uniform with respect to α, whereas it shrinks to zero when $\alpha \to 0$ in [13].

Added in proofs. After this paper was submitted, Christof Sparber has informed us of the preprint [29] where related results are obtained in the spirit of [13]. Pu does not provide an error estimate but performs the expansion at higher orders. As in [13], the methods used do not allow uniform estimates as $\alpha \to 0$.

Acknowledgements The three authors acknowledge the support of IMPA, the Brazilian-French program in Mathematics, and the MathAmSud Project "Propagation of Nonlinear Dispersive Equations." D. L. acknowledges support from the project ANR-08-BLAN-0301-01 and J.-C. S. from the project ANR-07-BLAN-0250 of the Agence Nationale de la Recherche.

References

1. Alvarez-Samaniego, B., Lannes, D.: Large time existence for 3D water waves and asymptotics. Invent. Math. **171**, 485–541 (2008)
2. de Bouard, A.: Stability and instability of some nonlinear dispersive solitary waves in higher dimension. Proc. Roy. Soc. Edinburgh Sect. A **126**(1), 89–112 (1996)
3. Bustamente, E., Isaza, P., Mejía, J.: On the support of solutions to the Zakharov–Kuznetsov equation. J. Differ. Equat. **251**, 2728–2736 (2011)
4. Chae, D., Tadmor, E.: On the finite time blow-up of the Euler–Poisson equations in $\mathbb{R}^2$. Comm. Math. Sci. **6**(3), 785–789 (2008)
5. Cheng, B., Tadmor, E.: A Sharp Local Blow-up Condition for Euler–Poisson Equations with Attractive Forcing. arXiv : 0902.1582v1
6. Colin, T., Ben Youssef, W.: Rigorous derivation of Korteweg-de Vries type systems from a general class of hyperbolic systems. Math. Mod. Num. Anal. (M2AN) **34**(4), 873–911 (2000)
7. Davidson, R.C.: Methods in Nonlinear Plasma Physics. Academic, New-York (1972)
8. Faminskii, A.: The Cauchy problem for the Zakharov–Kuznetsov equation. Differ. Equat. **31**(6), 1002–1012 (1995)
9. Farah, L.G., Linares F., Pastor A.: A note on the 2D generalized Zakharov-Kuznetsov equation: Local, global, and scattering results. *J. Differential Equations* **253**(8), 2558–2571 (2012)
10. Gérard-Varet, D., Han-Kwan, D., Rousset, F.: Quasineutral limit of the Euler-Poisson system for ions in a domain with boundaries, Indiana Univ. Math. J., to appear.
11. Grenier, E.: Pseudo-differential energy estimates of singular perturbations. Comm. Pure Appl. Math. **50**, 0821–0865 (1997)
12. Guo, Y., Pausader, B.: Global Smooth Ion Dynamics in the Euler–Poisson System, arXiv:1003.3653.
13. Guo, Y., Pu, X.: KdV Limit of the Euler–Poisson System, arXiv:1202.1830
14. Haragus, M., Nicholls, D.P., Sattinger, D.H.: Solitary wave interactions of the Euler–Poisson equations. J. Math. Fluid Mech. **5**, 92–118 (2003)
15. Haragus, M., Scheel, A.: Linear stability and instability of ion-acoustic plasma solitary waves. Physica D **170**, 13–30 (2002)
16. Himonas, A.A., Misiolek, G., Tiglay, F.: On unique continuation for the modified Euler–Poisson equations. Disc. Cont. Dyn. Syst. **19**(3), 515–529 (2007)
17. Infeld, E., Rowlands, G.: Nonlinear Waves, Solitons and Chaos. Cambridge University Press, Cambridge (1990)
18. Kato, T., Ponce, G.: Commutator estimates and the Euler and Navier–Stokes equations. Comm. Pure Appl. Math. **41**, 891–917 (1988)
19. Laedke, E.W., Spatschek, K.H.: Growth rates of bending solitons. J. Plasma Phys. **28**(3), 469–484 (1982)
20. Lannes, D.: Sharp estimates for pseudo-differential operators with symbols of limited smoothness and commutators. J. Funct. Anal. **232**, 495–539 (2006)
21. Larkin, N.A., Tronco, E.: Regular solutions of the 2D Zakharov–Kuznetsov equation in a half-space (2012, preprint)
22. Li, Y., Sattinger, D.H.: Solitons collisions in the ion-acoustic plasma equations. J. Math Fluid Mech. **1**, 117–130 (1999)
23. Linares, F., Pastor, A.: Well-posedness for the two-dimensional modified Zakharov–Kuznetsov equation. SIAM J. Math. Anal. **4**(4), 1329–1339 (2009)
24. Linares, F., Pastor, A.: Local and global well-posedness for the 2D generalized Zakharov–Kuznetsov equation. J. Funct. Anal. **260**, 1060–1085 (2011)
25. Linares, F., Saut, J.-C.: The Cauchy problem for the 3D Zakharov–Kuznetsov equation. Disc. Cont. Dyn. Syst. A **24**(2), 547–565 (2009)
26. Linares, F., Pastor, A., Saut, J.-C.: Well-posedness for the Zakharov–Kuznetsov equation in a cylinder and on the background of a KdV soliton. Comm. Part. Differ. Equat. **35**(9), 1674–1689 (2010)

27. Panthee, M.: A note on the unique continuation property for Zakharov–Kuznetsov equation. Nonlinear Anal. **59**, 425–438 (2004)
28. Perla-Menzala, G., Rosier, L., Saut, J.-C., Temam, R.: Boundary control of the Zakharov–Kuznetsov equation, submitted
29. Pu, X.: Dispersive Limit of the Euler–Poisson System in Higher Dimensions, arXiv:1204.5435v1 [math-ph] 24 April 2012
30. Ribaud, F., Vento, S.: Well-posedness results for the 3D Zakharov–Kuznetsov equation. SIAM J. Math. Anal. **44**(4), 2289–2304 (2012)
31. Ribaud, F., Vento, S.: A note on the Cauchy problem for the 2D generalized Zakharov–Kuznetsov equations. Comptes Rendus Mathématiques **350**, 499–503 (2012)
32. Saut, J.-C., Temam, R.: The initial boundary value problem for the Zakharov–Kuznetsov equation. Adv. Differ. Equat. **15**(11–12), 1001–1031 (2010)
33. Saut, J.-C., Temam, R., Wang, C.: An initial and boundary-value problem for the Zakharov-Kuznestov equation in a bounded domain. J. Math. Phys. **53**, 115612 (2012). http://dx.doi.org/10.1063/1.4752102
34. Tiğlay, F.: The Cauchy problem and integrability of a modified Euler–Poisson equation. Trans. A.M.S. **360**(4), 1861–1877 (2008)
35. Verheest, F., Mace, R.L., Pillay, S.R., Hellberg, M.A.: Unified derivation of Korteweg-de Vries-Zakharov–Kuznetsov equations in multispecies plasmas, J. Phys. A: Math. Gen. **35**, 795–806 (2002)
36. Zakharov, V.E., Kuznetsov, E.A.: On three dimensional solitons. Sov. Phys. JETP. **39**, 285–286 (1974)

Chapter 11
L^1 Estimates for Oscillating Integrals Related to Structural Damped Wave Models

Takashi Narazaki and Michael Reissig

Abstract The goal of this paper is to derive $L^p - L^q$ estimates away from the conjugate line for structural damped wave models. The damping term interpolates between exterior damping and viscoelastic damping. The crucial point is to derive at first $L^1 - L^1$ estimates. Depending on the behavior of the characteristic roots of the operator, one has to take into consideration oscillations in one part of the extended phase space. The radial symmetric behavior of the roots allows to apply the theory of modified Bessel functions. Oscillations may produce unbounded time-dependent constants (either for small times close to 0 or for large times close to infinity) in the $L^1 - L^1$ estimates. Some interpolation techniques imply the desired $L^p - L^q$ estimates away from the conjugate line.

Key words: Modified Bessel functions, L^1–L^1 estimates, Structural damping, Wave models

2010 Mathematics Subject Classification: 35L99, 35B40, 35A23.

11.1 Introduction

Let us consider the Cauchy problem for the structural damped wave equation

$$u_{tt} - \Delta u + \mu(-\Delta)^\sigma u_t = 0,\ u(0,x) = u_0(x),\ u_t(0,x) = u_1(x), \tag{11.1}$$

T. Narazaki
Department of Mathematical Sciences, Tokai University, Kitakaname, Kanagawa, 259 1292 Japan
e-mail: narazaki@tokai-u.jp

M. Reissig (✉)
Faculty for Mathematics and Computer Science, TU Bergakademie Freiberg, Prüferstr. 9, D-09596 Freiberg, Germany
e-mail: reissig@math.tu-freiberg.de

M. Cicognani et al. (eds.), *Studies in Phase Space Analysis with Applications to PDEs*, Progress in Nonlinear Differential Equations and Their Applications 84,
DOI 10.1007/978-1-4614-6348-1_11,

where μ is a positive constant and $\sigma \in (0,1]$. Here we have two limit cases. The case $\sigma = 0$ describes waves with external damping, and $\sigma = 1$ describes waves with viscoelastic damping. In both cases $L^p - L^q$ estimates for the solutions are well understood; see [4–7, 10] and [11] for the case $\sigma = 0$ (linear or semi-linear models) and [8] for the case $\sigma = 1$. Applying partial Fourier transformation to (11.1), we have the following representation of solution (let us assume $\lambda_1 \neq \lambda_2$):

$$u(t,x) = F^{-1}(v(t,\xi)) = F^{-1}\Big(\frac{\lambda_1 e^{\lambda_2 t} - \lambda_2 e^{\lambda_1 t}}{\lambda_1 - \lambda_2} v_0(\xi)\Big) + F^{-1}\Big(\frac{e^{\lambda_1 t} - e^{\lambda_2 t}}{\lambda_1 - \lambda_2} v_1(\xi)\Big),$$

where

$$\lambda_{1,2} = \frac{-\mu|\xi|^{2\sigma} \pm \sqrt{\mu^2|\xi|^{4\sigma} - 4|\xi|^2}}{2}, \; v_0(\xi) = F(u_0)(\xi), \; v_1(\xi) = F(u_1)(\xi). \tag{11.2}$$

We introduce the notations

$$J_0(t,x)(u_0) := F^{-1}\Big(\frac{\lambda_1 e^{\lambda_2 t} - \lambda_2 e^{\lambda_1 t}}{\lambda_1 - \lambda_2} v_0(\xi)\Big),$$
$$J_1(t,x)(u_1) := F^{-1}\Big(\frac{e^{\lambda_1 t} - e^{\lambda_2 t}}{\lambda_1 - \lambda_2} v_1(\xi)\Big).$$

Let $\chi = \chi(|\xi|)$ be a smooth function which localizes to small frequencies, then we shall estimate the Fourier multipliers

$$J_{01}(t,x)(u_0) := F^{-1}\Big(\frac{\lambda_1 e^{\lambda_2 t} - \lambda_2 e^{\lambda_1 t}}{\lambda_1 - \lambda_2} \chi(|\xi|) v_0(\xi)\Big), \tag{11.3}$$
$$J_{02}(t,x)(u_0) := F^{-1}\Big(\frac{\lambda_1 e^{\lambda_2 t} - \lambda_2 e^{\lambda_1 t}}{\lambda_1 - \lambda_2} (1-\chi(|\xi|)) v_0(\xi)\Big), \tag{11.4}$$
$$J_{11}(t,x)(u_1) := F^{-1}\Big(\frac{e^{\lambda_1 t} - e^{\lambda_2 t}}{\lambda_1 - \lambda_2} \chi(|\xi|) v_1(\xi)\Big), \tag{11.5}$$
$$J_{12}(t,x)(u_1) := F^{-1}\Big(\frac{e^{\lambda_1 t} - e^{\lambda_2 t}}{\lambda_1 - \lambda_2} (1-\chi(|\xi|)) v_1(\xi)\Big). \tag{11.6}$$

In every part of the phase space we have to take into consideration the asymptotic behavior of the characteristic roots from (11.2).

Remark 11.1. In the further considerations it is sufficient to study the Fourier multipliers (11.3)–(11.6) for sufficiently small and sufficiently large frequencies. This is arranged by the choice of χ. For the middle frequencies $\{\xi \in \mathbb{R}^n : |\xi| \in [\varepsilon, N]\}$, where ε is small and N is large we have $\mathbb{R}\lambda_k \leq C(\varepsilon, N) < 0$. So, we have no dominant influence on the desired estimates.

Different strategies appear because in one part of the phase space, we have oscillations in the characteristic roots. If $\sigma \in (0,1/2)$, this effect appears for large frequencies; if $\sigma \in (1/2,1)$, this effect appears for small frequencies. The oscillations produce some unbounded behavior of the time-dependent constants close to $t=0$ in the case $\sigma \in (0,1/2)$ and some unbounded behavior of the time-dependent constants close to $t=\infty$ in the case $\sigma \in (1/2,1)$. In our considerations we want to understand the critical case $\sigma = 1/2$. The estimates for the Fourier multipliers having no oscillations are quite satisfied; we estimate all of them by $\lesssim 1$. Here we apply heavily radial symmetry and theory of modified Bessel functions (see Appendix). More difficulties appear for Fourier multipliers with oscillations. We obtain time-dependent bounds in Lemmas 11.7–11.10 and in Lemmas 11.14 and 11.15 which are reasonable. Here we take into consideration the connection to Fourier multipliers appearing for wave models (see Appendix).

By using rules for modified Bessel functions (see Appendix) we will derive $L^p - L^q$ estimates for $J_{kl}(t,x)(u_k)$ in the next sections. To explain our motivation for this paper we apply Young's inequality to estimate $J_{01}(t,x)(u_0)$ in the following way:

$$\|J_{01}(t,x)(u_0)(t,\cdot)\|_{L^p} \le \Big\|F^{-1}\Big(\frac{\lambda_1 e^{\lambda_2 t} - \lambda_2 e^{\lambda_1 t}}{\lambda_1 - \lambda_2}\chi(|\xi|)\Big)(t,\cdot)\Big\|_{L^r}\|u_0\|_{L^q},$$

where $1 \le q \le p \le \infty$ and $1 + \frac{1}{p} = \frac{1}{r} + \frac{1}{q}$. Here we assume $u_0 \in S(\mathbb{R}^n)$. So, the interesting question is for L^r estimates of the oscillating integral

$$F^{-1}\Big(\frac{\lambda_1 e^{\lambda_2 t} - \lambda_2 e^{\lambda_1 t}}{\lambda_1 - \lambda_2}\chi(|\xi|)\Big).$$

In general, there is no problem to derive L^2 estimates. But, what about L^1 estimates? We will focus our attention to give answers to this question.

We shall divide our considerations into the cases $\sigma \in (0,1/2)$, $\sigma = 1/2$, and $\sigma \in (1/2,1]$. To get an idea about expected results, let us study the very special case $\sigma = 1/2$ and $\mu = 2$. In this case $\lambda_1 \equiv \lambda_2$ and we have

$$\begin{aligned} J_0(t,x)(u_0) &= F^{-1}\big(e^{-|\xi|t}v_0(\xi)\big) + tF^{-1}\big(e^{-|\xi|t}|\xi|v_0(\xi)\big),\\ J_1(t,x)(u_1) &= tF^{-1}\big(e^{-|\xi|t}v_1(\xi)\big). \end{aligned}$$

Due to [9] we have for all $t > 0$ the relation

$$\int_{\mathbb{R}^n} e^{-2\pi|\xi|t}e^{-2\pi i x\cdot\xi}d\xi = c_n\frac{t}{(t^2+|x|^2)^{\frac{n+1}{2}}}.$$

So, we can conclude immediately the following statement:

Lemma 11.1. *We have the following estimates*:

$$\|F^{-1}(e^{-c|\xi|t})\|_{L^r(\mathbb{R}^n)} \lesssim t^{-n(1-\frac{1}{r})} \text{ for } r \in [1,\infty],\ c > 0 \text{ and } t > 0. \tag{11.7}$$

Moreover, we can conclude

$$\big\|F^{-1}\big(e^{-c|\xi|t}|\xi|t\big)\big\|_{L^r(\mathbb{R}^n)} \lesssim t^{-n(1-\frac{1}{r})} \text{ for } r\in[1,\infty],\ c>0 \text{ and } t>0. \quad (11.8)$$

Using Young's inequality we conclude the next result.

Proposition 11.1. *Let us consider the Cauchy problem*

$$u_{tt}-\Delta u+2(-\Delta)^{1/2}u_t=0,\ u(0,x)=u_0(x),\ u_t(0,x)=u_1(x). \quad (11.9)$$

Then the solution satisfies the L^p-L^q estimate

$$\|u(t,\cdot)\|_{L^p} \lesssim t^{-n(1-\frac{1}{r})}\|u_0\|_{L^q}+t^{1-n(1-\frac{1}{r})}\|u_1\|_{L^q}, \quad (11.10)$$

where $1\le q\le p\le\infty$ and $1+\frac{1}{p}=\frac{1}{r}+\frac{1}{q}$.

The main goal of this paper is to derive L^1-L^1 estimates for the solutions to (11.1) of the following type:

$$\|u(t,\cdot)\|_{L^1} \le C_0(t)\|u_0\|_{L^1}+C_1(t)\|u_1\|_{L^1}.$$

We are interested to explain the behavior of the functions $C_0(t)$ and $C_1(t)$ for $t\to+0$ and for $t\to\infty$.

By using a rough interpolation strategy the statement of Proposition 11.1 for $\sigma=1/2$ is generalized in Propositions 11.8 and 11.11 to the cases $\sigma\in(0,1/2)$ and $\sigma\in(1/2,1)$. In this way we derive L^p-L^q estimates not necessarily on the conjugate line. Here we use the interpolation between L^1 and L^∞ estimates for the oscillating integrals appearing in (11.3) to (11.6) (see (11.81), (11.82), (11.113), and (11.114)). The case $\sigma=1$ is already studied in [8].

11.2 L^p Estimates for a Model Oscillating Integrals

In this section we derive L^p estimates for the oscillating integral

$$F^{-1}\big(e^{-c|\xi|^{2\kappa}t}\big). \quad (11.11)$$

Our main goal is to show how the theory of modified Bessel functions coupled with some new ideas can be used to prove the desired estimates. In Lemma 11.2 we study the 3d-case. The 2d-case is studied in Lemma 11.3. In Lemma 11.4 we will explain how the higher-dimensional case can be reduced to one of the basic cases from Lemmas 11.2 and 11.3.

Lemma 11.2. *The following estimate holds in $\mathbb{R}^3$:*

$$\|F^{-1}\big(e^{-c|\xi|^{2\kappa}t}\big)\|_{L^p(\mathbb{R}^3)} \lesssim t^{-\frac{3}{2\kappa}(1-\frac{1}{p})} \quad (11.12)$$

for $\kappa\in(0,1]$, $p\in[1,\infty]$ and $t>0$. Here c is supposed to be a positive constant.

Proof. Using the radial symmetry of $e^{-c|\xi|^{2\kappa}t}$ we have

$$F^{-1}(e^{-c|\xi|^{2\kappa}t}) = \int_0^\infty e^{-cr^{2\kappa}t} r^2 \tilde{J}_{\frac{1}{2}}(r|x|)dr, \tag{11.13}$$

where $\tilde{J}_{\frac{1}{2}}(r|x|)$ is the modified Bessel function. For $\kappa = \frac{1}{2}$ the explicit representation of $F^{-1}(e^{-c|\xi|t})$ for $t = 1$ gives

$$F^{-1}(e^{-c|\xi|}) \sim \frac{1}{\langle x\rangle^4}. \tag{11.14}$$

Our strategy is the following:
First we prove

$$|F^{-1}(e^{-c|\xi|^{2\kappa}})| \lesssim \frac{1}{\langle x\rangle^{3+2\kappa}} \quad \text{for all} \quad x \in \mathbb{R}^3. \tag{11.15}$$

Then after a change of variables we derive the representation

$$F^{-1}(e^{-c|\xi|^{2\kappa}t}) = \frac{1}{t^{\frac{3}{2\kappa}}} G\Big(\frac{x}{t^{\frac{1}{2\kappa}}}\Big), \tag{11.16}$$

where

$$G(y) = \int_{\mathbb{R}^3} e^{iy\eta} e^{-c|\eta|^{2\kappa}} d\eta. \tag{11.17}$$

So, from the first step we have

$$\|G\|_{L^p(\mathbb{R}^3_y)} \lesssim 1 \tag{11.18}$$

and after backward transformation

$$\|F^{-1}(e^{-c|\xi|^{2\kappa}t})\|_{L^1(\mathbb{R}^3_x)} = \frac{1}{t^{\frac{3}{2\kappa}}} \Big\|G\Big(\frac{x}{t^{\frac{1}{2\kappa}}}\Big)\Big\|_{L^1(\mathbb{R}^3_x)} \lesssim 1,$$
$$\|F^{-1}(e^{-c|\xi|^{2\kappa}t})\|_{L^p(\mathbb{R}^3_x)} = \frac{1}{t^{\frac{3}{2\kappa}}} \Big\|G\Big(\frac{x}{t^{\frac{1}{2\kappa}}}\Big)\Big\|_{L^p(\mathbb{R}^3_x)} \lesssim t^{-\frac{3}{2\kappa}(1-\frac{1}{p})}$$

for $p \in (1,\infty]$, respectively. Let us devote how to show the basic estimate

$$|F^{-1}(e^{-c|\xi|^{2\kappa}})| \lesssim \frac{1}{\langle x\rangle^{3+2\kappa}}. \tag{11.19}$$

If $|x| \le 1$, then

$$|F^{-1}(e^{-c|\xi|^{2\kappa}})| \le \int_{\mathbb{R}^n} e^{-c|\xi|^{2\kappa}} d\xi \lesssim 1.$$

Now let us restrict ourselves to $\{|x| \ge 1\}$. For this reason we take account of the radial symmetry representation and study for $t = 1$ the integral

$$\int_0^\infty e^{-cr^{2\kappa}t}\, r^2 \tilde{J}_{\frac{1}{2}}(r|x|)dr. \tag{11.20}$$

Using for the modified Bessel functions the relation

$$\tilde{J}_{\frac{1}{2}}(r|x|) = -\frac{1}{r|x|^2}\partial_r \tilde{J}_{-\frac{1}{2}}(r|x|) \tag{11.21}$$

and for $x \in \mathbb{R}^3$ the explicit representation $\tilde{J}_{-\frac{1}{2}}(r|x|) = \cos(r|x|)$, we arrive at

$$\begin{aligned} G(x) &= \frac{1}{|x|^2}\int_0^\infty d_r\big(e^{-cr^{2\kappa}} r\big)\cos(r|x|)dr \\ &= \frac{1}{|x|^2}\int_0^\infty \big(1-2\kappa c r^{2\kappa}\big)e^{-cr^{2\kappa}}\cos(r|x|)dr. \end{aligned}$$

To generate the decay rate $\langle x\rangle^{-(3+2\kappa)}$ we will apply two more steps of partial integration. First

$$\begin{aligned} G(x) &= -\frac{1}{|x|^3}\int_0^\infty d_r\big(\big(1-2\kappa c r^{2\kappa}\big)e^{-cr^{2\kappa}}\big)\sin(r|x|)dr \\ &= \frac{1}{|x|^3}\int_0^\infty \big(c2\kappa + 4c\kappa^2 - 4c^2\kappa^2 r^{2\kappa}\big) r^{2\kappa-1} e^{-cr^{2\kappa}}\sin(r|x|)dr. \end{aligned}$$

To estimate the integral $\int_0^\infty r^{2\kappa-1}e^{-cr^{2\kappa}}\sin(r|x|)dr$ we divide it into

$$\int_0^{\frac{1}{|x|}} r^{2\kappa-1}e^{-cr^{2\kappa}}\sin(r|x|)dr + \int_{\frac{1}{|x|}}^\infty r^{2\kappa-1}e^{-cr^{2\kappa}}\sin(r|x|)dr. \tag{11.22}$$

The first integral is estimated by $\langle x\rangle^{-2\kappa}$. In the second integral we carry out one more step of partial integration. Then we obtain

$$\begin{aligned} \int_{\frac{1}{|x|}}^\infty r^{2\kappa-1}e^{-cr^{2\kappa}}\sin(r|x|)dr &= -\frac{1}{|x|}\int_{\frac{1}{|x|}}^\infty r^{2\kappa-1}e^{-cr^{2\kappa}}\partial_r\cos(r|x|)dr \\ &= \frac{1}{|x|}\int_{\frac{1}{|x|}}^\infty \big((2\kappa-1)r^{2\kappa-2} - c2\kappa r^{4\kappa-2}\big)e^{-cr^{2\kappa}}\cos(r|x|)dr + \text{additional term}, \end{aligned}$$

where the additional term can be estimated by $\langle x\rangle^{-2\kappa}$. Dividing the last integral into,

$$\begin{aligned} &\int_{\frac{1}{|x|}}^\infty \big((2\kappa-1)r^{2\kappa-2} - c2\kappa r^{4\kappa-2}\big)e^{-cr^{2\kappa}}\cos(r|x|)dr \\ &= \int_{\frac{1}{|x|}}^1 \big((2\kappa-1)r^{2\kappa-2} - c2\kappa r^{4\kappa-2}\big)e^{-cr^{2\kappa}}\cos(r|x|)dr \\ &\quad + \int_1^\infty \big((2\kappa-1)r^{2\kappa-2} - c2\kappa r^{4\kappa-2}\big)e^{-cr^{2\kappa}}\cos(r|x|)dr \end{aligned}$$

straightforward estimates lead to

$$\int_0^\infty r^{2\kappa-1}e^{-cr^{2\kappa}}\sin(r|x|)dr \lesssim \begin{cases} \dfrac{1}{\langle x\rangle^{2\kappa}}, \ 0<\kappa<1/2, \\ \dfrac{\log\langle x\rangle}{\langle x\rangle}, \quad \kappa = 1/2, \\ \dfrac{1}{\langle x\rangle}, \ 1/2<\kappa\le 1. \end{cases} \tag{11.23}$$

Summarizing we have shown $\|G\|_{L^p(\mathbb{R}^3)} \lesssim 1$. This completes the proof. □

Following the same proof we are able to derive the following result:

Corollary 11.1. *The following estimate holds in* $\mathbb{R}^3$:

$$\|F^{-1}(|\xi|^a e^{-c|\xi|^{2\kappa}t})\|_{L^p(\mathbb{R}^3)} \lesssim t^{-\frac{a}{2\kappa}-\frac{3}{2\kappa}(1-\frac{1}{p})} \tag{11.24}$$

for $\kappa\in(0,1]$, $p\in[1,\infty]$ *and* $t>0$. *Here* c *and* a *are supposed to be positive constants.*

Lemma 11.3. *The following estimate holds in* $\mathbb{R}^2$:

$$\|F^{-1}(e^{-c|\xi|^{2\kappa}t})\|_{L^p(\mathbb{R}^2)} \lesssim t^{-\frac{2}{2\kappa}(1-\frac{1}{p})} \tag{11.25}$$

for $\kappa\in(0,1]$, $p\in[1,\infty]$ *and* $t>0$. *Here* c *is supposed as a positive constant.*

Proof. As in the proof to Lemma 11.2 we have finally to study the integral

$$\int_0^\infty e^{-cr^{2\kappa}}r\tilde{J}_0(r|x|)dr. \tag{11.26}$$

From the relation $J_0(s)=\frac{J_1(s)}{s}+\frac{d}{ds}J_1(s)$, it follows immediately $\tilde{J}_0(s)=2\tilde{J}_1(s)+s\frac{d}{ds}\tilde{J}_1(s)$. Instead of the last integral we will study

$$\int_0^\infty e^{-cr^{2\kappa}}r\left(2\tilde{J}_1(r|x|)+r\partial_r\tilde{J}_1(r|x|)\right)dr. \tag{11.27}$$

After partial integration this integral is equal to

$$2\kappa c\int_0^\infty e^{-cr^{2\kappa}}r^{2\kappa+1}\tilde{J}_1(r|x|)dr. \tag{11.28}$$

This integral we divide into

$$\int_0^{\frac{1}{|x|}} e^{-cr^{2\kappa}}r^{2\kappa+1}\tilde{J}_1(r|x|)dr+\int_{\frac{1}{|x|}}^\infty e^{-cr^{2\kappa}}r^{2\kappa+1}\tilde{J}_1(r|x|)dr. \tag{11.29}$$

Using the boundedness of $\tilde{J}_1(s)$ for $s \in [0,1]$ the first integral can be estimated by $\langle x\rangle^{-(2\kappa+2)}$. To estimate the second integral we apply the following asymptotic formula for $\tilde{J}_1(s)$ for $s \geq 1$:

$$\tilde{J}_1(s) = C_1 \frac{1}{s^{\frac{3}{2}}} \cos\Big(s - \frac{3}{4}\pi\Big) + O\Big(\frac{1}{|s|^{\frac{5}{2}}}\Big). \tag{11.30}$$

The integral can be estimated as follows:

$$\int_{\frac{1}{|x|}}^{\infty} e^{-cr^{2\kappa}} r^{2\kappa+1} O\Big(\frac{1}{(r|x|)^{\frac{5}{2}}}\Big) dr \lesssim \begin{cases} \frac{1}{\langle x\rangle^{2\kappa+2}}, & 0 < \kappa < 1/4, \\ \frac{\log\langle x\rangle}{\langle x\rangle^{\frac{5}{2}}}, & \kappa = 1/4, \\ \frac{1}{\langle x\rangle^{\frac{5}{2}}}, & 1/4 < \kappa \leq 1. \end{cases} \tag{11.31}$$

It remains to estimate the integrals

$$\frac{1}{|x|^{\frac{3}{2}}} \int_{\frac{1}{|x|}}^{\infty} e^{-cr^{2\kappa}} r^{2\kappa-\frac{1}{2}} \cos(r|x|) dr, \quad \frac{1}{|x|^{\frac{3}{2}}} \int_{\frac{1}{|x|}}^{\infty} e^{-cr^{2\kappa}} r^{2\kappa-\frac{1}{2}} \sin(r|x|) dr. \tag{11.32}$$

Here we proceed as in the proof to Lemma 11.2. We explain only the first integral which we split into

$$\frac{1}{|x|^{\frac{3}{2}}} \int_{\frac{1}{|x|}}^{1} e^{-cr^{2\kappa}} r^{2\kappa-\frac{1}{2}} \cos(r|x|) dr + \frac{1}{|x|^{\frac{3}{2}}} \int_{1}^{\infty} e^{-cr^{2\kappa}} r^{2\kappa-\frac{1}{2}} \cos(r|x|) dr. \tag{11.33}$$

The first integral is equal to

$$\frac{1}{|x|^{\frac{5}{2}}} \int_{\frac{1}{|x|}}^{1} e^{-cr^{2\kappa}} r^{2\kappa-\frac{1}{2}} \partial_r \sin(r|x|) dr. \tag{11.34}$$

After partial integration the limit terms behave as $\langle x\rangle^{-(2\kappa+2)}$. The new integral is estimated by

$$\frac{1}{|x|^{\frac{5}{2}}} \int_{\frac{1}{|x|}}^{1} e^{-cr^{2\kappa}} r^{2\kappa-\frac{3}{2}} dr \lesssim \begin{cases} \frac{1}{\langle x\rangle^{2\kappa+2}}, & 0 < \kappa < 1/4, \\ \frac{\log\langle x\rangle}{\langle x\rangle^{\frac{5}{2}}}, & \kappa = 1/4, \\ \frac{1}{\langle x\rangle^{\frac{5}{2}}}, & 1/4 < \kappa \leq 1. \end{cases} \tag{11.35}$$

The second integral is equal to

$$\frac{1}{|x|^{\frac{5}{2}}} \int_{1}^{\infty} e^{-cr^{2\kappa}} r^{2\kappa-\frac{1}{2}} \partial_r \sin(r|x|) dr. \tag{11.36}$$

It can be estimated by $\langle x\rangle^{-\frac{5}{2}}$. In the same way we treat the integral

$$\frac{1}{|x|^{\frac{3}{2}}}\int_{\frac{1}{|x|}}^{\infty} e^{-cr^{2\kappa}} r^{2\kappa-\frac{1}{2}}\sin(r|x|)dr. \tag{11.37}$$

This completes the proof. □

Corollary 11.2. *The following estimate holds in* $\mathbb{R}^2$:

$$\|F^{-1}(|\xi|^a e^{-c|\xi|^{2\kappa}t})\|_{L^p(\mathbb{R}^2)} \lesssim t^{-\frac{a}{2\kappa}-\frac{2}{2\kappa}(1-\frac{1}{p})} \tag{11.38}$$

for $\kappa\in(0,1]$, $p\in[1,\infty]$ *and* $t>0$. *Here* c *and* a *are supposed to be positive constants.*

The next goal is to prove the following generalization of Lemmas 11.2 and 11.3.

Lemma 11.4. *The following estimate holds in* $\mathbb{R}^n$ *for* $n\geq 4$:

$$\|F^{-1}(e^{-c|\xi|^{2\kappa}t})\|_{L^p(\mathbb{R}^n)} \lesssim t^{-\frac{n}{2\kappa}(1-\frac{1}{p})} \tag{11.39}$$

for $\kappa\in(0,1]$, $p\in[1,\infty]$, *and* $t>0$. *Here* c *is supposed to be a positive constant.*

Proof. If $n\geq 4$ is odd, then we carry out $\frac{n+1}{2}$ steps of partial integration. We can in $\frac{n-1}{2}$ steps apply the rules

$$\tilde{J}_{\mu+1}(r|x|) = -\frac{1}{r|x|^2}\partial_r\tilde{J}_\mu(r|x|),\ \ |\tilde{J}_\mu(s)|\leq C_\mu \tag{11.40}$$

for real nonnegative μ to conclude

$$\begin{aligned} F^{-1}(e^{-c|\xi|^{2\kappa}}) &= \int_0^\infty e^{-cr^{2\kappa}} r^{n-1}\tilde{J}_{\frac{n}{2}-1}(r|x|)dr \\ &= (-1)^{\frac{n-1}{2}}\frac{1}{|x|^{n-1}}\int_0^\infty\Big(\frac{\partial}{\partial r}\frac{1}{r}\Big)^{\frac{n-1}{2}}(e^{-cr^{2\kappa}}r^{n-1})\tilde{J}_{-\frac{1}{2}}(r|x|)dr. \end{aligned}$$

Among all integrals the integrals

$$\int_0^\infty e^{-cr^{2\kappa}}\cos(r|x|)dr,\quad \int_0^\infty e^{-cr^{2\kappa}}r^{2\kappa}\cos(r|x|)dr \tag{11.41}$$

have the dominant influence. The same approach as in the proof to Lemma 11.2 gives immediately $\|G\|_{L^p(\mathbb{R}^3)}\lesssim 1$. This completes the proof for odd $n\geq 4$. Let us devote to the case of even $n\geq 4$. Analogous to the odd case we can carry out $\frac{n}{2}-1$ steps of partial integration by using the rule

$$\tilde{J}_{\mu+1}(r|x|) = -\frac{1}{r|x|^2}\partial_r\tilde{J}_\mu(r|x|). \tag{11.42}$$

Among all integrals the integrals

$$\int_0^\infty e^{-cr^{2\kappa}} r\tilde{J}_0(r|x|)dr, \int_0^\infty e^{-cr^{2\kappa}} r^{1+2\kappa}\tilde{J}_0(r|x|)dr \tag{11.43}$$

have the dominant influence. The same approach as in the proof to Lemma 11.3 gives immediately $\|G\|_{L^p(\mathbb{R}^3)} \lesssim 1$. This completes the proof. □

Corollary 11.3. *The following estimate holds in* $\mathbb{R}^n$, $n \geq 4$:

$$\|F^{-1}(|\xi|^a e^{-c|\xi|^{2\kappa}t})\|_{L^p(\mathbb{R}^n)} \lesssim t^{-\frac{a}{2\kappa}-\frac{n}{2\kappa}(1-\frac{1}{p})} \tag{11.44}$$

for $\kappa \in (0,1]$, $p \in [1,\infty]$ *and* $t > 0$. *Here* c *and* a *are supposed to be positive constants.*

11.3 The General Case $\sigma = \frac{1}{2}$

Let us consider the model (11.1) with a positive $\mu \neq 2$.

If $\mu \in (2,\infty)$, then we get $\lambda_{1,2} = \frac{1}{2}(-\mu \pm \sqrt{\mu^2-4})|\xi|$. The estimate of $J_0(t,x)u_0$ can be reduced to the application of Lemma 11.1. To estimate $J_1(t,x)u_1$ we apply

$$J_1(t,x)u_1 = tF^{-1}\Big(\int_0^1 e^{\frac{1}{2}\left(-\mu+(2\theta-1)\sqrt{\mu^2-4}\right)|\xi|t} d\theta\, v_1(\xi)\Big). \tag{11.45}$$

Taking into consideration $-\mu + \sqrt{\mu^2-4} \leq -\frac{2}{\mu}$ we can reduce this case to the application of Lemma 11.1, too. We may conclude from (11.45)

$$\begin{aligned}\|J_1(t,x)u_1\|_{L^p} &\lesssim t\int_0^1 \|F^{-1}(e^{\frac{1}{2}\left(-\mu+(2\theta-1)\sqrt{\mu^2-4}\right)|\xi|t} v_1(\xi))\|_{L^p} d\theta\\ &\lesssim t^{1-n(1-\frac{1}{r})}\|u_1\|_{L^q}.\end{aligned}$$

In this way we have shown the following statement:

Proposition 11.2. *Let us consider for* $\mu \in (2,\infty)$ *the Cauchy problem*

$$u_{tt} - \Delta u + \mu(-\Delta)^{1/2}u_t = 0,\ u(0,x) = u_0(x),\ u_t(0,x) = u_1(x). \tag{11.46}$$

Then the solution satisfies the $L^p - L^q$ *estimate*

$$\|u(t,\cdot)\|_{L^p} \lesssim t^{-n(1-\frac{1}{r})}\|u_0\|_{L^q} + t^{1-n(1-\frac{1}{r})}\|u_1\|_{L^q}, \tag{11.47}$$

where $1 \leq q \leq p \leq \infty$ *and* $1 + \frac{1}{p} = \frac{1}{r} + \frac{1}{q}$.

Now let us devote to the case $\mu \in (0,2)$. The characteristic roots are $\lambda_{1,2} = \frac{1}{2}(-\mu \pm i\sqrt{4-\mu^2})|\xi|$. We feel some relation to wave models due to the term $i\sqrt{4-\mu^2})|\xi|$. As model Fourier multipliers we will study

$$F^{-1}\big(e^{-c_1|\xi|t}\cos(c_2|\xi|t)\big),\ F^{-1}\big(e^{-c_1|\xi|t}\sin(c_2|\xi|t)\big)$$

with a positive constant c_1 and a real constant $c_2 \neq 0$.

Lemma 11.5. *The following estimates hold in $\mathbb{R}^n$ for $n \geq 2$:*

$$\|F^{-1}(e^{-c_1|\xi|t}\cos(c_2|\xi|t))\|_{L^p(\mathbb{R}^n)} \lesssim t^{-n(1-\frac{1}{p})},$$
$$\|F^{-1}(e^{-c_1|\xi|t}\sin(c_2|\xi|t))\|_{L^p(\mathbb{R}^n)} \lesssim t^{-n(1-\frac{1}{p})}$$

for $p \in [1,\infty]$ and $t > 0$.

Proof. Here we can follow the proofs to the Lemmas 11.2–11.4. Let us only sketch how to obtain the estimate for $\|F^{-1}(e^{-c_1|\xi|t}\cos(c_2|\xi|t))\|_{L^p(\mathbb{R}^3)}$. The change of variables from the proof to Lemma 11.2 reduces the considerations to the Fourier multiplier $F^{-1}(e^{-c_1|\eta|}\cos(c_2|\eta|))$. We have

$$F^{-1}(e^{-c_1|\eta|}\cos(c_2|\eta|)) = \frac{1}{|x|^2}\int_0^\infty d_r\big(e^{-c_1 r}\cos(c_2 r)r\big)\cos(r|x|)dr. \tag{11.48}$$

In the next step of partial integration we use $\sin(r|x|) = 0$ for $r = 0$. A fourth step of partial integration gives immediately the estimate

$$\begin{aligned}|F^{-1}(e^{-c_1|\eta|}\cos(c_2|\eta|))| &= \Big|\frac{1}{|x|^2}\int_0^\infty d_r\big(e^{-c_1 r}\cos(c_2 r)r\big)\cos(r|x|)dr\Big| \\ &\lesssim \frac{1}{\langle x\rangle^4}.\end{aligned}$$

This implies $\|F^{-1}(e^{-c_1|\eta|}\cos(c_2|\eta|))\|_{L^1(\mathbb{R}^3)} \lesssim 1$. It follows the desired statement.□

The estimate of $J_0(t,x)u_0$ can be reduced to the application of Lemma 11.5. To estimate $J_1(t,x)u_1$ we apply

$$J_1(t,x)u_1 = tF^{-1}\Big(\int_0^1 e^{\frac{1}{2}(-\mu+i(2\theta-1)\sqrt{4-\mu^2})|\xi|t}\,d\theta\, v_1(\xi)\Big). \tag{11.49}$$

We may conclude from (11.49)

$$\begin{aligned}\|J_1(t,x)u_1\|_{L^p} &\lesssim t\int_0^1 \big\|F^{-1}\big(e^{\frac{1}{2}(-\mu+i(2\theta-1)\sqrt{4-\mu^2})|\xi|t}\,v_1(\xi)\big)\big\|_{L^p} d\theta \\ &\lesssim t^{1-n(1-\frac{1}{r})}\|u_1\|_{L^q}.\end{aligned}$$

In this way we have shown the following statement:

Proposition 11.3. *Let us consider for $\mu \in (0,2)$ the Cauchy problem*

$$u_{tt} - \Delta u + \mu(-\Delta)^{1/2}u_t = 0,\ u(0,x) = u_0(x),\ u_t(0,x) = u_1(x). \tag{11.50}$$

Then the solution satisfies the $L^p - L^q$ estimate

$$\|u(t,\cdot)\|_{L^p} \lesssim t^{-n(1-\frac{1}{r})}\|u_0\|_{L^q} + t^{1-n(1-\frac{1}{r})}\|u_1\|_{L^q}, \tag{11.51}$$

where $1 \leq q \leq p \leq \infty$ *and* $1+\frac{1}{p} = \frac{1}{r}+\frac{1}{q}$.

11.4 The Case $\sigma \in (0,1/2)$

In this section we will study for $\sigma \in (0,1/2)$ the Cauchy problem

$$u_{tt} - \Delta u + (-\Delta)^\sigma u_t = 0,\ u(0,x) = u_0(x),\ u_t(0,x) = u_1(x). \tag{11.52}$$

A change of variables in (11.1) implies without loss of generality $\mu = 1$. The characteristic roots are

$$\lambda_{1,2} = \frac{-|\xi|^{2\sigma} \pm \sqrt{|\xi|^{4\sigma} - 4|\xi|^2}}{2}.$$

Lemma 11.6. *We have the following asymptotic behavior of the characteristic roots*:

1. $\lambda_1 \sim -|\xi|^{2(1-\sigma)}$, $\lambda_2 \sim -|\xi|^{2\sigma}$ *and* $\lambda_1 - \lambda_2 \sim |\xi|^{2\sigma}$ *if* $|\xi|$ *is small,*
2. $\lambda_1 \sim -|\xi|^{2\sigma} + i|\xi|$, $\lambda_2 \sim -|\xi|^{2\sigma} - i|\xi|$ *and* $\lambda_1 - \lambda_2 \sim i|\xi|$ *if* $|\xi|$ *is large.*

11.4.1 L^1 Estimates for Small Frequencies

Taking into consideration Lemma 11.6 and the Fourier multipliers (11.3)–(11.6), then we shall estimate the L^1-norms of the following oscillating integrals for all $t > 0$:

$$F^{-1}\big(e^{\lambda_1 t}\chi(|\xi|)\big),\ F^{-1}\big(|\xi|^{2-4\sigma}e^{\lambda_2 t}\chi(|\xi|)\big),\ F^{-1}\big(e^{(\lambda_2-\lambda_1)t}\chi(|\xi|)\big).$$

Here we proceed as follows: at first we estimate the L^1-norm of $F^{-1}\big(e^{(\lambda_2-\lambda_1)t}\chi(|\xi|)\big)$. Then we give a remark about the estimate for the L^1-norm of

$$F^{-1}\big(|\xi|^a e^{(\lambda_1+\lambda_2)t}\chi(|\xi|)\big).$$

The last two estimates allow to estimate the L^1-norm of $F^{-1}\big(|\xi|^{2-4\sigma}e^{\lambda_2 t}\chi(|\xi|)\big)$. Finally, we devote to the L^1-estimate of $F^{-1}\big(e^{\lambda_1 t}\chi(|\xi|)\big)$.

Proposition 11.4. *The following estimate holds for all* $t > 0$:

$$\big\|F^{-1}\big(e^{(\lambda_2-\lambda_1)t}\chi(|\xi|)\big)\big\|_{L^1(\mathbb{R}^n)} \lesssim 1.$$

Proof. If we localize to small $|x|$, then we have immediately

$$\|F^{-1}(e^{(\lambda_2-\lambda_1)t}\chi(|\xi|))\chi(|x|)\|_{L^1(\mathbb{R}^n)} \lesssim 1.$$

For this reason we assume in the following $|x| \geq 1$. A lot of steps in the proof are similar to those from Lemmas 11.2–11.4, nevertheless, we shall give a proof in detail to feel *changes which are necessary due to nonhomogeneous function* $\lambda_2 - \lambda_1$.

At first we consider the case $t \in (0,1]$. Let us introduce the function

$$I(t,x) := F^{-1}(e^{t(\lambda_2-\lambda_1)}\chi(|\xi|))(x).$$

Using the radial symmetry of the integrand and modified Bessel functions we have to consider

$$I(t,x) = c\int_0^\infty e^{-t\sqrt{r^{4\sigma}-4r^2}}\chi(r)r^{n-1}\tilde{J}_{n/2-1}(r|x|)\,dr.$$

Introducing the vector field

$$Xf(r) := \frac{d}{dr}\left(\frac{1}{r}f(r)\right)$$

we get

$$X^k\left(f(r)r^l\right) = \sum_{j=0}^{k} c_{kj}d_r^j f(r)r^{l-2k+j} \text{ for all } r>0. \tag{11.53}$$

Let us choose an odd $n = 2m+1$, $m \geq 1$. Then

$$I(t,x) = c\int_0^\infty e^{-t\sqrt{r^{4\sigma}-4r^2}}\chi(r)r^{2m}\tilde{J}_{m-1/2}(r|x|)\,dr.$$

Since

$$X^k\left(e^{-t\sqrt{r^{4\sigma}-4r^2}}\chi(r)r^{2m}\right)\Big|_0^\infty = 0$$

when $k = 0,\cdots,m-1$, Lemma 11.22 from the Appendix and integration by parts imply

$$I(t,x) = \frac{c}{|x|^{2m}}\int_0^\infty X^m\left(e^{-t\sqrt{r^{4\sigma}-4r^2}}\chi(r)r^{2m}\right)\cos(r|x|)dr.$$

Now we can apply one more step of partial integration (we use $\sin 0 = 0$) and obtain

$$I(t,x) = -\frac{c}{|x|^n}\int_0^\infty \partial_r\left(X^m\left(e^{-t\sqrt{r^{4\sigma}-4r^2}}\chi(r)r^{2m}\right)\right)\sin(r|x|)dr.$$

A formal calculation gives with (11.53) that the right-hand side is equal to

$$\sum_{j=0}^{m}\sum_{k=0}^{j+1}\frac{c_{jk}}{|x|^n}\int_0^\infty \partial_r^{j+1-k}e^{-t\sqrt{r^{4\sigma}-4r^2}}\chi^{(k)}(r)r^j\sin(r|x|)dr$$
$$+\sum_{j=0}^{m}\sum_{k=0}^{j}\frac{c_{jk}}{|x|^n}\int_0^\infty \partial_r^{j-k}e^{-t\sqrt{r^{4\sigma}-4r^2}}\chi^{(k+1)}(r)r^j\sin(r|x|)dr$$
$$+\sum_{j=1}^{m}\sum_{k=0}^{j}\frac{c_{jk}}{|x|^n}\int_0^\infty \partial_r^{j-k}e^{-t\sqrt{r^{4\sigma}-4r^2}}\chi^{(k)}(r)r^{j-1}\sin(r|x|)dr$$

with universal constants c_{jk}. Using for $k\geq 1$ that

$$\left|\partial_r^j e^{-t\sqrt{r^{4\sigma}-4r^2}}\right|\leq C_j \text{ for all } j\geq 0 \text{ and } t>0$$

on the support of the derivatives of χ (which is away from $r=0$), one more step of partial integration leads to the bound $\lesssim |x|^{-(n+1)}$ for all integrals with $k\geq 1$. It remains to study for $j=0,\cdots,m$ the integrals

$$I_{j,0}(t,x):=\int_0^\infty \partial_r^{j+1}e^{-t\sqrt{r^{4\sigma}-4r^2}}\chi(r)r^j\sin(r|x|)dr.$$

We note that for $j=1,\cdots,m+1$

$$\left|\partial_r^j e^{-t\sqrt{r^{4\sigma}-4r^2}}\right|\leq C_j r^{2\sigma-j}$$

on the support of χ. Hence, the splitting of the integral $\int_0^\infty$ as in Sect. 11.2 gives on the one hand

$$\int_0^{\frac{1}{|x|}}\left|\partial_r^j e^{-t\sqrt{r^{4\sigma}-4r^2}}\chi(r)r^{j-1}\sin(r|x|)\right|dr\leq\frac{C_j}{|x|^{2\sigma}},$$

and on the other hand

$$\begin{aligned}
&\Big|\int_{\frac{1}{|x|}}^\infty \partial_r^j e^{-t\sqrt{r^{4\sigma}-4r^2}}\chi(r)r^{j-1}\sin(r|x|)dr\Big|\\
&\leq\frac{1}{|x|}\left|\partial_r^j e^{-t\sqrt{r^{4\sigma}-4r^2}}\chi(r)r^{j-1}\cos(r|x|)\right|_{r=\frac{1}{|x|}}\\
&+\frac{1}{|x|}\int_{\frac{1}{|x|}}^\infty\left|\partial_r\Big(\partial_r^j e^{-t\sqrt{r^{4\sigma}-4r^2}}\chi(r)r^{j-1}\Big)\cos(r|x|)\right|dr\\
&\lesssim\frac{1}{|x|^{2\sigma}}+\frac{1}{|x|}\int_{\frac{1}{|x|}}^\infty r^{2\sigma-2}dr\lesssim\frac{1}{|x|^{2\sigma}}.
\end{aligned}$$

Summarizing we have shown

$$\left|I_{j,0}(t,x)\right|\lesssim\frac{1}{|x|^{2\sigma}}.$$

From the above estimates we have the desired estimate

$$\left\|F^{-1}\left(e^{(\lambda_2-\lambda_1)t}\chi(|\xi|)\right)(1-\chi(|x|))\right\|_{L^1(\mathbb{R}^n)} \le C \quad \text{for all } t\in(0,1] \text{ and } n=2m+1. \tag{11.54}$$

Now we devote to the case $t\in(0,1]$ and $n=2m$, $m\ge 1$. In this case we shall study

$$I(t,x)=c\int_0^\infty e^{-t\sqrt{r^{4\sigma}-4r^2}}\chi(r)r^{2m-1}\tilde{J}_{m-1}(r|x|)dr.$$

Applying $m-1$ times the second rule from Lemma 11.22 we obtain

$$I(t,x)=\frac{c}{|x|^{2m-2}}\int_0^\infty X^{m-1}\left(e^{-t\sqrt{r^{4\sigma}-4r^2}}\chi(r)r^{2m-1}\right)J_0(r|x|)dr.$$

By (11.53) the right-hand side is equal to

$$\sum_{j=0}^{m-1}\frac{c_j}{|x|^{2m-2}}\int_0^\infty \partial_r^j\left(e^{-t\sqrt{r^{4\sigma}-4r^2}}\chi(r)\right)r^{j+1}J_0(r|x|)dr=:\sum_{j=0}^{m-1}c_jI_j(t,x).$$

Using the first rule of Lemma 11.22 from the Appendix for $\mu=1$ and the fifth rule for $\mu=0$ after partial integration we conclude

$$\begin{aligned}I_0(t,x)&=-\frac{1}{|x|^{2m-2}}\int_0^\infty \partial_r\left(e^{-t\sqrt{r^{4\sigma}-4r^2}}\chi(r)\right)r^2\tilde{J}_1(r|x|)dr\\&=-\frac{1}{|x|^{2m}}\int_0^\infty \partial_r\left(\partial_r\left(e^{-t\sqrt{r^{4\sigma}-4r^2}}\chi(r)\right)r\right)\tilde{J}_0(r|x|)dr.\end{aligned}$$

Taking into consideration

$$\left|\partial_r\left(\partial_r\left(e^{-t\sqrt{r^{4\sigma}-4r^2}}\chi(r)\right)r\right)\right|\le Cr^{2\sigma-1} \quad \text{for all } t\in(0,1]$$

on the support of χ, we have

$$\left|\int_0^{\frac{1}{|x|}}\partial_r\left(\partial_r\left(e^{-t\sqrt{r^{4\sigma}-4r^2}}\chi(r)\right)r\right)J_0(r|x|)dr\right|\lesssim\frac{1}{|x|^{2\sigma}}.$$

Here we have used the estimate $|J_0(s)|\le C$ for $s\in[0,1]$. We also have $|J_0(s)|\le Cs^{-1/2}$ for $s>1$ and $\chi(r)=0$ when $r\ge 1$. Hence,

$$\begin{aligned}&\left|\int_{\frac{1}{|x|}}^\infty \partial_r\left(\partial_r\left(e^{-t\sqrt{r^{4\sigma}-4r^2}}\chi(r)\right)r\right)J_0(r|x|)dr\right|\\&\le\frac{c}{|x|^{\frac12}}\int_{\frac{1}{|x|}}^1 r^{2\sigma-\frac32}dr\le\begin{cases}\dfrac{C}{|x|^{\frac12}}, & 1/4<\sigma<1/2,\\[2mm] \dfrac{C\log|x|}{|x|^{\frac12}}, & \sigma=1/4,\\[2mm] \dfrac{C}{|x|^{2\sigma}}, & 0<\sigma<1/4.\end{cases}\end{aligned}$$

From the above two estimates we have

$$\|I_0(t,\cdot)\|_{L^1(\mathbb{R}^{2m})} \le C \text{ for all } t\in(0,1] \text{ and } m\ge 1. \tag{11.55}$$

Let $j\in[1,m-1]$ be an integer. Then the same treatment as for $I_0(t,x)$ implies

$$\begin{aligned}
I_j(t,x) &= \frac{1}{|x|^{2m-2}}\int_0^\infty \partial_r^j\left(e^{-t\sqrt{r^{4\sigma}-4r^2}}\chi(r)\right)\\
&\qquad\times\left(\partial_r\left(r^{j+2}\tilde{J}_1(r|x|)\right)-jr^{j+1}\tilde{J}_1(r|x|)\right)dr\\
&= -\frac{1}{|x|^{2m}}\int_0^\infty \partial_r\left(\partial_r^{j+1}\left(e^{-t\sqrt{r^{4\sigma}-4r^2}}\chi(r)\right)r^{j+1}\right)J_0(r|x|)dr\\
&\quad -\frac{j}{|x|^{2m}}\int_0^\infty \partial_r\left(\partial_r^{j}\left(e^{-t\sqrt{r^{4\sigma}-4r^2}}\chi(r)\right)r^{j}\right)J_0(r|x|)dr.
\end{aligned}$$

The same splitting of the integral as for $I_0(t,x)$ leads to the same estimate

$$\|I_j(t,\cdot)\|_{L^1(\mathbb{R}^{2m})} \le C \text{ for all } t\in(0,1] \text{ and } m\ge 1. \tag{11.56}$$

From the above estimates (11.55) and (11.56) we have the desired estimate

$$\begin{aligned}
&\left\|F^{-1}\left(e^{(\lambda_2-\lambda_1)t}\chi(|\xi|)\right)(1-\chi(|x|))\right\|_{L^1(\mathbb{R}^n)} \le C\\
&\text{for all } t\in(0,1] \text{ and } n=2m,\ m\ge 1.
\end{aligned} \tag{11.57}$$

Now we consider the case $t\ge 1$. By the change of variables $\xi = t^{-\frac{1}{2\sigma}}\eta$ we have

$$F^{-1}\left(e^{(\lambda_2-\lambda_1)t}\chi(|\xi|)\right)(x) = t^{-\frac{n}{2\sigma}}F^{-1}\left(e^{-\sqrt{|\eta|^{4\sigma}-4t^{2-\frac{1}{\sigma}}|\eta|^2}}\chi\left(t^{-\frac{1}{2\sigma}}|\eta|\right)\right)\left(t^{-\frac{1}{2\sigma}}x\right).$$

It is clear that it suffices to show that

$$\|H(t,\cdot)\|_{L^1(\mathbb{R}^n)} \le C \text{ for all } t\in[1,\infty),$$

where

$$H(t,x) = F^{-1}\left(e^{-\sqrt{|\eta|^{4\sigma}-4t^{2-\frac{1}{\sigma}}|\eta|^2}}\chi\left(t^{-\frac{1}{2\sigma}}|\eta|\right)\right)(x). \tag{11.58}$$

We note that $|\eta|^{2\sigma}\sim\sqrt{|\eta|^{4\sigma}-4t^{2-\frac{1}{\sigma}}|\eta|^2}$ on the support (which is chosen small enough) of $\chi(t^{-\frac{1}{2\sigma}}|\eta|)$. We shall study

$$H(t,x) = c\int_0^\infty e^{-\sqrt{r^{4\sigma}-4t^{2-\frac{1}{\sigma}}r^2}}\chi\left(t^{-\frac{1}{2\sigma}}r\right)r^{n-1}\tilde{J}_{\frac{n}{2}-1}(r|x|)dr.$$

Let $n=2m+1$, $m\ge 1$. Then

$$\begin{aligned}H(t,x) &= \frac{c}{|x|^n}\int_0^\infty \partial_r\Big(X^m\Big(e^{-\sqrt{r^{4\sigma}-4t^{2-\frac{1}{\sigma}}r^2}}\chi\big(t^{-\frac{1}{2\sigma}}r\big)r^{2m}\Big)\Big)\sin(r|x|)dr\\ &=: \sum_{1\le j+k\le m+1, j\ge 0, k\ge 0} c_{jk}H_{jk}(t,x),\end{aligned}$$

whereby (11.53)

$$H_{jk}(t,x) := \frac{t^{-\frac{k}{2\sigma}}}{|x|^n}\int_0^\infty \partial_r^{j+1}e^{-\sqrt{r^{4\sigma}-4t^{2-\frac{1}{\sigma}}r^2}}\chi^{(k)}\big(t^{-\frac{1}{2\sigma}}r\big)r^{j+k}\sin(r|x|)dr.$$

For $k = 1,\cdots,m+1$ we may conclude as follows:

$$\begin{aligned}|H_{jk}(t,x)| &\le \frac{t^{-\frac{k}{2\sigma}}}{|x|^{n+1}}\Big|\int_0^\infty \partial_r\Big(\partial_r^{j+1}e^{-\sqrt{r^{4\sigma}-4t^{2-\frac{1}{\sigma}}r^2}}\chi^{(k)}\big(t^{-\frac{1}{2\sigma}}r\big)r^{j+k}\Big)\cos(r|x|)dr\Big|\\ &\le \frac{Ct^{-\frac{k}{2\sigma}}}{|x|^{n+1}}\int_0^\infty r^{2\sigma+k-2}e^{-\sqrt{r^{4\sigma}-4t^{2-\frac{1}{\sigma}}r^2}}dr \le \frac{C}{|x|^{n+1}}e^{-at}\end{aligned}$$

with some positive constant a. Here we used the location of the support of derivatives of $\chi(t^{-1/2\sigma}r)$, in particular, $r \sim t^{\frac{1}{2\sigma}}$ on this set. Now let us devote to $k = 0$. Taking account of

$$\Big|\partial_r^{j+1}e^{-\sqrt{r^{4\sigma}-4t^{2-\frac{1}{\sigma}}r^2}}\Big| \le C_j r^{2\sigma-j-1}$$

on the support of $\chi\big(t^{-\frac{1}{2\sigma}}r\big)$, we have the following two estimates after splitting the integral $\int_0^\infty$:

$$\int_0^{\frac{1}{|x|}}\Big|\partial_r^{j+1}e^{-\sqrt{r^{4\sigma}-4t^{2-\frac{1}{\sigma}}r^2}}\chi\big(t^{-\frac{1}{2\sigma}}r\big)r^j\sin(r|x|)\Big|dr \le \frac{C_j}{|x|^{2\sigma}},$$

and

$$\begin{aligned}&\Big|\int_{\frac{1}{|x|}}^\infty \partial_r^{j+1}e^{-\sqrt{r^{4\sigma}-4t^{2-\frac{1}{\sigma}}r^2}}\chi\big(t^{-\frac{1}{2\sigma}}r\big)r^j\sin(r|x|)dr\Big|\\ &\le \frac{1}{|x|}\Big|\partial_r^{j+1}e^{-\sqrt{r^{4\sigma}-4t^{2-\frac{1}{\sigma}}r^2}}\chi\big(t^{-\frac{1}{2\sigma}}r\big)r^j\cos(r|x|)\Big|_{r=1/|x|}\\ &\quad+\frac{1}{|x|}\int_{\frac{1}{|x|}}^\infty\Big|\partial_r\Big(\partial_r^{j+1}e^{-\sqrt{r^{4\sigma}-4t^{2-\frac{1}{\sigma}}r^2}}\chi\big(t^{-\frac{1}{2\sigma}}r\big)r^j\Big)\Big|dr\\ &\le C_j\Big(\frac{1}{|x|^{2\sigma}}+\frac{1}{|x|}\int_{\frac{1}{|x|}}^\infty r^{2\sigma-2}dr\Big) < \frac{C_j}{|x|^{2\sigma}}.\end{aligned}$$

Summarizing we have shown

$$|H_{jk}(t,x)| \le \frac{C_j}{|x|^{n+2\sigma}} \text{ for all } t\ge 1,\ |x|\ge 1.$$

But this gives us immediately

$$\left\| F^{-1}\left(e^{(\lambda_2-\lambda_1)t}\chi(|\xi|)\right)(1-\chi(|x|))\right\|_{L^1(\mathbb{R}^n)} \lesssim 1 \text{ for all } t \geq 1. \tag{11.59}$$

Let $n = 2m$. Then repeating the above calculations for large t, we have the same estimate (11.59). Here we have used the following inequalities:

$$\int_{\frac{1}{|x|}}^{\infty} r^{2\sigma-\frac{3}{2}} e^{-ar^{2\sigma}}\, dr$$

$$\leq \int_{\frac{1}{|x|}}^{1} r^{2\sigma-\frac{3}{2}}\, dr + \int_{1}^{\infty} e^{-ar^{2\sigma}}\, dr \leq C \begin{cases} |x|^{\frac{1}{2}-2\sigma}, & \sigma < 1/4, \\ \log(|x|+2), & \sigma = 1/4, \\ 1, & 1/4 < \sigma < 1/2. \end{cases}$$

In this way the proof is completed. □

Following the proof of Proposition 11.4 we expect the following L^1 estimate for the oscillating integral:

$$F^{-1}\big(|\xi|^{2-4\sigma}e^{(\lambda_1+\lambda_2)t}\chi(|\xi|)\big) = F^{-1}\big(|\xi|^{2-4\sigma}e^{-|\xi|^{2\sigma}t}\chi(|\xi|)\big).$$

Corollary 11.4. *The following estimate holds for all $t > 0$:*

$$\|F^{-1}\big(|\xi|^{2-4\sigma}e^{(\lambda_1+\lambda_2)t}\chi(|\xi|)\big)\|_{L^1(\mathbb{R}^n)} \lesssim 1.$$

Proof. If $t \in (0,1]$, then we can follow directly the proof to Proposition 11.4. The only change is that instead of $e^{-t\sqrt{r^{4\sigma}-4r^2}}$, we have the much better term $r^{2-4\sigma}e^{-tr^{2\sigma}}$. This term satisfies of course the same estimates we used for $e^{-t\sqrt{r^{4\sigma}-4r^2}}$ in the proof. Even the additional power of r is helpful.
If $t \in [1,\infty)$, then we have to prove the uniform L^1-boundedness for

$$H(t,x) = F^{-1}\left(e^{-|\eta|^{2\sigma}}\chi\left(t^{-\frac{1}{2\sigma}}|\eta|\right)\right)(x).$$

But this follows exactly by the same reasoning. □

From Proposition 11.4 and Corollary 11.4 we conclude the following corollary:

Corollary 11.5. *The following estimate holds for all $t > 0$:*

$$\|F^{-1}\big(|\xi|^{2-4\sigma}e^{\lambda_2 t}\chi(|\xi|)\big)\|_{L^1(\mathbb{R}^n)} \lesssim 1.$$

Proof. The statement follows immediately from the representation

$$F^{-1}(|\xi|^{2-4\sigma}e^{\lambda_2 t}\chi(|\xi|)^2) = F^{-1}(|\xi|^{2-4\sigma}e^{(\lambda_1+\lambda_2)\frac{t}{2}}\chi(|\xi|)) * F^{-1}(e^{(\lambda_2-\lambda_1)\frac{t}{2}}\chi(|\xi|))$$

together with the above-mentioned results. □

It remains to consider the oscillating integral

$$F^{-1}(e^{\lambda_1 t}\chi(|\xi|)) = F^{-1}\Big(e^{-\frac{1}{2}\big(|\xi|^{2\sigma}-\sqrt{|\xi|^{4\sigma}-4|\xi|^2}\big)t}\chi(|\xi|)\Big).$$

Proposition 11.5. *The following estimate holds for all* $t > 0$:

$$\|F^{-1}(e^{\lambda_1 t}\chi(|\xi|))\|_{L^1(\mathbb{R}^n)} \lesssim 1.$$

Proof. Here we can follow again the proof to Proposition 11.4 because from Lemma 11.6 we know that for small frequencies λ_1 is real and behaves as $-|\xi|^{2(1-\sigma)}$. So, if terms are estimated by $\langle x\rangle^{-2\sigma}$, then now these terms can be estimated by $\langle x\rangle^{-2(1-\sigma)}$ or by $\langle x\rangle^{-1}$. The basic estimates which are used in the proof are

$$\big|\partial_r^j e^{-\lambda_1 t}\big| \le C_j \text{ for all } j \ge 0 \text{ and } t > 0 \text{ on the support of } \chi^{(k)},\ k \ge 1;$$
$$\big|\partial_r^j e^{-\lambda_1 t}\big| \le C_j r^{2(1-\sigma)-j} \text{ for all } j = 1,\cdots,m+1, \text{ on the support of } \chi;$$
$$\Big|\partial_r^{j+1} e^{-\lambda_1\big(t^{-\frac{1}{2(1-\sigma)}}r\big)t}\Big| \le C_j r^{2(1-\sigma)-j-1} \text{ on the support of } \chi\big(t^{-\frac{1}{2(1-\sigma)}}r\big).$$

□

From Corollary 11.5 and Proposition 11.5 we conclude the following statement:

Corollary 11.6. *The following estimate holds for all* $t > 0$:

$$\Big\|F^{-1}\Big(\frac{\lambda_2 e^{\lambda_1 t} - \lambda_1 e^{\lambda_2 t}}{\lambda_1 - \lambda_2}\chi(|\xi|)\Big)\Big\|_{L^1(\mathbb{R}^n)} \lesssim 1.$$

From Propositions 11.4 and 11.5 we conclude the following statement:

Corollary 11.7. *The following estimate holds for all* $t > 0$:

$$\Big\|F^{-1}\Big(\frac{e^{\lambda_1 t} - e^{\lambda_2 t}}{\lambda_1 - \lambda_2}\chi^2(|\xi|)\Big)\Big\|_{L^1(\mathbb{R}^n)} \le Ct.$$

Proof. Since

$$\begin{aligned}\frac{e^{\lambda_1 t} - e^{\lambda_2 t}}{\lambda_1 - \lambda_2} &= t\int_0^1 e^{\theta\lambda_1 t + (1-\theta)\lambda_2 t}d\theta \\ &= te^{t\lambda_1}\int_0^1 e^{-t\theta\sqrt{|\xi|^{4\sigma}-4|\xi|^2}}d\theta\end{aligned}$$

the application of Propositions 11.4 and 11.5 allows the following conclusion:

$$
\begin{aligned}
&\Big\|F^{-1}\Big(\frac{e^{t\lambda_1}-e^{t\lambda_2}}{\lambda_1-\lambda_2}\chi^2(|\xi|)\Big)\Big\|_{L^1(\mathbb{R}^n)}\\
&\leq t\int_0^1\Big\|F^{-1}\Big(e^{t\lambda_1}\chi(|\xi|)e^{-t\theta\sqrt{|\xi|^{4\sigma}-4|\xi|^2}}\chi(|\xi|)\Big)\Big\|_{L^1(\mathbb{R}^n)}d\theta\\
&\leq Ct\int_0^1\Big\|F^{-1}\Big(e^{t\lambda_1}\chi(|\xi|)\Big)\Big\|_{L^1(\mathbb{R}^n)}\Big\|F^{-1}\Big(e^{-t\theta\sqrt{|\xi|^{4\sigma}-4|\xi|^2}}\chi(|\xi|)\Big)\Big\|_{L^1(\mathbb{R}^n)}d\theta\\
&\leq Ct.
\end{aligned}
$$

This proves the desired result. □

11.4.2 L^1 Estimates for Large Frequencies

In this section we shall estimate the following L^1 norms for all $t>0$:

$$\Big\|F^{-1}\Big(\frac{e^{\lambda_1 t}-e^{\lambda_2 t}}{\lambda_1-\lambda_2}(1-\chi(|\xi|))\Big)\Big\|_{L^1(\mathbb{R}^n)}, \tag{11.60}$$

$$\Big\|F^{-1}\Big(\frac{\lambda_1 e^{\lambda_2 t}-\lambda_2 e^{\lambda_1 t}}{\lambda_1-\lambda_2}(1-\chi(|\xi|))\Big)\Big\|_{L^1(\mathbb{R}^n)}. \tag{11.61}$$

To get those estimates we have to understand in a first step the following model oscillating integral:

$$
\begin{aligned}
&F^{-1}\Big(e^{-c_1|\xi|^{2\kappa}t}|\xi|^{2\beta}\frac{\sin(c_2|\xi|t)}{|\xi|}(1-\chi(|\xi|))\Big),\\
&F^{-1}\Big(e^{-c_1|\xi|^{2\kappa}t}\cos(c_2|\xi|t)(1-\chi(|\xi|))\Big).
\end{aligned}
$$

In a second step we have to estimate the oscillating integrals

$$F^{-1}\Big(e^{-c_1|\xi|^{2\kappa}t}|\xi|^{2\beta}\frac{\sin\big(c_2|\xi|f(|\xi|)t\big)}{|\xi|f(|\xi|)}(1-\chi(|\xi|))\Big), \tag{11.62}$$

$$F^{-1}\Big(e^{-c_1|\xi|^{2\kappa}t}\cos(c_2|\xi|f(|\xi|)t)\big(1-\chi(|\xi|)\big)\Big), \tag{11.63}$$

where $f(|\xi|)=\sqrt{1-|\xi|^{4\kappa-2}/4}$. If $\beta=0$, then an estimate for (11.62) implies an estimate for the first oscillating integral (11.60). If $\beta=\kappa=\sigma$, then estimates for (11.62) and (11.63) imply an estimate for the second oscillating integral (11.61).

Lemma 11.7. *Let us given the model oscillating integral*

$$F^{-1}\Big(e^{-c_1|\xi|^{2\kappa}t}|\xi|^{2\beta}\frac{\sin(c_2|\xi|t)}{|\xi|}(1-\chi(|\xi|))\Big) \tag{11.64}$$

with $\beta \geq 0$, $\kappa \in (0,1/2)$ and $t > 0$. Here c_1 is a positive and $c_2 \neq 0$ is a real constant. Then the following estimate holds in $\mathbb{R}^n$:

$$\Big\|F^{-1}\Big(e^{-c_1|\xi|^{2\kappa}t}|\xi|^{2\beta}\frac{\sin(c_2|\xi|t)}{|\xi|}(1-\chi(|\xi|))\Big)\Big\|_{L^1(\mathbb{R}^n)} \lesssim \begin{cases} t^{1+[(n-2)/2](1-\frac{1}{2\kappa})-\frac{\beta}{\kappa}}, & t \in (0,1], \\ e^{-ct}, & t \in [1,\infty). \end{cases}$$

Proof. Let us set

$$h(s,x) := F^{-1}\Big(e^{-c_1|\xi|^{2\kappa}s}|\xi|^{2\beta}(1-\chi(|\xi|))\Big)(x) \text{ for all } s > 0, \tag{11.65}$$

and

$$w(t,s,x) := F^{-1}\Big(\frac{\sin(c_2|\xi|t)}{|\xi|}e^{-c_1|\xi|^{2\kappa}s}|\xi|^{2\beta}(1-\chi(|\xi|))\Big)(x) \text{ for all } t,s > 0. \tag{11.66}$$

After using Lemma 11.23 from the Appendix, we have for $n \geq 2$ the estimate

$$\|w(t,s,\cdot)\|_{L^p} \leq \sum_{0\leq|\alpha|\leq[(n-2)/2]} |a_\alpha| t^{|\alpha|+1}\|\partial_x^\alpha h(s,\cdot)\|_{L^p} \quad \text{for all } p \in [1,\infty]. \tag{11.67}$$

Let $s > 0$ be arbitrary and fixed. Then due to (11.65) and (11.66), the function $w = w(t,s,x)$ is the solution to the Cauchy problem for the wave equation

$$w_{tt} - c_2^2\Delta w = 0, \quad w(0,s,x) = 0, \quad w_t(0,s,x) = h(s,x).$$

Setting $s = t$ in (11.67) yields

$$\|w(t,t,\cdot)\|_{L^p} \leq \sum_{0\leq|\alpha|\leq[(n-2)/2]} |a_\alpha| t^{|\alpha|+1}\|\partial_x^\alpha h(t,\cdot)\|_{L^p} \tag{11.68}$$

for any $p \in [1,\infty]$. It remains to estimate the following norms

$$t^{|\alpha|+1}\|\partial_x^\alpha h(t,\cdot)\|_{L^1(\mathbb{R}^n)} = t^{|\alpha|+1}\Big\|F^{-1}\Big(\xi^\alpha|\xi|^{2\beta}e^{-c_1|\xi|^{2\kappa}t}(1-\chi(|\xi|))\Big)\Big\|_{L^1(\mathbb{R}^n)} \text{ for all } |\alpha| \leq [(n-2)/2].$$

Since

$$F^{-1}\Big(\xi^\alpha|\xi|^{2\beta}e^{-c_1|\xi|^{2\kappa}t}(1-\chi(|\xi|))\Big)(x) = t^{-\frac{n}{2\kappa}-\frac{|\alpha|}{2\kappa}-\frac{\beta}{\kappa}}F^{-1}\Big(\xi^\alpha|\xi|^{2\beta}e^{-c_1|\xi|^{2\kappa}}\Big(1-\chi\Big(|\xi|t^{-\frac{1}{2\kappa}}\Big)\Big)\Big)\Big(xt^{-\frac{1}{2\kappa}}\Big),$$

and

$$\left\|F^{-1}\left(\xi^\alpha|\xi|^{2\beta}e^{-c_1|\xi|^{2\kappa}}\left(1-\chi\left(|\xi|t^{-\frac{1}{2\kappa}}\right)\right)\right)\left(xt^{-\frac{1}{2\kappa}}\right)\right\|_{L^1(\mathbb{R}^n)} \lesssim t^{\frac{n}{2\kappa}}\left\|F^{-1}\left(\xi^\alpha|\xi|^{2\beta}e^{-c_1|\xi|^{2\kappa}}\right)\right\|_{L^1(\mathbb{R}^n)},$$

we conclude

$$\|w(t,t,\cdot)\|_{L^1} \le \sum_{0\le|\alpha|\le[(n-2)/2]} |a_\alpha| t^{|\alpha|+1} t^{-\frac{|\alpha|}{2\kappa}-\frac{\beta}{\kappa}}.$$

This is the desired estimate for $t\in(0,1]$. To understand the exponential decay for $t\in[1,\infty)$, we consider

$$F^{-1}\Big(e^{-c_1|\xi|^{2\kappa}t}|\xi|^{2\beta}\frac{\sin(c_2|\xi|t)}{|\xi|}(1-\chi(|\xi|))\Big)\chi(|x|)$$

and

$$F^{-1}\Big(e^{-c_1|\xi|^{2\kappa}t}|\xi|^{2\beta}\frac{\sin(c_2|\xi|t)}{|\xi|}(1-\chi(|\xi|))\Big)(1-\chi(|x|)).$$

The L^1 norm of the first integral can be estimated by

$$\Big\|F^{-1}\Big(e^{-c_1|\xi|^{2\kappa}t}|\xi|^{2\beta}\frac{\sin(c_2|\xi|t)}{|\xi|}(1-\chi(|\xi|))\Big)\chi(|x|)\Big\|_{L^1(\mathbb{R}^n)} \le \int_{\mathbb{R}^n}\chi(|x|)dx\int_{\mathbb{R}^n}e^{-c_1|\xi|^{2\kappa}t}|\xi|^{2\beta-1}(1-\chi(|\xi|))d\xi \lesssim e^{-ct}.$$

To derive the exponential decay for the second integral, we use the representation by the aid of modified Bessel functions from the Appendix. So we have to consider

$$\int_0^\infty e^{-c_1r^{2\kappa}t}r^{2\beta}\frac{\sin(c_2rt)}{r}(1-\chi(r))r^{n-1}\tilde{J}_{\frac{n}{2}-1}(r|x|)dr \text{ for } |x|\ge 1,\ t\ge 1.$$

Now we can follow the procedure of partial integration to produce a term $|x|^{-(n+1)}$ which guarantees the L^1 property in x. Moreover, the integral over any derivative of the integrand with respect to r possesses an exponential decay. This completes the proof. □

Lemma 11.8. *Let us given the oscillating integral*

$$F^{-1}\Big(e^{-c_1|\xi|^{2\kappa}t}\cos(c_2|\xi|t)(1-\chi(|\xi|))\Big) \tag{11.69}$$

with $\kappa\in(0,1/2)$ and $t>0$. Here c_1 is a positive and $c_2\neq 0$ is a real constant. Then the following estimate holds in $\mathbb{R}^n$:

$$\Big\|F^{-1}\Big(e^{-c_1|\xi|^{2\kappa}t}\cos(c_2|\xi|t)\big(1-\chi(|\xi|)\big)\Big)\Big\|_{L^1(\mathbb{R}^n)} \lesssim \begin{cases} t^{[n/2](1-\frac{1}{2\kappa})}, & t\in(0,1], \\ e^{-ct}, & t\in[1,\infty). \end{cases}$$

Proof. The proof follows the same lines as the proof to the previous lemma. We have only to take account of the fact that now we shall use Lemma 11.23 from the Appendix for

$$\partial_t w(t,x) = \partial_t F^{-1}\Big(\frac{\sin(|\xi|t)}{|\xi|}F(h)\Big)(x), \quad t>0.$$

In this way we conclude the desired estimates. □

Now let us devote to the oscillating integrals (11.62) and (11.63).

Lemma 11.9. *Let us given the oscillating integral*

$$F^{-1}\Big(e^{-c_1|\xi|^{2\kappa}t}|\xi|^{2\beta}\frac{\sin\big(c_2|\xi|f(|\xi|)t\big)}{|\xi|f(|\xi|)}\big(1-\chi(|\xi|)\big)\Big) \tag{11.70}$$

with $\beta\geq 0$, $\kappa\in(0,1/2)$, $t>0$, *and* $f(|\xi|)=\sqrt{1-|\xi|^{4\kappa-2}/4}$. *Here* c_1 *is a positive and* $c_2\neq 0$ *is a real constant. Then the following estimate holds in* $\mathbb{R}^n$:

$$\Big\|F^{-1}\Big(e^{-c_1|\xi|^{2\kappa}t}|\xi|^{2\beta}\frac{\sin(c_2|\xi|f(|\xi|)t)}{|\xi|f(|\xi|)}\big(1-\chi(|\xi|)\big)\Big)\Big\|_{L^1(\mathbb{R}^n)} \lesssim \begin{cases} t^{1+[(n-2)/2](1-\frac{1}{2\kappa})-\frac{\beta}{\kappa}}, & t\in(0,1], \\ e^{-ct}, & t\in[1,\infty). \end{cases}$$

Proof. First we consider the oscillating integral for $t\in(0,1]$. It holds

$$|\xi|f(|\xi|)t = |\xi|t - |\xi|^{2\kappa}g(|\xi|^{2\kappa-1})t,$$

where

$$g(r)=\frac{r}{4(1+\sqrt{1-r^2/4})}.$$

We may assume that there exists a smooth radial function $\chi_1(\xi)$ such that $1-\chi_1(\xi)\neq 0$ when $|\xi|\leq 1$ and $(1-\chi_1(\xi))(1-\chi(|\xi|))=(1-\chi(|\xi|))$. Since $1-1/f(|\xi|)$ satisfies the condition in Lemma 11.24 from the Appendix, we have

$$\Big\|F^{-1}\Big(\frac{1}{f(|\xi|)}h(t,\xi)\big(1-\chi(|\xi|)\big)\Big)\Big\|_{L^1(\mathbb{R}^n)} \lesssim \Big\|F^{-1}\big(h(t,\xi)(1-\chi(|\xi|))\big)\Big\|_{L^1(\mathbb{R}^n)}$$

under the assumption that the right-hand side exists. Assuming this existence from Lemma 11.24 we have

$$
\begin{aligned}
&\Big\|F^{-1}\Big(e^{-c_1|\xi|^{2\kappa}t}|\xi|^{2\beta}\tfrac{\sin(c_2|\xi|f(|\xi|)t)}{|\xi|f(|\xi|)}\big(1-\chi(|\xi|)\big)\Big)\Big\|_{L^1(\mathbb{R}^n)}\\
&\quad\le\Big\|F^{-1}\Big(e^{-c_1|\xi|^{2\kappa}t}|\xi|^{2\beta}\tfrac{\sin(c_2|\xi|t)}{|\xi|}\cos\big(c_2|\xi|^{2\kappa}g(|\xi|^{2\kappa-1})t\big)\big(1-\chi(|\xi|)\big)\Big)\Big\|_{L^1(\mathbb{R}^n)}\\
&\qquad+\Big\|F^{-1}\Big(e^{-c_1|\xi|^{2\kappa}t}|\xi|^{2\beta}\cos(c_2|\xi|t)\tfrac{\sin\big(c_2|\xi|^{2\kappa}g(|\xi|^{2\kappa-1})t\big)}{|\xi|}\big(1-\chi(|\xi|)\big)\Big)\Big\|_{L^1(\mathbb{R}^n)}\\
&\quad\lesssim\textstyle\sum_{|\alpha|\le[(n-2)/2]}t^{|\alpha|+1}\\
&\qquad\times\Big\|F^{-1}\Big(\xi^\alpha e^{-c_1|\xi|^{2\kappa}t}|\xi|^{2\beta}\cos\big(c_2|\xi|^{2\kappa}g(|\xi|^{2\kappa-1})t\big)\big(1-\chi(|\xi|)\big)\Big)\Big\|_{L^1(\mathbb{R}^n)}\\
&\quad+\textstyle\sum_{|\alpha|\le[n/2]}t^{|\alpha|}\Big\|F^{-1}\Big(\xi^\alpha e^{-c_1|\xi|^{2\kappa}t}|\xi|^{2\beta}\tfrac{\sin\big(c_2|\xi|^{2\kappa}g(|\xi|^{2\kappa-1})t\big)}{|\xi|}\big(1-\chi(|\xi|)\big)\Big)\Big\|_{L^1(\mathbb{R}^n)}\\
&\quad=:I_1+I_2.
\end{aligned}
\tag{11.71}
$$

Moreover, we use

$$
\begin{aligned}
\cos\big(c_2|\xi|^{2\kappa}g(|\xi|^{2\kappa-1})t\big)&=1+\textstyle\sum_{k=1}^{N-1}\frac{(-1)^{k-1}}{(2k)!}\big(c_2|\xi|^{2\kappa}g(|\xi|^{2\kappa-1})t\big)^{2k}\\
&\quad+R_N(t,\xi),
\end{aligned}
\tag{11.72}
$$

where

$$
\begin{aligned}
&R_N(t,\xi)\\
&=\frac{(-1)^N}{(2N-1)!}\big(c_2|\xi|^{2\kappa}g(|\xi|^{2\kappa-1})t\big)^{2N}\int_0^1(1-\theta)^{2N-1}\cos\big(\theta c_2|\xi|^{2\kappa}g(|\xi|^{2\kappa-1})t\big)\,d\theta.
\end{aligned}
$$

Now we can choose $N=N(\kappa)$ large enough such that the following estimates hold for all admissible α in formula (11.71):

$$
\Big|\partial_\xi^\gamma\Big(\xi^\alpha e^{-c_1|\xi|^{2\kappa}t}|\xi|^{2\beta}R_N(t,\xi)\Big)\Big|\le C_{\alpha,\beta,\gamma}|\xi|^{-(n+1)}\ \text{ for all } t\in(0,1] \text{ and } |\gamma|\le n+1.
$$

The application of Lemma 11.24 yields

$$
\Big\|F^{-1}\Big(\xi^\alpha e^{-c_1|\xi|^{2\kappa}t}|\xi|^{2\beta}R_N(t,\xi)\big(1-\chi(|\xi|)\big)\Big)\Big\|_{L^1(\mathbb{R}^n)}\lesssim 1\quad\text{for all } t\in(0,1].
\tag{11.73}
$$

The functions $g(|\xi|^{2\kappa-1})^{2k}$ for $k=1,\cdots,N-1$ satisfy the assumptions in Lemma 11.24 with positive $\nu=\nu(k)$. So, we may apply it. Together with the Corollaries 11.1–11.3 we conclude

$$
\begin{aligned}
&\Big\|F^{-1}\Big(\xi^\alpha e^{-c_1|\xi|^{2\kappa}t}|\xi|^{2\beta}\big(|\xi|^{2\kappa}g(|\xi|^{2\kappa-1})t\big)^{2k}\big(1-\chi(|\xi|)\big)\Big)\Big\|_{L^1(\mathbb{R}^n)}\\
&\quad\lesssim\Big\|F^{-1}\Big(\xi^\alpha e^{-c_1|\xi|^{2\kappa}t}|\xi|^{2\beta}\big(|\xi|^{2\kappa}t\big)^{2k}\Big)\Big\|_{L^1(\mathbb{R}^n)}\lesssim t^{-\frac{|\alpha|+2\beta}{2\kappa}}
\end{aligned}
\tag{11.74}
$$

for any $t\in(0,1]$. It follows from (11.73) and (11.74) that

$$\Big\| F^{-1}\Big(\xi^\alpha e^{-c_1|\xi|^{2\kappa}t}|\xi|^{2\beta}\cos\big(c_2|\xi|^{2\kappa}g(|\xi|^{2\kappa-1})t\big)\big(1-\chi(|\xi|)\big)\Big)\Big\|_{L^1(\mathbb{R}^n)} \lesssim t^{-\frac{|\alpha|+2\beta}{2\kappa}} \tag{11.75}$$

for any $t \in (0,1]$. Therefore we obtain

$$I_1 \lesssim \sum_{|\alpha|\le[(n-2)/2]} t^{1+|\alpha|(1-\frac{1}{2\kappa})-\frac{\beta}{\kappa}} \lesssim t^{1+[(n-2)/2](1-\frac{1}{2\kappa})-\frac{\beta}{\kappa}} \text{ for any } t \in (0,1]. \tag{11.76}$$

By the same reasoning we estimate I_2. Using the formula

$$\begin{aligned}
&\frac{\sin\big(c_2|\xi|^{2\kappa}g(|\xi|^{2\kappa-1})t\big)}{|\xi|}\\
&= |\xi|^{2\kappa-1}t\Big(c_2 g(|\xi|^{2\kappa-1}) + \sum_{k=1}^{N-1}\frac{(-1)^k}{(2k+1)!}\big(c_2|\xi|^{2\kappa}t\big)^{2k} g(|\xi|^{2\kappa-1})^{2k+1} + R_N'(t,\xi)\Big),
\end{aligned}$$

where

$$\begin{aligned}
R_N'(t,\xi) := \frac{(-1)^N}{(2N)!}\big(c_2|\xi|^{2\kappa}t\big)^{2N} g(|\xi|^{2\kappa-1})^{2N+1}&\\
\cdot\int_0^1 (1-\theta)^{2N-1}\cos\big(\theta c_2|\xi|^{2\kappa}g(|\xi|^{2\kappa-1})t\big)\,d\theta&,
\end{aligned}$$

the oscillating integral from (11.73) with R_N replaced by R_N' can be estimated in the same way. To derive an estimate of type (11.74) we may apply again Lemma 11.24 together with the Corollaries 11.1–11.3. So we get for all $|\alpha| \ge 0$ the estimates

$$\begin{aligned}
&\Big\| F^{-1}\Big(\xi^\alpha e^{-c_1|\xi|^{2\kappa}t}|\xi|^{2\beta}|\xi|^{2\kappa-1}t\big(|\xi|^{2\kappa}t\big)^{2k} g(|\xi|^{2\kappa-1})^{2k+1}\big(1-\chi(|\xi|)\big)\Big)\Big\|_{L^1(\mathbb{R}^n)}\\
&\quad\lesssim t\Big\| F^{-1}\Big(\xi^\alpha|\xi|^{2\beta+2\kappa-1}e^{-c_1|\xi|^{2\kappa}t}\big(|\xi|^{2\kappa}t\big)^{2k}\Big)\Big\|_{L^1(\mathbb{R}^n)} \lesssim t^{-\frac{|\alpha|}{2\kappa}-\frac{2\beta-1}{2\kappa}}.
\end{aligned}$$

From the last inequalities it follows

$$\begin{aligned}
&\Big\| F^{-1}\Big(\xi^\alpha e^{-c_1|\xi|^{2\kappa}t}|\xi|^{2\beta}\frac{\sin\big(c_2|\xi|^{2\kappa}g(|\xi|^{2\kappa-1})t\big)}{|\xi|}\big(1-\chi(|\xi|)\big)\Big)\Big\|_{L^1(\mathbb{R}^n)}\\
&\lesssim t^{-\frac{|\alpha|}{2\kappa}-\frac{2\beta-1}{2\kappa}} \text{ for } t\in(0,1].
\end{aligned}$$

Summarizing we may conclude

$$\begin{aligned}
I_2 &\lesssim \textstyle\sum_{|\alpha|\le[n/2]} t^{|\alpha|(1-\frac{1}{2\kappa})+\frac{1}{2\kappa}-\frac{\beta}{\kappa}} \lesssim t^{[n/2](1-\frac{1}{2\kappa})+\frac{1}{2\kappa}-\frac{\beta}{\kappa}}\\
&\lesssim t^{1+[(n-2)/2](1-\frac{1}{2\kappa})-\frac{\beta}{\kappa}} \text{ for any } t\in(0,1].
\end{aligned} \tag{11.77}$$

The estimates (11.76) and (11.77) imply the statement of the lemma for $t \in (0,1]$. To prove the exponential decay for large t we can proceed as in the proof to Lemma 11.7. The proof is completed. □

Lemma 11.10. *Let us given the oscillating integral*

$$F^{-1}\Big(e^{-c_1|\xi|^{2\kappa}t}\cos(c_2|\xi|f(|\xi|)t)(1-\chi(|\xi|))\Big) \tag{11.78}$$

with $\kappa \in (0,1/2)$, $t > 0$, and $f(|\xi|) = \sqrt{1-|\xi|^{4\kappa-2}/4}$. Here c_1 is a positive and $c_2 \neq 0$ is a real constant. Then the following estimate holds in $\mathbb{R}^n$:

$$\Big\|F^{-1}\Big(e^{-c_1|\xi|^{2\kappa}t}\cos(c_2|\xi|f(|\xi|)t)(1-\chi(|\xi|))\Big)\Big\|_{L^1(\mathbb{R}^n)} \lesssim \begin{cases} t^{[n/2](1-\frac{1}{2\kappa})}, & t \in (0,1], \\ e^{-ct}, & t \in [1,\infty). \end{cases}$$

Proof. The proof follows the arguments from the proof to Lemma 11.9. □

Taking into consideration the statements from Lemmas 11.7 to 11.10 we have the following results:

Corollary 11.8. *The oscillating integrals*

$$F^{-1}\Big(\frac{\lambda_2 e^{\lambda_1 t} - \lambda_1 e^{\lambda_2 t}}{\lambda_1 - \lambda_2}(1-\chi(|\xi|))\Big) \tag{11.79}$$

with $\sigma \in (0,1/2)$ and $t > 0$ satisfy the following estimates in $\mathbb{R}^n$, $n \geq 2$:

$$\Big\|F^{-1}\Big(\frac{\lambda_2 e^{\lambda_1 t} - \lambda_1 e^{\lambda_2 t}}{\lambda_1 - \lambda_2}(1-\chi(|\xi|))\Big)\Big\|_{L^1(\mathbb{R}^n)} \lesssim \begin{cases} t^{[n/2](1-\frac{1}{2\sigma})}, & t \in (0,1], \\ e^{-ct}, & t \in [1,\infty). \end{cases}$$

Corollary 11.9. *The oscillating integrals*

$$F^{-1}\Big(\frac{e^{\lambda_1 t} - e^{\lambda_2 t}}{\lambda_1 - \lambda_2}(1-\chi(|\xi|))\Big) \tag{11.80}$$

with $\sigma \in (0,1/2)$ and $t > 0$ satisfy the following estimates in $\mathbb{R}^n$, $n \geq 2$:

$$\Big\|F^{-1}\Big(\frac{e^{\lambda_1 t} - e^{\lambda_2 t}}{\lambda_1 - \lambda_2}(1-\chi(|\xi|))\Big)\Big\|_{L^1(\mathbb{R}^n)} \lesssim \begin{cases} t^{1+[(n-2)/2](1-\frac{1}{2\sigma})}, & t \in (0,1], \\ e^{-ct}, & t \in [1,\infty). \end{cases}$$

11.4.3 L^∞ Estimates

Using the properties of the Fourier transformation and Lemma 11.6 we conclude the following results after standard calculations.

Proposition 11.6. *We have the following L^∞ estimates for $t > 0$ and $\sigma \in (0,1/2)$:*

$$\Big\|F^{-1}\Big(\frac{\lambda_1 e^{\lambda_2 t}-\lambda_2 e^{\lambda_1 t}}{\lambda_1-\lambda_2}\chi(|\xi|)\Big)\Big\|_{L^\infty(\mathbb{R}^n)} \lesssim (1+t)^{-\frac{n}{2(1-\sigma)}},$$

$$\Big\|F^{-1}\Big(\frac{\lambda_1 e^{\lambda_2 t}-\lambda_2 e^{\lambda_1 t}}{\lambda_1-\lambda_2}(1-\chi(|\xi|))\Big)\Big\|_{L^\infty(\mathbb{R}^n)} \lesssim t^{-\frac{n}{2\sigma}},$$

$$\Big\|F^{-1}\Big(\frac{e^{\lambda_1 t}-e^{\lambda_2 t}}{\lambda_1-\lambda_2}\chi(|\xi|)\Big)\Big\|_{L^\infty(\mathbb{R}^n)} \lesssim (1+t)^{-\frac{n-2\sigma}{2(1-\sigma)}},$$

$$\Big\|F^{-1}\Big(\frac{e^{\lambda_1 t}-e^{\lambda_2 t}}{\lambda_1-\lambda_2}(1-\chi(|\xi|))\Big)\Big\|_{L^\infty(\mathbb{R}^n)} \lesssim \begin{cases} t^{1-\frac{n}{2\sigma}}, & t\in(0,1],\\ t^{-\frac{n-1}{2\sigma}}, & t\in[1,\infty).\end{cases}$$

11.4.4 $L^p - L^q$ Estimates not Necessarily on the Conjugate Line

From Corollaries 11.6–11.9 and Proposition 11.6 we have the following statements:

Proposition 11.7. *The following estimates hold for $t > 0$:*

$$\Big\|F^{-1}\Big(\frac{\lambda_2 e^{\lambda_1 t}-\lambda_1 e^{\lambda_2 t}}{\lambda_1-\lambda_2}\Big)\Big\|_{L^1(\mathbb{R}^n)} \lesssim \begin{cases} t^{[n/2](1-\frac{1}{2\sigma})}, & t\in(0,1],\\ 1, & t\in[1,\infty),\end{cases}$$

$$\Big\|F^{-1}\Big(\frac{e^{\lambda_1 t}-e^{\lambda_2 t}}{\lambda_1-\lambda_2}\Big)\Big\|_{L^1(\mathbb{R}^n)} \lesssim \begin{cases} t^{1+[(n-2)/2](1-\frac{1}{2\sigma})}, & t\in(0,1],\\ t, & t\in[1,\infty),\end{cases}$$

$$\Big\|F^{-1}\Big(\frac{\lambda_2 e^{\lambda_1 t}-\lambda_1 e^{\lambda_2 t}}{\lambda_1-\lambda_2}\Big)\Big\|_{L^\infty(\mathbb{R}^n)} \lesssim \begin{cases} t^{-\frac{n}{2\sigma}}, & t\in(0,1],\\ t^{-\frac{n}{2(1-\sigma)}}, & t\in[1,\infty),\end{cases}$$

$$\Big\|F^{-1}\Big(\frac{e^{\lambda_1 t}-e^{\lambda_2 t}}{\lambda_1-\lambda_2}\Big)\Big\|_{L^\infty(\mathbb{R}^n)} \lesssim \begin{cases} t^{1-\frac{n}{2\sigma}}, & t\in(0,1],\\ t^{-\frac{n-2\sigma}{2(1-\sigma)}}, & t\in[1,\infty).\end{cases}$$

By interpolation we conclude from the last proposition

$$\Big\|F^{-1}\Big(\frac{\lambda_2 e^{\lambda_1 t}-\lambda_1 e^{\lambda_2 t}}{\lambda_1-\lambda_2}\Big)\Big\|_{L^r(\mathbb{R}^n)} \lesssim \begin{cases} t^{[n/2](1-\frac{1}{2\sigma})\frac{1}{r}-\frac{n}{2\sigma}(1-\frac{1}{r})}, & t\in(0,1],\\ t^{-\frac{n}{2(1-\sigma)}(1-\frac{1}{r})}, & t\in[1,\infty),\end{cases} \tag{11.81}$$

$$\Big\|F^{-1}\Big(\frac{e^{\lambda_1 t}-e^{\lambda_2 t}}{\lambda_1-\lambda_2}\Big)\Big\|_{L^r(\mathbb{R}^n)} \lesssim \begin{cases} t^{1+[(n-2)/2](1-\frac{1}{2\sigma})\frac{1}{r}-\frac{n}{2\sigma}(1-\frac{1}{r})}, & t\in(0,1],\\ t^{1-\frac{n+2-4\sigma}{2(1-\sigma)}(1-\frac{1}{r})}, & t\in[1,\infty),\end{cases} \tag{11.82}$$

for all $r\in[1,\infty]$.

Proposition 11.8. *Let us consider with $\sigma \in (0,1/2)$ the Cauchy problem*

$$u_{tt} - \Delta u + \mu(-\Delta)^{\sigma} u_t = 0,\ u(0,x) = u_0(x),\ u_t(0,x) = u_1(x). \tag{11.83}$$

Then the solution satisfies the following $L^p - L^q$ estimates:

$$\|u(t,\cdot)\|_{L^p} \lesssim t^{[n/2](1-\frac{1}{2\sigma})\frac{1}{r} - \frac{n}{2\sigma}(1-\frac{1}{r})} \|u_0\|_{L^q} + t^{1+[(n-2)/2](1-\frac{1}{2\sigma})\frac{1}{r} - \frac{n}{2\sigma}(1-\frac{1}{r})} \|u_1\|_{L^q}$$
$$\text{for } t \in (0,1],$$
$$\|u(t,\cdot)\|_{L^p} \lesssim t^{-\frac{n}{2(1-\sigma)}(1-\frac{1}{r})} \|u_0\|_{L^q} + t^{1-\frac{n+2-4\sigma}{2(1-\sigma)}(1-\frac{1}{r})} \|u_1\|_{L^q} \text{ for } t \in [1,\infty),$$

where $1 \le q \le p \le \infty$ and $1 + \frac{1}{p} = \frac{1}{r} + \frac{1}{q}$.

Proof. The statement follows immediately from (11.81) and (11.82) by using the explicit representation of the solution by Fourier multipliers and the convolution rule. □

Remark 11.2. Setting $\sigma = 1/2$ in the last statement we see that Proposition 11.8 coincides with Proposition 11.3.

11.5 The Case $\sigma \in (1/2,1)$

In this section we will study for $\sigma \in (1/2,1)$ the Cauchy problem

$$u_{tt} - \Delta u + (-\Delta)^{\sigma} u_t = 0,\ u(0,x) = u_0(x),\ u_t(0,x) = u_1(x). \tag{11.84}$$

A change of variables in (11.1) implies without loss of generality $\mu = 1$. The characteristic roots are

$$\lambda_{1,2} = \frac{-|\xi|^{2\sigma} \pm \sqrt{|\xi|^{4\sigma} - 4|\xi|^2}}{2}.$$

Lemma 11.11. *We have the following asymptotic behavior of the characteristic roots:*

1. *$\lambda_1 \sim -|\xi|^{2\sigma} + i|\xi|$, $\lambda_2 \sim -|\xi|^{2\sigma} - i|\xi|$ and $\lambda_1 - \lambda_2 \sim i|\xi|$ if $|\xi|$ is small,*
2. *$\lambda_1 \sim -|\xi|^{2(1-\sigma)}$, $\lambda_2 \sim -|\xi|^{2\sigma}$ and $\lambda_1 - \lambda_2 \sim |\xi|^{2\sigma}$ if $|\xi|$ is large.*

11.5.1 L^1 Estimates for Small Frequencies

In this section we shall estimate the following L^1 norms for all $t > 0$:

$$\Big\| F^{-1}\Big(\frac{\lambda_1 e^{\lambda_2 t} - \lambda_2 e^{\lambda_1 t}}{\lambda_1 - \lambda_2} \chi(|\xi|) \Big) \Big\|_{L^1(\mathbb{R}^n)}, \tag{11.85}$$

$$\Big\| F^{-1}\Big(\frac{e^{\lambda_1 t} - e^{\lambda_2 t}}{\lambda_1 - \lambda_2} \chi(|\xi|) \Big) \Big\|_{L^1(\mathbb{R}^n)}. \tag{11.86}$$

To get those estimates we shall estimate the oscillating integrals

$$F^{-1}\Big(e^{-c_1|\xi|^{2\kappa}t}|\xi|^{2\beta}\frac{\sin(c_2|\xi|f(|\xi|)t)}{|\xi|f(|\xi|)}\chi(|\xi|)\Big), \qquad (11.87)$$

$$F^{-1}\Big(e^{-c_1|\xi|^{2\kappa}t}\cos(c_2|\xi|f(|\xi|)t)\chi(|\xi|)\Big), \qquad (11.88)$$

where $f(|\xi|)=\sqrt{1-|\xi|^{4\kappa-2}/4}$. The localizing function χ is chosen in such a way that $f(|\xi|)$ is positive on the support of χ. If $\beta=\kappa=\sigma$, then estimates for (11.87) and (11.88) imply an estimate for the first oscillating integral (11.85). If $\beta=0$, then an estimate for (11.87) implies an estimate for the second oscillating integral (11.86).

Lemma 11.12. *Let us given the oscillating integral*

$$F^{-1}\Big(e^{-c_1|\xi|^{2\kappa}t}|\xi|^{2\beta}\frac{\sin\big(c_2|\xi|f(|\xi|)t\big)}{|\xi|f(|\xi|)}\chi(|\xi|)\Big) \qquad (11.89)$$

with $\beta\geq 0$, $\kappa\in(1/2,1)$, $t\in(0,1]$ and $f(|\xi|)=\sqrt{1-|\xi|^{4\kappa-2}/4}$. Here c_1 is a positive and $c_2\neq 0$ is a real constant. Then the following estimate holds in $\mathbb{R}^n$:

$$\Big\|F^{-1}\Big(e^{-c_1|\xi|^{2\kappa}t}|\xi|^{2\beta}\frac{\sin\big(c_2|\xi|f(|\xi|)t\big)}{|\xi|f(|\xi|)}\chi(|\xi|)\Big)\Big\|_{L^1(\mathbb{R}^n)}\lesssim 1 \text{ for } t\in(0,1].$$

Proof. We use the radial symmetry and study the integral

$$\int_0^b e^{-c_1r^{2\kappa}t}r^{2\beta}\frac{\sin\big(c_2rf(r)t\big)}{rf(r)}\chi(r)\tilde{J}_{\frac{n}{2}-1}(r|x|)dr.$$

Then we can proceed as in Sect. 11.2; this means we can carry out a sufficient number of steps of partial integration to ensure the desired L^1 behavior with respect to x. Here the asymptotic behavior of $\frac{\sin(c_2rf(r)t)}{rf(r)}$ and its first three derivatives near to $r=0$ helps. □

Lemma 11.13. *Let us given the oscillating integral*

$$F^{-1}\Big(e^{-c_1|\xi|^{2\kappa}t}\cos\big(c_2|\xi|f(|\xi|)t\big)\chi(|\xi|)\Big) \qquad (11.90)$$

with $\kappa\in(1/2,1)$, $t\in(0,1]$, and $f(|\xi|)=\sqrt{1-|\xi|^{4\kappa-2}/4}$. Here c_1 is a positive and $c_2\neq 0$ is a real constant. Then the following estimate holds in $\mathbb{R}^n$with $n\geq 2$:

$$\Big\|F^{-1}\Big(e^{-c_1|\xi|^{2\kappa}t}\cos\big(c_2|\xi|f(|\xi|)t\big)\chi(|\xi|)\Big)\Big\|_{L^1(\mathbb{R}^n)}\lesssim 1 \text{ for } t\in(0,1].$$

Proof. Repeating the arguments in the proof of the previous lemma we conclude the desired estimate. □

Lemma 11.14. *Let us given the oscillating integral*

$$F^{-1}\Big(e^{-c_1|\xi|^{2\kappa}t}|\xi|^{2\beta}\frac{\sin\big(c_2|\xi|f(|\xi|)t\big)}{|\xi|f(|\xi|)}\chi(|\xi|)\Big) \tag{11.91}$$

with $\beta\geq 0$, $\kappa\in(1/2,1)$, $t\geq 1$, *and* $f(|\xi|)=\sqrt{1-|\xi|^{4\kappa-2}/4}$. *Here* c_1 *is a positive and* $c_2\neq 0$ *is a real constant. Then the following estimate holds in* $\mathbb{R}^n$:

$$\Big\|F^{-1}\Big(e^{-c_1|\xi|^{2\kappa}t}|\xi|^{2\beta}\frac{\sin\big(c_2|\xi|f(|\xi|)t\big)}{|\xi|f(|\xi|)}\chi(|\xi|)\Big)\Big\|_{L^1(\mathbb{R}^n)}\lesssim t^{1+[(n-2)/2](1-\frac{1}{2\kappa})-\frac{\beta}{\kappa}}$$

for $t\geq 1$.

Proof. We introduce the abbreviations $\nu:=1-1/2\kappa$ and $g(t,|\xi|):=t^\nu|\xi|-\sqrt{t^{2\nu}|\xi|^2-|\xi|^{4\kappa}/4}$. In the first step we will show for $\alpha\geq 0$ the auxiliary estimates

$$\Big\|F^{-1}\Big(\xi^\alpha e^{-c_1|\xi|^{2\kappa}}\frac{\sin(g(t,|\xi|))}{|\xi|}\chi\big(t^{-\frac{1}{2\kappa}}|\xi|\big)\Big)\Big\|_{L^1(\mathbb{R}^n)}\lesssim 1,\quad t\in[1,\infty), \tag{11.92}$$

$$\Big\|F^{-1}\Big(\xi^\alpha e^{-c_1|\xi|^{2\kappa}}\cos(g(t,|\xi|))\chi\big(t^{-\frac{1}{2\kappa}}|\xi|\big)\Big)\Big\|_{L^1(\mathbb{R}^n)}\lesssim 1,\quad t\in[1,\infty). \tag{11.93}$$

Let $h(\xi)=h(r)$, $r:=|\xi|$ be a radial smooth function. Then we have

$$\Big|\partial_\xi^\alpha h(\xi)\Big|\lesssim\sum_{m=1}^{|\alpha|}\frac{1}{r^{|\alpha|-m}}\big|d_r^m h(r)\big|, \tag{11.94}$$

and, in particular,

$$\big|\partial_\xi^\beta g(t,|\xi|)\big|\lesssim|\xi|^2\quad\text{if}\quad(1-\chi(|\xi|))\chi\big(t^{-\frac{1}{2\kappa}}|\xi|\big)\neq 0. \tag{11.95}$$

Here we use $1\lesssim|\xi|, t^{-\nu}\lesssim|\xi|^{2\kappa-1}$ and $1\lesssim\sqrt{4-t^{-2\nu}|\xi|^{4\kappa-2}}$. Hence, we obtain

$$\begin{aligned}
&\Big|x^\beta F^{-1}\Big(\xi^\alpha e^{-c_1|\xi|^{2\kappa}}\frac{\sin(g(t,|\xi|))}{|\xi|}(1-\chi(|\xi|))\chi\big(t^{-\frac{1}{2\kappa}}|\xi|\big)\Big)(x)\Big|\\
&\sim\Big|F^{-1}\Big(\partial_\xi^\beta\Big(\xi^\alpha e^{-c_1|\xi|^{2\kappa}}\frac{\sin(g(t,|\xi|))}{|\xi|}(1-\chi(|\xi|))\chi\big(t^{-\frac{1}{2\kappa}}|\xi|\big)\Big)\Big)\Big|\\
&\lesssim\int_{\mathbb{R}^n}|\xi|^{|\alpha|+2}e^{-c_1|\xi|^{2\kappa}}d\xi\lesssim 1
\end{aligned}$$

if $|\beta|\leq n+1$. This implies

$$\Big\|F^{-1}\Big(\xi^\alpha e^{-c_1|\xi|^{2\kappa}}\frac{\sin(g(t,|\xi|))}{|\xi|}(1-\chi(|\xi|))\chi\big(t^{-\frac{1}{2\kappa}}|\xi|\big)\Big)\Big\|_{L^1(\mathbb{R}^n)}\lesssim 1\text{ for all }\alpha. \tag{11.96}$$

Moreover, we have

$$\big|\partial_\xi^\beta g(t,|\xi|)\big|\lesssim|\xi|^{2\kappa-|\beta|}\quad\text{if}\quad\chi(|\xi|)\neq 0. \tag{11.97}$$

Hence, it follows for large $|x|$, only these are interesting for the further considerations,

$$\begin{aligned}
&\Big\|F^{-1}\Big(\xi^\alpha e^{-c_1|\xi|^{2\kappa}}\frac{\sin(g(t,|\xi|))}{|\xi|}\chi(|\xi|)\chi(t^{-\frac{1}{2\kappa}}|\xi|)\Big)\Big\|_{L^1(\mathbb{R}^n)}\\
&\lesssim \textstyle\sum_{|\beta|\le n+1}\Big\|x^\beta F^{-1}\Big(\xi^\alpha e^{-c_1|\xi|^{2\kappa}}\frac{\sin(g(t,|\xi|))}{|\xi|}\chi(|\xi|)\chi(t^{-\frac{1}{2\kappa}}|\xi|)\Big)\Big\|_{L^\infty(\mathbb{R}^n)}\\
&\lesssim \textstyle\sum_{|\beta|\le n+1}\Big\|F^{-1}\Big(\partial_\xi^\beta\Big(\xi^\alpha e^{-c_1|\xi|^{2\kappa}}\frac{\sin(g(t,|\xi|))}{|\xi|}\chi(|\xi|)\Big)\Big)\Big\|_{L^\infty(\mathbb{R}^n)}\\
&\lesssim \textstyle\int_{\operatorname{supp}\chi}|\xi|^{2\kappa+|\alpha|-n-2}d\xi\lesssim 1 \quad\text{when}\quad |\alpha|\ge 1.
\end{aligned} \tag{11.98}$$

Now we show that (11.98) holds for $\alpha=0$, too. Therefore we note that $(\sin\theta-\theta)/\theta^3=:h(\theta)\in C^\infty(\mathbb{R})$ and $|h^{(k)}(\theta)|\le C_k$ when $|\theta|\lesssim 1$. Hence, from (11.97) and direct calculations we conclude

$$\Big|\partial_\xi^\beta\frac{\sin(g(t,|\xi|))-g(t,|\xi|)}{|\xi|}\Big|=\Big|\partial_\xi^\beta\frac{g(t,|\xi|)^3h(g(t,|\xi|))}{|\xi|}\Big|\lesssim|\xi|^{6\kappa-n-1}$$

when $|\beta|\le n+1$ and $\chi(|\xi|)\ne 0$. Since

$$\Big|\partial_\xi^\beta\Big(\frac{g(t,|\xi|)}{|\xi|}-\frac{t^{-\nu}|\xi|^{4\kappa-2}}{8}\Big)\Big|\lesssim|\xi|^{2\kappa-n-1}$$

when $|\beta|\le n+1$ and $\chi(|\xi|)\ne 0$, similar arguments to those which lead to (11.98) imply

$$\Big|F^{-1}\Big(e^{-c_1|\xi|^{2\kappa}}\frac{\sin(g(t,|\xi|))-g(t,|\xi|)}{|\xi|}\chi(|\xi|)\Big)\Big|\lesssim\langle x\rangle^{-n-1},$$

$$\Big|F^{-1}\Big(e^{-c_1|\xi|^{2\kappa}}\Big(\frac{g(t,|\xi|)}{|\xi|}-\frac{t^{-\nu}|\xi|^{4\kappa-2}}{8}\Big)\chi(|\xi|)\Big)\Big|\lesssim\langle x\rangle^{-n-1}.$$

The considerations from Sect. 11.2 give

$$\Big\|F^{-1}\Big(e^{-c_1|\xi|^{2\kappa}}\frac{t^{-\nu}|\xi|^{4\kappa-2}}{8}\chi(|\xi|)\Big)\Big\|_{L^1(\mathbb{R}^n)}\lesssim 1.$$

Therefore we have (11.98) for $\alpha=0$, that is,

$$\Big\|F^{-1}\Big(e^{-c_1|\xi|^{2\kappa}}\frac{\sin(g(t,|\xi|))}{|\xi|}\chi(|\xi|)\chi(t^{-\frac{1}{2\kappa}}|\xi|)\Big)\Big\|_{L^1(\mathbb{R}^n)}\lesssim 1 \tag{11.99}$$

because $\chi(|\xi|)\chi(t^{-1/2\kappa}|\xi|)=\chi(|\xi|)$ when $t\ge 1$. Summarizing we obtain from (11.96), (11.98), and (11.99) the desired estimate (11.92). In the same way we prove (11.93). Applying the change of variables $\eta:=t^{1/2\kappa}\xi$ and $y:=t^{1/2\kappa}x$ we have

$$\begin{aligned}
&\Big\|F^{-1}\Big(e^{-c_1|\xi|^{2\kappa}t}|\xi|^{2\beta}\frac{\sin(t|\xi|f(|\xi|))}{|\xi|}\chi(|\xi|)\Big)\Big\|_{L^1(\mathbb{R}^n)}\\
&\lesssim t^{\frac{1}{2\kappa}-\frac{\beta}{\kappa}}\Big\|F^{-1}\Big(e^{-c_1|\xi|^{2\kappa}}|\xi|^{2\beta}\frac{\sin(t^\nu|\xi|f(|\xi|))}{|\xi|}\chi(t^{-\frac{1}{2\kappa}}|\xi|)\Big)\Big\|_{L^1(\mathbb{R}^n)}.
\end{aligned} \tag{11.100}$$

Since $t^\nu|\xi|f(|\xi|) = t^\nu|\xi| - g(t,|\xi|)$ and $F^{-1}(e^{-c_1|\xi|^{2\kappa}}|\xi|^{2\beta}) \in L^1(\mathbb{R}^n)$ for any positive constant c_1 and nonnegative β, the inequality (11.100) gives

$$\begin{aligned}
&\Big\|F^{-1}\Big(e^{-c_1|\xi|^{2\kappa}t}|\xi|^{2\beta}\frac{\sin(t|\xi|f(|\xi|))}{|\xi|}\chi(|\xi|)\Big)\Big\|_{L^1(\mathbb{R}^n)} \\
&\lesssim t^{\frac{1}{2\kappa}-\frac{\beta}{\kappa}}\Big\|F^{-1}\Big(\frac{\sin(t^\nu|\xi|)}{|\xi|}e^{-c_1|\xi|^{2\kappa}}|\xi|^{2\beta}\cos(g(t,|\xi|))\chi(t^{-\frac{1}{2\kappa}}|\xi|)\Big)\Big\|_{L^1(\mathbb{R}^n)} \\
&+t^{\frac{1}{2\kappa}-\frac{\beta}{\kappa}}\Big\|F^{-1}\Big(\cos(t^\nu|\xi|)e^{-c_1|\xi|^{2\kappa}}|\xi|^{2\beta}\frac{\sin(g(t,|\xi|))}{|\xi|}\chi(t^{-\frac{1}{2\kappa}}|\xi|)\Big)\Big\|_{L^1(\mathbb{R}^n)}.
\end{aligned} \tag{11.101}$$

Using Lemma 11.23 from the Appendix and the estimates (11.92) and (11.93), we arrive at

$$\begin{aligned}
&t^{\frac{1}{2\kappa}-\frac{\beta}{\kappa}}\Big\|F^{-1}\Big(\tfrac{\sin(t^\nu|\xi|)}{|\xi|}e^{-c_1|\xi|^{2\kappa}}|\xi|^{2\beta}\cos(g(t,|\xi|))\chi(t^{-\frac{1}{2\kappa}}|\xi|)\Big)\Big\|_{L^1(\mathbb{R}^n)} \\
&\lesssim t^{\frac{1}{2\kappa}-\frac{\beta}{\kappa}}\textstyle\sum_{|\alpha|\le[(n-2)/2]}t^{(|\alpha|+1)\nu} \\
&\quad\times\Big\|F^{-1}\Big(\xi^\alpha e^{-c_1|\xi|^{2\kappa}}|\xi|^{2\beta}\cos(g(t,|\xi|))\chi(t^{-\frac{1}{2\kappa}}|\xi|)\Big)\Big\|_{L^1(\mathbb{R}^n)} \\
&\lesssim t^{\nu[(n-2)/2]+1-\frac{\beta}{\kappa}} \quad \text{for all } t\in[1,\infty),
\end{aligned} \tag{11.102}$$

and

$$\begin{aligned}
&t^{\frac{1}{2\kappa}-\frac{\beta}{\kappa}}\Big\|F^{-1}\Big(\cos(t^\nu|\xi|)e^{-c_1|\xi|^{2\kappa}}|\xi|^{2\beta}\frac{\sin(g(t,|\xi|))}{|\xi|}\chi(t^{-\frac{1}{2\kappa}}|\xi|)\Big)\Big\|_{L^1(\mathbb{R}^n)} \\
&\lesssim t^{\frac{1}{2\kappa}-\frac{\beta}{\kappa}}\sum_{|\alpha|\le[n/2]}t^{|\alpha|\nu}\Big\|F^{-1}\Big(\xi^\alpha e^{-c_1|\xi|^{2\kappa}}|\xi|^{2\beta}\frac{\sin(g(t,|\xi|))}{|\xi|}\chi(t^{-\frac{1}{2\kappa}}|\xi|)\Big)\Big\|_{L^1(\mathbb{R}^n)} \\
&\lesssim t^{\nu[(n-2)/2]+1-\frac{\beta}{\kappa}} \quad \text{for all } t\in[1,\infty).
\end{aligned} \tag{11.103}$$

From (11.92), (11.93), and (11.100)–(11.103) we have

$$\begin{aligned}
&\Big\|F^{-1}\Big(e^{-c_1|\xi|^{2\kappa}t}|\xi|^{2\beta}\frac{\sin(t|\xi|f(|\xi|))}{|\xi|}\chi(|\xi|)\Big)\Big\|_{L^1(\mathbb{R}^n)} \\
&\lesssim t^{1+[(n-2)/2](1-\frac{1}{2\kappa})-\frac{\beta}{\kappa}} \quad \text{for all } t\in[1,\infty).
\end{aligned} \tag{11.104}$$

Let $\chi_1 = \chi_1(|\xi|) \in C^\infty(\mathbb{R}^n)$ be a radial function such that $\chi_1(|\xi|)\chi(|\xi|) = \chi(|\xi|)$ and $f(|\xi|) \ge a > 0$ on the support of χ_1 with some constant a. Then

$$F^{-1}\Big(\frac{\chi_1(|\xi|)}{f(|\xi|)}\Big) \in L^1(\mathbb{R}^n).$$

Applying Young's inequality we obtain the desired estimate

$$\Big\|F^{-1}\Big(e^{-c_1|\xi|^{2\kappa}t}|\xi|^{2\beta}\frac{\sin(t|\xi|f(|\xi|))}{f(|\xi|)|\xi|}\chi(|\xi|)\Big)\Big\|_{L^1(\mathbb{R}^n)}$$
$$\lesssim t^{1+[(n-2)/2](1-\frac{1}{2\kappa})-\frac{\beta}{\kappa}} \quad \text{for all } t\in[1,\infty).$$

This completes the proof. □

Lemma 11.15. *Let us given the oscillating integral*

$$F^{-1}\Big(e^{-c_1|\xi|^{2\kappa}t}\cos\big(c_2|\xi|f(|\xi|)t\big)\chi(|\xi|)\Big) \tag{11.105}$$

with $\kappa\in(1/2,1)$, $t>0$, *and* $f(|\xi|)=\sqrt{1-|\xi|^{4\kappa-2}/4}$. *Here* c_1 *is a positive and* $c_2\neq 0$ *is a real constant. Then the following estimate holds in* $\mathbb{R}^n$ *with* $n\geq 2$:

$$\Big\|F^{-1}\Big(e^{-c_1|\xi|^{2\kappa}t}\cos\big(c_2|\xi|f(|\xi|)t\big)\chi(|\xi|)\Big)\Big\|_{L^1(\mathbb{R}^n)}\lesssim(1+t)^{[n/2](1-\frac{1}{2\kappa})} \text{ for } t>0.$$

Proof. Repeating the arguments in the proof of the previous lemma we conclude the desired estimate. □

Taking into consideration the statements from Lemmas 11.12 to 11.15 we have the following results:

Corollary 11.10. *The oscillating integrals*

$$F^{-1}\Big(\frac{\lambda_2 e^{\lambda_1 t}-\lambda_1 e^{\lambda_2 t}}{\lambda_1-\lambda_2}\chi(|\xi|)\Big) \tag{11.106}$$

with $\sigma\in(1/2,1)$ *and* $t>0$ *satisfy the following estimates in* $\mathbb{R}^n$, $n\geq 2$:

$$\Big\|F^{-1}\Big(\frac{\lambda_2 e^{\lambda_1 t}-\lambda_1 e^{\lambda_2 t}}{\lambda_1-\lambda_2}\chi(|\xi|)\Big)\Big\|_{L^1(\mathbb{R}^n)}\lesssim(1+t)^{[n/2](1-\frac{1}{2\sigma})} \text{ for } t>0.$$

Corollary 11.11. *The oscillating integrals*

$$F^{-1}\Big(\frac{e^{\lambda_1 t}-e^{\lambda_2 t}}{\lambda_1-\lambda_2}\chi(|\xi|)\Big) \tag{11.107}$$

with $\sigma\in(1/2,1)$ *and* $t>0$ *satisfy the following estimates in* $\mathbb{R}^n$, $n\geq 2$:

$$\Big\|F^{-1}\Big(\frac{e^{\lambda_1 t}-e^{\lambda_2 t}}{\lambda_1-\lambda_2}\chi(|\xi|)\Big)\Big\|_{L^1(\mathbb{R}^n)}\lesssim(1+t)^{1+[(n-2)/2](1-\frac{1}{2\sigma})} \text{ for } t>0.$$

11.5.2 L^1 Estimates for Large Frequencies

Lemma 11.16. *The following estimate holds for all* $t>0$:

$$\|F^{-1}(|\xi|^a e^{(\lambda_1+\lambda_2)t}(1-\chi(|\xi|)))\|_{L^1(\mathbb{R}^n)} \lesssim 1.$$

Here $a<1$ is supposed to be a real constant.

Proof. We only prove the result for odd $n \geq 3$. Using modified Bessel functions, we have to consider for large $|x|$ the integral

$$\begin{aligned} I(t,x) &:= \int_1^\infty e^{-cr^{2\sigma}t}(1-\chi(r))r^{n-1+a}\tilde{J}_{1/2}(r|x|)dr \\ &= \frac{1}{|x|^2}\int_1^\infty \partial_r\Big(e^{-cr^{2\sigma}t}(1-\chi(r))r^{n-2+a}\Big)\cos(r|x|)dr. \end{aligned}$$

We carry out $n-1$ more steps of partial integration (in each step boundary values vanish) and obtain modulo a factor

$$I_n(t,x) = \frac{1}{|x|^{n+1}}\int_1^\infty \partial_r^n\Big(e^{-cr^{2\sigma}t}(1-\chi(r))r^{n-2+a}\Big)\cdots(r|x|)dr,$$

where $\cdots$ is staying for sin or cos. Integrals containing derivatives of χ are not of interest. Any derivative of $e^{-cr^{2\sigma}t}$ gives $r^{-1}r^{2\sigma}t$. The second factor is linked with $e^{-cr^{2\sigma}t}$. For this reason we have a term r^{-2+a} which is integrable over $(1,\infty)$ due to the assumption $a<1$. □

In this section we shall estimate the following L^1 norms for all $t>0$:

$$\Big\|F^{-1}\Big(e^{\lambda_1 t}(1-\chi(|\xi|))\Big)\Big\|_{L^1(\mathbb{R}^n)}, \Big\|F^{-1}\Big(|\xi|^{2-4\sigma}e^{\lambda_2 t}(1-\chi(|\xi|))\Big)\Big\|_{L^1(\mathbb{R}^n)},$$
$$\Big\|F^{-1}\Big(e^{(\lambda_2-\lambda_1)t}(1-\chi(|\xi|))\Big)\Big\|_{L^1(\mathbb{R}^n)}, \Big\|F^{-1}\Big(\frac{e^{\lambda_1 t}-e^{\lambda_2 t}}{\lambda_1-\lambda_2}(1-\chi(|\xi|))\Big)\Big\|_{L^1(\mathbb{R}^n)}.$$

Lemma 11.17. *Let us given the oscillating integral*

$$F^{-1}\Big(e^{(\lambda_2-\lambda_1)t}(1-\chi(|\xi|))\Big) \tag{11.108}$$

with $\sigma \in (1/2,1]$ and $t>0$. Then the following estimate holds in $\mathbb{R}^n$:

$$\Big\|F^{-1}\Big(e^{(\lambda_2-\lambda_1)t}(1-\chi(|\xi|))\Big)(1-\chi(|x|))\Big\|_{L^1(\mathbb{R}^n)} \lesssim 1.$$

Proof. We only prove the statement for $n=3$. As in the proof to Lemma 11.2 we carry out several steps of partial integration, now three steps. Then we reduce our considerations to the integral

$$\frac{1}{|x|^4}\int_0^\infty \partial_r^3\Big(e^{-t\sqrt{r^{4\sigma}-4r^2}}(1-\chi(r))r\Big)\cos(r|x|)dr.$$

Taking account of

$$\left|\partial_r^j\left(e^{-t\sqrt{r^{4\sigma}-4r^2}}\right)\right| \le \frac{C_j}{r^j} \text{ for } t>0 \text{ and } r\ge 1 \text{ in the case } j=3,$$

we conclude the desired statement. □

Lemma 11.18. *Let us given the oscillating integral*

$$F^{-1}\left(e^{(\lambda_2-\lambda_1)t}(1-\chi(|\xi|))\right) \tag{11.109}$$

with $\sigma \in (1/2,1]$ and $t>0$. Then the following estimate holds in $\mathbb{R}^n$:

$$\left\|F^{-1}\left(e^{(\lambda_2-\lambda_1)t}(1-\chi(|\xi|))\right)\chi(|x|)\right\|_{L^1(\mathbb{R}^n)} \lesssim 1.$$

Proof. If $t \ge 1/2$, then we conclude as follows:

$$\begin{aligned}&\left\|F^{-1}\left(e^{(\lambda_2-\lambda_1)t}(1-\chi(|\xi|))\right)\chi(|x|)\right\|_{L^1(\mathbb{R}^n)}\\ &\le \int_{\mathbb{R}^n}\chi(|x|)dx\int_{\mathbb{R}^n} e^{-c_1|\xi|^{2\sigma}}(1-\chi(|\xi|))d\xi \lesssim 1.\end{aligned}$$

If $t\in(0,1/2]$, then we split the integral

$$\begin{aligned}&\int_{\mathbb{R}^n} F^{-1}\left(e^{(\lambda_2-\lambda_1)t}(1-\chi(|\xi|))\right)\chi(|x|)dx\\ &=\int_{\mathbb{R}^n} F^{-1}\left(e^{(\lambda_2-\lambda_1)t}(1-\chi(|\xi|))\right)\chi(|x|)\chi\left(t^{-\frac{1}{2\sigma}}x\right)dx\\ &\quad+\int_{\mathbb{R}^n} F^{-1}\left(e^{(\lambda_2-\lambda_1)t}(1-\chi(|\xi|))\right)\chi(|x|)\left(1-\chi\left(t^{-\frac{1}{2\sigma}}x\right)\right)dx.\end{aligned}$$

To estimate both integrals we use as in the proof to Lemma 11.2 the transformations $\eta := \xi t^{\frac{1}{2\sigma}}$ and $x =: yt^{\frac{1}{2\sigma}}$. We may conclude for the first integral

$$\begin{aligned}&\int_{\mathbb{R}^n}\left|F^{-1}\left(e^{(\lambda_2-\lambda_1)t}(1-\chi(|\xi|))\right)\chi(|x|)\chi\left(t^{-\frac{1}{2\sigma}}r\right)\right|dx\\ &\le C\int_{\mathbb{R}^n}\chi(y)dy\int_{\mathbb{R}^n} e^{-c_1|\eta|^{2\sigma}}d\eta \lesssim 1.\end{aligned}$$

We show only for $n=3$ how to estimate the second integral. The integral of interest is

$$\int_{\mathbb{R}^3}(1-\chi(y))\left(\int_0^\infty e^{-r^{2\sigma}\sqrt{1-t^{\frac{2\sigma-1}{\sigma}}r^{2-4\sigma}}}r^2\left(1-\chi\left(t^{-\frac{1}{2\sigma}}r\right)\right)\tilde{J}_{\frac{1}{2}}(r|y|)dr\right)dy.$$

We apply the approach from the proof to Lemma 11.2 and observe the following after three steps of partial integration: our starting integral is transformed to

$$\frac{1}{|y|^4}\int_0^\infty \partial_r^3\left(e^{-r^{2\sigma}\sqrt{1-\left(\frac{t^{\frac{1}{2\sigma}}}{r}\right)^{4\sigma-2}}}r\left(1-\chi\left(t^{-\frac{1}{2\sigma}}r\right)\right)\right)\cos(r|y|)dy.$$

Integrals containing derivatives of $\chi\big(\frac{r}{t^{\frac{1}{2\sigma}}}\big)$ are not of interest because on its support we have $t^{\frac{1}{2\sigma}} \sim r$, so a power of r is produced. Differentiation of $e^{-r^{2\sigma}\sqrt{1-\big(\frac{t^{\frac{1}{2\sigma}}}{r}\big)^{4\sigma-2}}}$ gives the factor $r^{2\sigma-1}$. Taking into consideration the support of $1-\chi\big(\frac{r}{t^{\frac{1}{2\sigma}}}\big)$, we see that all integrands appearing after three steps of partial integration have at least the asymptotic behavior $r^{2\sigma-2}$ near 0. So, the integral $\int_0^1$ exists for $\sigma \in (1/2,1]$. Due to the term $e^{-c_1 r^{2\sigma}}$ the integral $\int_1^\infty$ exists, too. The asymptotic behavior $O(|y|^{-4})$ implies the desired estimate

$$\int_{\mathbb{R}^3} \Big|F^{-1}\Big(e^{(\lambda_2-\lambda_1)t}(1-\chi(|\xi|))\Big)\chi(|x|)\Big(1-\chi\Big(t^{-\frac{1}{2\sigma}}x\Big)\Big)\Big|dx \lesssim 1.$$

This completes the proof. □

From Lemmas 11.17 and 11.18 we obtain the following corollary:

Corollary 11.12. *Let us given the oscillating integral*

$$F^{-1}\Big(e^{(\lambda_2-\lambda_1)t}(1-\chi(|\xi|))\Big) \tag{11.110}$$

with $\sigma \in (1/2,1]$ and $t>0$. Then the following estimate holds in $\mathbb{R}^n$:

$$\Big\|F^{-1}\Big(e^{(\lambda_2-\lambda_1)t}(1-\chi(|\xi|))\Big)\Big\|_{L^1(\mathbb{R}^n)} \lesssim 1.$$

From Lemma 11.16 and Corollary 11.12 we conclude the following statement:

Corollary 11.13. *The following estimate holds for all $t>0$:*

$$\Big\|F^{-1}\Big(|\xi|^{2-4\sigma}e^{\lambda_2 t}(1-\chi(|\xi|))\Big)\Big\|_{L^1(\mathbb{R}^n)} \lesssim 1.$$

Proof. The statement follows immediately from the representation

$$\begin{aligned}
&F^{-1}\Big(|\xi|^{2-4\sigma}e^{\lambda_2 t}(1-\chi(|\xi|))^2\Big)\\
&\quad = F^{-1}\Big(|\xi|^{2-4\sigma}e^{(\lambda_1+\lambda_2)\frac{t}{2}}(1-\chi(|\xi|))\Big) * F^{-1}\Big(e^{(\lambda_2-\lambda_1)\frac{t}{2}}(1-\chi(|\xi|))\Big)
\end{aligned}$$

together with the above-mentioned results. □

It remains to consider the oscillating integral $F^{-1}\big(e^{\lambda_1 t}(1-\chi(|\xi|))\big)$ for $\sigma \in (1/2,1]$.

Lemma 11.19. *Let us given the oscillating integral*

$$F^{-1}\Big(e^{\lambda_1 t}(1-\chi(|\xi|))\Big) \tag{11.111}$$

with $\sigma \in (1/2, 1]$ *and* $t > 0$. *Then the following estimate holds in* $\mathbb{R}^n$:

$$\left\| F^{-1}\left(e^{\lambda_1 t}(1-\chi(|\xi|))\right)(1-\chi(|x|))\right\|_{L^1(\mathbb{R}^n)} \lesssim 1.$$

Proof. We prove the statement only for $n = 3$. As in the proof to Lemma 11.2 we carry out several steps of partial integration, now three steps. Then we reduce our considerations to the integral

$$\frac{1}{|x|^4}\int_0^\infty \partial_r^3\left(e^{\lambda_1(r)t}(1-\chi(r))r\right)\cos(r|x|)dr.$$

Taking account of

$$\left|\partial_r^j e^{\lambda_1(r)t}\right| \le C_j r^{2-2\sigma-j} \text{ for } t > 0 \text{ and } r \ge 1 \text{ in the case } j = 3,$$

we conclude the desired statement. □

Lemma 11.20. *Let us given the oscillating integral*

$$F^{-1}\left(e^{\lambda_1 t}(1-\chi(|\xi|))\right) \tag{11.112}$$

with $\sigma \in (1/2, 1)$ *and* $t > 0$. *Then the following estimate holds in* $\mathbb{R}^n$:

$$\left\| F^{-1}\left(e^{\lambda_1 t}(1-\chi(|\xi|))\right)\chi(|x|)\right\|_{L^1(\mathbb{R}^n)} \lesssim 1.$$

Proof. If $t \ge 1/2$, then we conclude as follows:

$$\begin{aligned}&\left\| F^{-1}(e^{\lambda_1 t}(1-\chi(|\xi|)))\chi(|x|)\right\|_{L^1(\mathbb{R}^n)}\\&\quad\le \int_{\mathbb{R}^n}\chi(|x|)dx\int_{\mathbb{R}^n} e^{-c_1|\xi|^{2-2\sigma}}(1-\chi(|\xi|))\,d\xi \lesssim 1.\end{aligned}$$

If $t \in (0, 1/2]$, then we split the integral

$$\begin{aligned}&\int_{\mathbb{R}^n} F^{-1}\left(e^{\lambda_1 t}(1-\chi(|\xi|))\right)\chi(|x|)dx\\&\quad= \int_{\mathbb{R}^n} F^{-1}\left(e^{\lambda_1 t}(1-\chi(|\xi|))\right)\chi(|x|)\chi\left(t^{-\frac{1}{2-2\sigma}}x\right)dx\\&\quad+\int_{\mathbb{R}^n} F^{-1}\left(e^{\lambda_1 t}(1-\chi(|\xi|))\right)\chi(|x|)\left((1-\chi)t^{-\frac{1}{2-2\sigma}}x\right)dx.\end{aligned}$$

To estimate both integrals we use as in the proof to Lemma 11.2 the transformations $\eta := \xi t^{\frac{1}{2-2\sigma}}$ and $x =: y t^{\frac{1}{2-2\sigma}}$. We may conclude for the first integral

$$\int_{\mathbb{R}^n} \Big| F^{-1}\Big(e^{\lambda_1 t}(1-\chi(|\xi|))\Big)\chi(|x|)\chi\Big(t^{-\frac{1}{2-2\sigma}}x\Big)\Big| dx \\ \le C\int_{\mathbb{R}^n} \chi(y)dy \int_{\mathbb{R}^n} e^{-c_1|\eta|^{2-2\sigma}} d\eta \lesssim 1.$$

We show only for the case $n=3$ how to estimate the second integral. The integral of interest is

$$\int_{\mathbb{R}^3} (1-\chi(y))\Big(\int_0^\infty e^{\lambda_1\left(rt^{-\frac{1}{2-2\sigma}}\right)t} r^2\Big(1-\chi\Big(t^{-\frac{1}{2-2\sigma}}r\Big)\Big)\tilde{J}_{\frac{1}{2}}(r|y|)dr\Big)dy.$$

We apply the approach from the proof to Lemma 11.2 and observe the following after two steps of partial integration: our starting integral is transformed to

$$\frac{1}{|y|^3}\int_0^\infty \partial_r^2\Big(e^{\lambda_1\left(rt^{-\frac{1}{2-2\sigma}}\right)t} r\Big(1-\chi\Big(t^{-\frac{1}{2-2\sigma}}r\Big)\Big)\Big)\sin(r|y|)dy.$$

Integrals containing derivatives of $\chi\big(\frac{r}{t^{\frac{1}{2-2\sigma}}}\big)$ are not of interest because on its support we have $t^{\frac{1}{2-2\sigma}} \sim r$, so a power of r is produced. Differentiation of $e^{\lambda_1\left(rt^{-\frac{1}{2-2\sigma}}\right)t}$ gives the factor $r^{1-2\sigma}$. Taking into consideration the support of $1-\chi\big(\frac{r}{t^{\frac{1}{2-2\sigma}}}\big)$ we see that all integrands appearing after two steps of partial integration have at least the asymptotic behavior $r^{1-2\sigma}$ near 0. Now we apply the splitting

$$\int_0^\infty = \int_0^{\frac{1}{|y|}} + \int_{\frac{1}{|y|}}^1 + \int_1^\infty.$$

The weak singularity $r^{1-2\sigma}$ implies for the first integral the behavior $|y|^{2\sigma-5}$. To estimate the other integrals we carry out a third step of partial integration. Then the second integral is estimated by $|y|^{2\sigma-5}$, too. Finally, the last integral is estimated by $|y|^{-4}$. Summarizing we obtain the asymptotic behavior $O(|y|^{2\sigma-5})$. This implies the desired estimate

$$\int_{\mathbb{R}^3} \Big| F^{-1}\Big(e^{\lambda_1 t}(1-\chi(|\xi|))\Big)\chi(|x|)\Big(1-\chi\Big(t^{-\frac{1}{2-2\sigma}}x\Big)\Big)\Big| dx \lesssim 1.$$

This completes the proof. □

Remark 11.3. In Lemma 11.20 we exclude $\sigma=1$. In this case $\lambda_1(|\xi|)+1 \sim |\xi|^{-2}$. Consequently, we get e^{-t} in the estimate; on the other hand, we cannot expect the existence of $F^{-1}(e^{c|\xi|^{-2}}(1-\chi(|\xi|)))$ for $c>0$.

Finally, it remains to estimate

$$\Big\| F^{-1}\Big(\frac{e^{\lambda_1 t}-e^{\lambda_2 t}}{\lambda_1-\lambda_2}(1-\chi(|\xi|))\Big)\Big\|_{L^1(\mathbb{R}^n)}.$$

Here we follow the proof of Corollary 11.7.

Corollary 11.14. *The following estimate holds for all $t > 0$:*

$$\left\| F^{-1}\left(\frac{e^{\lambda_1 t} - e^{\lambda_2 t}}{\lambda_1 - \lambda_2}(1 - \chi(|\xi|))\right)\right\|_{L^1(\mathbb{R}^n)} \leq Ct.$$

11.5.3 L^∞ Estimates

Using the properties of the Fourier transformation and Lemma 11.11 we conclude the following results after standard calculations.

Proposition 11.9. *We have the following L^∞ estimates for $t > 0$ and $\sigma \in (1/2, 1)$:*

$$\left\|F^{-1}\Big(\frac{\lambda_1 e^{\lambda_2 t} - \lambda_2 e^{\lambda_1 t}}{\lambda_1 - \lambda_2}\chi(|\xi|)\Big)\right\|_{L^\infty(\mathbb{R}^n)} \lesssim (1+t)^{-\frac{n}{2\sigma}},$$
$$\left\|F^{-1}\Big(\frac{\lambda_1 e^{\lambda_2 t} - \lambda_2 e^{\lambda_1 t}}{\lambda_1 - \lambda_2}(1-\chi(|\xi|))\Big)\right\|_{L^\infty(\mathbb{R}^n)} \lesssim t^{-\frac{n+2-4\sigma}{2\sigma}},$$
$$\left\|F^{-1}\Big(\frac{e^{\lambda_1 t} - e^{\lambda_2 t}}{\lambda_1 - \lambda_2}\chi(|\xi|)\Big)\right\|_{L^\infty(\mathbb{R}^n)} \lesssim (1+t)^{-\frac{n-2\sigma}{2\sigma}},$$
$$\left\|F^{-1}\Big(\frac{e^{\lambda_1 t} - e^{\lambda_2 t}}{\lambda_1 - \lambda_2}(1-\chi(|\xi|))\Big)\right\|_{L^\infty(\mathbb{R}^n)} \lesssim t^{-\frac{n-2\sigma}{2\sigma}}.$$

11.5.4 $L^p - L^q$ Estimates not Necessarily on the Conjugate Line

From Corollaries 11.10–11.14, Lemmas 11.19 and 11.20, and Proposition 11.9, we have the following statements:

Proposition 11.10. *The following estimates hold for $t > 0$, $\sigma \in (1/2, 1)$ and $n \geq 2$:*

$$\left\|F^{-1}\Big(\frac{\lambda_2 e^{\lambda_1 t} - \lambda_1 e^{\lambda_2 t}}{\lambda_1 - \lambda_2}\Big)\right\|_{L^1(\mathbb{R}^n)} \lesssim (1+t)^{[n/2](1-\frac{1}{2\sigma})},$$
$$\left\|F^{-1}\left(\frac{e^{\lambda_1 t} - e^{\lambda_2 t}}{\lambda_1 - \lambda_2}\right)\right\|_{L^1(\mathbb{R}^n)} \lesssim t(1+t)^{[(n-2)/2](1-\frac{1}{2\sigma})},$$
$$\left\|F^{-1}\Big(\frac{\lambda_2 e^{\lambda_1 t} - \lambda_1 e^{\lambda_2 t}}{\lambda_1 - \lambda_2}\Big)\right\|_{L^\infty(\mathbb{R}^n)} \lesssim t^{-\frac{n+2-4\sigma}{2\sigma}},$$
$$\left\|F^{-1}\left(\frac{e^{\lambda_1 t} - e^{\lambda_2 t}}{\lambda_1 - \lambda_2}\right)\right\|_{L^\infty(\mathbb{R}^n)} \lesssim t^{-\frac{n-2\sigma}{2\sigma}}.$$

By interpolation we conclude from the last proposition

$$\left\|F^{-1}\Big(\frac{\lambda_2 e^{\lambda_1 t}-\lambda_1 e^{\lambda_2 t}}{\lambda_1-\lambda_2}\Big)\right\|_{L^r(\mathbb{R}^n)} \lesssim \begin{cases} t^{-\frac{n+2-4\sigma}{2\sigma}(1-\frac{1}{r})}, \ t\in(0,1], \\ t^{-\frac{n+2-4\sigma}{2\sigma}(1-\frac{1}{r})+[n/2]\frac{2\sigma-1}{2\sigma}\frac{1}{r}}, \ t\in[1,\infty), \end{cases} \tag{11.113}$$

$$\left\|F^{-1}\left(\frac{e^{\lambda_1 t}-e^{\lambda_2 t}}{\lambda_1-\lambda_2}\right)\right\|_{L^r(\mathbb{R}^n)} \lesssim \begin{cases} t^{1-\frac{n}{2\sigma}(1-\frac{1}{r})}, \ t\in(0,1], \\ t^{1-\frac{n}{2\sigma}(1-\frac{1}{r})+[(n-2)/2]\frac{2\sigma-1}{2\sigma}\frac{1}{r}}, \ t\in[1,\infty), \end{cases} \tag{11.114}$$

for all $r\in[1,\infty]$.

Proposition 11.11. *Let us consider with $\sigma\in(1/2,1)$ the Cauchy problem*

$$u_{tt}-\Delta u+\mu(-\Delta)^\sigma u_t=0,\ u(0,x)=u_0(x),\ u_t(0,x)=u_1(x). \tag{11.115}$$

Then the solution satisfies the following L^p-L^q estimates :

$$\|u(t,\cdot)\|_{L^p}\lesssim t^{-\frac{n+2-4\sigma}{2\sigma}(1-\frac{1}{r})}\|u_0\|_{L^q}+t^{1-\frac{n}{2\sigma}(1-\frac{1}{r})}\|u_1\|_{L^q} \ \textit{for } t\in(0,1],$$
$$\|u(t,\cdot)\|_{L^p}\lesssim t^{-\frac{n+2-4\sigma}{2\sigma}(1-\frac{1}{r})+[n/2]\frac{2\sigma-1}{2\sigma}\frac{1}{r}}\|u_0\|_{L^q}+t^{1-\frac{n}{2\sigma}(1-\frac{1}{r})+[(n-2)/2]\frac{2\sigma-1}{2\sigma}\frac{1}{r}}\|u_1\|_{L^q}$$
$$\textit{for } t\in[1,\infty),$$

where $1\le q\le p\le\infty$ and $1+\frac{1}{p}=\frac{1}{r}+\frac{1}{q}$.

Proof. The statement follows immediately from (11.113) and (11.114) by using the explicit representation of the solution by Fourier multipliers and the convolution rule. □

Remark 11.4. Setting $\sigma=1/2$ in the last statement we see that Proposition 11.11 coincides with Proposition 11.3. If we set formally $\sigma=1$, then we conclude

$$\|u(t,\cdot)\|_{L^p}\lesssim t^{-\frac{n-2}{2}(1-\frac{1}{r})}\|u_0\|_{L^q}+t^{1-\frac{n}{2}(1-\frac{1}{r})}\|u_1\|_{L^q} \ \text{for } t\in(0,1],$$
$$\|u(t,\cdot)\|_{L^p}\lesssim t^{-\frac{n-2}{2}(1-\frac{1}{r})+[n/2]\frac{1}{2}\frac{1}{r}}\|u_0\|_{L^q}+t^{1-\frac{n}{2}(1-\frac{1}{r})+[(n-2)/2]\frac{1}{2}\frac{1}{r}}\|u_1\|_{L^q}$$
$$\text{for } t\in[1,\infty),$$

where $1\le q\le p\le\infty$ and $1+\frac{1}{p}=\frac{1}{r}+\frac{1}{q}$. Our main goal of this paper is to prove the basic L^1-L^1 estimates for $\sigma\in(0,1)$. So, setting $p=q=r=1$ in these estimates we may conclude

$$\|u(t,\cdot)\|_{L^1}\lesssim\|u_0\|_{L^1}+t\|u_1\|_{L^1} \ \text{for } t\in(0,1],$$
$$\|u(t,\cdot)\|_{L^1}\lesssim t^{[n/2]\frac{1}{2}}\|u_0\|_{L^1}+t^{1+[(n-2)/2]\frac{1}{2}}\|u_1\|_{L^1} \ \text{for } t\in[1,\infty).$$

In [8] the following result was proved for the case $\sigma=1$:

1. If $n=2k$ is an even number and $n\ge2$, then

$$\|u(t,\cdot)\|_{L^1}\lesssim(1+t)^{\frac{k}{2}}\|u_0\|_{L^1}+(1+t)^{\frac{1}{2}+\frac{k}{2}}\|u_1\|_{L^1} \ \text{for } t>0.$$

2. If $n = 2k+1$ is an odd number and $n \geq 3$, then

$$\|u(t,\cdot)\|_{L^1} \lesssim (1+t)^{\frac{k}{2}} \|u_0\|_{L^1} + (1+t)^{\frac{1}{2}+\frac{k}{2}} \|u_1\|_{L^1} \text{ for } t > 0.$$

We see that in the estimates for $\|u(t,\cdot)\|_{L^1}$ our upper bounds for $t \to \infty$ coincide with those from [8].

Remark 11.5. In the future we will use the $L^1 - L^1$ estimates to treat semi-linear structural damped wave models (see [2] and [3]). A first step is to derive optimal $L^p - L^q$ estimates not necessarily on the conjugate line for Fourier integrals with amplitudes localized to small or to large frequencies [2]. The interested reader can study such results for semi-linear classical damped waves in the recent paper [1].

Appendix

Modified Bessel Functions

Here we summarize some rules for modified Bessel functions.

Let $J_\mu = J_\mu(s)$ be the Bessel function of order $\mu \in (-\infty,\infty)$, and let $\tilde{J}_\mu(s) := J_\mu(s)/s^\mu$ when μ is not a negative integer.

Lemma 11.21. *Let $f \in L^p(\mathbb{R}^n)$, $p \in [1,2]$, be a radial function. Then the Fourier transform $F(f)$ is also a radial function and it satisfies*

$$F(f)(\xi) = c\int_0^\infty g(r) r^{n-1} \tilde{J}_{n/2-1}(r|\xi|)\,dr, \quad g(|x|) := f(x).$$

Lemma 11.22. *Assume that μ is not a negative integer. Then the following rules hold:*

(1) $s d_s \tilde{J}_\mu(s) = \tilde{J}_{\mu-1}(s) - 2\mu \tilde{J}_\mu(s)$.
(2) $d_s \tilde{J}_\mu(s) = -s\tilde{J}_{\mu+1}(s)$.
(3) $\tilde{J}_{-1/2}(s) = \sqrt{\frac{\pi}{2}}\cos s$.
(4) *We have for any μ the relations*

$$|\tilde{J}_\mu(s)| \leq Ce^{\pi|\mathrm{Im}\mu|} \text{ if } |s| \leq 1,$$
$$J_\mu(s) = Cs^{-1/2}\cos\left(s - \frac{\mu}{2}\pi - \frac{\pi}{4}\right) + O(|s|^{-3/2}) \text{ if } |s| \geq 1.$$

(5) $\tilde{J}_{\mu+1}(r|x|) = -\frac{1}{r|x|^2}\partial_r \tilde{J}_\mu(r|x|), \quad r \neq 0,\ x \neq 0.$

Solutions to Wave Equations

Lemma 11.23 (Lemma 3.2 from [8]). *Put*

$$w(t,x) = F^{-1}\Big(\frac{\sin(|\xi|t)}{|\xi|}F(h)\Big)(x), \quad t>0.$$

Then, for suitable constants a_α we have

$$w(t,x) = \sum_{0\le|\alpha|\le(n-3)/2} a_\alpha t^{|\alpha|+1}\int_{|y|=1} y^\alpha\big(\partial_x^\alpha h\big)(x+ty)\,d\sigma_y$$

for odd $n\ge 3$;
and

$$w(t,x) = \sum_{0\le|\alpha|\le(n-2)/2} a_\alpha t^{|\alpha|+1}\int_{|y|\le 1} y^\alpha\big(\partial_x^\alpha h\big)(x+ty)\frac{1}{\sqrt{1-|y|^2}}\,dy$$

for even $n\ge 2$.

Estimates for Fourier Multipliers

Lemma 11.24. *Let ν be positive constant. Let $h=h(|\xi|)$ be a smooth radial function on $\{\xi\in\mathbb{R}^n : |\xi|\ge 1\}$ such that*

$$\Big|\partial_\xi^\alpha h(|\xi|)\Big| \lesssim |\xi|^{-\nu-|\alpha|} \quad \textit{for all } |\alpha|\le n+1.$$

Then $F^{-1}\Big(h(|\xi|)(1-\chi(|\xi|))\Big)\in L^1(\mathbb{R}^n)$ and

$$\Big\|F^{-1}\Big(h(|\xi|)(1-\chi(|\xi|))\Big)\Big\|_{L^1(\mathbb{R}^n)} \le C\sum_{|\alpha|\le n+1}\sup_{\{|\xi|\ge 1\}}|\xi|^{\nu+|\alpha|}|\partial_\xi^\alpha h(|\xi|)|.$$

Here the function $1-\chi(|\xi|)$ localizes to $\{|\xi|\ge 1\}$.

Proof. We put for all $j\in\mathbb{N}$

$$\varphi(x) = F^{-1}\big(h(|\xi|)(1-\chi(|\xi|))\big),\ \varphi_j(x) = F^{-1}\big(h(|\xi|)(1-\chi(|\xi|))\chi(2^{-j}|\xi|)\big).$$

Then we have $\varphi_j\in\mathscr{S}(\mathbb{R}^n)\subset L^1(\mathbb{R}^n)\cap L^\infty(\mathbb{R}^n)$. On the one hand we have

$$\begin{aligned}|\varphi_j(x)-\varphi_{j+1}(x)| &\le c\int_{\mathbb{R}^n}|h(|\xi|)|(1-\chi(|\xi|))\big|\chi(2^{-j}|\xi|)-\chi(2^{-j-1}|\xi|)\big|\,d\xi\\ &\le c\int_{2^{j-1}\le|\xi|\le 2^{j+1}}|\xi|^{-\nu}\,d\xi \le c\int_{2^{j-1}}^{2^{j+1}} r^{-\nu+n-1}\,dr \le C2^{j(n-\nu)},\end{aligned}$$

and, on the other hand, we have for $n \leq |\alpha| \leq n+1$ the estimates

$$\begin{aligned}
&\left|x^\alpha \left(\varphi_{j+1}(x) - \varphi_j(x)\right)\right| \\
&\leq c \int_{\mathbb{R}^n} \left|\partial_\xi^\alpha \Big(h(\xi)\,(1-\chi(|\xi|))\,\big(\chi(2^{-j}|\xi|) - \chi(2^{-j-1}|\xi|)\big)\Big)\right| d\xi \\
&\leq c \sum_{k=0}^{|\alpha|} \int_{\mathbb{R}^n} |\xi|^{-\nu-n+k} 2^{-jk} \left|\chi(2^{-j}|\xi|) - 2^{-k}\chi(2^{-j-1}|\xi|)\right| d\xi \\
&\leq c \sum_{k=0}^{|\alpha|} \int_{2^{j-1}}^{2^{j+1}} r^{-\nu+k-1} 2^{-jk}\, dr \leq C 2^{-j\nu}.
\end{aligned}$$

Hence, we have

$$|x|^{n+1} \left|\varphi_{j+1}(x) - \varphi_j(x)\right| \leq C 2^{-j\nu}. \tag{11.116}$$

Now let us distinguish the cases $|x| \geq 1$ and $|x| \leq 1$. If $|x| \geq 1$, then we have

$$|x|^{(n+1)(1-\theta)} \left|\varphi_{j+1}(x) - \varphi_j(x)\right| \leq C 2^{-j\nu(1-\theta)+\theta(n-\nu)}. \tag{11.117}$$

If $\nu \geq n$, then we choose $\theta < \frac{1}{n+1}$ and obtain from the last inequality

$$|x|^{(n+1)(1-\theta)} \left|\varphi_{j+1}(x) - \varphi_j(x)\right| \leq C 2^{-j\nu(1-\theta)}. \tag{11.118}$$

If $\nu \in (0,n)$, then we choose $\theta < \min\{\frac{1}{n+1}; \frac{\nu}{2n-\nu}\}$. Then we obtain

$$|x|^{(n+1)(1-\theta)} \left|\varphi_{j+1}(x) - \varphi_j(x)\right| \leq C 2^{-j\nu(1-\theta)/2}. \tag{11.119}$$

Both inequalities (11.118) and (11.119) show that $\{\varphi_j\}_j$ is a Cauchy sequence in $L^1(\mathbb{R}^n \setminus \{|x| \leq 1\})$. This implies that

$$\varphi = \lim_{j\to\infty} \varphi_j \in L^1(\mathbb{R}^n \setminus \{|x| \leq 1\}).$$

In the case $|x| \leq 1$ we choose from the above estimates

$$|x|^n \left|\varphi_{j+1}(x) - \varphi_j(x)\right| \leq C 2^{-j\nu} \quad \text{and} \quad \left|\varphi_j(x) - \varphi_{j+1}(x)\right| \leq C 2^{j(n-\nu)}. \tag{11.120}$$

As above we conclude

$$|x|^{n(1-\theta)} \left|\varphi_{j+1}(x) - \varphi_j(x)\right| \leq C 2^{-j\nu(1-\theta)+j(n-\nu)\theta}. \tag{11.121}$$

If $\nu \geq n$, then we choose $\theta > 0$ small. If $\nu \in (0,n)$, then we choose as above $\theta \leq \frac{\nu}{2n-\nu}$. Both choices imply that $\{\varphi_j\}_j$ is a Cauchy sequence in $L^1(\{|x| \leq 1\})$. This implies that

$$\varphi = \lim_{j\to\infty} \varphi_j \in L^1(\{|x| \leq 1\}).$$

This completes the proof. □

Acknowledgements This project started during a sabbatical of the first author at TU Bergakademie Freiberg from September 2010 to March 2011. The first author thanks the faculty of Mathematics and Computer Science for the hospitality.

References

1. D'Abbicco, M., Lucente, S., Reissig, M.: Semi-linear wave equations with effective damping, 28 A4, accepted for publication in Chinese Annals of Mathematics, Ser. B.
2. D'Abbicco, M., Reissig, M.: Semi-linear structural damped waves, 23 A4, submitted for publication
3. Lu, X., Reissig, M.: Rates of decay for structural damped models with decreasing in time coefficients. Int. J. of Dyn. Syst. Differ. Equat. **2**, 1–2, 21–55 (2009)
4. Matsumura, A.: On the asymptotic behavior of solutions of semi-linear wave equations. Publ. Res. Inst. Math. Sci. Kyoto Univ. **12**, 169–189 (1976)
5. Narazaki, T.: L^p-L^q estimates for damped wave equations and their applications to semilinear problem. J. Math. Soc. Jpn. **56**, 585–626 (2004)
6. Narazaki, T.: Global solutions to the Cauchy problem for a system of damped wave equations. J. Differ. Int. **24**, 569–600 (2011)
7. Nishihara, K.: L^p-L^q estimates of solutions to the damped wave equation in 3-dimensional space and their application. Math. Z. **244**, 631–649 (2003)
8. Shibata, Y.: On the rate of decay of solutions to linear viscoelastic equation. MMAS **23**, 203–226 (2000)
9. Stein, E.M., Weiss, G.: In: Introduction to Fourier Analysis on Euclidean Spaces. Princeton Mathematical Series. University press, Princeton (1971)
10. Todorova, Y., Yordanov, B.: Critical exponent for a nonlinear wave equation with damping. J. Differ. Equat. **174**, 464–489 (2001)
11. Wang, W., Wang, W.: The pointwise estimates of solutions for semilinear dissipative wave equation in multi-dimensions. J. Math. Anal. Appl. **366**, 226–241 (2010)

Chapter 12
On the Cauchy Problem for Noneffectively Hyperbolic Operators, a Transition Case

Tatsuo Nishitani

Abstract We discuss the well-posedness of the Cauchy problem for noneffectively hyperbolic operators assuming that the spectral structure of the Hamilton map changes across a submanifold of codimension 1 of the double characteristic manifold. Under the assumption that there is no null bicharacteristic tangent to the submanifold where the spectral transition occurs, we derive microlocal a priori estimates assuming the strict Ivrii-Petkov-Hörmander condition.

Key words: Cauchy problem, Hamilton map and flow, Transition case, Noneffectively hyperbolic operator

2010 Mathematics Subject Classification: Primary: 35L15; Secondary: 35B30.

12.1 Introduction

Let

$$P(x,D) = -D_0^2 + \sum_{|\alpha|\leq 2, \alpha_0<2} a_\alpha(x) D^\alpha = p(x,D) + P_1 + P_0$$

be a second order differential operator defined in a neighborhood of the origin of $\mathbb{R}^{n+1}$ where $x = (x_0, x') = (x_0, x_1, ..., x_n)$. We assume that the principal symbol $p(x, \xi)$ vanishes exactly of order 2 on a C^∞ manifold Σ and

$$\text{rank}\Big(\sum_{j=0}^{n} d\xi_j \wedge dx_j\Big|_\Sigma\Big) = constant. \tag{12.1}$$

T. Nishitani (✉)
Department of Mathematics, Osaka University, Machikaneyama 1-1, Toyonaka, 560-0043, Osaka, Japan
e-mail: nishitani@math.sci.osaka-u.ac.jp

M. Cicognani et al. (eds.), *Studies in Phase Space Analysis with Applications to PDEs*, Progress in Nonlinear Differential Equations and Their Applications 84,
DOI 10.1007/978-1-4614-6348-1_12,

We also assume that p is noneffectively hyperbolic, that is, the Hamilton map F_p (the linearization of the Hamilton flow H_p on Σ, see [7, 9]) has no nonzero real eigenvalues on Σ. Assuming that the spectral structure of F_p is stable on Σ, that is, either $\operatorname{Ker} F_p^2 \cap \operatorname{Im} F_p^2 = \{0\}$ or $\operatorname{Ker} F_p^2 \cap \operatorname{Im} F_p^2 \neq \{0\}$ throughout Σ, the well-posedness of the Cauchy problem is discussed in [3–5]. In this note we study the Cauchy problem for noneffectively hyperbolic operators when the spectral structure of the Hamilton map changes. We restrict ourselves to the case $\operatorname{codim} \Sigma = 3$, and the spectral structure of F_p changes across a codimension 1 submanifold of Σ. Our aim is studying this simplest transition case, to elucidate the effects of the spectral transition on the well-posedness of the Cauchy problem.

More general geometrical settings of transitions are studied in [6]. They proved microlocal a priori estimates assuming no null bicharacteristic tangent to Σ, while in this note discussions are mainly devoted to the case that there may be a null bicharacteristic tangent to Σ (but not to the submanifold where the transition occurs). Some transition cases from effectively hyperbolic to noneffectively hyperbolic are studied in [2].

By assumption, near any point $\rho \in \Sigma$, one can write

$$p(x,\xi) = -\xi_0^2 + \varphi_1(x,\xi')^2 + \varphi_2(x,\xi')^2$$

where $d\varphi_1$ and $d\varphi_2$ are linearly independent at ρ and $\Sigma = \{\xi_0 = 0, \varphi_1 = 0, \varphi_2 = 0\}$. We assume that

$$\begin{cases} \text{the spectral structure of } F_p \text{ changes across} \\ \text{a submanifold } S \text{ of } \Sigma \text{ of codimension 1.} \end{cases} \tag{12.2}$$

Let $\rho \in \Sigma$. Then the Taylor expansion of $p(x,\xi)$ around ρ starts with $p_\rho(X)$, a second order homogeneous polynomial in $X = (x,\xi)$, called the localization of p at ρ which is a hyperbolic polynomial in the direction $\theta = (0,...,0,1,0,...,0)$ (see [9]). Let $Q(X,Y)$ be the quadratic form associated to p_ρ then the Hamilton map at ρ is given by

$$\sigma(X, F_p Y) = Q(X,Y), \quad X, Y \in \mathbb{R}^{2(n+1)}.$$

Since p_ρ is hyperbolic in the direction θ then the hyperbolicity cone Γ_ρ is given by

$$\Gamma_\rho = \text{the connected component of } \theta \text{ in } \{X \mid p_\rho(X) \neq 0\}$$

and the propagation cone C_ρ is defined as

$$C_\rho = \{X \mid \sigma(X,Y) \leq 0, \forall Y \in \Gamma_\rho\}$$

which is the "minimal" cone containing all bicharacteristics with limit point ρ (Lemma 1.1.1 in [12]). Recall that (Lemma 1.1.3 in [12])

$$p \text{ is noneffectively hyperbolic at } \rho \Longleftrightarrow C_\rho \cap T_\rho \Sigma \neq \{0\}. \tag{12.3}$$

Lemma 12.1. *Assume (12.1) and (12.2). Then we have either* $\{\xi_0, \varphi_1\} \neq 0$ *or* $\{\xi_0, \varphi_2\} \neq 0$ *on S.*

Proof. Suppose that $\{\xi_0,\varphi_j\}=0$ at $\rho\in S$ for $j=1,2$. Then in a suitable symplectic basis in $T_\rho\Sigma$ we can write

$$p_\rho=-\xi_0^2+\xi_1^2+\xi_2^2, \tag{12.4}$$
$$p_\rho=-\xi_0^2+\mu(x_1^2+\xi_1^2),\quad \mu>0 \tag{12.5}$$

according to $\{\varphi_1,\varphi_2\}(\rho)=0$ or $\{\varphi_1,\varphi_2\}(\rho)\neq 0$. If the first case occurs then from (12.1) we have $\mathrm{rank}\,(\sigma|_\Sigma)=0$ and hence p_ρ takes the form (12.4) everywhere on Σ in a suitable symplectic basis. In the second case we note that $\pm i\mu$ are the eigenvalues of F_p. From the continuity of $F_p(\rho)$ with respect to $\rho\in\Sigma$, it follows that F_p has still nonzero pure imaginary eigenvalues near ρ on Σ, and hence p_ρ takes the form (12.5) in a suitable symplectic basis. Therefore, in both cases, the spectral structure of F_p does not change near ρ. This proves the assertion. □

We now assume that $\{\xi_0,\varphi_j\}(\rho)\neq 0$ with some j. Considering

$$\tilde\varphi_i=\sum_{j=1}^{2}O_{ij}\varphi_j$$

with a smooth orthogonal matrix (O_{ij}), we may assume without restrictions that

$$\{\xi_0,\varphi_2\}\neq 0,\quad \{\xi_0,\varphi_1\}=O(|\varphi|) \tag{12.6}$$

near ρ. Here and in what follows $f=O(|\varphi|)$, $\varphi=(\varphi_1,\varphi_2)$ means that f is a linear combination of φ_1 and φ_2. We look for nonzero eigenvalues μ of F_p; $F_pX=\mu X$. Since $\mu\neq 0$ it suffices to consider F_p on the image of F_p, that is, on the vector space $\langle H_{\xi_0},H_{\varphi_1},H_{\varphi_2}\rangle$ spanned by H_{ξ_0}, H_{φ_1}, H_{φ_2}. Let $X=\alpha H_{\xi_0}+\beta H_{\varphi_1}+\gamma H_{\varphi_2}$ and consider $F_pX=\mu X$. Since

$$F_pX=-\gamma\{\varphi_2,\xi_0\}H_{\xi_0}+\gamma\{\varphi_2,\varphi_1\}H_{\varphi_1}+(\alpha\{\xi_0,\varphi_2\}+\beta\{\varphi_1,\varphi_2\})H_{\varphi_2}$$

we have

$$-\gamma\{\varphi_2,\xi_0\}=\mu\alpha,\ \gamma\{\varphi_2,\varphi_1\}=\mu\beta,\ \alpha\{\xi_0,\varphi_2\}+\beta\{\varphi_1,\varphi_2\}=\mu\gamma$$

so that

$$\begin{pmatrix}0&0&\{\xi_0,\varphi_2\}\\0&0&-\{\varphi_1,\varphi_2\}\\\{\xi_0,\varphi_2\}&\{\varphi_1,\varphi_2\}&0\end{pmatrix}\begin{pmatrix}\alpha\\\beta\\\gamma\end{pmatrix}=\mu\begin{pmatrix}\alpha\\\beta\\\gamma\end{pmatrix}.$$

Thus the characteristic equation is

$$\mu\left(\mu^2-(\{\xi_0,\varphi_2\}^2-\{\varphi_1,\varphi_2\}^2)\right)=0. \tag{12.7}$$

Lemma 12.2. *If the spectral structure of F_p changes across S, then we have*

$$\{\xi_0,\varphi_2\}^2-\{\varphi_1,\varphi_2\}^2=0\ \text{ on }\ S.$$

Therefore we have $\{\xi_0,\varphi_2\}^2-\{\varphi_1,\varphi_2\}^2<0$ *in* $\Sigma\setminus S$*, and hence* p *is noneffectively hyperbolic in* $\Sigma\setminus S$ *with* $\mathrm{Ker}\,F_p^2\cap\mathrm{Im}\,F_p^2=\{0\}$ *and noneffectively hyperbolic on* S *with* $\mathrm{Ker}\,F_p^2\cap\mathrm{Im}\,F_p^2\neq\{0\}$.

Proof. Note that if $\{\xi_0,\varphi_2\}^2-\{\varphi_1,\varphi_2\}^2\neq 0$ at $\rho\in S$, then we have $\{\xi_0,\varphi_2\}^2-\{\varphi_1,\varphi_2\}^2<0$ because otherwise $F_p(\rho)$ would have nonzero real eigenvalues. By continuity F_p has still nonzero pure imaginary eigenvalues near ρ on Σ, and then the spectral structure of F_p does not change near ρ on Σ. This proves the first assertion. Since

$$F_p^2H_{\varphi_2}=(\{\xi_0,\varphi_2\}^2-\{\varphi_1,\varphi_2\}^2)H_{\varphi_2}$$

it is clear that $0\neq H_{\varphi_2}\in\mathrm{Ker}\,F_p^2\cap\mathrm{Im}\,F_p^2$ if $\{\xi_0,\varphi_2\}^2-\{\varphi_1,\varphi_2\}^2=0$ and we conclude the assertion. □

Remark 12.1. If $\{\xi_0,\varphi_2\}^2-\{\varphi_1,\varphi_2\}^2>0$ in $\Sigma\setminus S$ and $\{\xi_0,\varphi_2\}^2-\{\varphi_1,\varphi_2\}^2=0$ on S, that is, p is effectively hyperbolic in $\Sigma\setminus S$ and noneffectively hyperbolic on S with $\mathrm{Ker}\,F_p^2\cap\mathrm{Im}\,F_p^2\neq\{0\}$, the well-posedness question seems to be much harder to answer.

Let $i\mu(\rho)$ ($\mu(\rho)>0$), $\rho\in\Sigma\setminus S$ be the pure imaginary eigenvalue of $F_p(\rho)$. Then from (12.7) one has

$$\mu(\rho)^2=\{\varphi_1,\varphi_2\}^2-\{\xi_0,\varphi_2\}^2 \tag{12.8}$$

and hence $\mu(\rho)^2$ extends smoothly on Σ. From (12.7) and Lemma 12.2, one can write

$$\{\xi_0,\varphi_2\}^2-\{\varphi_1,\varphi_2\}^2=-\theta^{2m}+c_1\varphi_1+c_2\varphi_2 \tag{12.9}$$

with some $m\in\mathbb{N}$ in a neighborhood of ρ where S is defined by $\{\theta=0\}\cap\Sigma$ and $d\theta\neq 0$ on S. Recall that

$$\{\xi_0,\varphi_2\}^2-\{\varphi_1,\varphi_2\}^2=\{\xi_0+\varphi_1,\varphi_2\}\{\xi_0-\varphi_1,\varphi_2\} \tag{12.10}$$

and hence we have either $\{\xi_0+\varphi_1,\varphi_2\}=0$ or $\{\xi_0-\varphi_1,\varphi_2\}=0$ on S. Since the arguments are completely parallel we assume that

$$\{\xi_0-\varphi_1,\varphi_2\}=0\quad\text{on}\quad S$$

and hence one can assume $\{\xi_0,\varphi_2\}=\{\varphi_1,\varphi_2\}>0$ on S without restrictions. Thus we can write

$$\{\xi_0-\varphi_1,\varphi_2\}=-\tilde{\theta}^{2m}+c_1'\varphi_1+c_2'\varphi_2 \tag{12.11}$$

near ρ. From (12.8), (12.10) and (12.11) we have

$$2\{\varphi_1,\varphi_2\}\tilde{\theta}^{2m}=\mu^2\quad\text{on}\quad S$$

then for any $\varepsilon>0$, there is a neighborhood of ρ where we have

$$(1-\varepsilon)\mathrm{Tr}^+F_p\leq\sqrt{2\{\varphi_1,\varphi_2\}}|\tilde{\theta}|^m\leq(1+\varepsilon)\mathrm{Tr}^+F_p. \tag{12.12}$$

Define

$$\hat{\theta} = \tilde{\theta} - \frac{\{\varphi_1, \tilde{\theta}\}}{\{\varphi_1, \varphi_2\}}\varphi_2 + \frac{\{\varphi_2, \tilde{\theta}\}}{\{\varphi_1, \varphi_2\}}\varphi_1 \tag{12.13}$$

so that

$$S = \{\hat{\theta} = 0\} \cap \Sigma, \ \{\hat{\theta}, \varphi_j\} = O(|\varphi|), \ j = 1, 2.$$

Note that (12.11) and (12.12) still holds with $\hat{\theta}$. Here we note

Lemma 12.3. *We have $C_\rho \cap T_\rho S = \{0\}$ if and only if $\{\xi_0, \hat{\theta}\}(\rho) \neq 0$. Hence there is no null bicharacteristic tangent to S if $\{\xi_0, \hat{\theta}\} \neq 0$.*

Proof. Note that

$$C_\rho \subset (T_\rho \Sigma)^\sigma = \{X \mid \sigma(X,Y) = 0, \forall Y \in \Gamma_\rho\} = \langle H_{\xi_0}, H_{\varphi_1}, H_{\varphi_2} \rangle. \tag{12.14}$$

Indeed otherwise there were $X \in C_\rho$ such that $\sigma(X, W) \neq 0$ with some $W \in T_\rho \Sigma$. On the other hand since $\Gamma_\rho + T_\rho \Sigma \subset \Gamma_\rho$ we have

$$\sigma(X, Z + tW) = \sigma(X, Z) + t\sigma(X, W) \leq 0$$

for any $Z \in \Gamma_\rho$ and any $t \in \mathbb{R}$. This would give a contradiction and hence (12.14).

Let $X \in C_\rho \cap T_\rho S$ so that $X = \alpha H_{\xi_0} + \beta H_{\varphi_1} + \gamma H_{\varphi_2} \in T_\rho S$. Since $d\xi_0(X) = 0$, $d\varphi_j(X) = 0$, $d\hat{\theta}(X) = 0$, we have $X = 0$ if $\{\xi_0, \hat{\theta}\}(\rho) \neq 0$. Conversely assume that $\{\xi_0, \hat{\theta}\}(\rho) = 0$. Then for any $X = \alpha H_{\xi_0} + \beta H_{\varphi_1} + \gamma H_{\varphi_2} + \delta H_{\hat{\theta}} \in (T_\rho S)^\sigma$, one has

$$d\xi_0(X)^2 \leq d\varphi_1(X)^2 + d\varphi_2(X)^2$$

and hence $X \notin \Gamma_\rho$ so that $\Gamma_\rho \cap (T_\rho S)^\sigma = \emptyset$. This proves $C_\rho \cap T_\rho S \neq \{0\}$. □

12.2 Case $C \cap TS = \{0\}$

In this section we discuss the C^∞ well-posedness of the Cauchy problem assuming $C \cap TS = \{0\}$. We also assume that $|\mu(\rho)|$ vanishes "simply" on S, that is with some constant $C > 0$ we have

$$C^{-1}\mathrm{dist}(\rho, S) \leq |\mu(\rho)| \leq C\mathrm{dist}(\rho, S) \tag{12.15}$$

on Σ where $\mathrm{dist}(\rho, S)$ denotes the distance on Σ from ρ to S. From (12.12) the condition (12.15) is equivalent to $m = 1$. From Lemma 12.3 we may assume

$$\{\xi_0, \hat{\theta}\}(\rho) > 0 \tag{12.16}$$

without restrictions.

We make a dilation of the coordinate x_0; $x_0 \to \mu x_0$ with small $\mu > 0$ so that we have

$$P(x,\xi,\mu)=\mu^2P(\mu x_0,x',\mu^{-1}\xi_0,\xi')$$
$$=p(\mu x_0,x',\xi_0,\mu\xi')+\mu P_1(\mu x_0,x',\xi_0,\mu\xi')+\mu^2P_0(\mu x_0,x')$$
$$=p(x,\xi,\mu)+P_1(x,\xi,\mu)+P_0(x,\mu).$$

In what follows we often write such symbols dropping μ. Let

$$g_0=\mu^2dx_0^2+|dx'|^2+\langle\xi'\rangle_\mu^{-2}|d\xi'|^2,\ \ \langle\xi'\rangle_\mu^2=\mu^{-2}+|\xi'|^2$$

then it is clear that $a(\mu x_0,x',\mu\xi')=a(x,\xi',\mu)\in S(\langle\mu\xi'\rangle^m,g_0)$ if $a(x,\xi')\in S^m_{1,0}$. Note that

$$\langle\mu\xi'\rangle=\mu\langle\xi'\rangle_\mu,\ \ \langle\xi'\rangle_\mu^{-1}=\mu\langle\mu\xi'\rangle^{-1}.$$

We assume that our assumptions are satisfied globally and hence one can assume

$$\begin{cases}p(x,\xi)=-\xi_0^2+\varphi_1(x,\xi')^2+\varphi_2(x,\xi')^2,\ \ \varphi_j\in S(\langle\mu\xi'\rangle,g_0)\\ \{\xi_0-\varphi_1,\varphi_1\}=c_1\varphi_1+c_2\varphi_2,\ \ c_j\in\mu S(1,g_0),\\ \{\xi_0-\varphi_1,\varphi_2\}=-\mu\hat\theta^2\langle\mu\xi'\rangle+c_1\varphi_1+c_2\varphi_2,\ \ c_j\in\mu S(1,g_0),\\ \{\varphi_1,\varphi_2\}\ge c\mu\langle\mu\xi'\rangle,\ \ c>0\end{cases}\tag{12.17}$$

where $\hat\theta\in S(1,g_0)$ is defined by (12.13) so that

$$\{\hat\theta,\varphi_j\}=c_{j1}\varphi_1+c_{j2}\varphi_2,\ \ c_{jk}\in\mu S(\langle\mu\xi'\rangle^{-1},g).\tag{12.18}$$

Recall that

$$P(x,D)=(p+P_{sub})^w+S(1,g_0)$$

where $p=-\xi_0^2+\varphi_1^2+\varphi_2^2$. We assume that the strict Ivrii-Petkov-Hörmander condition is satisfied:

$$\mathsf{Im}\,P_{sub}=0,\quad |\mathsf{Re}\,P_{sub}|\le\mu(1-\varepsilon^*)\mathrm{Tr}^+F_p\tag{12.19}$$

on Σ with some $\varepsilon^*>0$. Since

$$\varepsilon(\hat\theta,\varphi)=\sup\left(|\hat\theta|+|\varphi_1\langle\mu\xi'\rangle^{-1}|+|\varphi_2\langle\mu\xi'\rangle^{-1}|\right)$$

can be assumed to be sufficiently small working in a small neighborhood of ρ and then extending the reference symbols globally, taking (12.18) and (12.12) into account, one can assume

$$\begin{cases}\{\xi_0\pm\lambda,\hat\theta\}\ge c\mu,\ \ c>0\\ \sqrt{\mu}(1-\varepsilon^*/4)\mathrm{Tr}^+F_p\le\langle\mu\xi'\rangle^{1/2}\sqrt{2\{\varphi_1,\varphi_2\}}|\hat\theta|\\ \qquad\le\sqrt{\mu}(1+\varepsilon^*/4)\mathrm{Tr}^+F_p,\\ 60|\{\varphi_j,\hat\theta\}|^2\le\varepsilon^*\{\xi_0+\lambda,\hat\theta\}\{\xi_0+\lambda,\hat\theta\}.\end{cases}\tag{12.20}$$

Let us put $P_{sub} = T_1 + iT_2$ with real $T_i \in \mu S(\langle \mu\xi' \rangle, g_0)$; then from (12.19) one can write

$$T_1 = \mu c_0 \hat{\theta} \langle \mu\xi' \rangle + c_{11}\varphi_1 + c_{12}\varphi_2, \quad T_2 = c_{21}\varphi_1 + c_{22}\varphi_2$$

with $c_0 \in S(1, g_0)$, $c_{ij} \in \mu S(1, g_0)$. Let us rewrite p as

$$\begin{aligned}
p &= -(\xi_0 + \varphi_1)(\xi_0 - \varphi_1) + \varphi_2^2 \\
&= -(\xi_0 + \varphi_1 - a\hat{\theta}^2\varphi_1 - \varphi_1^3 \langle \mu\xi' \rangle^{-2})(\xi_0 - \varphi_1 + a\hat{\theta}^2\varphi_1 + \varphi_1^3 \langle \mu\xi' \rangle^{-2}) \\
&\quad + \varphi_2^2 + 2a\hat{\theta}^2\varphi_1^2(1 - a\hat{\theta}^2/2) + 2\varphi_1^4 \langle \mu\xi' \rangle^{-2}(1 - a\hat{\theta}^2 - \varphi_1^2 \langle \mu\xi' \rangle^{-2}/2) \\
&= -(\xi_0 + \lambda)(\xi_0 - \lambda) + \varphi_2^2 + \tilde{a}^2\hat{\theta}^2\varphi_1^2 + \tilde{b}^2\varphi_1^4 \langle \mu\xi' \rangle^{-2} \qquad (12.21) \\
&= -(\xi_0 + \lambda)(\xi_0 - \lambda) + q
\end{aligned}$$

where

$$a = \frac{\mu \langle \mu\xi' \rangle}{\{\varphi_1, \varphi_2\}} \geq c_1 > 0 \qquad (12.22)$$

and

$$\begin{cases}
\lambda = \varphi_1 - a\hat{\theta}^2\varphi_1 - \varphi_1^3 \langle \mu\xi' \rangle^{-2}, \\
q = \varphi_2^2 + \tilde{a}^2\hat{\theta}^2\varphi_1^2 + \tilde{b}^2\varphi_1^4 \langle \mu\xi' \rangle^{-2}, \\
\tilde{a} = \sqrt{2a(1 - a\hat{\theta}^2/2)}, \quad \tilde{b} = \sqrt{2(1 - a\hat{\theta}^2 - \varphi_1^2 \langle \mu\xi' \rangle^{-2}/2)}.
\end{cases}$$

As in [10], we move $(c_{11} + ic_{21})\varphi_1$ into the principal part:

$$\begin{aligned}
&-(\xi_0 + \lambda)(\xi_0 - \lambda) + (c_{11} + ic_{21})\varphi_1 \\
&= -(\xi_0 + \lambda + \alpha)(\xi_0 - \lambda - \alpha) - 2\alpha(\hat{\theta}^2\varphi_1 + \varphi_1^3 \langle \mu\xi' \rangle^{-2}) - \alpha^2
\end{aligned}$$

where $\alpha = (c_{11} + ic_{21}) \in \mu S(1, g_0)$. Thus we are led to consider $P = -M\Lambda + Q$ with $M = \xi_0 + \lambda + \alpha$, $\Lambda = \xi_0 - \lambda - \alpha$ and $Q = q + T_1 + iT_2$ where

$$\begin{cases}
T_1 = \mu c_0 \hat{\theta} \langle \mu\xi' \rangle + d_0\hat{\theta}\varphi_1 + d_1\varphi_1^2 \langle \mu\xi' \rangle^{-1} + d_2\varphi_2, \\
T_2 = d_0'\hat{\theta}\varphi_1 + d_1'\varphi_1^2 \langle \mu\xi' \rangle^{-1} + d_2'\varphi_2.
\end{cases} \qquad (12.23)$$

Lemma 12.4. *We have*

$$\begin{aligned}
&\{\xi_0 - \varphi_1 + a\hat{\theta}^2\varphi_1, \hat{\theta}\} \geq \mu c > 0, \ c > 0, \\
&\{\xi_0 - \varphi_1 + a\hat{\theta}^2\varphi_1, \varphi_1\} = c_1\varphi_1 + c_2\varphi_2, \ c_j \in \mu S(1, g_0), \\
&\{\xi_0 - \varphi_1 + a\hat{\theta}^2\varphi_1, \varphi_2\} = c_1'\varphi_1 + c_2'\varphi_2, \ c_j' \in \mu S(1, g_0).
\end{aligned}$$

Proof. The first assertion is obvious because $\{\varphi_1, \hat{\theta}\} = c_1\varphi_1 + c_2\varphi_2$ with $c_j \in \mu S(\langle \mu\xi' \rangle^{-1}, g_0)$ and $\varepsilon(\hat{\theta}, \varphi)$ is small enough. The second assertion is also clear. Note

$$\{\xi_0-\varphi_1+a\hat{\theta}^2\varphi_1,\varphi_2\}=\{\xi_0-\varphi_1,\varphi_2\}+a\hat{\theta}^2\{\varphi_1,\varphi_2\}+\{a\hat{\theta}^2,\varphi_2\}\varphi_1$$
$$=-\mu\hat{\theta}^2\langle\mu\xi'\rangle+a\hat{\theta}^2\{\varphi_1,\varphi_2\}+O(|\varphi|)=\sum c_j\varphi_j$$

by (12.22) and hence the third assertion. □

Remark 12.2. In the case $\{\varphi_1,\varphi_2\}<0$ we take

$$a=-\frac{\mu\langle\mu\xi'\rangle}{\{\varphi_1,\varphi_2\}}>0$$

so that

$$\{\xi_0+\varphi_1-a\hat{\theta}^2\varphi_1,\varphi_2\}=\{\xi_0+\varphi_1,\varphi_2\}-a\hat{\theta}^2\{\varphi_1,\varphi_2\}+O(|\varphi|)$$
$$=-\mu\hat{\theta}^2\langle\mu\xi'\rangle+O(|\varphi|)-a\hat{\theta}^2\{\varphi_1,\varphi_2\}=O(|\varphi|).$$

Throughout the paper we use the same letter to denote a symbol and the operator if there is no confusion. We denote by $\varphi\psi$ the product of symbols φ and ψ and by $\varphi(\psi u)$ the composition of the operators φ and ψ:

$$\varphi\psi u=(\varphi\psi)^w u,\quad \varphi(\psi u)=(\varphi\#\psi)^w u.$$

Here we check that

Lemma 12.5. *We have*

$$\mu^{3/2}\|\langle\mu\xi'\rangle^{1/4}u\|^2\le c^{-1}(1+c^{-1})\big(\|\varphi_2 u\|^2+\|\tilde{b}\varphi_1^2\langle\mu\xi'\rangle^{-1}u\|^2\big)+C\|u\|^2$$

with some $C>0$.

Proof. Note that

$$\{\varphi_1\langle\mu\xi'\rangle^{-1/2},\varphi_2\}=\{\varphi_1,\varphi_2\}\langle\mu\xi'\rangle^{-1/2}+\varphi_1\langle\mu\xi'\rangle^{-1/2}T,\ \ T\in\mu S(1,g_0)$$

which proves

$$(i[\varphi_1\langle\mu\xi'\rangle^{-1/2},\varphi_2]u,u)\ge(\{\varphi_1,\varphi_2\}\langle\mu\xi'\rangle^{-1/2}u,u)$$
$$-C\mu\|\varphi_1\langle\mu\xi'\rangle^{-1/2}u\|^2-C\mu\|u\|^2.$$

Since $\{\varphi_1,\varphi_2\}\langle\mu\xi'\rangle^{-1/2}\ge c\mu\langle\mu\xi'\rangle^{1/2}$ this shows

$$c\mu^{3/2}\|\langle\mu\xi'\rangle^{1/4}u\|^2\le\|\varphi_2 u\|^2+\mu\|\varphi_1\langle\mu\xi'\rangle^{-1/2}u\|^2+C\mu^{3/2}\|u\|^2.$$

Note

$$\{\varphi_1^3\langle\mu\xi'\rangle^{-2},\varphi_2\}=3\varphi_1^2\{\varphi_1,\varphi_2\}\langle\mu\xi'\rangle^{-2}+\varphi_1^3\langle\mu\xi'\rangle^{-2}T,\ \ T\in\mu S(1,g_0)$$

and hence

$$(i[\varphi_1^3\langle\mu\xi'\rangle^{-2},\varphi_2]u,u)\geq 3(\varphi_1^2\{\varphi_1,\varphi_2\}\langle\mu\xi'\rangle^{-2}u,u)$$
$$-C\mu\|\varphi_1^3\langle\mu\xi'\rangle^{-2}u\|^2-C\mu\|u\|^2.$$

Then we get

$$3c\mu(\varphi_1^2\langle\mu\xi'\rangle^{-1}u,u)\leq\|\varphi_2 u\|^2+\|\varphi_1^3\langle\mu\xi'\rangle^{-2}u\|^2$$
$$+C\mu(\|\varphi_1^3\langle\mu\xi'\rangle^{-2}u\|^2+\|u\|^2).$$

Since

$$\|\varphi_1^3\langle\mu\xi'\rangle^{-2}u\|^2=([(\varphi_1^3\langle\mu\xi'\rangle^{-2})\#(\varphi_1^3\langle\mu\xi'\rangle^{-2})]u,u)$$
$$\leq([(\tilde{b}\varphi_1^2\langle\mu\xi'\rangle^{-1})\#(\tilde{b}\varphi_1^2\langle\mu\xi'\rangle^{-1})]u,u)+C\mu^2\|u\|^2$$

which follows from the Fefferman–Phong (Theorem 18.6.8 in [8]) inequality because

$$\tilde{b}^2\varphi_1^4\langle\mu\xi'\rangle^{-2}-\varphi_1^6\langle\mu\xi'\rangle^{-4}\geq 0$$

can be assumed, thus we get the assertion. □

Let us define

$$w=\sqrt{\hat{\theta}^2+\langle\xi'\rangle_\mu^{-3/4}},\;\;\varphi=\hat{\theta}+w,\;\;W=e^{\ell\log\varphi}=\varphi^\ell$$

where $\ell\in\mathbb{R}$ will be determined later. Then it is easy to check that

$$w\in S(w,w^{-2}(\mu^2dx_0^2+|dx'|^2+\langle\xi'\rangle_\mu^{-2}|d\xi'|^2))$$

and

$$w^{-2}(\mu^2dx_0^2+|dx'|^2+\langle\xi'\rangle_\mu^{-2}|d\xi'|^2)\leq\langle\xi'\rangle_\mu^{3/4}|dx|^2+\langle\xi'\rangle_\mu^{-5/4}|d\xi'|^2=\bar{g}.$$

Lemma 12.6. *We have*

$$\varphi\in S(\varphi,g)$$

and hence $\varphi^k\in S(\varphi^k,g)$ *for any* $k\in\mathbb{R}$.

Proof. Note that

$$\partial_{x,\xi'}^\alpha\varphi=\frac{\partial_{x,\xi'}^\alpha\hat{\theta}}{w}\varphi+\frac{\partial_{x,\xi'}^\alpha\langle\xi'\rangle_\mu^{-3/4}}{2w}$$

for $|\alpha|=1$ and $\langle\xi'\rangle_\mu^{-3/4}\leq 2w\varphi$. Since $\hat{\theta}\in S(1,g_0)\subset S(1,g)$ and $w^{-1}\in S(w^{-1},g)$ we get the desired estimates for higher derivatives by induction on $|\alpha|$. □

12.2.1 A Priori Estimates

We follow the arguments in [13] to derive weighted a priori etimates. Consider

$$PW = W\tilde{P}, \quad \tilde{P} = -\tilde{M}\tilde{\Lambda} + \tilde{Q}.$$

We note that

Lemma 12.7. *Let $a \in S(m, g_0)$. Then we have*

$$a\#W - W\#a = i\ell w^{-1}\{\hat{\theta}, a\}W + W\#R$$

with some $R \in S(m\langle\xi'\rangle_\mu^{-1}, \bar{g}) \subset \mu S(m\langle\mu\xi'\rangle^{-1}, \bar{g})$.

Proof. Since $a \in S(m, g_0)$ and $\varphi^\ell \in S(\varphi^\ell, g)$ we have

$$a\#W - W\#a = \frac{1}{i}\{a, \varphi^\ell\} + R_1, \;\; R_1 \in S(m\varphi^\ell\langle\xi'\rangle_\mu^{-15/8}, g).$$

Note that

$$\begin{aligned}\{a, \varphi^\ell\} = \ell\{a, \varphi\}\varphi^{\ell-1} &= \ell\varphi^\ell\Big(w^{-1}\{a, \hat{\theta}\} + \frac{\{a, \langle\xi'\rangle_\mu^{-3/4}\}}{2w\varphi}\Big)\\ &= \ell w^{-1}\varphi^\ell\{a, \hat{\theta}\} + R_2, \;\; R_2 \in S(m\varphi^\ell\langle\xi'\rangle_\mu^{-1}, g)\end{aligned}$$

since $2\varphi w \geq \langle\xi'\rangle_\mu^{-3/4}$ so that

$$a\#W - W\#a = i\ell w^{-1}\{\hat{\theta}, a\}W + \tilde{R}, \;\; \tilde{R} \in S(m\varphi^\ell\langle\xi'\rangle_\mu^{-1}, g).$$

Since $\varphi^{-\ell} \in S(\varphi^{-\ell}, g)$ we have $WW^{-1} = 1 - R$ with $R \in \mu^{1/4}S(\langle\mu\xi'\rangle^{-1/4}, g) \subset \mu^{1/4}S(\langle\mu\xi'\rangle^{-1/4}, \bar{g})$. Then it follows from [1] that there exists $\tilde{W} \in S(\varphi^{-\ell}, \bar{g})$ for $0 < \mu \leq \mu_0$ such that

$$W\#\tilde{W} = \tilde{W}\#W = 1.$$

Thus we can write $\tilde{R} = W\#(\tilde{W}\#\tilde{R}) = W\#R$ with $R \in S(m\langle\xi'\rangle_\mu^{-1}, \bar{g})$ which proves the assertion. □

We now study $\Lambda\#W$.

Lemma 12.8. *We have*

$$\Lambda\#W = W\#\big(\Lambda - i\ell w^{-1}\{\xi_0 - \lambda, \hat{\theta}\} + R_1 + R_2\big)$$

where $R_1 \in \mu S(\langle\xi'\rangle_\mu^{1/8}, \bar{g})$ is real and $R_2 \in \mu^{9/8}S(1, \bar{g})$.

Proof. From Lemma 12.7 we have

$$\Lambda\#W - W\#\Lambda = -i\ell w^{-1}\{\Lambda, \hat{\theta}\}W + W\#R, \;\; R \in \mu S(1, \bar{g}).$$

Noting that

$$W\#(w^{-1}\{\Lambda,\hat{\theta}\}) = w^{-1}\{\Lambda,\hat{\theta}\}W + \tilde{R}_1 + \tilde{R}_2$$

where $\tilde{R}_1 \in \mu S(\varphi^\ell \langle \xi' \rangle_\mu^{1/8}, g)$ is pure imaginary and $\tilde{R}_2 \in \mu^{9/8} S(\langle \mu\xi' \rangle^{-1/8}, g)$ because $w^{-1}\{\Lambda,\hat{\theta}\} \in \mu S(w^{-1}, g)$. Writing $\tilde{R}_j = W\#(\tilde{W}\#\tilde{R}_j) = W\#R_j$ we get the desired assertion. □

We turn to study $q\#W$.

Lemma 12.9. *We have*

$$\begin{aligned} q\#W - W\#q = W\#\big[& i\{\ell w^{-1}\{q,\hat{\theta}\} + c_1'\varphi_1 + c_2'\varphi_2 + c_3'\hat{\theta}\varphi_2 + c_4'\varphi_1^2\langle \mu\xi' \rangle^{-1}\} \\ & + \{\tilde{c}\varphi_1\varphi_2 + \tilde{c}_1\varphi_1 + \tilde{c}_2\varphi_2 + \tilde{c}_3\hat{\theta}\varphi_1 + \tilde{c}_4\varphi_1^2\langle \mu\xi' \rangle^{-1}\} + r\big] \end{aligned}$$

where

$$c_1' \in \mu S(\langle \xi' \rangle_\mu^{-1/2}, g),\ c_j' \in \mu S(1,g),\ \tilde{c} \in S(w^{-3}\langle \xi' \rangle_\mu^{-2}, g),\ \tilde{c}_j \in \mu S(w^{-2}\langle \xi' \rangle_\mu^{-1}, g)$$

which are real and $r \in \mu^2 S(w^{-1}, \bar{g})$.

Proof. Note that $q\#W - W\#q = -i\{q,\varphi^\ell\} + r$ with $r \in \mu^2 S(\varphi^\ell \langle \xi' \rangle_\mu^{1/8}, g)$ because $q \in S(\langle \mu\xi' \rangle^2, g_0)$ and $\varphi^\ell \in S(\varphi^\ell, g)$. Hence we have

$$q\#W = W\#(q+\tilde{r}) - i\{q,\varphi^\ell\}$$

with $\tilde{r} \in \mu S(\langle \xi' \rangle_\mu^{1/8}, \bar{g})$. Recall that

$$\begin{aligned} \{q,\varphi^\ell\} &= \ell\frac{\{q,\hat{\theta}\}}{w}\varphi^\ell + \ell\frac{\{q,\langle \xi' \rangle_\mu^{-3/4}\}}{2w}\varphi^{\ell-1} \\ &= \ell\varphi^\ell\Big(\frac{\{q,\hat{\theta}\}}{w} + \frac{\{q,\langle \xi' \rangle_\mu^{-3/4}\}}{2w\varphi}\Big), \\ q &= \varphi_2^2 + \tilde{a}^2\hat{\theta}^2\varphi_1^2 + \tilde{b}^2\varphi_1^4\langle \mu\xi' \rangle^{-2} \end{aligned}$$

with $\tilde{a}, \tilde{b} \in S(1, g_0)$. Since $\{\varphi_j, \hat{\theta}\} = c_{j1}\varphi_1 + c_{j2}\varphi_2$ with $c_{ji} \in S(\langle \xi' \rangle_\mu^{-1}, g)$ it is easy to see that

$$\{q,\hat{\theta}\} = c_1\varphi_1\varphi_2 + c_2\varphi_2^2 + c_3\hat{\theta}^2\varphi_1^2 + c_4\varphi_1^4\langle \mu\xi' \rangle^{-2}$$

with real $c_j \in S(\langle \xi' \rangle_\mu^{-1}, g)$. □

We prepare a lemma which will be used frequently in the following.

Lemma 12.10. *Let* $a_i \in S(m_i, g)$ *and* $\psi_i \in S(\langle \mu\xi' \rangle, g_0)$ *be real. Then*

$$a_1\#(a_2\psi_1\psi_2),\ \ (a_1\psi_1)\#(a_2\psi_2),\ \ a_1\#(\psi_1\psi_2)\#a_2$$

can be written as

$$a_1a_2\psi_1\psi_2 + c\psi_1\psi_2 + c_1\psi_1 + c_2\psi_2 + r \\ + i\{c'\psi_1\psi_2 + c_1'\psi_1 + c_2'\psi_2 + r'\} + R$$

where

$c \in S(m_1m_2w^{-4}\langle\xi'\rangle_\mu^{-2}, g)$, $c_i \in \mu S(m_1m_2w^{-3}\langle\xi'\rangle_\mu^{-1}, g)$ *and* $r \in \mu^2 S(m_1m_2w^{-2}, g)$,

$c' \in S(m_1m_2w^{-2}\langle\xi'\rangle_\mu^{-1}, g)$, $c_j' \in \mu S(m_1m_2w^{-1}, g)$ *and* $r' \in \mu^2 S(m_1m_2w^{-2}\langle\xi'\rangle_\mu^{-1/4}, g)$

which are real and $R \in \mu^2 S(m_1m_2\langle\xi'\rangle_\mu, g)$.

Proof. Recall

$$a_1\#(a_2\psi_1\psi_2) = a_1a_2\psi_1\psi_2 \\ + \sum_{j=1}^{3} \frac{1}{(2i)^j} \sum_{|\beta'+\beta''+\alpha'+\alpha''|=j} \frac{(-1)^{|\beta'+\beta''|}}{\alpha'!\beta'!\alpha''!\beta''!} a_{1(\beta'+\beta'')}^{(\alpha'+\alpha'')} a_{2(\alpha')}^{(\beta')} (\psi_1\psi_2)_{(\alpha'')}^{(\beta'')} + r$$

where $r \in \mu^2 S(m_1m_2\langle\xi'\rangle_\mu, g)$. It is easy to see that the second and fourth terms in the right-hand side can be written

$$i\{c\psi_1\psi_2 + c_1\psi_1 + c_2\psi_2 + r\}$$

with $r \in \mu^2 S(m_1m_2w^{-2}\langle\xi'\rangle_\mu^{-1/4}, g)$ and

$$c \in S(m_1m_2w^{-2}\langle\xi'\rangle_\mu^{-1}, g), \quad c_j \in \mu S(m_1m_2w^{-1}, g)$$

and the third term in the right-hand side is written

$$\{c\psi_1\psi_2 + c_1\psi_1 + c_2\psi_2 + r\}$$

where $r \in \mu^2 S(m_1m_2w^{-2}, g)$ and

$$c \in S(m_1m_2w^{-4}\langle\xi'\rangle_\mu^{-2}, g), \; c_j \in \mu S(m_1m_2w^{-3}\langle\xi'\rangle_\mu^{-1}, g).$$

These prove the assertion.

Let $a \in S(\langle\xi'\rangle_\mu^{-1}, g)$. From Lemma 12.10 we get

$$\varphi^\ell\#(w^{-1}a\varphi_1\varphi_2) = \varphi^\ell\{w^{-1}a\varphi_1\varphi_2 + c\varphi_1\varphi_2 + c_1\varphi_1 + c_2\varphi_2\} \\ + i\varphi^\ell\{c'\varphi_1\varphi_2 + c_1'\varphi_1 + c_2'\varphi_2\} + R$$

where $c \in S(w^{-1}\langle\xi'\rangle_\mu^{-3/2}, g)$, $c_j \in \mu S(\langle\xi'\rangle_\mu^{-1/2}, g)$, $c' \in S(w^{-3}\langle\xi'\rangle_\mu^{-2}, g)$, $c_j' \in \mu S(w^{-2}\langle\xi'\rangle_\mu^{-1}, g)$ and $R \in \mu^2 S(\varphi^\ell w^{-1}, g)$. From Lemma 12.10 again it follows that

$$\varphi^\ell\#(c'\varphi_1\varphi_2) = \varphi^\ell\{c'\varphi_1\varphi_2 + c_1\varphi_1 + c_2\varphi_2\} \\ + i\varphi^\ell\{c''\varphi_1\varphi_2 + c_1'\varphi_1 + c_2'\varphi_2\} + R$$

where $c_j \in \mu S(w^{-2}\langle\xi'\rangle_\mu^{-9/8}, g)$, $c'' \in \mu S(w^{-1}\langle\xi'\rangle_\mu^{-3/2}, g)$, $c'_j \in \mu^2 S(\langle\xi'\rangle_\mu^{-1/2}, g)$, $R \in \mu^2 S(\varphi^\ell w^{-1}\langle\xi'\rangle_\mu^{-1/4}, g)$, and

$$\begin{aligned}\varphi^\ell \#(c\varphi_1\varphi_2) = \varphi^\ell\{c\varphi_1\varphi_2 + c_1\varphi_1 + c_2\varphi_2\}\\ + i\varphi^\ell\{c'_1\varphi_1 + c'_2\varphi_2\} + R\end{aligned}$$

with $c_j \in \mu S(\langle\xi'\rangle_\mu^{-5/8}, g)$, $c'_j \in \mu S(w^{-2}\langle\xi'\rangle_\mu^{-9/8}, g)$, and $R \in \mu^2 S(\varphi^\ell\langle\xi'\rangle_\mu^{-1/8}, g)$. On the other hand it is easy to see that for $c_j \in \mu S(w^{-2}\langle\xi'\rangle_\mu^{-1}, g)$

$$\varphi^\ell \#(c\varphi_j) = \varphi^\ell\left[c_j\varphi_j + ic'_j\varphi_j\right] + r$$

with $c'_j \in S(\langle\xi'\rangle_\mu^{-1/2}, g)$, $r \in \mu^2 S(\varphi^\ell w^{-1}\langle\xi'\rangle_\mu^{-1/8}, g)$ and for $c_j \in \mu S(\langle\xi'\rangle_\mu^{-1/4}, g)$

$$\varphi^\ell \#(c_j\varphi_j) = c_j\varphi^\ell\varphi_j + r$$

where $r \in \mu^2 S(\varphi^\ell w^{-1}\langle\xi'\rangle_\mu^{-1/8}, g)$. Thus we can write

$$\begin{aligned}w^{-1}\varphi^\ell a\varphi_1\varphi_2 &= \varphi^\ell \#\big[\{w^{-1}a\varphi_1\varphi_2 + c\varphi_1\varphi_2 + c_1\varphi_1 + c_2\varphi_2\}\\ &\qquad + i\{\tilde{c}\varphi_1\varphi_2 + \tilde{c}_1\varphi_1 + \tilde{c}_2\varphi_2\} + r\big]\\ &= \varphi^\ell \#\big[\{w^{-1}a\varphi_1\varphi_2 + c_1\varphi_1 + c'_2\varphi_2\}\\ &\qquad + i\{\tilde{c}\varphi_1\varphi_2 + \tilde{c}_1\varphi_1 + \tilde{c}_2\varphi_2\} + r\big]\end{aligned} \tag{12.24}$$

where we have $c_1 \in \mu S(\langle\xi'\rangle_\mu^{-1/2}, g)$, $c'_2 \in \mu S(1, g)$, $\tilde{c} \in S(w^{-3}\langle\xi'\rangle_\mu^{-2}, g)$, $\tilde{c}_j \in \mu S(w^{-2}\langle\xi'\rangle_\mu^{-1}, g)$, and $r \in \mu^2 S(w^{-1}, \bar{g})$. Applying the same reasoning we obtain similar estimates as (12.24) for φ_2^2, $(\hat{\theta}\varphi_1)^2$, $(\varphi_1^2\langle\mu\xi'\rangle^{-1})^2$.

Since it is easy to see

$$\{q, \langle\xi'\rangle_\mu^{-3/4}\} = c_1\hat{\theta}\varphi_1 + c_2\varphi_2 + c_3\varphi_1^2\langle\mu\xi'\rangle^{-1}$$

with $c_j \in \mu S(w\varphi, g)$, then we have

$$\begin{aligned}\frac{\{q, \langle\xi'\rangle_\mu^{-3/4}\}}{2w}\varphi^{\ell-1} &= \varphi^\ell \#\big[(c'_1\hat{\theta}\varphi_1 + c'_2\varphi_2 + c'_3\varphi_1^2\langle\mu\xi'\rangle^{-1})\\ &\qquad + i(\tilde{c}_1\hat{\theta}\varphi_1 + \tilde{c}_2\varphi_2 + \tilde{c}_3\varphi_1^2\langle\mu\xi'\rangle^{-1}) + r\big]\end{aligned}$$

with $c'_j \in \mu S(1, g)$, $\tilde{c}_j \in \mu S(w^{-2}\langle\xi'\rangle_\mu^{-1}, g)$ and $r \in \mu^2 S(w^{-1}, \bar{g})$. This together with (12.24) proves the assertion. □

Let us write $\tilde{M} = D_0 - \tilde{m}(x, D')$, $\tilde{\Lambda} = D_0 - \tilde{\lambda}(x, D')$; then we have

Proposition 12.1. *With $\tilde{\Lambda}_\theta = \tilde{\Lambda} - i\theta$, $\tilde{M}_\theta = \tilde{M} - i\theta$, we have*

$$2\mathsf{Im}(\tilde{P}_\theta u, \tilde{\Lambda}_\theta u) \geq \frac{d}{dx_0}(\|\tilde{\Lambda}_\theta u\|^2 + ((\mathsf{Re}\,\tilde{Q})u,u) + \theta^2\|u\|^2) + \theta\|\tilde{\Lambda}_\theta u\|^2$$
$$+2\theta\mathsf{Re}(\tilde{Q}u,u) + 2((\mathsf{Im}\,\tilde{m})\tilde{\Lambda}_\theta u, \tilde{\Lambda}_\theta u) + 2\mathsf{Re}(\tilde{\Lambda}_\theta u, (\mathsf{Im}\,\tilde{Q})u)$$
$$+\mathsf{Im}([D_0 - \mathsf{Re}\tilde{\lambda}, \mathsf{Re}\,\tilde{Q}]u,u) + 2\mathsf{Re}((\mathsf{Re}\,\tilde{Q})u, (\mathsf{Im}\tilde{\lambda})u)$$
$$+\theta^3\|u\|^2 + 2\theta^2((\mathsf{Im}\tilde{\lambda})u,u).$$

Proof. See Proposition 4.1 in [3]. □

Recall that

$$\mathsf{Im}\,\tilde{\lambda} = \ell w^{-1}\{\xi_0 - \lambda, \hat{\theta}\} + R, \quad \mathsf{Im}\,\tilde{m} = \ell w^{-1}\{\xi_0 + \lambda, \hat{\theta}\} + R'$$

with $R, R' \in \mu S(1,\bar{g})$. Since $\{\xi_0 \pm \lambda, \hat{\theta}\} \geq c\mu$ with some $c > 0$, we see that

$$((\mathsf{Im}\,\tilde{\lambda})u,u) \geq c\ell\mu(w^{-1}u,u) - C\mu\|u\|^2. \tag{12.25}$$

The same arguments show

$$((\mathsf{Im}\,\tilde{m})\tilde{\Lambda}_\theta u, \tilde{\Lambda}_\theta u) \geq \ell(w^{-1}\{\xi_0 + \lambda, \hat{\theta}\}\tilde{\Lambda}_\theta u, \tilde{\Lambda}_\theta u) - C\mu\|\tilde{\Lambda}_\theta u\|^2$$
$$\geq c\ell\mu(w^{-1}\tilde{\Lambda}_\theta u, \tilde{\Lambda}_\theta u) - C\mu\|\tilde{\Lambda}_\theta u\|^2 \tag{12.26}$$

Let us consider $(w^{-1}\tilde{\Lambda}_\theta u, \tilde{\Lambda}_\theta u)$. Note that

$$-\mathsf{Im}\,(\tilde{\Lambda}_\theta u, u) = \frac{d}{dx_0}\|u\|^2 + \theta\|u\|^2 + (\mathsf{Im}\,\tilde{\lambda}\,u,u)$$
$$\geq \frac{d}{dx_0}\|u\|^2 + \frac{\theta}{2}\|u\|^2 + c\ell\mu(w^{-1}u,u) \tag{12.27}$$

for $\theta \geq \theta_0$. In particular

$$\theta\|\tilde{\Lambda}_\theta u\|^2 \geq \theta^2\frac{d}{dx_0}\|u\|^2 + c\theta^3\|u\|^2 + c\mu\ell\theta^2(w^{-1}u,u)$$

with some $c > 0$.

Proposition 12.2. *We have*

$$((\mathsf{Im}\,\tilde{m})\tilde{\Lambda}_\theta u, \tilde{\Lambda}_\theta u) + C\mu\ell^2\|\tilde{\Lambda}_\theta u\|^2 \geq c\mu^2\ell^2\frac{d}{dx_0}\|w^{-1}u\|^2$$
$$+c\mu^2\ell^2\theta\|w^{-1}u\|^2 + c\mu^3\ell^3(w^{-3}u,u) - C\|u\|^2,$$

$$\theta\|\tilde{\Lambda}_\theta u\|^2 \geq \theta^2\frac{d}{dx_0}\|u\|^2 + c\theta^3\|u\|^2 + c\mu\ell\theta^2(w^{-1}u,u)$$

with $c > 0$ and $C = C(\ell,\mu)$.

Proof. Replacing u by $w^{-1}u$ in (12.27), we have

$$-\operatorname{Im}(\tilde{\Lambda}_\theta w^{-1}u, w^{-1}u) \geq \frac{d}{dx_0}\|w^{-1}u\|^2$$
$$+\frac{\theta}{2}\|w^{-1}u\|^2 + c\ell\mu(w^{-1}(w^{-1}u), w^{-1}u). \tag{12.28}$$

Here we prepare a lemma.

Lemma 12.11. *Let $T \in S(w^p, g)$, $p \leq 0$. Then one can write*

$$T = w^{p/2}\#\tilde{T}\#w^{p/2} + r, \quad \tilde{T}, r \in S(1, g).$$

In particular we have

$$|(Tu, u)| \leq C\|w^{p/2}u\|^2 + C\|u\|^2$$

with $C > 0$.

Proof. Since $w^{p/2}\#(w^{-p}T)\#w^{p/2} = T + R_1$ with $R_1 \in S(w^{p-2}\langle\xi'\rangle_\mu^{-1}, g)$. Write $R_1 = w^{p/2}\#T_1\#w^{p/2} + R_2$ with $T_1 \in S(w^{-2}\langle\xi'\rangle_\mu^{-1}, g)$, $R_2 \in S(w^{p-4}\langle\xi'\rangle_\mu^{-2}, g)$. Repeating this argument we conclude that one can write

$$T = w^{p/2}\#\tilde{T}\#w^{p/2} + R, \quad \tilde{T}, R \in S(1, g)$$

which is the assertion. □

Write

$$(\tilde{\Lambda}_\theta w^{-1}u, w^{-1}u) = (w^{-1}\tilde{\Lambda}_\theta u, w^{-1}u) + ([\tilde{\Lambda}_\theta, w^{-1}]u, w^{-1}u)$$

and note that

$$[\tilde{\Lambda}_\theta, w^{-1}] = c + r, \quad r \in \mu S(w^{-1}\langle\mu\xi'\rangle^{-1/4}, g)$$

where $c \in \mu S(w^{-2}, g)$ is independent of ℓ. Applying Lemma 12.11 one has

$$-\operatorname{Im}(\tilde{\Lambda}_\theta w^{-1}u, w^{-1}u) \leq -\operatorname{Im}(w^{-1}\tilde{\Lambda}_\theta u, w^{-1}u)$$
$$+c_1\mu\|w^{-3/2}u\|^2 + C(\|w^{-1}u\|^2 + \|u\|^2).$$

We now estimate $\operatorname{Im}(w^{-1}\tilde{\Lambda}_\theta u, w^{-1}u)$. Note

$$-\operatorname{Im}(w^{-1}\tilde{\Lambda}_\theta u, w^{-1}u) \leq -\operatorname{Im}(w^{-1/2}\tilde{\Lambda}_\theta u, w^{-1/2}w^{-1}u)$$
$$+C\mu^{9/8}(\|\tilde{\Lambda}_\theta u\|^2 + \|u\|^2)$$
$$\leq (c\mu\ell)^{-1}\|w^{-1/2}\tilde{\Lambda}_\theta u\|^2 + \frac{c\mu\ell}{2}\|w^{-1/2}w^{-1}u\|^2$$
$$+C\mu^{9/8}(\|\tilde{\Lambda}_\theta u\|^2 + \|u\|^2).$$

Since $w^{-1}\#w^{-1/2}\#w^{-1/2}\#w^{-1} = w^{-3} + r$ with $r \in \mu^{9/8}S(\langle\mu\xi'\rangle^{-1/8}, g)$ then $-\operatorname{Im}(\tilde{\Lambda}_\theta w^{-1}u, w^{-1}u)$ is bounded by

$$(c\mu\ell)^{-1}\|w^{-1/2}\tilde{\Lambda}_\theta u\|^2+(\frac{c\mu\ell}{2}+c_1\mu)(w^{-3}u,u)$$
$$+C\big(\|\tilde{\Lambda}_\theta u\|^2+\|w^{-1}u\|^2+\|u\|^2\big) \tag{12.29}$$

with some $C=C(\ell,\mu)$. We turn to check $(w^{-1}(w^{-1}u),w^{-1}u)$. Repeating similar arguments we see

$$(w^{-1}(w^{-1}u),w^{-1}u)\geq(w^{-3}u,u)-C\mu^{9/8}(\|w^{-1}u\|^2+\|u\|^2).$$

This together with (12.29) and (12.28) shows

$$(c\mu\ell)^{-1}\|w^{-1/2}\tilde{\Lambda}_\theta u\|^2+C\|\tilde{\Lambda}_\theta u\|^2\geq\frac{d}{dx_0}\|w^{-1}u\|^2$$
$$+c\theta\|w^{-1}u\|^2+\frac{c\mu\ell}{3}(w^{-3}u,u)-C\|u\|^2 \tag{12.30}$$

for $\ell\geq\ell_0$, $\theta\geq\theta_0(\mu,\ell)$ with $C=C(\mu,\ell)$. Remark that

$$\|w^{-1/2}\tilde{\Lambda}_\theta u\|^2\leq(w^{-1}\tilde{\Lambda}_\theta u,\tilde{\Lambda}_\theta u)+C\mu^{1/8}\|\tilde{\Lambda}_\theta u\|^2$$

and multiply $\mu^2\ell^2$ to both sides of (12.30), we get the assertion from (12.26). □

We now estimate the term $\mathrm{Re}(\mathrm{Re}\,\tilde{Q}u,\mathrm{Im}\tilde{\lambda}u)$.

Lemma 12.12. *We have*

$$\mathrm{Re}(\mathrm{Re}\,\tilde{Q}u,\mathrm{Im}\tilde{\lambda}u)+C\mu^3\ell(w^{-3}u,u)$$
$$\geq((\mathrm{Im}\tilde{\lambda}\,\mathrm{Re}\,\tilde{Q})u,u)-C\big(\|w^{-1}u\|^2+\|u\|^2\big)$$

with $C=C(\mu,\ell)$.

Proof. Recall that

$$\mathrm{Re}\,\tilde{Q}=q+Q_1+T_1+R$$

where $q=\varphi_2^2+\tilde{a}^2\hat{\theta}^2\varphi_1^2+\tilde{b}^2\varphi_1^4\langle\mu\xi'\rangle^{-2}\in S(\langle\mu\xi'\rangle^2,g_0)$, $R\in\mu^2S(w^{-1},\bar{g})$ and

$$T_1=\mu c_0\hat{\theta}\langle\mu\xi'\rangle+c_2\varphi_2+c_3\hat{\theta}\varphi_1+c_4\varphi_1^2\langle\mu\xi'\rangle^{-1} \tag{12.31}$$

where $c_j\in\mu S(1,g_0)$, $j\geq 2$ and

$$Q_1=\tilde{c}\varphi_1\varphi_2+\tilde{c}_1\varphi_1+\tilde{c}_2\varphi_2+\tilde{c}_3\hat{\theta}\varphi_1+\tilde{c}_4\varphi_1^2\langle\mu\xi'\rangle^{-1} \tag{12.32}$$

where $\tilde{c}\in\mu S(w^{-3}\langle\xi'\rangle_\mu^{-2},g)$ and $\tilde{c}_j\in\mu S(\langle\xi'\rangle_\mu^{-1/4},g)$. Note $\mathrm{Re}(\mathrm{Im}\tilde{\lambda}\#q)=\mathrm{Im}\tilde{\lambda}\,q+r$ where $r\in\ell\mu^3S(w^{-3},g)$ because $\mathrm{Im}\tilde{\lambda}\in\mu\ell S(w^{-1},g)$. Hence we see

$$\mathrm{Re}(qu,\mathrm{Im}\tilde{\lambda}u)=((\mathrm{Im}\tilde{\lambda}\,q)u,u)+\mathrm{Re}(ru,u). \tag{12.33}$$

Let us consider

$$K\mu^3\ell w^{-3}+r=\mu^3\ell[w^{-3/2}(K+\ell^{-1}\mu^{-3}w^{-3}r)^{1/2}]$$
$$\#[w^{-3/2}(K+\ell^{-1}\mu^{-3}w^{-3}r)^{1/2}]+R$$

with $R\in S(w^{-7}\langle\xi'\rangle_\mu^{-2},g)\subset\mu^{1/8}S(w^{-2}\langle\mu\xi'\rangle^{-1/8},g)$ where we choose $K>0$ large so that $K+\ell^{-1}\mu^{-3}w^{-3}r\ge c>0$. Thanks to Lemma 12.11, we get

$$K\mu^3\ell(w^{-3}u,u)+(ru,u)\ge -C\big(\|w^{-1}u\|^2+\|u\|^2\big). \tag{12.34}$$

Since

$$\mathsf{Re}(\mathsf{Im}\tilde{\lambda}\#T_1)=\mathsf{Im}\tilde{\lambda}\,T_1+R,\;\; R\in\mu^3S(w^{-1}\langle\xi'\rangle_\mu^{-1/4},g),$$
$$\mathsf{Re}(\mathsf{Im}\tilde{\lambda}\#Q_1)=\mathsf{Im}\tilde{\lambda}\,Q_1+c+R,\;\; R\in\mu^3S(w^{-2},g)$$

with $c\in\mu^3\ell S(w^{-3}\langle\xi'\rangle_\mu^{-1/8},g)$, it follows from Lemma 12.11 that

$$\mathsf{Re}(Q_1u,\mathsf{Im}\tilde{\lambda}u)\ge(\mathsf{Im}\tilde{\lambda}Q_1u,u)-c\mu^{25/8}\ell(w^{-3}u,u)-C(\|w^{-1}u\|^2+\|u\|^2),$$
$$\mathsf{Re}(T_1u,\mathsf{Im}\tilde{\lambda}u)\ge(\mathsf{Im}\tilde{\lambda}\,T_1u,u)-C(\|w^{-1}u\|^2+\|u\|^2).$$

Thus from (12.33) and (12.34) we conclude the assertion. □

We now estimate $(\mathsf{Re}\tilde{Q}u,u)$ from below. From the assumption we have

$$|T_1|\le\mu(1-\varepsilon^*)\mathrm{Tr}^+F_p\quad\text{on}\quad\Sigma.$$

Since $T_1=\mu c_0\hat{\theta}\langle\mu\xi'\rangle+c_2\varphi_2+c_3\hat{\theta}\varphi_1+c_4\varphi_1^2\langle\mu\xi'\rangle^{-1}$ we have

$$|\mu c_0\hat{\theta}\langle\mu\xi'\rangle|\le\mu(1-\varepsilon^*)\mathrm{Tr}^+F_p\le\sqrt{\mu}\frac{1-\varepsilon^*}{1-\varepsilon^*/4}\langle\mu\xi'\rangle^{1/2}\sqrt{2\{\varphi_1,\varphi_2\}}|\hat{\theta}|$$

that is,

$$|c_0|\le(1-\frac{\varepsilon^*}{2})\mu^{-1/2}\langle\mu\xi'\rangle^{-1/2}\sqrt{2\{\varphi_1,\varphi_2\}}.$$

Lemma 12.13. *There exists $C>0$ such that*

$$\mu|(c_0\hat{\theta}\langle\mu\xi'\rangle u,u)|\le(1-\varepsilon^*/4)\big(\|\varphi_2u\|^2+\|\tilde{a}\hat{\theta}\varphi_1u\|^2$$
$$+C\mu\|\tilde{b}\varphi_1^2\langle\mu\xi'\rangle^{-1}u\|^2\big)+C\|u\|^2.$$

Proof. Let us set

$$k=\frac{c_0\mu\langle\mu\xi'\rangle}{\tilde{a}\{\varphi_1,\varphi_2\}}.$$

Then it is clear that one can assume $|k|\le(1-\varepsilon^*/2)$ because $1-a\hat{\theta}^2/2$ is enough close to 1. Note that

$$\{\tilde{a}\hat{\theta}\varphi_1,k\varphi_2\}=\mu c_0\hat{\theta}\langle\mu\xi'\rangle+C_0\hat{\theta}\varphi_1+C_1\varphi_1^2\langle\mu\xi'\rangle^{-1}+C_2\varphi_2$$

with $C_j \in \mu S(1,g_0)$, thanks to (12.18). Thus we get

$$(1+\nu)\|k\varphi_2 u\|^2+(1+\nu)^{-1}\|\tilde{a}\hat{\theta}\varphi_1 u\|^2 \geq |\mu(c_0\hat{\theta}\langle\mu\xi'\rangle u,u)|$$
$$-C\mu\left(\|\varphi_2 u\|^2+\|\tilde{a}\hat{\theta}\varphi_1 u\|^2+\|\tilde{b}\varphi_1^2\langle\mu\xi'\rangle^{-1}u\|^2\right)-C\|u\|^2.$$

Choosing $\nu>0$ so that $(1+\nu)(1-\varepsilon^*/2)=(1+\nu)^{-1}$, we get the desired assertion.□

Proposition 12.3. *There exists $c>0$ such that we have*

$$(\mathsf{Re}\tilde{Q}u,u)\geq\frac{\varepsilon^*}{8}\big(\|\varphi_2 u\|^2+\|\tilde{a}\hat{\theta}\varphi_1 u\|^2+\|\tilde{b}\varphi_1^2\langle\mu\xi'\rangle^{-1}u\|^2$$
$$+c\mu^{3/2}\|\langle\mu\xi'\rangle^{1/4}u\|^2\big)-C\|u\|^2.$$

Proof. Since q is a sum of terms ψ^2 with $\psi\in S(\langle\mu\xi'\rangle,g_0)$, the estimates for q is easy. With real $c\in S(w^{-3}\langle\xi'\rangle_\mu^{-2},g)$, we have

$$c\varphi_1\varphi_2=\mathsf{Re}((c\varphi_1)\#\varphi_2)+r,\ \ r\in\mu^2 S(\langle\xi'\rangle_\mu^{-1/8},g)$$

where $c\varphi_1\in\mu S(\langle\xi'\rangle_\mu^{1/8},g)$ and hence

$$(c\varphi_1\varphi_2 u,u)\geq-C\mu\|\varphi_2 u\|^2-C\mu^2\|\langle\mu\xi'\rangle^{1/4}u\|^2-C\|u\|^2.$$

On the other hand as for real $c\in\mu S(\langle\xi'\rangle_\mu^{-1/4},g)$, we have

$$(c\varphi_1 u,u)\geq-C\mu^{1/2}\|\varphi_1^2\langle\mu\xi'\rangle^{-1}u\|^2-C\mu^{5/2}\|\langle\mu\xi'\rangle^{1/4}u\|^2-C\|u\|^2. \quad (12.35)$$

Indeed one has $c\varphi_1=\mathsf{Re}(c\langle\mu\xi'\rangle^{1/2})\#(\langle\mu\xi'\rangle^{-1/2}\varphi_1)+r$ with $r\in\mu^2 S(1,g)$ and

$$\|\varphi_1\langle\mu\xi'\rangle^{-1/2}u\|^2\leq C\|\varphi_1^2\langle\mu\xi'\rangle^{-1}u\|^2+C\|u\|^2,$$
$$\|c\langle\mu\xi'\rangle^{1/2}u\|^2\leq C\mu^{5/2}\|\langle\mu\xi'\rangle^{1/4}u\|^2.$$

Therefore one has

$$\mathsf{Re}(Q_1 u,u)\geq-C\mu^{1/2}(\|\varphi_2 u\|^2+\|\tilde{a}\hat{\theta}\varphi_1 u\|^2+\|\tilde{b}\varphi_1^2\langle\mu\xi'\rangle^{-1}u\|^2)$$
$$-C\mu^{5/2}\|\langle\mu\xi'\rangle^{1/4}u\|^2-C\|u\|^2.$$

Writing

$$((\mathsf{Re}\tilde{Q})u,u)=([(1-\frac{\varepsilon^*}{4})q+Q_1+T_1]u,u)+\frac{\varepsilon^*}{4}(qu,u)$$

we get the assertion from Lemmas 12.5 and 12.13. □

We next estimate $((\mathsf{Im}\tilde{\lambda}\,\mathsf{Re}\tilde{Q})u,u)$ from below. To simplify notations let us set

$$\rho=\sqrt{(\mu\ell)^{-1}\mathsf{Im}\tilde{\lambda}}\in S(w^{-1/2},g)$$

where we have $c_1 \le w^{1/2}\rho \le c_2$ with some $c_i > 0$ and

$$\rho \text{ is uniformly bounded in } S(w^{-1/2}, g) \text{ with respect to } \ell \text{ and } \mu.$$

Remark that

Lemma 12.14. *Let $p \ge 0$. Then we have*

$$c_1' \|w^{-p/2}u\|^2 - C\|u\|^2 \le \|\rho^p u\|^2 \le c_2' \|w^{-p/2}u\|^2 + C\|u\|^2$$

with some $c_i' > 0$, $C > 0$.

Proof. Since one can write

$$\mathrm{Re}(w^{-p/2}\# w^{-p/2}) = \rho^p \#(\rho^{-2p}w^{-p} + R_1)\#\rho^p + R_2$$

with $R_1 \in \mu^{1/2}S(1,g)$ and $R_2 \in S(1,g)$, then the proof is immediate. □

Then we have

Lemma 12.15. *We have*

$$\begin{aligned}\mathrm{Re}(\rho^2 \mathrm{Re}\tilde{Q}u, u) \ge (1 - C\mu^{1/2})(\mathrm{Re}\tilde{Q}(\rho u), \rho u) - C\mu^2 \|w^{-3/2}u\|^2 \\ -C\mu^{3/2}(\|w^{-1}u\|^2 + \|\langle \mu\xi'\rangle^{1/4}u\|^2) \\ -C(\|\varphi_2 u\|^2 + \|\hat{\theta}\varphi_1 u\|^2 + \|\varphi_1^2 \langle \mu\xi'\rangle^{-1}u\|^2 + \|u\|^2).\end{aligned}$$

Proof. Recalling $\mathrm{Re}\tilde{Q} = q + Q_1 + T_1 + R$, we first consider q. Since q is a sum of terms $\rho^2\psi^2$ with $\psi \in S(\langle \mu\xi'\rangle, g_0)$, one can write from Lemma 12.10

$$\rho^2\psi^2 = \rho\#\psi^2\#\rho + c_1\psi^2 + c_2\#(\psi\#\rho) + r$$

with $c_1 \in S(\langle \xi'\rangle_\mu^{-1/8}, g)$, $c_2 \in \mu S(w^{-1/2}\langle \xi'\rangle_\mu^{1/8}, g)$, and $r \in \mu^2 S(w^{-3}, g)$. Here we note

$$(c_1\psi^2 u, u) \le C\|\psi u\|^2 + C\mu^2(\|w^{-1}u\|^2 + \|\langle \mu\xi'\rangle^{1/4}u\|^2) + C\|u\|^2$$

and

$$\|c_2 u\|^2 \le (c_2^2 u, u) + C\|u\|^2 \le C\mu^{7/4}(\|w^{-1}u\|^2 + \|\langle \mu\xi'\rangle^{1/4}u\|^2) + C\|u\|^2.$$

To estimate $\mathrm{Re}(ru, u)$ it suffices to apply Lemma 12.11. We turn to estimate the term $(\rho^2 Q_1 u, u)$. Let $c \in S(w^{-3}\langle \xi'\rangle_\mu^{-2}, g)$ be real. Then

$$\rho^2 c\varphi_1\varphi_2 = \rho\#(c\varphi_1\varphi_2)\#\rho + r, \;\; r \in \mu^2 S(w^{-3}\langle \xi'\rangle_\mu^{-1/8}, g)$$

and hence

$$(\rho^2 c\varphi_1\varphi_2 u, u) \ge (c\varphi_1\varphi_2(\rho u), \rho u) - C\mu^{17/8}\|w^{-3/2}u\|^2 - C\|u\|^2.$$

With real $c \in \mu S(\langle\xi'\rangle_\mu^{-1/4}, g)$ and $\psi \in S(\langle\mu\xi'\rangle, g_0)$, it is easy to see that $c\psi\rho^2 = \rho\#(c\psi)\#\rho) + r$ with $r \in \mu^2 S(w^{-2}\langle\xi'\rangle_\mu^{-1/8}, g)$ so that

$$(c\varphi_1\rho^2 u, u) \geq (c\varphi_1(\rho u), \rho u) - C\mu^2(\|w^{-1}u\|^2 + \|u\|^2)$$

thanks to Lemma 12.11. Repeating similar arguments we get

$$(\rho^2 T_1 u, u) \geq (T_1(\rho u), \rho u) - C\mu^2(\|w^{-1}u\|^2 + \|u\|^2)$$

and hence we obtain the desired assertion. □

Since $\mathrm{Im}\tilde{\lambda}\,\mathrm{Re}\tilde{Q} = \mu\ell\rho^2\mathrm{Re}\tilde{Q}$ we get from Lemmas 12.12 and 12.15

Proposition 12.4. *We have*

$$\begin{aligned}&\mathrm{Re}(\mathrm{Re}\tilde{Q}u, \mathrm{Im}\tilde{\lambda}u) + C\mu^3\ell(w^{-3}u, u)\\ &\geq \mu\ell(1 - C\mu^{1/4})(\mathrm{Re}\tilde{Q}(\rho u), \rho u) - C\mu^{5/2}(\|w^{-1}u\|^2 + \|\langle\mu\xi'\rangle^{1/4}u\|^2)\\ &-C(\|\varphi_2 u\|^2 + \|\tilde{a}\hat{\theta}\varphi_1 u\|^2 + \|\tilde{b}\varphi_1^2\langle\mu\xi'\rangle^{-1}u\|^2 + \|u\|^2).\end{aligned}$$

We proceed to estimate the term $\mathrm{Im}([D_0 - \mathrm{Re}\tilde{\lambda}, \mathrm{Re}\tilde{Q}]u, u)$. Recall that $\mathrm{Re}\tilde{\lambda} = \xi_0 - \lambda + r$ with $r \in \mu S(\langle\xi'\rangle_\mu^{1/8}, g)$ where $\lambda = \varphi_1 - a\hat{\theta}^2\varphi_1$. It is easy to see that

$$\begin{aligned}|(\{r, \mathrm{Re}\tilde{Q}\}u, u)| \leq C\mu^2(&\|\rho\varphi_2 u\|^2 + \|\rho\tilde{a}\hat{\theta}\varphi_1 u\|^2 + \|\rho\tilde{b}\varphi_1^2\langle\mu\xi'\rangle^{-1}u\|^2\\ &+ \|\rho\langle\mu\xi'\rangle^{1/4}u\|^2) + C(\|w^{-1/2}u\|^2 + \|u\|^2).\end{aligned}$$

We now estimate $\mathrm{Im}([D_0 - \mathrm{Re}\tilde{\lambda}, T_1 + Q_1]u, u)$. It is enough to study $(\{\xi_0 - \lambda, T_1 + Q_1\}u, u)$. Recalling (12.31) one can write

$$\{\xi_0 - \lambda, T_1\} = c_0'\hat{\theta}\langle\mu\xi'\rangle + c_1'\varphi_1 + c_2\varphi_2',\ \ c_j' \in \mu^2 S(1, g_0)$$

Thanks to (12.17) and (12.18). Applying the same arguments proving Lemma 12.13, one gets

$$\begin{aligned}|(c_0'\hat{\theta}\langle\mu\xi'\rangle u, u)| \leq C\mu(&\|\varphi_2 u\|^2 + \|\tilde{a}\hat{\theta}\varphi_1 u\|^2 + \|\tilde{b}\varphi_1^2\langle\mu\xi'\rangle^{-1}u\|^2)\\ &+ C\mu^3(\|w^{-1/2}u\|^2 + \|u\|^2).\end{aligned}$$

Similar arguments shows that

$$|(c_1'\varphi_1 u, u)| \leq C\mu(\|\varphi_2 u\|^2 + \|\tilde{b}\varphi_1^2\langle\mu\xi'\rangle^{-1}u\|^2 + \|u\|^2).$$

Thus we have

$$\begin{aligned}|(\{\xi_0 - \lambda, T_1\}u, u)| \leq C\mu(&\|\varphi_2 u\|^2 + \|\tilde{a}\hat{\theta}\varphi_1 u\|^2\\ &+ \|\tilde{b}\varphi_1^2\langle\mu\xi'\rangle^{-1}u\|^2 + \|w^{-1/2}u\|^2 + \|u\|^2).\end{aligned}$$

In virtue of (12.32) and (12.17) we can write

$$\begin{aligned}\{\xi_0-\lambda,Q_1\}&=c\varphi_1\varphi_2+c_0\hat{\theta}\langle\mu\xi'\rangle+c_1\varphi_1\\&\quad+c_2\varphi_2+c_3\hat{\theta}\varphi_1+c_4\varphi_1^2\langle\mu\xi'\rangle^{-1}\end{aligned}$$

with $c\in\mu S(w^{-1}\langle\xi'\rangle_\mu^{-5/4},g)$, $c_0\in\mu^2S(\langle\xi'\rangle_\mu^{-1/4},g)$, $c_j\in\mu^2S(w^{-1}\langle\xi'\rangle_\mu^{-1/4},g)$, $j=1,...,4$. Write

$$c\varphi_1\varphi_2=\mathsf{Re}((c\varphi_1)\#\varphi_2)+r,\;\; r\in\mu^3S(\langle\xi'\rangle_\mu^{-1/8},g)$$

where $c\varphi_1\in\mu^2S(w^{-1}\langle\xi'\rangle_\mu^{-1/4},g)$ and hence

$$|(c\varphi_1\varphi_2u,u)|\le C\mu^2(\|\varphi_2u\|^2+\|w^{-1/2}u\|^2)+C\|u\|^2.$$

We consider $(c_1\varphi_1u,u)$ with real $c_1\in\mu^2S(w^{-1}\langle\xi'\rangle_\mu^{-1/4},g)$. Write

$$c_1\varphi_1=\mathsf{Re}[(\rho^{3/2}\langle\mu\xi'\rangle^{-1/2}\varphi_1)\#(\rho^{-3/2}c_1\langle\mu\xi'\rangle^{1/2})]+r$$

where $r\in\mu^3S(w^{-2},g)$. Here note that

$$\begin{aligned}&\mathsf{Re}[(\rho^{3/2}\langle\mu\xi'\rangle^{-1/2}\varphi_1)\#(\rho^{3/2}\langle\mu\xi'\rangle^{-1/2}\varphi_1)]\\&=\mathsf{Re}[(\rho\tilde{b}\langle\mu\xi'\rangle^{-1}\varphi_1^2)\#c]+r,\;\; r\in\mu S(w^{-2},g),\;\; c\in S(w^{-1},g)\end{aligned}$$

and

$$\begin{aligned}&\mathsf{Re}[(\rho^{-3/2}c_1\langle\mu\xi'\rangle^{1/2})\#(\rho^{-3/2}c_1\langle\mu\xi'\rangle^{1/2})]\\&=(\rho\langle\mu\xi'\rangle^{1/4})\#(\rho^{-4}c_1^2\langle\mu\xi'\rangle^{3/4})+r,\;\; r\in\mu^3S(w^{-2},g).\end{aligned}$$

Since $\rho^{-4}c_1^2\langle\mu\xi'\rangle^{3/4}\in\mu^4S(\langle\mu\xi'\rangle^{1/4},g)$ we conclude that

$$\begin{aligned}|(c_1\varphi_1u,u)|&\le C\mu^3(\|\rho\tilde{b}\langle\mu\xi'\rangle^{-1}\varphi_1^2u\|^2+\|\rho\langle\mu\xi'\rangle^{1/4}u\|^2)\\&\quad+C(\|w^{-1}u\|^2+\|\langle\mu\xi'\rangle^{1/4}u\|^2).\end{aligned}$$

Then we conclude that

$$\begin{aligned}|(\{\xi_0-\lambda,Q_1\}u,u)|&\le C\mu(\|\varphi_2u\|^2+\|\tilde{a}\theta\varphi_1u\|^2+\|\tilde{a}\hat{\theta}\varphi_1u\|^2\\&+\|\tilde{b}\varphi_1^2\langle\mu\xi'\rangle^{-1}u\|^2)+C\mu^3(\|\rho\tilde{b}\varphi_1^2\langle\mu\xi'\rangle^{-1}u\|^2+\|\rho\langle\mu\xi'\rangle^{1/4}u\|^2)\\&+C(\|w^{-1}u\|^2+\|u\|^2).\end{aligned}$$

It remains to consider $(\{\xi_0-\lambda,q\}u,u)$. From Lemma 12.4 we see that

$$\{\xi_0-\lambda,q\}=c_1\varphi_1\varphi_2+c_2\hat{\theta}\varphi_1\varphi_1,\;\; c_j\in\mu S(1,g_0).$$

modulo $\mu O(|(\varphi_2,\hat{\theta}\varphi_1,\varphi_1^2\langle\mu\xi'\rangle^{-1})|^2)$. We only estimate the term $(c_1\varphi_1\varphi_2u,u)$ because $(c_2\hat{\theta}\varphi_1\varphi_1u,u)$ is estimated similarly. We first note that

$$\mathsf{Re}(c_1\varphi_1\varphi_2)=\mathsf{Re}((\rho\varphi_2)\#c_1'\#(w^{1/2}\varphi_1))+\sum\tilde{c}_j\varphi_j+r$$

where $c_1' \in \mu S(1,g_0)$, $\tilde{c}_j \in \mu^2 S(w^{-1}\langle\xi'\rangle_\mu^{-1/4},g)$, $r \in \mu^2 S(w^{-2},g)$. To estimate the terms $(\tilde{c}_j\varphi_j u,u)$ it is enough to apply the same arguments as above. To estimate $\|w^{1/2}\varphi_1 u\|$ we apply the following lemma.

Lemma 12.16. *We have*

$$\|w^{1/2}\varphi_1 u\|^2,\ (w\varphi_1^2 u,u) \le C\big([\varphi_2^2+\tilde{a}^2\hat{\theta}^2\varphi_1^2+\tilde{b}^2\varphi_1^4\langle\mu\xi'\rangle^{-2}](\rho u),\rho u\big)$$
$$+C\big(\|w^{-1/2}u\|^2+\|\langle\mu\xi'\rangle^{1/4}u\|^2+\|\varphi_1^2\langle\mu\xi'\rangle^{-1}u\|^2\big)$$

with some $C>0$.

Proof. Since $w^2=\hat{\theta}^2+\langle\xi'\rangle_\mu^{-3/4}=\hat{\theta}^2+\mu^{3/4}\langle\mu\xi'\rangle^{-3/4}$, it follows that

$$w^2\varphi_1^2 \le C(\hat{\theta}^2\varphi_1^2+\varphi_1^4\langle\mu\xi'\rangle^{-2}+\mu^{3/2}\langle\mu\xi'\rangle^{1/2})$$

and hence

$$([\tilde{a}^2\hat{\theta}^2\varphi_1^2+\tilde{b}^2\varphi_1^4\langle\mu\xi'\rangle^{-2}+\mu^{3/2}\langle\mu\xi'\rangle^{1/2}]u,u)-(w^2\varphi_1^2 u,u)\ge -C\|u\|^2$$

by the Fefferman–Phong inequality because $w^2\varphi_1^2\in S(\langle\mu\xi'\rangle^2,g_0)$. Replacing u by ρu we get

$$C([\tilde{a}^2\hat{\theta}^2\varphi_1^2+\tilde{b}^2\varphi_1^4\langle\mu\xi'\rangle^{-2}+\mu^{3/2}\langle\mu\xi'\rangle^{1/2}](\rho u),\rho u)$$
$$\ge (w^2\varphi_1^2(\rho u),\rho u)-C\|\rho u\|^2.$$

Let us study $(w^2\varphi_1^2(\rho u),\rho u)$. Write

$$\mathsf{Re}(\rho\#(w^2\varphi_1^2)\#\rho)=\mathsf{Re}((\rho w\varphi_1)\#(\rho w\varphi_1))+c\varphi_1+r$$

where $c\in\mu S(\langle\xi'\rangle_\mu^{-1/4},g)$, $r\in\mu^2 S(w^{-1},g)$ which proves

$$(w^2\varphi_1^2(\rho u),\rho u)\ge\|\rho w\varphi_1 u\|^2$$
$$-C(\|\varphi_1^2\langle\mu\xi'\rangle^{-1}u\|^2+\|\langle\mu\xi'\rangle^{1/4}u\|^2+\|w^{-1/2}u\|^2).$$

Note that one can write

$$w^{1/2}\varphi_1=c\#(\rho w\varphi_1)+r,\quad c\in S(1,g),\ r\in\mu S(w^{-1/2},g).$$

On the other hand by Lemma 12.5, one has

$$\mu^{3/2}(\langle\mu\xi'\rangle^{1/2}(\rho u),\rho u)$$
$$\le C([\varphi_2^2+\tilde{b}^2\varphi_1^4\langle\mu\xi'\rangle^{-2}](\rho u),\rho u)+C\mu\|w^{-1/2}u\|^2$$

and hence we get the assertion for $\|w^{1/2}\varphi_1 u\|^2$. To examine the assertion for $(w\varphi_1^2 u,u)$, it suffices to note

$$\mathsf{Re}[(w^{1/2}\varphi_1)\#(w^{1/2}\varphi_1)] = w\varphi_1^2 + c\varphi_1 + r$$

with some $c \in \mu S(\langle\xi'\rangle_\mu^{-1/4}, g)$ and $r \in \mu^2 S(w^{-1}, g)$. □

Thus we obtain

$$\begin{aligned}&|\mathsf{Re}(c_1\varphi_1\varphi_2 u, u)|,\ |\mathsf{Re}(c_2\hat{\theta}\varphi_1\varphi_1 u, u)|\\ &\le C\mu\big([\varphi_2^2 + \tilde{a}^2\hat{\theta}^2\varphi_1^2 + \tilde{b}^2\varphi_1^4\langle\mu\xi'\rangle^{-2}](\rho u), \rho u\big)\\ &+C\big(\|w^{-1}u\|^2 + \|\langle\mu\xi'\rangle^{1/4}u\|^2 + \|\tilde{b}\varphi_1^2\langle\mu\xi'\rangle^{-1}u\|^2\big)\end{aligned}$$

with some $C > 0$. Indeed it is enough to note that for $\psi \in S(\langle\mu\xi'\rangle, g_0)$, one has

$$\begin{aligned}C_1(\psi^2(\rho u), \rho u) &\ge \|\rho\psi u\|^2 - C_2\mu^2\|w^{-1}u\|^2 - C_2\mu\|u\|^2,\\ C_1(\langle\mu\xi'\rangle^{1/2}(\rho u), \rho u) &\ge \|\rho\langle\mu\xi'\rangle^{1/4}u\|^2 - C_2\mu^{1/2}\|u\|^2.\end{aligned}$$

We summarize in

Proposition 12.5. *We have*

$$\begin{aligned}&|\mathsf{Im}([D_0 - \mathsf{Re}\tilde{\lambda}, \mathsf{Re}\tilde{Q}]u, u)|\\ &\le C\mu\big(\mathsf{Re}\tilde{Q}(\rho u), \rho u\big) + C(\mathsf{Re}\tilde{Q}u, u) + C\big(\|w^{-1}u\|^2 + \|u\|^2\big).\end{aligned}$$

We finally estimate $(\tilde{\Lambda}_\theta u, (\mathsf{Im}\tilde{Q})u)$. From Lemma 12.9 and (12.23) we have

$$\mathsf{Im}\tilde{Q} = \ell w^{-1}\{q, \hat{\theta}\} + c_1\varphi_1 + c_2\varphi_2 + c_3\hat{\theta}\varphi_1 + c_4\varphi_1^2\langle\mu\xi'\rangle^{-1}$$

$c_1 \in \mu S(\langle\xi'\rangle_\mu^{-1/4}, g)$, $c_j \in \mu S(1, g)$ for $j \ge 2$. Since

$$\begin{aligned}|(\tilde{\Lambda}_\theta u, (\mathsf{Im}\tilde{Q} - \ell w^{-1}\{q, \hat{\theta}\} - c_1\varphi_1)u)| &\le C(\|\tilde{\Lambda}_\theta u\|^2 + \|\varphi_2 u\|^2\\ &+\|\tilde{a}\hat{\theta}\varphi_1 u\|^2 + \|\tilde{b}\varphi_1^2\langle\mu\xi'\rangle^{-1}u\|^2 + \|u\|^2)\end{aligned}$$

it is enough to estimate $(\tilde{\Lambda}_\theta u, \ell w^{-1}\{q, \hat{\theta}\}u)$, $(\tilde{\Lambda}_\theta u, c_1\varphi_1 u)$. Noting

$$c_1\varphi_1 = \tilde{c}\#(w^{1/2}\varphi_1) + r$$

with $\tilde{c} \in \mu S(1, g)$ and $r \in \mu^2 S(w^{-1/2}, g)$, we obtain

$$|(\tilde{\Lambda}_\theta u, c_1\varphi_1 u)| \le C\mu^2\|w^{1/2}\varphi_1 u\|^2 + C(\|\tilde{\Lambda}_\theta u\|^2 + \|w^{-1/2}u\|^2).$$

We turn to estimate $\ell|(\tilde{\Lambda}_\theta u, w^{-1}\{q, \hat{\theta}\}u)|$. We estimate the case $q = \varphi_2^2$ since the other cases are similar. With $a = \sqrt{\ell w^{-1}\{\xi_0 + \lambda, \hat{\theta}\}}$, consider

$$\begin{aligned}2\ell|(\tilde{\Lambda}_\theta u, w^{-1}\{\varphi_2, \hat{\theta}\}\varphi_2 u)| &= 2|(a\tilde{\Lambda}_\theta u, a^{-1}(\ell w^{-1}\{\varphi_2, \hat{\theta}\}\varphi_2 u))|\\ &\le \delta\|a\tilde{\Lambda}_\theta u\|^2 + \delta^{-1}\|a^{-1}(\ell w^{-1}\{\varphi_2, \hat{\theta}\}\varphi_2 u)\|^2\end{aligned}$$

with a $\delta > 0$. Let us study the second term in the right-hand side. Applying the same arguments as in the proof of Lemma 12.10 one can show

$$(\ell w^{-1}\{\varphi_2,\hat{\theta}\}\varphi_2)\#a^{-1}\#a^{-1}\#(\ell w^{-1}\{\varphi_2,\hat{\theta}\}\varphi_2)$$
$$=\mu\ell\rho\#(b\varphi_2^2)\#\rho+\mu\ell c_0\varphi_2^2+\mu^2\ell c_1\varphi_1+\mu^3\ell c_2$$

where

$$b=\frac{\{\varphi_2,\hat{\theta}\}^2}{\{\xi_0-\lambda,\hat{\theta}\}\{\xi_0+\lambda,\hat{\theta}\}},\ c_0\in S(w^{-5}\langle\xi'\rangle_\mu^{-2},g),$$
$$c_1\in S(w^{-4}\langle\xi'\rangle_\mu^{-1},g),\ c_2\in S(w^{-3},g).$$

From the assumption and the Fefferman–Phong inequality we get

$$(b\varphi_2^2(\rho u),\rho u)\le\frac{\varepsilon^*}{60}(\varphi_2^2(\rho u),\rho u)+C\|w^{-1/2}u\|^2.$$

Repeating the same arguments as in the proof of Lemma 12.10 one can write

$$c_0\varphi_2^2=\rho\#\varphi_2\#(c_0\rho^{-2})\#\varphi_2\#\rho+c_{00}\varphi_2+c_{01},$$
$$c_1\varphi_2=\mathrm{Re}\big((\rho^{-1}c_1)\#\varphi_2\#\rho\big)+c_{10}\varphi_2+c_{11}$$

where

$$c_{00}\in\mu S(w^{-1},g),\ c_{01}\in\mu^2S(w^{-2}\langle\xi'\rangle_\mu^{-1/16},g),$$
$$c_{10}\in S(1,g),\ c_{11}\in\mu S(w^{-2}\langle\xi'\rangle_\mu^{-1/8},g).$$

This proves that

$$|(c_0\varphi_2u,u)|\le C\mu^{1/4}(\varphi_2^2(\rho u),\rho u)+C\|w^{-1}u\|^2,$$
$$\mu^2|(c_1\varphi_2u,u)|\le C\mu^{3/2}(\varphi_2^2(\rho u),\rho u)+C\mu^{5/2}\|\langle\mu\xi'\rangle^{1/4}(\rho u)\|^2$$
$$+C(\|\varphi_2u\|^2+\|w^{-1}u\|^2).$$

Thus we obtain

$$2\ell|(\tilde{\Lambda}_\theta u,w^{-1}\{\varphi_2,\hat{\theta}\}\varphi_2u)|\le\delta^{-1}\mu\ell\big(\frac{\varepsilon^*}{60}+C\mu^{1/4}\big)\big[(\varphi_2^2(\rho u),\rho u)$$
$$+c\mu^{3/2}\|\langle\mu\xi'\rangle^{1/4}(\rho u)\|^2\big]+C\delta^{-1}\mu^3\ell(w^{-3}u,u)$$
$$+C\big[\|\varphi_2^2u\|^2+\|w^{-1}u\|^2+\|u\|^2\big]$$
$$+\delta\ell(w^{-1}\{\xi_0+\lambda,\hat{\theta}\}\tilde{\Lambda}_\theta u,\tilde{\Lambda}_\theta u)+C\|\tilde{\Lambda}_\theta u\|^2.$$

Since we get similar estimates for $\hat{\theta}^2\varphi_1^2$ and $\varphi_1^4\langle\mu\xi'\rangle^{-2}$, taking $\delta=1/2$ for instance, we conclude

Proposition 12.6. *We have*

$$|(\tilde{\Lambda}_\theta u, (\mathsf{Im}\,\tilde{Q})u)| \le \frac{1}{10}\mu\ell(\varepsilon^* + C\mu^{1/4})(\mathsf{Re}\,\tilde{Q}(\rho u), \rho u)$$
$$+C\mu^3\ell(w^{-3}u,u) + C(\mathsf{Re}\,\tilde{Q}u,u) + \frac{1}{2}\ell(w^{-1}\{\xi_0+\lambda,\hat{\theta}\}\tilde{\Lambda}_\theta u, \tilde{\Lambda}_\theta u)$$
$$+C\big(\|\tilde{\Lambda}_\theta u\|^2 + \|w^{-1}u\|^2 + \|u\|^2\big).$$

To obtain a priori estimates we first take $\ell \ge \ell_0$ large so that

$$\mu\ell(\mathsf{Re}\,\tilde{Q}(\rho u), \rho u)$$

in Proposition 12.4 controls $C\mu(\mathsf{Re}\,\tilde{Q}(\rho u), \rho u)$ in Proposition 12.5 and then

$$\frac{c}{2}\mu^3\ell^3(w^{-3}u,u)$$

in Proposition 12.2 absorbs $C\mu^3\ell(w^{-3}u,u)$ in Proposition 12.4 and $C\mu^3\ell(w^{-3}u,u)$ in Proposition 12.6. We next choose μ small $0 < \mu \le \mu_0 = \mu_0(\ell)$ and finally take $\theta \ge \theta_0 = \theta_0(\ell_0, \mu_0)$ large. Then from Propositions 12.2, 12.3, 12.4, 12.5, and 12.6, we obtain a priori estimates which proves

Theorem 12.1. *Assume (12.1) and (12.2). We also assume that the global version of the assumptions (12.15), (12.16), (12.19) are satisfied. Then the Cauchy problem is C^∞ well posed.*

12.3 Case $C \cap TS \neq \{0\}$

In this section we study the C^∞ well-posedness of the Cauchy problem assuming $C \cap TS \neq \{0\}$. From Lemma 12.3 we see that

$$C_\rho \cap T_\rho S \neq \{0\} \Longleftrightarrow \{\xi_0, \hat{\theta}\}(\rho) = 0$$

and hence one can write

$$\{\xi_0, \hat{\theta}\} = c\hat{\theta} + c_1\varphi_1 + c_2\varphi_2. \tag{12.36}$$

Let ψ be a smooth function vanishing on Σ such that

$$F_p^2 H_\psi = -\mu^2 H_\psi, \;\; H_\psi \notin T\Sigma \tag{12.37}$$

where $i\mu(\rho)$ ($\mu(\rho) > 0$) is the pure imaginary eigenvalue of $F_p(\rho)$ and hence $\mu(\rho)^2$ is smooth up to S on Σ as observed in Sect. 12.1. We introduce a condition

$$|H_\psi^3 p| \le C\mathrm{Tr}^+ F_p \quad \text{on} \quad \Sigma \tag{12.38}$$

with some $C > 0$.

Lemma 12.17. *The condition (12.38) is equivalent to*

$$|\{\varphi_2, \Delta\}| \le C\mathrm{Tr}^+ F_p \quad on \ \Sigma$$

where $\Delta = \{\xi_0, \varphi_2\}^2 - \{\varphi_1, \varphi_2\}^2$.

Proof. Let $\psi = \alpha\xi_0 + \beta\varphi_1 + \gamma\varphi_2$. From (12.37) it follows that $\gamma \neq 0$ and (α, β) is proportional to $(\{\varphi_2, \xi_0\}, \{\varphi_1, \varphi_2\})$ with proportionality k. Then we have

$$\{\psi, \{\psi, p\}\} = \Delta(-\gamma^2 + k^2\Delta).$$

Then it is not difficult to examine that

$$H_\psi^3 p = -\gamma^2\{\varphi_2, \Delta\} + O(\hat{\theta}^{2m-1}) \tag{12.39}$$

on Σ. This proves the assertion because of (12.12). □

Remark 12.3. When $m = 1$ the condition (12.38) is equivalent to

$$H_\psi^3 p = 0 \quad \text{on} \quad S.$$

Lemma 12.18. *We assume that there is $C > 0$ such that $|H_\psi^3 p| \le C\mathrm{Tr}^+ F_p$ on Σ. Then we have*

$$\{\xi_0 - \varphi_1, \varphi_2\} = -\hat{\theta}^{2m} + c_0\hat{\theta}^m\varphi_1 + c_1\varphi_1^2 + c_2\varphi_2. \tag{12.40}$$

Proof. Recall that one can write $\{\xi_0 - \varphi_1, \varphi_2\} = -\hat{\theta}^{2m} + c_1\varphi_1 + c_2\varphi_2$ and $\{\xi_0 + \varphi_1, \varphi_2\} \neq 0$. Thus by Lemma 12.17 (or rather by (12.39)), one gets

$$|\{\varphi_2, \{\varphi_2, \xi_0 - \varphi_1\}\}| \le C\mathrm{Tr}^+ F_p.$$

Since $\{\varphi_1, \varphi_2\} \neq 0$ it follows that c_1 is a linear combination of $\hat{\theta}^m$, φ_1 and φ_2 which proves the assertion. □

We assume that our assumptions are satisfied globally:

$$\begin{cases} p(x,\xi) = -\xi_0^2 + \varphi_1(x,\xi')^2 + \varphi_2(x,\xi')^2, \ \ \varphi_j \in S(\langle\xi'\rangle, g_0) \\ \{\xi_0 - \varphi_1, \varphi_2\} = -\hat{\theta}^{2m}\langle\xi'\rangle + c_0\hat{\theta}^m\varphi_1 + c_1\varphi_1^2\langle\xi'\rangle^{-1} + c_2\varphi_2, \\ \{\xi_0, \hat{\theta}\} = c_0'\hat{\theta} + c_1'\varphi_1\langle\xi'\rangle^{-1} + c_2'\varphi_2\langle\xi'\rangle^{-1}, \\ \{\varphi_1, \varphi_2\} \ge c\langle\xi'\rangle, \ \ c > 0, \\ (1-\varepsilon)\mathrm{Tr}^+ F_p \le \langle\xi'\rangle^{1/2}\sqrt{2\{\varphi_1, \varphi_2\}}|\hat{\theta}|^m \le (1+\varepsilon)\mathrm{Tr}^+ F_p. \end{cases} \tag{12.41}$$

We also assume that the strict Ivrii-Petkov-Hörmander condition is satisfied:

$$\mathsf{Im}\, P_{sub} = 0, \quad |\mathsf{Re} P_{sub}| \le (1-\varepsilon^*)\mathrm{Tr}^+ F_p \tag{12.42}$$

on Σ with some $\varepsilon^* > 0$ where we can assume $\varepsilon < \varepsilon^*$ working in a small neighborhood of ρ and then extending the reference symbols globally.

12.3.1 Elementary Decomposition

Let us rewrite p as

$$\begin{aligned} p &= -(\xi_0 + \varphi_1 - a\hat{\theta}^{2m}\varphi_1 - \varphi_1^3\langle\xi'\rangle^{-2})(\xi_0 - \varphi_1 + a\hat{\theta}^{2m}\varphi_1 + \varphi_1^3\langle\xi'\rangle^{-2}) \\ &\quad + \varphi_2^2 + 2a\hat{\theta}^{2m}\varphi_1^2(1 - a\hat{\theta}^{2m}/2) + 2\varphi_1^4\langle\xi'\rangle^{-2}(1 - a\hat{\theta}^{2m} - \varphi_1^2\langle\xi'\rangle^{-2}/2) \\ &= -(\xi_0 + \lambda)(\xi_0 - \lambda) + q \end{aligned} \tag{12.43}$$

where

$$a = \frac{\langle\xi'\rangle}{\{\varphi_1, \varphi_2\}} \geq c_1 > 0$$

and

$$\begin{aligned} \lambda &= \varphi_1 - a\hat{\theta}^{2m}\varphi_1 - \varphi_1^3\langle\xi'\rangle^{-2}, \\ q &= \varphi_2^2 + 2a\hat{\theta}^{2m}\varphi_1^2(1 - a\hat{\theta}^{2m}/2) + 2\varphi_1^4\langle\xi'\rangle^{-2}(1 - a\hat{\theta}^{2m} - \varphi_1^2\langle\xi'\rangle^{-2}/2) \\ &= \varphi_2^2 + \tilde{a}^2\hat{\theta}^{2m}\varphi_1^2 + \tilde{b}^2\varphi_1^4\langle\xi'\rangle^{-2}. \end{aligned}$$

Lemma 12.19. *We have*

$$\begin{aligned} &\{\xi_0 - \varphi_1, \varphi_1\} = O(|(\varphi_1, \varphi_2)|), \\ &\{\xi_0 - \varphi_1 + a\hat{\theta}^{2m}\varphi_1, \hat{\theta}\} = O(|(\hat{\theta}, \varphi_1\langle\xi'\rangle^{-1}, \varphi_2\langle\xi'\rangle^{-1})|), \\ &\{\xi_0 + \varphi_1 - a\hat{\theta}^{2m}\varphi_1, \hat{\theta}\} = O(|(\hat{\theta}, \varphi_1\langle\xi'\rangle^{-1}, \varphi_2\langle\xi'\rangle^{-1})|), \\ &\{\xi_0 - \varphi_1 + a\hat{\theta}^{2m}\varphi_1, \varphi_2\} = O(\hat{\theta}^m\varphi_1) + O(\varphi_1^2\langle\xi'\rangle^{-1}) + O(\varphi_2). \end{aligned}$$

Proof. It is enough to note that

$$\begin{aligned} \{\xi_0 - \varphi_1 + a\hat{\theta}^{2m}\varphi_1, \varphi_2\} &= \{\xi_0 - \varphi_1, \varphi_2\} + a\hat{\theta}^{2m}\{\varphi_1, \varphi_2\} + O(\hat{\theta}^m\varphi_1) \\ &= O(\hat{\theta}^m\varphi_1) + O(\varphi_1^2\langle\xi'\rangle^{-1}) + O(\varphi_2) \end{aligned}$$

because $a\hat{\theta}^{2m}\{\varphi_1, \varphi_2\} = \hat{\theta}^{2m}\langle\xi'\rangle$. □

Then we have

Proposition 12.7. *The decomposition (12.43) is an elementary decomposition, that is,*

$$\{\xi_0 - \lambda, q\} = O(q).$$

Corollary 12.1. *Assume (12.38). Then there is no null bicharacteristic tangent to Σ.*

Proof. It suffices to repeat the proof in [11]. □

Proof (Proposition 12.7). It is clear that

$$\begin{aligned} \{\xi_0 - \varphi_1 + a\hat{\theta}^{2m}\varphi_1, \varphi_2^2\} &= [O(\hat{\theta}^m\varphi_1) + O(\varphi_1^2\langle\xi'\rangle^{-1}) + O(\varphi_2)]\varphi_2 \\ &\leq C(\varphi_2^2 + \hat{\theta}^{2m}\varphi_1^2 + \varphi_1^4\langle\xi'\rangle^{-2}) \leq C'q. \end{aligned}$$

As for the terms $\{\xi_0-\varphi_1+a\hat{\theta}^{2m}\varphi_1,\tilde{a}^2\hat{\theta}^{2m}\varphi_1^2\}$, $\{\xi_0-\varphi_1+a\hat{\theta}^{2m}\varphi_1,\tilde{b}^2\varphi_1^4\langle\xi'\rangle^{-2}\}$, the assertion is easily checked, thanks to Lemma 12.19. □

12.3.2 A Priori Estimates

Recalling $P=(p+P_{sub})^w$ let us consider

$$P=-(\xi_0+\lambda)(\xi_0-\lambda)+q+T_1+iT_2$$

with

$$q=\varphi_2^2+\tilde{a}^2\hat{\theta}^{2m}\varphi_1^2+\tilde{b}^2\varphi_1^4\langle\xi'\rangle^{-2}$$

where $T_i\in S(\langle\xi'\rangle,g_0)$ are real and

$$|T_1|\le(1-\varepsilon^*)\mathrm{Tr}^+F_p,\ \ T_2=c_1\varphi_1+c_2\varphi_2$$

with real $c_j'\in S(1,g_0)$. From this and (12.41) one can write

$$T_1=c_0\hat{\theta}^m\langle\xi'\rangle+c_1'\varphi_1+c_2'\varphi_2,\ \ c_j'\in S(1,g_0)$$

where

$$|c_0\hat{\theta}^m\langle\xi'\rangle|\le(1-\varepsilon^*)\mathrm{Tr}^+F_p\le\frac{1-\varepsilon^*}{1-\varepsilon}\langle\xi'\rangle^{1/2}\sqrt{2\{\varphi_1,\varphi_2\}}|\hat{\theta}^m|$$

which implies

$$|c_0|\le(1-\varepsilon')\langle\xi'\rangle^{-1/2}\sqrt{2\{\varphi_1,\varphi_2\}}$$

with some $\varepsilon'>0$.

As in Sect. 12.2 we move $(c_1'+ic_1)\varphi_1$ into the principal part:

$$\begin{aligned}&-(\xi_0+\lambda)(\xi_0-\lambda)+(c_1'+ic_1)\varphi_1\\&=-(\xi_0+\lambda+\alpha)(\xi_0-\lambda-\alpha)-2\alpha(\hat{\theta}^{2m}\varphi_1+\varphi_1^3\langle\xi'\rangle^{-2})-\alpha^2\end{aligned}$$

where $\alpha=(c_1'+ic_1)/2\in S(1,g_0)$. Thus we are led to consider $P=-M\Lambda+Q$ where $M=\xi_0+\lambda+\alpha$, $\Lambda=\xi_0-\lambda-\alpha$ and $Q=q+T_1'+iT_2'$ with

$$\begin{aligned}&T_1'=c_0\hat{\theta}^m\langle\xi'\rangle+d_0\hat{\theta}^m\varphi_1+d_1\varphi_1^2\langle\xi'\rangle^{-1}+d_2\varphi_2,\\&T_2'=d_0'\hat{\theta}^m\varphi_1+d_1'\varphi_1^2\langle\xi'\rangle^{-1}+d_2'\varphi_2.\end{aligned}$$

where c_0, d_j, $d_j'\in S(1,g_0)$. With $\Lambda_\theta=\Lambda-i\theta$, $M_\theta=M-i\theta$, we have

$$
\begin{aligned}
2\mathsf{Im}(P_\theta u,\Lambda_\theta u) \geq \frac{d}{dx_0}(\|\Lambda_\theta u\|^2 + ((\mathsf{Re}\,Q)u,u) + \theta^2\|u\|^2) + \theta\|\Lambda_\theta u\|^2 \\
+2\theta\mathsf{Re}(Qu,u) - (c_1\Lambda_\theta u,\Lambda_\theta u) + 2\mathsf{Re}(\Lambda_\theta u,(\mathsf{Im}\,Q)u) \\
+\mathsf{Im}([D_0-\lambda-c_1',\mathsf{Re}\,Q]u,u) + \mathsf{Re}((\mathsf{Re}\,Q)u,c_1u) \\
+\theta^3\|u\|^2 + \theta^2(c_1u,u).
\end{aligned}
$$

Lemma 12.20. *There are* $\varepsilon_1 > 0$ *and* $C > 0$ *such that*

$$
\begin{aligned}
|(c_0\hat{\theta}^m\langle\xi'\rangle u,u)| \leq (1-\varepsilon_1)\big(\|\varphi_2 u\|^2 + \|\tilde{a}\hat{\theta}^m\varphi_1 u\|^2 \\
+\|\tilde{b}\varphi_1^2\langle\xi'\rangle^{-1}u\|^2\big) + C\|u\|^2.
\end{aligned}
$$

Proof. We repeat the proof of Lemma 12.13. Let us set

$$
k = \frac{c_0\langle\xi'\rangle}{\tilde{a}\{\varphi_1,\varphi_2\}}
$$

Then it is clear that one can assume $|k| \leq (1-\varepsilon'/2)$ working in a small neighborhood of ρ. Note that

$$
\{\tilde{a}\hat{\theta}^m\varphi_1, k\varphi_2\} = c_0\hat{\theta}^m\langle\xi'\rangle + C_0\hat{\theta}^m\varphi_1 + C_1\varphi_1^2\langle\xi'\rangle^{-1} + C_2\varphi_2.
$$

Thus we get

$$
\begin{aligned}
(1+\nu)\|k\varphi_2 u\|^2 + (1+\nu)^{-1}\|\tilde{a}\hat{\theta}^m\varphi_1 u\|^2 \geq |(c_0\hat{\theta}^m\langle\xi'\rangle u,u)| \\
-\delta\big(\|\varphi_2 u\|^2 + \|\tilde{a}\hat{\theta}^m\varphi_1 u\|^2 + \|\tilde{b}\varphi_1^2\langle\xi'\rangle^{-1}u\|^2\big) - C_\delta\|u\|^2
\end{aligned}
$$

for any $\delta > 0$. Choosing $\nu > 0$ and $\delta > 0$ so that $\delta + (1+\nu)(1-\varepsilon'/2) < 1$ and $\delta + (1+\nu)^{-1} < 1$, we get the desired assertion. □

Theorem 12.2. *Assume (12.1) and (12.2). We also assume that the global version of the assumptions (12.36), (12.38), and (12.42) are satisfied. Then the Cauchy problem is* C^∞ *well-posed.*

12.3.3 Geometric Observations

In this subsection we discuss about geometric aspects of (12.38).

Proposition 12.8. *Assume that* $H_\psi^3 p \neq 0$ *at* $\rho \in S$. *Then there is a null bicharacteristic landing at* ρ *tangentially to* S.

Proof. We follow the arguments given in [14]. To simplify notations let us set $\Xi_0 = \xi_0 - \varphi_1$, $X_0 = x_0$ and extend (X_0,Ξ_0) to a full symplectic coordinates (X,Ξ). Switching the notation from (X,Ξ) to (x,ξ) and writing θ for $\hat{\theta}$, one can write

$$
p = -\xi_0(\xi_0 + 2\varphi_1) + \varphi_2^2
$$

where we have

$$\{\xi_0, \varphi_1\} = O(|\varphi|), \quad \{\xi_0, \varphi_2\} = -\theta^2 + O(|\varphi|),$$
$$\{\theta, \varphi_j\} = O(|\varphi|), \quad \{\xi_0, \theta\} = O(|(\theta, \varphi)|)$$

with $\varphi = (\varphi_1, \varphi_2)$. Let us take

$$\xi_0, x_0, \varphi_1, \varphi_2, \theta, \psi_1, \ldots, \psi_r$$

to be a system of local coordinates around ρ. Note that we can assume that ψ_j are independent of x_0 taking $\psi_j(0, x', \xi')$ as new ψ_j. Moreover we can assume that

$$\{\psi_j, \varphi_k\} = O(|\varphi|), \quad k = 1, 2$$

taking $\psi_j - \{\psi_j, \varphi_1\}\varphi_2/\{\varphi_2, \varphi_1\} - \{\psi_j, \varphi_2\}\varphi_1/\{\varphi_1, \varphi_2\}$ as new ψ_j. Recall the Hamilton system

$$\begin{cases} \dot{x} = \dfrac{\partial}{\partial \xi} p(x, \xi) \\ \dot{\xi} = -\dfrac{\partial}{\partial x} p(x, \xi). \end{cases} \tag{12.44}$$

Let $\gamma(s) = (x(s), \xi(s))$ be a solution to the Hamilton system (12.44) and we consider $\xi_0(s)$, $x_0(s)$, $\varphi_j(\gamma(s))$, $\theta(\gamma(s))$, $\psi_j(\gamma(s))$. Note that

$$\frac{d}{ds} f(\gamma(s)) = \{p, f\}(\gamma(s)).$$

Let us change the parameter from s to t:

$$t = \frac{1}{s}$$

so that we have

$$\frac{d}{ds} = -tD, \quad D = t\frac{d}{dt}$$

and hence $tD(t^p F) = t^{p+1}(DF + pF)$. Let us introduce new unknowns:

$$\begin{cases} \xi_0(s) = t^4 \Xi_0(t), \ x_0(s) = tX_0(t), \\ \varphi_1(\gamma(s)) = t^2 \Phi_1(t), \ \varphi_2(\gamma(s)) = t^3 \Phi_2(t), \\ \theta(\gamma(s)) = t^2 \Theta(t), \ \psi_j(\gamma(s)) = t^2 \Psi_j(t). \end{cases}$$

Let us set

$$\{\varphi_2, \xi_0\} = \theta^2 + C_1^{20}\varphi_1 + C_2^{20}\varphi_2,$$
$$\{\varphi_1, \xi_0\} = C_1^{10}\varphi_1 + C_2^{10}\varphi_2$$

then it is not difficult to see

$$\begin{cases} DX_0 = -X_0 + 2\Phi_1 + O(t^2), \\ D\Phi_1 = -2\Phi_1 + 2\{\varphi_1, \varphi_2\}\Phi_2 + O(t), \\ D\Xi_0 = -4\Xi_0 - 2C_1^{20}\Phi_1\Phi_2 + O(t), \\ D\Phi_2 = -3\Phi_2 - 2C_1^{20}\Phi_1^2 + 2\{\varphi_1, \varphi_2\}\Xi_0 + O(t), \\ D\Theta = -2\Theta + O(t), \\ D\Psi_j = -2\Psi_j + O(t^2). \end{cases}$$

Let us put $\kappa = C_1^{20}(\rho)$, $\delta = \{\varphi_1, \varphi_2\}(\rho)$. From (12.39) and (12.41) it follows that

$$\kappa = \{\varphi_2, \{\varphi_2, \xi_0\}\}(\rho) \neq 0, \;\; \delta \neq 0.$$

Let us set

$$V = (X_0, \Phi_2, \Xi_0, \Phi_1, \Theta, \Psi), \;\; \Psi = (\Psi_1, ..., \Psi_r)$$

and we summarize what we have obtained:

$$\begin{cases} DX_0 = -X_0 + 2\Phi_1 + tG(t,V), \\ D\Xi_0 = -4\Xi_0 - 2\kappa\Phi_1\Phi_2 + tG(t,V), \\ D\Phi_1 = -2\Phi_1 + 2\delta\Phi_2 + tG(t,V), \\ D\Phi_2 = -3\Phi_2 - 2\kappa\Phi_1^2 + 2\delta\Xi_0 + tG(t,V), \\ D\Theta = -2\Theta + tG(t,V), \\ D\Psi_j = -2\Psi_j + tG(t,V) \end{cases} \tag{12.45}$$

where $G(t,V)$ denotes a smooth function in (t,V) such that $G(t,0) = 0$.

Let us define the class of formal series in t and $\log 1/t$ in which we look for our formal solutions to the reduced system (12.45):

$$\mathscr{E} = \{ \sum_{0 \le j \le i} t^i (\log 1/t)^j V_{ij} \mid V_{ij} \in \mathbb{C}^N \}.$$

Lemma 12.21. *Assume that $V \in \mathscr{E}$ satisfies (12.45) formally and $\Phi_2(0) \neq 0$. Then $X_0(0)$, $\Xi_0(0)$, $\Phi_1(0)$, $\Theta(0)$, $\Psi_j(0)$ are uniquely determined. In particular $X_0(0) \neq 0$.*

Repeating similar (much simpler) arguments as in [14], one can show

Lemma 12.22. *There exists a formal solution $V \in \mathscr{E}$ to (12.45) such that $\Phi_2(0) \neq 0$.*

Repeating a similar but much simpler argument as in [14], we can conclude that there is a solution to (12.45) which is asymptotic to the formal solution given in Lemma 12.22. Thus we get a solution $(x(s), \xi(s))$ to the Hamilton system (12.44). Since

$$\frac{d\theta}{dx_0}\Big|_{x_0=0} = \Big(\frac{d\theta}{dt}/\frac{dx_0}{dt}\Big)_{x_0=0} = \frac{t\Theta(t)}{X_0(t)}\Big|_{t=0} = 0$$

and hence we see that $(x(s), \xi(s))$ is actually tangent to S, thus we end the proof. □

From Theorems 12.1, 12.2 and Proposition 12.8 we conclude

Theorem 12.3. *Assume (12.1) and (12.2) and that the pure imaginary eigenvalue of F_p vanishes simply on S and there is no null bicharacterisic tangent to S. Then under the strict Ivrii-Petkov-Hörmander condition, the Cauchy problem is C^∞ well-posed.*

References

1. Beals, R.: Characterization of pseudodifferential operators and applications. Duke Math. J. **44**, 45–57 (1977)
2. Bernardi, E., Bove, A.: Geometric transition for a class of hyperbolic operators with double characteristics. Jpn. J. Math. **23**, 1–87 (1997)
3. Bernardi, E., Nishitani, T.: On the Cauchy problem for non-effectively hyperbolic operators, the Gevrey 5 well-posedness. J. Anal. Math. **116**, 197–240 (2008)
4. Bernardi, E., Nishitani, T.: On the Cauchy problem for noneffectively hyperbolic operators: the Gevrey 3 well-posedness. JHDE **8**, 615–650 (2011)
5. Bernardi, E., Nishitani, T.: On the Cauchy problem for noneffectively hyperbolic operators: the Gevrey 4 well-posedness. Kyoto J. Math. **51**, 767–810 (2011)
6. Bernardi, E., Parenti, C., Parmeggiani, A.: The Cauchy problem for hyperbolic operators with double characteristics in presence of transition. Comm. Part. Differ. Equat. **37**, 1315–1356 (2012)
7. Hörmander, L.: The Cauchy problem for differential equations with double characteristics. J. Analyse Math. **32**, 118–196 (1977)
8. Hörmander, L.: The Analysis of Linear Partial Differential Operators, vol. III. Springer, Berlin (1985)
9. Ivrii, V.Ja., Petkov, V.M.: Necessary conditions for the Cauchy problem for non strictly hypebolic equations to be well posed. Uspehi Mat. Nauk. **29**, 3–70 (1974)
10. Ivrii, V.Ja.: The well posedness of the Cauchy problem for non strictly hyperbolic operators III: The energy integral. Trans. Moscow Math. Soc. **34**, 149–168 (1978)
11. Ivrii, V.Ja.: Wave fronts of solutions of certain pseudo-differential equations. Trans. Moscow Math. Soc. **39**, 49–86 (1981)
12. Kajitani, K., Nishitani, T.: In: The Hyperbolic Cauchy Problem. Lecture Notes in Mathematics, vol. 1505. Springer, Berlin (1991)
13. Nishitani, T.: Local energy integrals for effectively hyperbolic operators, I. J. Math. Kyoto Univ. **24**, 625–658; II, 659–666 (1984)
14. Nishitani, T.: Non effectively hyperbolic operators, Hamilton map and bicharacteristics. J. Math. Kyoto Univ. **44**, 55–98 (2004)

Chapter 13
A Note on Unique Continuation for Parabolic Operators with Singular Potentials

Takashi Okaji

Abstract We consider a class of heat-type differential operators. The coefficients in the lower-order terms are allowed to have critical singularities. These operators can be viewed as a perturbation of a simple model operator by a subcritical one. Under some conditions on the vector and scalar potentials in the critical part, we establish strong unique continuation theorems for such operators. For proof, we use a two-stage Carleman method. Firstly we derive a Carleman inequality for the model operators with critical potentials through an analysis of spectrum of some Schrödinger operators with compact resolvent. The obtained Carleman inequality at the first stage guarantees us to choose a weight function with higher singularity in a Carleman inequality at the second stage for the perturbed operators.

Key words: Carleman inequality, Parabolic operators, Strong unique continuation

2010 Mathematics Subject Classification: 35B60, 35K10, 35R45.

13.1 Introduction

In this paper we are concerned with strong unique continuation for a class of heat-type differential equations

$$\partial_t u + \Delta u + \Sigma_{j=1}^d A_j \partial_{x_j} u + Vu = 0$$

and inequalities

$$|\partial_t u + \Delta u + \Sigma_{j=1}^d A_j \partial_{x_j} u| \leq W|\nabla u| + |V||u|.$$

T. Okaji (✉)
Graduate School of Science, Kyoto University, Kyoto 606-8502, Japan
e-mail: okaji@math.kyoto-u.ac.jp

M. Cicognani et al. (eds.), *Studies in Phase Space Analysis with Applications to PDEs*, Progress in Nonlinear Differential Equations and Their Applications 84,
DOI 10.1007/978-1-4614-6348-1_13,

Here the coefficients $A = (A_1, \cdots, A_d)$, W, and V in the lower-order terms are allowed to have critical singularities.

Let P be a differential operator on a domain $\Omega \subset \mathbf{R}^d$:

$$P = \sum_{|\alpha| \le m} a_\alpha(x) D_x^\alpha, \; D = (-i\partial_x)^\alpha, \; x \in \Omega.$$

When all coefficients of evolution operators $L = \partial_t + P(t,x,D_x)$ are independent of time, then solutions to $P(x,D)u = 0$ satisfy the evolution equation $Lu = 0$. Before going to a detail of parabolic problems, we review some known results on strong unique continuation for elliptic operators. Hereafter, we use a standard notation for several function spaces. $L^p_{\mathrm{loc}}(\mathbf{R}^d)$ denotes the space of functions whose p-th power is locally integrable over $\mathbf{R}^d$, and $H^s_{\mathrm{loc}}(\mathbf{R}^d)$ $(= W^{s,2}_{loc})$ stands for the locally Sobolev space over $\mathbf{R}^d$.

We say that a function $u \in L^2_{\mathrm{loc}}(\Omega)$ vanishes of infinite order at a point x_0 if for any natural number N, there exists a constant C_N such that

$$\int_{B(x_0,r)} |u(x)|^2 dx \le C_N r^N, \; \forall N \in \mathbf{N}$$

for all small positive number r. Here $B(a,R) = \left\{ x \in \mathbf{R}^d \mid |x-a| \le R \right\}$. Throughout this paper the above condition is abbreviated to

$$u(x) = \mathscr{O}(|x|^\infty).$$

Definition 13.1. We say that P has (weak) unique continuation property if and only if any solution u to $Pu = 0$ in Ω is identically zero in Ω provided that u vanishes in a nonempty open subset of Ω.

Definition 13.2. We say that P has strong unique continuation property if and only if any solution u to $Pu = 0$ in Ω is identically zero in Ω provided that u vanishes of infinite order at a point of Ω.

For simplicity we only consider the following type equation:

$$Pu = \Delta u + \sum_{j=1}^{d} A_j(x) \partial_j u + V(x) u = 0, \; u \in H^1_{\mathrm{loc}}(\mathbf{R}^d).$$

When the coefficients A_j and V are smooth, P has the strong unique continuation property. This is a consequence of results by Carleman [4], Aronszajn [3], and Cordes [6].

As for non-smooth coefficients case, there are many works. Jerison–Kenig [10] proved the strong unique continuation when $d \ge 3$ and $A = (A_1, \cdots, A_d) \equiv 0$, $V \in L^{d/2}_{\mathrm{loc}}(\mathbf{R}^d)$. Later Sogge [24] extended the previous result to the following subcritical case; there exists a positive number δ such that

$$|x - x_0|^{1-\delta} A \in L^\infty_{\mathrm{loc}}(\mathbf{R}^d), \; V \in L^{d/2}_{\mathrm{loc}}(\mathbf{R}^d).$$

Moreover, Wolff [28] showed that the strong unique continuation is still valid for differential inequalities

$$|\Delta u| \le |A||\nabla u| \quad \text{for} \quad A \in L^{\max(d,(3d-4)/2)}_{\mathrm{loc}}(\mathbf{R}^d), \text{ when } d \ge 3.$$

As for critical cases, Regbaoui [20] and Grammatico [9] proved strong unique continuation property provided that

$$\| |x-x_0||A(x)| \|_{L^\infty(\mathbf{R}^d)} \le 1/\sqrt{2},\ |x-x_0|^2 V(x) \in L^\infty(\mathbf{R}^d).$$

We note that their results are almost optimal because Alinhac–Baouendi [2] has succeeded in constructing a function $u \in C^\infty(\mathbf{R}^2)$ with $\operatorname{supp} u = \mathbf{R}^2$ such that $C > 1$

$$|\Delta u| \le \frac{C}{|x|}|\nabla u|$$

and u vanishes of infinite order at the origin. Wolff [28] also obtained a similar result when $d \ge 4$.

Now we return to parabolic operators. Consider backward heat-type equations

$$\text{(P)} \quad \partial_t u + \Delta u + A\cdot\nabla u + Vu = 0,\ (t,x) \in \Omega_T = [0,T]\times\Omega$$

and inequalities

$$\text{(Q)} \quad |\partial_t u + \Delta u| \le |A||\nabla u| + |V||u|),\ (t,x) \in \Omega_T.$$

Here, $A = (A_1, \cdots, A_d)$ and $A\cdot\nabla = \sum_{j=1}^{d} A_j(t,x)\partial_{x_j}$.

We want to seek the weakest condition on vector potential $A(t,x)$ and scalar potential $V(t,x)$ in order that strong unique continuation is valid.

Since the principal symbols of parabolic operators degenerate at the time direction, it is natural to adjust the notion of unique continuation to parabolic operators. Let Ω be a domain of $\mathbf{R}^d$ and I an interval of $\mathbf{R}$. For an open subset ω of $I\times\Omega$, we define a horizontal set of ω as follows:

$$\mathscr{H}(\omega) = \{(t,x) \in I\times\Omega \mid \exists (t,y) \in \omega\}.$$

Definition 13.3. We say that $L = \partial_t + P(t,x,D_x)$ has (weak) unique continuation property if and only if any solution u to $Lu = 0$ in $I\times\Omega$ is identically zero on $\mathscr{H}(\omega)$ provided that u vanishes in a nonempty open subset ω of $I\times\Omega$.

Definition 13.4. We say that (local) strong unique continuation property for (P) [resp. Q] under space-like vanishing condition is valid at a point $(t_0,x_0) \in \Omega_T$ if and only if any solution $u(t,x) \in H^1_{\mathrm{loc}}(\Omega_T))$ to (P) [resp. (Q)] vanishes on $\{t = t_0\}\times\Omega$ provided that for any $N \in \mathbf{N}$, there exists a constant $C_N > 0$ such that for any small positive number r

$$\int_{|x-x_0|\le r} |u(t_0,x)|^2 dx \le C_N r^N \quad [\text{abbr. } u(t_0,x) = \mathscr{O}((|x-x_0|)^\infty)].$$

Definition 13.5. We say that (local) strong unique continuation property for (P) [resp. Q] under space–time vanishing condition is valid at a point $(t_0,x_0) \in \Omega_T$ if and only if any solution $u(t,x) \in H^1_{\mathrm{loc}}(\Omega_T))$ to (P) [resp. (Q)] vanishes on $\{t=t_0\} \times \Omega$ provided that for any $N \in \mathbf{N}$, there exists a constant $C_N > 0$ such that for any small number r,

$$\int_{Q_r(t_0,x_0)} |u(t,x)|^2 dtdx \le C_N r^N \quad [\text{abbr. } u(t,x) = \mathscr{O}((|t-t_0|+|x-x_0|^2)^\infty)],$$

where $Q_r(t_0,x_0) = \left\{(t,x) \in \mathbf{R}^{d+1} \,|\, 0 < t-t_0 < \sqrt{r},\ |x-x_0| \le r\right\}$.

It is known that these two notions of strong unique continuation properties for parabolic problems coincide in some cases [1]. In this talk we are not going into the detail in this direction.

We also consider a global version of strong unique continuation on the space–time $[0,T] \times \mathbf{R}^d$ which means that we impose an additional condition on the behavior of solutions around infinity in the space direction (global strong unique continuation property).

There is a long history of weak unique continuation property for parabolic operators of second order. In the case of smooth coefficients, we cite Nirenberg [17], Yamabe [29], Mizohata [16], and Lees and Protter [14]. In the case of non-smooth coefficients, we mention that Sogge [25] proved a weak unique continuation theorem for differential inequalities

$$|\partial_t u + \Delta u| \le |V(t,x)u|$$

for some $V \in L^{(d+2)/2}_{\mathrm{loc}}(\Omega_T)$.

As for strong unique continuation problem, after the pioneering work by Landis-Oleinik [13] who considered second-order parabolic operators with time-independent coefficients, many researches are known. We begin with the case that the vector potential A is identically equal to zero. Let us consider

$$L_0 u = \partial_t u + \Delta u + Vu = 0.$$

Lin [15] proved that under the condition that $V(t,x) = V(x) \in L^p(\Omega;\ \mathbf{R})$ with $p > d/2$, if $L_0 u = 0$ and $u(0,x) = \mathscr{O}(|x|^\infty)$, then $u(0,x) \equiv 0$ (local s.u.c.p.). Poon [19] proved that under the condition that $d \ge 3$ and $V = K(x/|x|)|x|^{-2}$ with $K \in L^\infty(\mathbf{S}^{d-1};\ \mathbf{R})$, if $L_0 u = 0$, $|u| + |\nabla_x u| \in L^\infty((0,\infty) \times \mathbf{R}^d)$ and $u(t,x) = \mathscr{O}((t+|x|^2)^\infty)$, then $u(0,x) \equiv 0$ (global s.u.c.p.).

As for differential inequalities, Escauriaza [7] proved global s.u.c.p. and Escauriaza and Vega [8] proved local s.u.c.p. for continuous solutions.

Theorem 13.1. *[7] Let $d \ge 2$. Suppose that either*

(i) $\|t^{1-(d/2p)} V\|_{L^\infty((0,T);\ L^p(\mathbf{R}^d))} \ll 1$, $d/2 < p \le \infty$
or
(ii) $\|t^{1-(d/2p)-(1/s)} V\|_{L^s((0,T);\ L^p(\mathbf{R}^d))} < +\infty$, $d/2 < p \le \infty$, $1 \le s < \infty$ *and* $(d/2p) + (1/s) \le 1$.

Then, any solution u to

$$|\partial_t u + \Delta u| \le |Vu|$$

in $(\mathbf{R}^d)_T$ is identically zero provided that there exists a positive number $a > 0$ such that $|u(t,x)| \le C_k(\sqrt{t} + |x|)^k e^{a|x|^2}$ in $[0,T] \times \mathbf{R}^d$.

For more general case where the vector potential A is not identically zero, we consider the equation

$$L_A u = \partial_t u + \Delta u + A \cdot \nabla u + Vu = 0, \text{ in } (\mathbf{R}^d)_T.$$

Chen [5] proved that under the hypothesis that $|A|$ and $|V| \in L^\infty$, if $L_A u = 0$, $u(0,x) = \mathscr{O}(|x|^\infty)$ and $|u(t,x)| \le Ce^{a|x|^2}$ with $a > 0$, then $u(t,x) \equiv 0$ (global s.u.c.p.). Poon [19] proved that under the hypothesis that $|A|$ and $|V| \in L^\infty$, if $L_A u = 0$, $u \in L^\infty$ and $|u(t,x)| = \mathscr{O}((t + |x|^2)^\infty)$, then $u(t,x) \equiv 0$ (global s.u.c.p.).

When both $|A|$ and $|V|$ are allowed to be unbounded, Koch and Tataru [12] proved local s.u.c.p. under two kinds of vanishing conditions for operators with general principal part. We only present a part of their results in a simpler version. Let

$$\widetilde{L}_A u = \partial_t u + \Delta u + A_1 \cdot \nabla u + \nabla \cdot A_2 u + Vu = 0, \text{ on } \mathbf{R} \times \mathbf{R}^d.$$

Let us use the notation $L^p L^q = L^p([0,T];\ L^q(\mathbf{R}^d)))$.

Theorem 13.2 ([12]). *Suppose that $d \ge 3$ and $\|A_j\|_{L^{d+2}} + \|V\|_{L^1 L^\infty + L^\infty L^{d/2}}$ is small enough. If $\widetilde{L}_A u = 0$ and $u(t,x) = \mathscr{O}((t + |x|^2)^\infty)$, then $u(0,x) = 0$ near the origin.*

13.2 Main Results

We are concerned with the critical case. Let $t_0 = 0$, $x_0 = 0$, and $\Omega_T = (0,T) \times \Omega$. First, let us consider the following heat-type differential operator:

$$L_0 u = \partial_t u + \Delta u + \frac{\kappa}{t} x \cdot \nabla u + \mu \left(\frac{x}{|x|^2} \cdot \nabla + \nabla \cdot \frac{x}{|x|^2} \right) u - \frac{K(\omega)}{|x|^2} u,$$

where κ and μ are real numbers and $K(\omega) \in L^\infty(\mathbf{S}^{d-1})$ is real valued. Let

$$H^{1,2}([0,T] \times \Omega) = \Big\{ u(t,x) \in L^2([0,T] \times \Omega) \mid \sum_{2j+|\alpha| \le 2} \partial_t^j \partial_x^\alpha u \in L^2([0,T] \times \Omega) \Big\}.$$

Then we have the following global strong unique continuation theorem:

Theorem 13.3. *Let $\Omega = \mathbf{R}^d$, $d \ge 3$, $\mu \ge 0$, and $m \ge 0$. Suppose that*

$$\operatorname*{ess\,inf}_{\omega \in \mathbf{S}^{d-1}} K(\omega) > -(d-2)^2/4 - \mu^2.$$

If $\kappa < 1/2$ and m is small enough, then any function $u \in H^{1,2}_{\mathrm{loc}}([0,T]\times \mathbf{R}^d)$ satisfying

$$|L_0 u| \le m t^{-1}|u| \tag{13.1}$$

is identically equal to zero on $[0,T]\times \mathbf{R}^d$ provided that for a positive number $\varepsilon < 1-2\kappa$ and any nonnegative integer k, there exists a positive constant C_k such that

$$|u(t,x)| \le C_k(\sqrt{t}+|x|)^k \exp\{(1-2\kappa-\varepsilon)\frac{|x|^2}{8t}\}.$$

Remark 13.1. When $\mu = 0$ and $K(\omega) = 0$, the same conclusion holds for $0 \le m < \sqrt{1-2\kappa}/4$.

Remark 13.2. Theorem 13.3 remains true when L_0 is altered to a perturbed operator $L = L_0 + t^{-1}q(x/\sqrt{t})$, where $q(y) \in L^{d/2}(\mathbf{R}^d) + L^\infty(\mathbf{R}^d)$ is a real-valued function (cf. Remark 13.5).

Next we shall extend the above result to a perturbed heat-type differential inequality

$$\left|L_0 u + i(A\cdot\nabla + \nabla\cdot A)u + Vu\right| \le M t^{(-1+\tau)/2}|\nabla u| + (m t^{-1} + |W|)|u|, \tag{13.2}$$

where $M,\ m \ge 0$, $\tau > 0$, $A(t,x)$ is $\mathbf{R}^d$-valued and $V(t,x)$ is real valued.

To state our hypothesis on the coefficients A, V and W, we shall use the terminology of relative boundedness.

Definition 13.6. Let R and S be linear operators in $\mathscr{H} = L^2(\mathbf{R}^d)$. We recall that R is said to be S-bounded with relative bound zero, if $D(S) \subset D(R)$ and if for any $\varepsilon > 0$, there exists a positive number δ such that

$$\|Rv\|_{\mathscr{H}} \le \varepsilon\|Sv\|_{\mathscr{H}} + \delta\|v\|_{\mathscr{H}}$$

for all $v \in C_0^\infty(\mathbf{R}^d)$.

Furthermore we shall need a more precise version of relative boundedness.

Definition 13.7. Let k be a positive number and σ a nonnegative number. We say that a function $q(t,x) \in L^2_{\mathrm{loc}}((0,T)\times\mathbf{R}^d)$ satisfies $(\mathrm{C})_{k,\sigma}$ if there exists a positive constant C independent of $t \in [0,T]$ such that for any $u \in C_0^\infty(\mathbf{R}^d)$ and any $t \in (0,T)$

$$\|t^{k/2}q(t,\cdot)u\|_{L^2(\mathbf{R}^d)} \le C t^\sigma \sum_{|\alpha|\le k} \|(\sqrt{t}\partial_x)^\alpha u\|_{L^2(\mathbf{R}^d)}. \tag{13.3}$$

We define $B^\sigma(|D|^k)$ to be a set of all functions satisfying $(\mathrm{C})_{k,\sigma}$.

Theorem 13.4. *Let $\Omega = \mathbf{R}^d$, $d \ge 5$, and $\mu \ge 0$. Suppose that*

$$\operatorname*{ess\,inf}_{\omega\in\mathbf{S}^{d-1}} K(\omega) > -(d-2)(d-4)/4 - \mu^2,$$

and that for some $\sigma > 0$, $A_j \in B^\sigma(|D|)$ $(1 \le j \le d)$, $W \in B^\sigma(|D|)$, $V \in B^\sigma(|D|^2)$ *and* $\mathrm{div}A \in B^\sigma(|D|^2)$. *Furthermore assume that all components of* $2t\partial_t A + x\cdot\nabla A$ *are* $\Delta^{1/2}$*-bounded with relative bound zero and* $2t\partial_t V + x\cdot\nabla V$ *is* Δ*-bounded with relative bound zero. If* $\kappa < 1/2$ *and* m *is small enough, then any solution* $u \in H^{1,2}_{\mathrm{loc}}([0,T]\times\mathbf{R}^d)$ *to (13.2) is identically equal to zero on* $[0,T]\times\mathbf{R}^d$ *provided that for a positive number* ε *and any nonnegative integer* k*, there exists a positive constant* C_k *such that*

$$|u(t,x)| \le C_k(\sqrt{t}+|x|)^k \exp\{(1-2\kappa-\varepsilon)\frac{|x|^2}{8t}\}. \tag{13.4}$$

Remark 13.3. When $\mu = 0$ and $K(\omega) = 0$, the same conclusion holds for any dimension.

As for local strong unique continuation properties, we have the following results.

Theorem 13.5. *Let* $d \ge 3$, $\Omega \ni 0$, $\mu \ge 0$, $\kappa < 1/2$, *and* $m \ge 0$. *Suppose that*

$$\operatorname*{ess\,inf}_{\omega\in\mathbf{S}^{d-1}} K(\omega) > -(d-2)^2/4-\mu^2.$$

Let $u \in H^{1,2}_{\mathrm{loc}}(\Omega_T)$ *be a solution to 13.1. If* m *is small enough and* $u(t,x) = \mathscr{O}((t+|x|^2)^\infty)$, *then there exists a neighborhood* Ω' *of the origin such that* $u(0,x)$ *is identically equal to zero in* Ω'.

Theorem 13.6. *Let* $d \ge 5$, $\Omega \ni 0$, $\mu \ge 0$, $\kappa < 1/2$, *and* $m \ge 0$ *Suppose that* $A_0 \in B^1(|D|)$ *and* $W \in B^1(|D|^2)$ *are nonnegative functions and that*

$$\operatorname*{ess\,inf}_{\omega\in\mathbf{S}^{d-1}} K(\omega) > -(d-2)(d-4)/4-\mu^2.$$

Let $u \in H^{1,2}_{\mathrm{loc}}(\Omega_T)$ *be a solution to*

$$|L_0 u| \le A_0|\nabla u| + W|u|.$$

If m *is small enough and* $u(t,x) = \mathscr{O}((t+|x|^2)^\infty)$, *then there exists a neighborhood* Ω' *of the origin such that* $u(0,x)$ *is identically equal to zero in* Ω'.

In a particular circumstance, a strong unique continuation theorem does not require any smallness assumption on scalar potentials $V(t,x) \in L^\infty((0,T);\, L^{d/2}(\mathbf{R}^d))$.

Corollary 13.1. *Let* $d \ge 5$, $\Omega \ni 0$, *and* $V(t,x) \in L^1_{\mathrm{loc}}((0,T)\times\mathbf{R}^d)$ *be real valued. Suppose that there exist a real-valued function* $U_0 \in L^{d/2}(\mathbf{R}^d)$, *a nonnegative function* $U_1 \in L^{d/2}(\mathbf{R}^d)$, *and a positive constant* C *such that*

$$|V(t,x) - t^{-1}U_0(x/\sqrt{t})| \le CU_1(x/\sqrt{t})$$

for any $t \in (0,T]$. *Let* $u \in H^{1,2}_{\mathrm{loc}}(\Omega_T)$ *be a solution to*

$$\partial_t u + \Delta u + Vu = 0.$$

If $u(t,x) = \mathscr{O}\left((t+|x|^2)^\infty\right)$, *then* $u(0,x) = 0$ *near the origin.*

We emphasize that $t^{-1}U_0(x/\sqrt{t}) \in L^\infty((0,T);\ L^{d/2}(\mathbf{R}^d))$

13.3 Sufficient Conditions for $(\mathrm{C})_{k,\sigma}$

In this section we shall give sufficient conditions which ensure the hypothesis $(\mathrm{C})_{k,\sigma}$.

For a positive number ρ we consider the following function:

$$\omega_\rho(x) = |x|^{\rho-d} \text{ for } 0<\rho<d,\ 1-\log|x|^2 \text{ for } \rho = d, \text{ and } 1 \text{ for } \rho > d.$$

Definition 13.8. For $\rho > 0$ and a measurable function q over $\mathbf{R}^d$, we define a non-negative number (including $+\infty$)

$$M_\rho(q) = \sup_{x\in\mathbf{R}^d}\left[\int_{|x-y|\le 1} |q(y)|^2\omega_\rho(x-y)dy\right]^{1/2}$$

and define a class of functions [23, 26]

$$M_\rho(\mathbf{R}^d) = \left\{q \in L^2_{\mathrm{loc}}(\mathbf{R}^d)|\ M_\rho(q) < +\infty\right\}.$$

For a real number τ, we also define

$$L^\infty_{-\tau} = \left\{q(x) \in L^1_{\mathrm{loc}}(\mathbf{R}^d) \mid |x|^\tau q(x) \in L^\infty(\mathbf{R}^d)\right\}.$$

Especially when ρ is equal to two or four, the spaces M_ρ are very important for us because for any $\varepsilon > 0$

$$L^\infty_{-2+\varepsilon}(\mathbf{R}^d) \subset M_4(\mathbf{R}^d) \text{ if } d \ge 5, \quad L^\infty_{-1+\varepsilon}(\mathbf{R}^d) \subset M_2(\mathbf{R}^d) \text{ if } d \ge 3.$$

We remark that there are the following relations for L^p spaces. From Hölder's inequality it follows that if $\rho > 2d/p$ and $2 < p \le \infty$, then

$$L^p(\mathbf{R}^d) \subset M_\rho(\mathbf{R}^d).$$

We consider function spaces with parameter $t \in I = (0,T)$. For $\varepsilon \ge 0$, we define spaces of functions $q(t,x)$ on $I \times \mathbf{R}^d$ as follows:

$$L^\infty M_\rho = L^\infty(I;\ M_\rho(\mathbf{R}^d)),\ C^\varepsilon M_\rho = \left\{q \mid t^{-\varepsilon}q(t,x) \in L^\infty M_\rho\right\}$$
$$L^\infty L^p = L^\infty(I;\ L^p(\mathbf{R}^d)),\ C^\varepsilon L^p = \left\{q \mid t^{-\varepsilon}q(t,x) \in L^\infty(I;\ L^p(\mathbf{R}^d))\right\}$$
$$C^\varepsilon L^\infty_{-\tau} = \left\{q(t,x) \mid t^{-\varepsilon}q(t,x) \in L^\infty L^\infty_{-\tau}\right\}.$$

Proposition 13.1. *Suppose that $d \geq 5$. Let ε and δ be nonnegative numbers such that $\varepsilon + \delta > 0$. If*

$$q \in C^{\varepsilon} M_{2k-\delta} + C^{\varepsilon} L^{\max(d/k,2)+\varepsilon+\delta} + C^{\varepsilon+\delta} L^{d/k} + C^{\varepsilon+\delta} L^{\infty}_{-k},$$

then there exists a positive number σ such that $q(t,x)$ satisfies $(C)_{k,\sigma}$ for each $k = 1,2$.

The next lemma is useful to prove Proposition 13.1.

Lemma 13.1. *Let $0 \leq s < k$. There exists a positive constant C such that for any $u \in C_0^{\infty}(\mathbf{R}^d)$ and $t > 0$,*

$$t^{k/2} \| |D|^s u \| = t^{(k-s)/2} \| |\sqrt{t} D|^s u \| \leq C t^{(k-s)/2} (\| |\sqrt{t} D|^k u \| + \|u\|).$$

Here $|D|^s$ is the Fourier multiplier whose symbol is equal to $|\xi|^s$.

In view of Lemma 13.1, it is not difficult to prove Proposition 13.1 by use of the following well-known inequalities.

Lemma 13.2 (Stummel and Schechter (Theorem 7.1 in [23])). *If $0 < \rho/2 < s$, then there exists a positive constant C such that*

$$\|qu\|_{L^2(\mathbf{R}^d)} \leq C M_{\rho}(q) \|u\|_{H^s(\mathbf{R}^d)}$$

for any $u \in C_0^{\infty}(\mathbf{R}^d)$.

Hölder's inequality and Sobolev's embedding theorem yield the following well-known inequality.

Lemma 13.3. *If $p > \max(d/s, 2)$, then there exists a positive constant C such that*

$$\|qu\|_{L^2(\mathbf{R}^d)} \leq C \|q\|_{L^p(\mathbf{R}^d)} \|u\|_{H^s(\mathbf{R}^d)}$$

for any $u \in C_0^{\infty}(\mathbf{R}^d)$.

As a consequence of Hardy–Littlewood–Sobolev inequality, we know the following inequalities.

Lemma 13.4. *Let $d \geq 5$. Then there exists a positive constant C such that*

$$\|qu\|_{L^2(\mathbf{R}^d)} \leq C \|q\|_{L^{d/2}(\mathbf{R}^d)} \|u\|_{H^2(\mathbf{R}^d)}$$

and

$$\|q \nabla u\|_{L^2(\mathbf{R}^d)} \leq C \|q\|_{L^d(\mathbf{R}^d)} \|u\|_{H^2(\mathbf{R}^d)}$$

for any $u \in C_0^{\infty}(\mathbf{R}^d)$.

Remark 13.4. The stronger version of Lemma 13.4 is also known as Strichartz's theorem (Theorem X.21 in [22]).

Lemma 13.5. *When $d \geq 3$,*

$$\| |x|^{-1} u(x) \|_{L^2(\mathbf{R}^d)} \leq \frac{2}{(d-2)} \| \nabla u(x) \|_{L^2(\mathbf{R}^d)}.$$

In addition, when $d \geq 5$,

$$\| |x|^{-2} u(x) \|_{L^2(\mathbf{R}^d)} \leq \frac{4}{(d-2)(d-4)} \| \Delta u(x) \|_{L^2(\mathbf{R}^d)}$$

for any $u \in C_0^\infty(\mathbf{R}^d)$.

Proof. The first part of Lemma 13.5 is well known as Hardy's inequality (cf. Auxiliary theorem 10.35 in [27]). For proving the second part, it suffices to apply Hardy's inequality twice. Let $\chi \in C^\infty(\mathbf{R}^d)$ be a nonnegative function such that

$$\chi(x) = 1 \text{ if } |x| \geq 1, \text{ and } = 0 \text{ if } |x| \leq 1/2.$$

With $\chi_\varepsilon(x) = \chi(x/\varepsilon)$ for a positive parameter ε, we see that for $u \in C_0^\infty(\mathbf{R}^d)$

$$\begin{aligned}\| |x|^{-2} \chi_\varepsilon u \| &\leq \frac{2}{d-2} \| \nabla |x|^{-1} \chi_\varepsilon u(x) \| \\ &\leq \frac{2}{d-2} \left\{ \| |x|^{-1} \chi_\varepsilon \nabla u \| + \| |x|^{-2} \chi_\varepsilon u \| + \| |x|^{-1} [\nabla, \chi_\varepsilon] u \| \right\}.\end{aligned}$$

Since it holds that

$$\lim_{\varepsilon \to +0} \| |x|^{-1} [\nabla, \chi_\varepsilon] u \| = 0,$$

we can conclude that

$$\frac{d-4}{d-2} \| |x|^{-2} u \| \leq \frac{2}{d-2} \| |x|^{-1} \nabla u \| \leq \left(\frac{2}{d-2}\right)^2 \| \Delta u \|.$$

□

13.4 Outline of Proof of Theorem 13.3

There are two major approaches to unique continuation problem. One is the so-called frequency function method, introduced by Garofalo-Lin, and the other is our approach based on an L^2 inequality with weight function, which is called Carleman inequality ([4]).

One of the most important ingredients in [7] is deformation of equations by means of a classical parabolic change of variables. In our settings, we have the following identity.

Lemma 13.6. *Let $t = s^2$ and $x = sy$. Then*

$$\begin{aligned} tL_0u &= \frac{1}{2}s\partial_s u + \left(\nabla_y - (1-2\kappa)\frac{y}{4} + \mu\frac{y}{|y|^2}\right)^2 u \\ &\quad - \left((1-2\kappa)\frac{y}{4} - \mu\frac{y}{|y|^2}\right)^2 u - \frac{K(\omega)}{|y|^2}u + (1-2\kappa)\frac{d}{4}u. \end{aligned}$$

Here we have used the following notation. For $A(y) = (A_1(y), \cdots, A_d(y))$,

$$(\nabla_y + A(y))^2 = \sum_{j=1}^{d} \left(\partial_{y_j} + A_j(y)\right)^2,\ A^2 = \sum_{j=1}^{d} A_j^2.$$

One can see that the transformed operator tL_0 is a sum of the first derivative in *time* and a formally self-adjoint operator with compact resolvent. This structure enables us to treat parabolic operators by a refined method for elliptic operators.

Now we introduce a new unknown function. Define

$$v(s,y) = e^{\psi_0(y)}u(s,y),\ \psi_0(y) = -(1-2\kappa)|y|^2/8 + \mu\log|y|.$$

It holds that $tL_0u = 0$ is equivalent to $\tilde{L}_0 v = 0$, where

$$\begin{aligned} \tilde{L}_0 &= e^{\psi_0(y)}tL_0 e^{-\psi_0(y)} \\ &= \frac{1}{2}s\partial_s + \Delta_y - \left((1-2\kappa)\frac{y}{4} - \mu\frac{y}{|y|^2}\right)^2 - \frac{K(\omega)}{|y|^2} + (1-2\kappa)\frac{d}{4}. \end{aligned}$$

Now we see that the infinitely vanishing condition on u in our theorems implies that

$$\exp\left\{-(1-2\kappa)\frac{|x|^2}{8t}\right\}|u(t,x)| \le C_k t^{k/2}(1+|x|/\sqrt{t})^k \exp\left\{-\delta\frac{|x|^2}{8t}\right\} \le C_k' t^{k/2}.$$

It follows that the function in the left-hand side vanishes of infinite order at $t = 0$. Hence we can use a weight function φ independent of x.

Let H be a self-adjoint operator in $L^2(\mathbf{R}^d)$ such that its domain $D(H)$ contains the space $C_0^\infty(\mathbf{R}^d)$. In what follows we use a notation $\|\cdot\|$ which stands for the standard norm of $L^2(\mathbf{R}^d)$. We seek a Carleman inequality for an operator $L = s\partial_s + \frac{1}{2} + H$.

Lemma 13.7. *Let $\varphi(s) = \log(1/s)$. Suppose that the resolvent set of H contains infinitely many compact intervals on the real axis, $I_n = [a_n, b_n]$ ($n \in \mathbf{N}$), such that for any $n \in \mathbf{N}$,*

$$a_n < b_n < a_{n+1} < b_{n+1},\ b_{n+1} - a_{n+1} = b_n - a_n > 0,\ \lim_{n\to\infty} a_n = +\infty.$$

Then, it holds that there exist a monotone increasing sequence of numbers, γ_n, and a positive constant C such that

$$\int_0^T \|s^{-\gamma_n}v\|^2 ds \le C\int_0^T \|s^{-\gamma_n}Lv\|^2 ds \tag{13.5}$$

holds if $v(s,\cdot) \in C_0^0((0,T);\ L^2(\mathbf{R}^d))$, $s\partial_s v(s,\cdot) \in L^2((0,T);\ L^2(\mathbf{R}^d))$ and $Hv(s,\cdot) \in L^2((0,T);\ L^2(\mathbf{R}^d))$.

Proof. It can be verified that for any given $v(s,\cdot)$ in Lemma 13.7,

$$\int_0^T \|s^{-\gamma}Lv\|^2 ds = \int_0^T \|s^{-\gamma}(s\partial_s + \frac{1}{2})v\|^2 ds + \int_0^T \|s^{-\gamma}(H-\gamma)v\|^2 ds. \tag{13.6}$$

It holds that for any $z \in \mathbf{C}$,

$$\|(H-z)v\| \geq \text{dist}(z, \sigma(H))\|v\|, \ v \in D(H),$$

where $\text{dist}(z, \sigma(H))$ denotes the distance between z and the spectrum of H.

Define $\gamma_n = (b_n + a_n)/2$ and $C_0 = (b_n - a_n)/2$. Then we see that

$$\|(H-\gamma_n)v\| \geq C_0\|v\|, \ v \in D(H). \tag{13.7}$$

In view of (13.6) and (13.7), we arrive at the desired inequality (13.5). □

We would like to apply Lemma 13.7 to the operator

$$\frac{1}{2}H = \Delta_y - \left((1-2\kappa)\frac{y}{4} - \mu\frac{y}{|y|^2}\right)^2 - \frac{K(\omega)}{|y|^2} + (1-2\kappa)\frac{d}{4} - \frac{1}{2}.$$

Consider a bilinear form over $C_0^\infty(\mathbf{R}^d) \times C_0^\infty(\mathbf{R}^d)$;

$$h(u,v) = \sum_{j=1}^d \langle \partial_{y_j}u, \partial_{y_j}v\rangle + \sum_{j=1}^d \langle((1-2\kappa)\frac{y_j}{4} - \mu\frac{y_j}{|y|^2})u, ((1-2\kappa)\frac{y_j}{4} - \mu\frac{y_j}{|y|^2})v\rangle$$
$$+\langle K(\omega)|y|^{-1}u, |y|^{-1}v\rangle,$$

where $\langle\cdot,\cdot\rangle$ denotes the standard inner product of $L^2(\mathbf{R}^d)$. This form is associated with an operator

$$H_0 = -\Delta + \left((1-2\kappa)\frac{y}{4} - \mu\frac{y}{|y|^2}\right)^2 + \frac{K(\omega)}{|y|^2}.$$

When $\mu = 0$ and $K(\omega) = 0$, H_0 is just a harmonic oscillator, so that it is known that its eigenvalues consist of $\lambda_n = \dfrac{1-2\kappa}{4}(2|\omega| + d)$, $\omega \in \mathbf{Z}_+^d$ which is distributed at even intervals.

Using a perturbation theory (Lemmas 13.13 and 13.14 in the appendix), we can obtain the following lemma.

Lemma 13.8. *Suppose that $d \geq 3$, $\mu \in \mathbf{R}$ and $K \in L^\infty(\mathbf{S}^{d-1})$ is real valued. If*

$$\operatorname*{ess\ inf}_{\omega \in \mathbf{S}^{d-1}} K(\omega) > -\frac{(d-2)^2}{4},$$

then H_0 has a self-adjoint Friedrichs extension H whose spectrum consists of eigenvalues $\{\lambda_j\}_{j=1}^\infty$. Moreover, a counting function

$$N(n) = \#\{j \in \mathbf{N} |\ \lambda_j \in [n, n+1]\}$$

satisfies $N_0 = \sup_{n \in \mathbf{N}} N(n) < +\infty$.

Remark 13.5. Let $V_0(y) \in L^{d/2}(\mathbf{R}^d)$ be a real-valued function. Then, Lemma 13.8 remains true if H_0 is altered to $H_0 + V_0$ because $|V_0|^{1/2}$ is $|D|$-bounded with relative bound zero (cf. Lemma 13.4).

Proof. We see that

$$H_0 = -\Delta + \left((1-2\kappa)\frac{y}{4}\right)^2 + \frac{\mu^2 + K(\omega)}{|y|^2} - (1-2\kappa)\mu/2.$$

Define

$$K_+(\omega) = \frac{1}{2}\left(|\mu^2 + K(\omega)| + \mu^2 + K(\omega)\right),$$
$$K_-(\omega) = \frac{1}{2}\left(|\mu^2 + K(\omega)| - \mu^2 - K(\omega)\right).$$

Let

$$S(\zeta) = -\Delta + \left((1-2\kappa)\frac{y}{4}\right)^2 + \zeta\frac{K_+(\omega)}{|y|^2}.$$

Then

$$H_0 = S(1) - K_-(\omega)|y|^{-2} - \frac{\mu(1-2\kappa)}{2}.$$

If $|\zeta|\|K_+\|_{L^\infty(\mathbf{S}^{d-1})} < \frac{(d-2)^2}{4}$, Hardy's inequality implies that there exists a positive number $b < 1$ such that

$$\langle |\zeta| K_+(\omega)|y|^{-2}u, u\rangle \le b\langle \delta u, u\rangle.$$

Let

$$\xi_0 = \min\left(\frac{(d-2)^2}{4\|K_+\|_{L^\infty(\mathbf{S}^{d-1})}}, \frac{1}{1+b}\right).$$

First, when $K_+(\omega) = 0$, then from Lemma 13.14, it follows that the spectrum of $S(\zeta)$ consists of real eigenvalues if $\zeta \in \mathbf{R}$ and $|\zeta| \le \xi_0$. We arrange them in increasing order

$$\lambda_1(\zeta) \le \lambda_2(\zeta) \le \cdots,$$

by counting their multiplicity. Then it holds that there exists a positive constant N_0 such that for any $n \in \mathbf{N}$, $\lambda_j(\zeta)$ satisfies

$$\lambda_j(\xi_0) \le N_0 \lambda_j(0).$$

From this inequality, it follows that the upper bound of the number of eigenvalues belonging to $[n, n+1]$ is finite. Next for $k \in \mathbf{N}$, we consider

$$S_k(\zeta) = -\Delta + \left((1-2\kappa)\frac{y}{4}\right)^2 + k\xi_0 \frac{K_+(\omega)}{|y|^2} + \zeta \frac{K_+(\omega)}{|y|^2}.$$

We can apply the same procedure repeatedly to $S_k(\zeta) + \zeta K_+(\omega)|y|^{-2}$ until we arrive at $S(1)$. Finally, applying the same procedure to $\widetilde{S}(\zeta) = S(1) - \zeta K_-(\omega)|y|^{-2}$, we obtain the conclusion for $\widetilde{S}(1)$. □

Lemma 13.9. *There exist a sequence $\{\gamma_n\}$ and a positive constant C_0 such that $\lim_{n\to\infty} \gamma_n = +\infty$ and*

$$\begin{aligned} C_0 \int_{\mathbf{R}_+ \times \mathbf{R}^d} & t^{-\gamma_n} \left(\frac{|x|}{\sqrt{t}}\right)^{2\mu} e^{-(1-2\kappa)|x|^2/4t} |v|^2 \frac{dt}{t^{(d+2)/2}} dx \\ \le \int_{\mathbf{R}_+ \times \mathbf{R}^d} & t^{-\gamma_n+2} \left(\frac{|x|}{\sqrt{t}}\right)^{2\mu} e^{-(1-2\kappa)|x|^2/4t} |L_0 v|^2 \frac{dt}{t^{(d+2)/2}} dx \end{aligned} \tag{13.8}$$

holds for any $n \in \mathbf{N}$ and $v(s,\cdot) \in C_0^0((0,T);\, L^2(\mathbf{R}^d))$, $s\partial_s v(s,\cdot) \in L^2((0,T);\, L^2(\mathbf{R}^d))$ and $H_0 v(s,\cdot) \in L^2((0,T);\, L^2(\mathbf{R}^d))$.

Here we recall

$$L_0 = \left[\partial_t u + \Delta_x u + \tfrac{\kappa}{t} x \cdot \nabla u + \mu \left(\tfrac{x}{|x|^2} \cdot \nabla + \nabla \cdot \tfrac{x}{|x|^2}\right) u - \tfrac{K(\omega)}{|x|^2} u\right].$$

Proof. Let $\sigma(H_0)$ be a set of all eigenvalues of H_0. From Lemma 13.8, it follows that each interval $[n, n+1]$ contains a subinterval J_n such that

$$J_n \cap \sigma(H_0) = \emptyset, \ |J_n| = 1/(N_0+1).$$

We can see that a collection of intervals, J_n, satisfies the hypothesis in Lemma 13.5, so that we arrive at the Carleman inequality (13.8). □

Theorem 13.3 follows from the Carleman inequality (13.8) by a standard way (cf. Sect. 3 of [7]).

13.5 Outline of Proof of Theorem 13.4

We are confronted with a difficulty when we treat our perturbed terms, because the left-hand side of the estimate in Lemma 13.9 contains no derivatives terms.

However, since hypothesis in Theorem 13.4 guarantees that H_0 is an essentially self-adjoint operator on $C_0^\infty(\mathbf{R}^d)$, we have the following modified Carleman inequality. Let $\psi(t,x) = \psi_0(x/\sqrt{t})$ and $\tilde{\gamma}_n = \gamma_n + (d+2)/2$.

Lemma 13.10. *There exists a positive constant C such that*

$$\int_{\mathbf{R}_+} t^{-\tilde{\gamma}_n}\|e^{\psi(t,x)}v\|^2dt +\gamma_n^{-1}\int_{\mathbf{R}_+} t^{-\tilde{\gamma}_n}\|e^{\psi(t,x)}\sqrt{t}\nabla_x v\|^2dt$$
$$+\gamma_n^{-2}\int_{\mathbf{R}_+} t^{-\tilde{\gamma}_n}\|e^{\psi(t,x)}t\Delta v\|^2dt$$
$$\leq C\int_{\mathbf{R}_+} t^{-\tilde{\gamma}_n+2}\|e^{\psi(t,x)}L_0 v\|^2dt,\ \forall n\in\mathbf{N} \tag{13.9}$$

holds for any $v\in C_0^\infty((0,T)\times\mathbf{R}^d)$.

Proof. Let

$$w=e^{\psi(t,x)}v,\ v\in C_0^\infty(0,T)\times\mathbf{R}^d$$

and recall

$$\tilde{L}_0=e^{-\psi_0}L_0e^{\psi_0}=\frac{s}{2}\partial_s+H_0(y,\partial_y).$$

It holds that

$$\int_0^\infty\left(\|\nabla_y w\|^2+\|(\nabla\psi_0(y))w\|^2\right)\frac{ds}{s}\leq\int_0^\infty\langle-\tilde{L}_0 w,w\rangle\frac{ds}{s}.$$

In view of

$$-\tilde{L}_0=-s^{-\gamma}\tilde{L}_0 s^{\gamma}+\gamma,$$

we see that

$$\int_0^\infty s^{-2\gamma}\left(\|\nabla_y w\|^2+\|(\nabla\psi_0(y))w\|^2\right)\frac{ds}{s}\leq\int_0^\infty s^{-2\gamma}(\|\tilde{L}_0 w\|^2+\gamma\|w\|^2)\frac{ds}{s}.$$

Then from Lemma 13.9 it follows that there exists a positive constant C such that

$$\gamma_n^{-1}\int_0^\infty s^{-2\gamma}\left(\|\nabla_y w\|^2+\|(\nabla\psi_0(y))w\|^2\right)\frac{ds}{s}\leq C\int_0^\infty s^{-2\gamma_n}\|\tilde{L}_0 w\|^2\frac{ds}{s}.$$

From the assumption $\mu^2+K(\omega)>-(d-2)(d-4)/4$, it follows that there exists a positive constant C such that

$$\int_0^\infty\|\Big(-\Delta+\frac{(1-2\kappa)^2|y|^2}{16}\Big)w\|^2\frac{ds}{s}\leq C\int_0^\infty\|\tilde{L}_0 w\|^2\frac{ds}{s}.$$

Hence

$$\gamma_n^{-2}\int_0^\infty s^{-2\gamma}\|(-\Delta+\frac{(1-2\kappa)^2|y|^2}{16})w\|^2\frac{ds}{s}\leq C\int_0^\infty s^{-2\gamma_n}\|\tilde{L}_0 w\|^2\frac{ds}{s}.$$

In the dimension $d \geq 5$, the inequalities in Lemma 13.5 imply that there exists a positive constant C such that

$$\|e^{\psi(t,x)}\sqrt{t}\partial_x v\|^2 = s^{-d}\|(\nabla_y - \nabla_y \psi_0)w\|^2 \leq Cs^{-d}\left(\|\nabla_y w\|^2 + \|\frac{(1-2\kappa)|y|}{4}w\|^2\right)$$

and

$$\begin{aligned}\|e^{\psi(t,x)}t\Delta v\|^2 &= s^{-d}\|(\partial_y - \nabla_y \psi_0)^2 w\|^2 \\ &\leq Cs^{-d}\left(\|\Delta_y w\|^2 + \|\frac{(1-2\kappa)^2|y|^2}{16}w\|^2 + \|w\|^2\right)\end{aligned}$$

for any $w = e^{\psi}v \in C_0^\infty((0,T)\times \mathbf{R}^d)$. We have arrive at the conclusion. □

Thanks to the modified Carleman estimate, we can improve vanishing order of thc solution.

Lemma 13.11. *Suppose $d \geq 5$. If m is small enough, then it holds that there exists a positive number ε such that for any solution $u \in H^{1,2}_{\mathrm{loc}}([0,T]\times \mathbf{R}^d)$ to (13.2) and for any small positive number τ,*

$$\int_0^\tau \int_{\mathbf{R}^d} \left(\frac{|x|}{\sqrt{t}}\right)^\mu e^{-(1-2\kappa)|x|^2/(4t)} \left\{|u(t,x)|^2 + |\sqrt{t}\nabla_x u|^2\right\} dt dx \leq C\exp\left[-\varepsilon\tau^{-1/\sigma}\right]$$

holds provided that u satisfies a vanishing condition (13.4). Here C stands for a positive constant independent of τ.

Proof. Let $R(t,x,\partial_x) = i(A\cdot\nabla + \nabla\cdot A) + V$. Then from the hypothesis on R it follows that there exists a positive constant $C > 0$ such that for any $t \in (0,T)$,

$$\begin{aligned}&\|e^{\psi(t,x)}tR(t,x,\partial_x)v\| = \|tR(t,x,\partial_x - \nabla_x\psi(t,x))w\| \\ &\leq Cs^{2\sigma-d}\left(\|\Delta_y w\|^2 + \||\nabla_y\psi_0|^2 w\|^2 + \|w\|^2\right)^{1/2}.\end{aligned} \tag{13.10}$$

Let $u \in H^{1,2}_{\mathrm{loc}}([0,T]\times\mathbf{R}^d)$ be a solution to (13.2) satisfying the vanishing condition (13.4). Applying the modified Carleman estimate (13.9) to $v = \chi(M\gamma_n t^\sigma)u$, where χ is a smooth cutoff function around the origin and M is a large parameter, we can see, by virtue of (13.10), that there exists a positive constant C such that

$$\int_0^{1/(2M\gamma_n^\sigma)} t^{-\gamma_n}\left\{\|e^\Psi u\|^2 + \|e^\Psi\sqrt{t}\nabla_x u\|^2\right\} dt \leq C\gamma_n \int_{1/(M\gamma_n^\sigma)}^{2/(M\gamma_n^\sigma)} t^{-\gamma_n}\|e^\Psi u\|^2 dt.$$

Since $t^{-\gamma_n}$ is a monotone decreasing function, it holds that for some positive constant C',

$$\int_0^{1/(2M\gamma_n^\sigma)} \left\{\|e^\Psi u\|^2 + \|e^\Psi\sqrt{t}\nabla_x u\|^2\right\} dt \leq C'\gamma_n 2^{-\gamma_n}.$$

Taking $1/(2M\gamma_n^\sigma) = R_n$, we obtain

$$\int_0^{R_n} \left\{ \|e^\Psi u\|^2 + \|e^\Psi \sqrt{t}\nabla_x u\|^2 \right\} dt \le C'(2MR_n)^{-1/\sigma} \exp\left(-(\log 2)(2MR_n)^{-1/\sigma}\right)$$

(cf. Proposition 5.4 in [18, 21]). □

Once we have established the above strong degeneracy of u, we can derive a nice Carleman inequality, which can control the first derivatives term ∇u, with a weight function with stronger singularity.

Lemma 13.12. *Let $0 < \sigma' < 1/\sigma$ and $\varphi(t) = t^{-\sigma'}$. Under the same hypothesis as in Theorem 13.4 it holds that there exist a positive number T and a constant $C > 0$ such that for any sufficiently large γ, the following inequality*

$$\int_0^T e^{2\gamma\varphi} \left\{ \gamma t(t\varphi')'(t)\|e^\Psi u\|^2 + \|e^\Psi \sqrt{t}\nabla u\|^2 + \gamma^{-1} t^{\sigma'} \|e^\Psi t\Delta u\|^2 \right\} \frac{dt}{t^{(d+2)/2}}$$
$$\le C \int_0^\infty e^{2\gamma\varphi} \left\| e^\Psi \left(t\partial_t u + t\Delta u + tRu\right)\right\|^2 \frac{dt}{t^{(d+2)/2}} \tag{13.11}$$

holds for any $u \in C_0^\infty((0,T) \times \mathbf{R}^d)$.

Here, we remark that

$$t(t\varphi')'(t) = (\sigma')^2 t^{-\sigma'} > 0.$$

Proof. Consider the operators

$$\tilde{L}_0 = tL_0 + tR, \text{ and } L_1 = e^{\psi(t,x)}\tilde{L}_0 e^{-\psi(t,x)}.$$

We see that

$$L_1 = \frac{s}{2}\partial_s + H + J$$

where J is a formally self-adjoint operator derived from tR. We use a standard notation for the commutator $[A,B] = AB - BA$. In view of the relation

$$[s\partial_s, V(s^2, sy)] = (2t + x\cdot\nabla_x)V(t,x), \text{ if } t = s^2,\ x = sy,$$

our hypothesis on R yields

$$\|Jv\| + \|[s\partial_s, J]v\| \le Cs^{2\sigma}\left(\|\Delta_y v\| + \||\nabla_y\psi_0|^2 v\| + \|v\|\right). \tag{13.12}$$

Take $\varphi_1(s) = s^{-2\sigma}$ and consider $w = e^{\gamma\varphi_1(s)}v(s,y)$ for $v(s,y) \in C_0^\infty((0,\sqrt{T}) \times \mathbf{R}^d)$. Denoting $L_{1,\gamma} = e^{\gamma\varphi_1(s)}L_1 e^{-\gamma\varphi_1(s)}$, we see that

$$\int_0^{\sqrt{T}} \|e^{\gamma\varphi_1(s)}L_1 v\|^2 \frac{ds}{s} = \int_0^{\sqrt{T}} \|L_{1,\gamma}w\|^2 \frac{ds}{s}$$
$$= \int_0^{\sqrt{T}} \|\frac{1}{2}(s\partial_s - \gamma s\varphi_1'(s))w + (H+J)w\|^2 \frac{ds}{s}$$

$$
\begin{aligned}
&= \int_0^{\sqrt{T}} \|\frac{s}{2}\partial_s w\|^2 \frac{ds}{s} + \int_0^{\sqrt{T}} \|\big(H+J-\frac{\gamma}{2}s\varphi_1'(s)\big)w\|^2 \frac{ds}{s} \\
&\quad +2\mathrm{Re}\int_0^{\sqrt{T}} \langle \frac{s}{2}\partial_s w, \big(H+J-\frac{\gamma}{2}s\varphi_1'(s)\big)w\rangle \frac{ds}{s} \\
&\geq \int_0^{\sqrt{T}} \frac{\gamma}{4} s\partial_s\big(s\varphi_1'(s)\big)\|w\|^2\frac{ds}{s} - \frac{1}{2}\int_0^{\sqrt{T}} \langle w, [s\partial_s, J]w\rangle \frac{ds}{s}. \qquad (13.13)
\end{aligned}
$$

On the other hand we see that

$$
\begin{aligned}
\int_0^{\sqrt{T}} \langle -Hw, w\rangle \frac{ds}{s} &= -\mathrm{Re}\int_0^{\sqrt{T}} \langle (\frac{s}{2}\partial_s + H)w, w\rangle \frac{ds}{s} \\
&= \mathrm{Re}\int_0^{\sqrt{T}} \langle -L_{1,\gamma} w, w\rangle \frac{ds}{s} - \int_0^{\sqrt{T}} \frac{\gamma}{2} s\varphi_1'(s)\|w\|^2 \frac{ds}{s} \\
&\quad + \int_0^{\sqrt{T}} \langle Jw, w\rangle \frac{ds}{s} \qquad (13.14)
\end{aligned}
$$

and

$$
\begin{aligned}
\int_0^{\sqrt{T}} \|Hw\|^2\frac{ds}{s} &\leq 2\int_0^{\sqrt{T}} \|\big(H+J-\frac{\gamma}{2}s\varphi_1'(s)\big)w\|^2\frac{ds}{s} \\
&\quad +4\int_0^{\sqrt{T}} \|Jw\|^2\frac{ds}{s} + 4\int_0^{\sqrt{T}} \|\frac{\gamma}{2}s\varphi_1'(s)w\|^2\frac{ds}{s}. \qquad (13.15)
\end{aligned}
$$

If m is small enough, it holds that there exist positive constants C_1, C_2, and C_3 independent of t such that for $v(x) = e^{\Psi}u(x) \in C_0^\infty(\mathbf{R}^d)$,

$$
\begin{aligned}
\frac{t}{2}\|e^{\Psi}\nabla_x u\|^2 &= \frac{t}{2}\|(\nabla_x - \nabla_x\psi)v\|^2 \\
&\leq \|\sqrt{t}\partial_x w\|^2 + \|\sqrt{t}\nabla_x\psi v\|^2 \\
&\leq C_1 s^d \langle -Hv, v\rangle_{L^2(\mathbf{R}_y^d)} \qquad (13.16)
\end{aligned}
$$

and

$$
\begin{aligned}
\|te^{\Psi}\Delta u\|^2 &= \|t(\partial_x - \nabla_x\psi)^2 v\|^2 \\
&\leq C_2\big(\|t\Delta v\|^2 + \||\sqrt{t}\nabla_x\psi|^2 v\|^2 + \|tv\|^2\big) \\
&\leq C_3 s^d\big(\|Hw\|^2_{L^2(\mathbf{R}_y^d)} + \|w\|^2_{L^2(\mathbf{R}_y^d)}\big). \qquad (13.17)
\end{aligned}
$$

In view of (13.12)–(13.17), if we take T as small enough, it holds that there exist positive constants C_0 and γ_0 such that for any $v \in C_0^\infty((0,T)\times\mathbf{R}^d)$ and $\gamma \geq \gamma_0$,

$$
\begin{aligned}
&\int_0^{\sqrt{T}} \|e^{\gamma\varphi_1(s)}L_1 v\|^2\frac{ds}{s} \\
&\geq C_0\left\{\gamma\int_0^{\sqrt{T}} t(t\varphi')'(t)\|e^{\gamma\varphi_1(s)}v\|^2\frac{ds}{s} + \int_0^{\sqrt{T}} \langle -Hw, w\rangle ds\right\}\frac{ds}{s}.
\end{aligned}
$$

□

Theorem 13.4 can be derived by following an argument in Sect. 3 of [7] with a slight modification caused by the presence of the vector potential A. Let χ_1 and χ_2 be nonnegative cutoff functions in $C^\infty(\mathbf{R}_+)$ such that $0 \le \chi_j \le 1$, $j = 1,2$,

$$\chi_1(t) = 1 \text{ if } t \in [2,\infty),\ \chi_1(t) = 0 \text{ if } t \in [0,1]$$

and

$$\chi_2(t) = 1 \text{ if } t \in [0,1],\ \chi_2(t) = 0 \text{ if } t \in [2,\infty).$$

Let k, R, and M be large positive parameters. Applying the strong Carleman inequality (13.11) to $v(t,x) = \chi_1(kt)\chi_2(Mt)\chi_2(|x|/R)u(t,x)$ and taking the limit as k and R tend to ∞, we conclude that

$$\int_0^{1/(2M)} \|e^{\Psi} u\|^2 \frac{dt}{t^{(d+2)/2}} = \mathscr{O}(e^{-2\gamma\{\varphi(1/2M)-\varphi(1/M)\}}), \qquad (13.18)$$

which shows that $u(t,x) = 0$ in $(0,(2M)^{-1}) \times \mathbf{R}^d$ because the right-hand side tends to 0 as $\gamma \to \infty$.

13.6 Outline of Proofs of Theorems 13.5 and 13.6

Both Theorems 13.5 and 13.6 are proved by the same method, so that we shall only consider Theorem 13.6.

Let $u \in H^{1,2}_{\text{loc}}([0,T] \times \Omega)$ be a solution to (13.2) satisfying the vanishing condition (13.4). Consider two cutoff functions $\chi_1 \in C_0^\infty(\mathbf{R})$ and $\chi_2 \in C_0^\infty(\mathbf{R}^d)$ such that χ_1 and χ_2 are nonnegative functions equal to 1 near the origin, respectively. Applying the modified Carleman estimate (13.9) to $v = \chi_1(M_1\gamma_n t^\sigma)\chi_2(M_2 x)u$, where M_1 and M_2 are large parameters, we can see that for any compact set of Ω and any small positive number τ, there exist positive number δ and constant C such that

$$\int_0^{\tau} \|e^{\Psi} u(t,x)\|^2_{L^2(K)} \frac{dt}{t^{d/2}} \le Ce^{-\frac{\delta}{\tau}}.$$

It is not difficult to see that the last inequality implies

$$\int_0^{\tau} \|e^{\frac{\delta}{t}} e^{\Psi} u(t,x)\|^2_{L^2(K)} \frac{dt}{t^{d/2}} < +\infty.$$

Since $u \in H^{1,2}([0,T] \times K)$ implies $u \in C([0,T];\ L^2(K))$, we can use a technique of convolution which is similar to one given in [8]. It holds that if $t > 0$ and $x \in K$ and $|y| \le \sqrt{\frac{4\delta}{(1-2\kappa)}}$,

$$\frac{\delta}{4t} \ge (1-2\kappa)\left(\frac{|x|^2}{8t} - \frac{|y-x|^2}{4t}\right).$$

Hence from the inequality

$$\int_0^\tau \|e^{\delta/(4t)}e^{\Psi}u(t,x)\|^2_{L^2(K)}\frac{dt}{t^{d/2}} \le Ce^{-\frac{\delta}{2\tau}},$$

it follows that

$$\int_{|y|\le r_0}\int_0^\tau \|e^{-(1-2\kappa)|y-x|^2/(4t)}u(t,x)\|^2_{L^2(K)}\frac{dt}{t^{d/2}}dy \le C|B(0,r_0)|e^{-\frac{\delta}{2\tau}}, \tag{13.19}$$

where $r_0 = \sqrt{\frac{4\delta}{(1-2\kappa)}}$. Let χ_K be a characteristic function of K. Define a positive constant c_0 by

$$c_0 = \int_{\mathbf{R}^d} \eta(x)dx,$$

where $\eta(x) = e^{-(1-2\kappa)|x|^2/2}$. We see that

$$\lim_{t\to+0}\int_{B(0,r_0)}\left|\int_{\mathbf{R}^d} t^{-d/2}\eta(\frac{y-x}{\sqrt{t}})|\chi_K(x)u(t,x)|^2dx - c_0|\chi_K(y)u(0,y)|^2\right|dy = 0. \tag{13.20}$$

Combining (13.19) with (13.20), we can conclude that $\chi_K(y)u(0,y) = 0$ on $B(0,r_0)$.

Appendix

Perturbation Theory

For a complex parameter ζ, we consider a family of operators

$$S(\zeta)u = Su + \zeta S^{(1)}u + \zeta^2 S^{(2)}u + \cdots,$$

where S is a densely defined and sectorial operator with domain $D(S)$ and the domain of each operator $S^{(n)}$ contains $D(S)$. Suppose that for some $a \ge 0$, $0 \le b < 1$

$$|(S^{(n)}u,u)| \le c^{n-1}\left(a\|u\|^2 + b\mathrm{Re}(Su,u)\right),\ u \in D(S).$$

Then

Lemma 13.13 (Theorem 4.12 in [11]). *When* $|\zeta| < (b+c)^{-1}$,

$$S(\zeta)u = Su + \zeta S^{(1)}u + \zeta^2 S^{(2)}u + \cdots,\ u \in D(S)$$

has a Friedrichs extension $T(\zeta)$ *which is a holomorphic family of type (B).*

Let $S(\zeta)$ be a self-adjoint holomorphic family of type (B) defined for $\zeta \in D_0 \subset \mathbf{C}$ and satisfies

$$\left|\frac{d}{d\zeta}(S(\zeta)u,u)\right| \le a'(u,u)+b'(S(\zeta)u,u),\ u\in D,\ \zeta\in I,$$

where $I \subset D_0$ is a compact interval of the real axis. Let $\mu(\zeta)$ be any continuous, piecewise holomorphic eigenvalue of $T(\zeta)$.

Lemma 13.14 (Theorem 4.21 in [11]).

$$|\mu(\zeta)-\mu(0)| \le \frac{1}{b'}\left(a'+b'\mu(0)\right)\left(e^{b'|\zeta|}-1\right)$$

as long as $\zeta \in I$ *and* $\lambda(\zeta)$ *is defined.*

Acknowledgements This research was partially supported by Grants-in-Aid for Scientific Research (No. 22540185), Japan Society for the Promotion of Science.

References

1. Alessandrini, G., Vessella, S.: Remark on the strong unique continuation property for parabolic operators. Proc. AMS **132**, 499–501 (2003)
2. Alinhac, S., Baouendi, M.S.: A counterexample to strong uniqueness for partial differential equations of Schrödinger's type. Comm. P.D.E. **19**, 1727–1733 (1994)
3. Aronszajn, N.: A unique continuation theorem for solutions of elliptic partial differential equations or inequalities of second order. J. Math. Pure. Appl. **36**, 235–249 (1957)
4. Carleman, T.: Sur un problème d'unicité pour les systèmes d'équations aux dérivées partielles à deux variables indépendantes. Ark. Mat. Astr. Fys. **26**,1–9 (1939)
5. Chen, X.Y.: A strong unique continuation theorem for parabolic equations. Math. Ann. **311**, 603–630 (1998)
6. Cordes, H.: Über die Besstimmtheit der Lösungen elliptischer Differentialgleichungen durch Anfangsvorgaben. Nachr. Akad. Wiss. Göttingen II a, 230–258 (1956)
7. Escauriaza, L.: Carleman inequalities and the heat operator. Duke Math. J. **104**, 113–127 (2000)
8. Escauriaza, L., Vega, L.: Carleman inequalities and the heat operator II. Indiana U. Math. J. **50**, 1149–1169 (2001)
9. Grammatico, C.: A result on strong unique continuation for the Laplace operator. Comm. P. D. E. **22**, 1475–1491 (1997)
10. Jerison, D., Kenig, C.E.: Unique continuation and absence of positive eigenvalues for Schrödinger operators. Ann. Math. **121**, 463–494 (1985)
11. Kato, T.: Perturbation Theory for Linear Operators. Springer (1976)
12. Koch, H., Tataru, D.: Carleman estimates and unique continuation for second order parabolic equations with non-smooth coefficients. Comm. P. D. E. **34**, 305–366 (2009)
13. Landis, E.M., Oleinik, O.A.: Generalized analyticity and some related properties of solutions of elliptic and parabolic equations. Uspekhi Mat. Nauk **29**(176):2, 190–206 (1974) (Russian). English transl.: Russian Math. Surveys **29**, 195–212 (1974)
14. Lees, M., Protter, M.H.: Unique continuation for parabolic differential equations and inequalities. Duke Math. J. **28**, 369–382 (1961)
15. Lin, F.H.: A uniqueness theorem for parabolic equations. CPAM **43**, 127–136 (1990)
16. Mizohata, S.: Unicité du prolongement des solutions pour quelques opérateurs différentiels paraboliques. Mem. Coll. Sci. Univ. Kyoto Ser. A, **31**(3), 219–239 (1958)
17. Nirenberg, L.: Uniqueness in Cauchy problems for differential equations with constant leading coefficients. Comm. Pure Appl. Math. **10**, 89–105 (1957)

18. Okaji, T.: Strong unique continuation property for time harmonic Maxwell equations. J. Math. Soc. Japan. **54-1**, 87–120 (2002)
19. Poon, C.C.: Unique continuation for parabolic operators. Comm. P. D. E. **21**, 521–539 (1996)
20. Regbaoui, R.: Prolongement unique pour les opérateurs de Schrödinger. Thesis, University of Rennes (1993)
21. Regbaoui, R.: Strong uniqueness for second order differential operators. J. Differ. Equat. **141**, 201–217 (1997)
22. Reed, M., Simon, B.: Methods of Modern Mathematical Physics, vol. 2. Academic, New York (1975)
23. Schechter, M.: Spectra of Partial Differential Operators, 2nd edn. Elsevier, Amsterdam (1986)
24. Sogge, C.D.: Strong uniqueness theorems for second order elliptic differential equations. Am. J. Math. **112**, 943–984 (1990)
25. Sogge, C.D.: A unique continuation theorem for second order parabolic differential operators. Ark. Mat. **28**, 159–182 (1990)
26. Stummel, F.: Singuläre elliptische Differentialoperatoren in Hilbertschen Räumen. Math. Ann. **132**, 150–176 (1956)
27. Weidmann, J.: Lincar Opcrators in Hilbcrt Spaces (English translation). Springer, Berlin (1980)
28. Wolff, T.H.: Recent work on sharp estimates in second order elliptic unique continuation problems. In: Fourier Analysis and partial Differential Equations, pp. 99–128. Studies in Advanced Mathematics. CRC, Boca Raton (1995)
29. Yamabe, H.: A unique continuation theorem of a diffusion equation. Ann. Math. (2), **69**, 462–466 (1959)

Chapter 14
On the Problem of Positivity of Pseudodifferential Systems

Alberto Parmeggiani

Abstract I will give here in the first place a survey of the problem of positivity estimates for systems of ψdos, such as the Sharp-Gårding and the Fefferman–Phong inequalities. Next, I will generalize Fujiwara's construction of the approximate positive/negative part of a first-order scalar ψdo to certain 2×2 systems of first-order ψdos.

Key words: Fefferman-Phong inequality, Positivity estimates, Positive/negative part of a first-order system, Systems of PDEs, Sharp-Gårding inequality

2010 Mathematics Subject Classification: Primary: 35S05; Secondary: 35B45, 35A30.

14.1 Introduction

Positivity estimates such as the Gårding and the Sharp-Gårding inequalities, the Fefferman–Phong inequality, and the Melin and the Hörmander inequalities (see Hörmander's third volume in [12]) are basic tools for studying, e.g., the Friedrichs extension of a formally self-adjoint operator, energy bounds of evolution equations, a priori estimates concerning hypoellipticity or solvability of PDEs. I will be concerned here with the Sharp-Gårding and the Fefferman–Phong inequalities (respectively, (SG) and (FP), for short) that I will first recall in the scalar setting. Next, I will give an overview on what is known about these estimates in the case of systems, and finally, after recalling Fujiwara's result on the approximate positive/negative part of a first-order *scalar* ψdo (pseudodifferential operator)

A. Parmeggiani (✉)
Dipartimento di Matematica, Università di Bologna, Piazza di Porta San Donato 5, I-40126 Bologna, Italy
e-mail: alberto.parmeggiani@unibo.it

M. Cicognani et al. (eds.), *Studies in Phase Space Analysis with Applications to PDEs*, Progress in Nonlinear Differential Equations and Their Applications 84, DOI 10.1007/978-1-4614-6348-1_14,

in the Hörmander class $S^1 = S^1_{1,0}(\mathbb{R}^n \times \mathbb{R}^n)$, I will state and prove an extension of such a construction to a first-order 2×2 system with Hermitian symbol in $S^1(\mathbb{R}^n \times \mathbb{R}^n; \mathsf{M}_2) =: S^1(\mathbb{R}^n \times \mathbb{R}^n) \otimes \mathsf{M}_2$, where M_N denotes the algebra of $N \times N$ complex matrices. (Analogous notation will be used in the case of S^m.)

My main aim in exploring these estimates is to understand the interplay between the phase-space geometry and the "frame" geometry, the latter being the geometry locally given by the matrix which gives the symbol of the operator. The main point in this approach is to avoid explicit requests on the eigenvalues of the symbol, which are seldom known explicitly.

I will use throughout the Weyl–Hörmander pseudodifferential calculus that I will briefly recall in the next section, addressing the reader to Hörmander's fundamental paper [11], or Hörmander's book [12], or also Lerner's book [13].

14.2 Background on the Weyl–Hörmander Calculus

Let $X = (x,\xi)$, $Y = (y,\eta)$, and $Z = (z,\zeta) \in \mathbb{R}^n \times \mathbb{R}^n$. Let $\sigma = \sum_{j=1}^n d\xi_j \wedge dx_j$ be the canonical symplectic 2-form in $\mathbb{R}^{2n}_X = \mathbb{R}^n_x \times \mathbb{R}^n_\xi$. Hence $\sigma(X,Y) = \langle \xi, y\rangle - \langle \eta, x\rangle$.

Definition 14.1. An *admissible metric* in $\mathbb{R}^{2n}$ is a function $\mathbb{R}^{2n} \ni X \longmapsto g_X$ where g_X is a positive-definite quadratic form on $\mathbb{R}^{2n}$ such that

- *Slowness:* There exists $C_0 > 0$ (the constant of *slowness*) such that for any given $X, Y \in \mathbb{R}^{2n}$ one has
$$g_X(Y - X) \le C_0^{-1} \Longrightarrow C_0^{-1} g_Y \le g_X \le C_0 g_Y.$$
- *Uncertainty:* For any given $X \in \mathbb{R}^{2n}$ one has
$$g_X \le g_X^\sigma,$$
where g_X^σ is the *dual* metric defined by
$$g_X^\sigma(Y) = \sup_{Z \neq 0} \frac{\sigma(Y,Z)^2}{g_X(Z)}.$$
- *Temperateness:* There exists $C_1 > 0$ and $N_1 \in \mathbb{Z}_+$ such that for all $X, Y \in \mathbb{R}^{2n}$, one has
$$g_X \le C_1 g_Y \big(1 + g_X^\sigma(X - Y)\big)^{N_1}.$$

I shall call the constants appearing above the *structural constants of the metric* g. The *Planck function* associated with g is by definition
$$h(X)^2 = \sup_{Z \neq 0} \frac{g_X(Z)}{g_X^\sigma(Z)}.$$
Hence, by the uncertainty property one always has $h \le 1$. ◇

Definition 14.2. Given an admissible metric g, a *g-admissible weight* is a positive function m on $\mathbb{R}^{2n}$ for which there exist constants $c, C, C' > 0$ and $N' \in \mathbb{Z}_+$ such that for all $X, Y \in \mathbb{R}^{2n}$,

$$g_X(X-Y) \le c \Longrightarrow C^{-1} \le \frac{m(X)}{m(Y)} \le C,$$

and

$$\frac{m(X)}{m(Y)} \le C'\big(1 + g_X^\sigma(Y-X)\big)^{N'}.$$

◇

In particular, given an admissible metric g, one always has that the Planck function h associated with g is a *g-admissible weight.*

Definition 14.3. Let g be an admissible metric and m be a g-admissible weight. Let $a \in C^\infty(\mathbb{R}^{2n})$. Denote by $a^{(k)}(X; v_1, \ldots, v_k)$ the k-th differential of a at X in the directions $v_1, \ldots, v_k$ of $\mathbb{R}^{2n}$. Define

$$|a|_k^g(X) := \sup_{0 \ne v_1, \ldots, v_k \in \mathbb{R}^{2n}} \frac{|a^{(k)}(X; v_1, \ldots, v_k)|}{\prod_{j=1}^k g_X(v_j)^{1/2}}.$$

We say that $a \in S(m,g)$ if for any given integer $k \in \mathbb{Z}_+$ the following seminorms are finite:

$$\|a\|_{k,S(m,g)} := \sup_{\ell \le k,\, X \in \mathbb{R}^{2n}} \frac{|a|_\ell^g(X)}{m(X)} < +\infty. \tag{14.1}$$

With $B^g_{X_0,r} = \{X;\, g_{X_0}(X - X_0) < r^2\}$, following Bony and Lerner [2], we say that $a \in C^\infty(\mathbb{R}^{2n})$ is a *symbol of weight m confined* to the ball $B^g_{X_0,r}$, and write $a \in \mathrm{Conf}(m, g, X_0, r)$, if for all $k \in \mathbb{Z}_+$

$$\|a\|_{k,\mathrm{Conf}(m,g,X_0,r)} := \sup_{\ell \le k,\, X \in \mathbb{R}^{2n}} \frac{|a|_\ell^{g_{X_0}}(X)}{m(X_0)} \big(1 + g^\sigma_{X_0}(X - B_{X_0,r})\big)^{k/2} < +\infty, \tag{14.2}$$

where $g^\sigma_Y(X - B) = \inf_{Z \in B} g^\sigma_Y(X - Z)$. Hence the space of symbols confined to the ball $B^g_{X_0,r}$ coincides with $\mathscr{S}(\mathbb{R}^{2n})$ endowed with the seminorms (14.2). Any given $\varphi \in C_0^\infty(B^g_{X_0,r})$ is automatically confined to the ball $B^g_{X_0,r}$. ◇

Given $p \in S(m,g)$, the ψdo associated with p through Weyl quantization is given by

$$p^{\mathrm{w}}(x,D)u(x) = (2\pi)^{-n} \iint e^{i\langle x-y,\xi\rangle} p\Big(\frac{x+y}{2}, \xi\Big) u(y)\, dy\, d\xi, \quad u \in \mathscr{S}(\mathbb{R}^n).$$

I shall denote by $\Psi(m,g)$ the class of ψdos obtained by Weyl quantization of symbols in $S(m,g)$.

As for the composition in the Weyl calculus, one has the following result:

Theorem 14.1. *Given* $a \in S(m_1,g)$, $b \in S(m_2,g)$ *then*

$$a^{\mathrm{w}}(x,D)b^{\mathrm{w}}(x,D) = (a\#b)^{\mathrm{w}}(x,D) \in \Psi(m_1m_2,g),$$

where for any given $N \in \mathbb{Z}_+$

$$(a\#b)(X) = \sum_{j=0}^{N} \frac{1}{j!}\Big(\frac{i}{2}\sigma(D_X,D_Y)\Big)^j a(X)b(Y)\Big|_{X=Y} + r_{N+1}(X), \tag{14.3}$$

with $r_{N+1} \in S(h^{N+1}m_1m_2,g)$.

Associated with an admissible metric g, one has a partition of unity as follows (see [2, 12, 13]):

Lemma 14.1. *Let* g *be an admissible metric, and let* $r^2 < C_0^{-1}$. *Then there exists a sequence of centers* $\{X_\nu\}_{\nu\in\mathbb{Z}_+}$, *a covering of* $\mathbb{R}^{2n}$ *made of g-balls* $B^g_{\nu,r} = \{X; g_{X_\nu}(X-X_\nu) < r^2\}$ *centered at* X_ν *and radius r, and a sequence of functions* $\{\varphi_\nu\}_{\nu\in\mathbb{Z}_+}$ uniformly *in* $S(1,g)$, *with* $\operatorname{supp}\varphi_\nu \subset B^g_{\nu,r}$, *such that* $\sum_{\nu\in\mathbb{Z}_+}\varphi_\nu^2 = 1$. *Moreover, for any given* r_* *such that* $r^2 \le r_*^2 < C_0^{-1}$, *there exists an integer* N_{r_*} *such that no more than* N_{r_*} *balls* B^g_{ν,r_*} *can intersect at each time (i.e., one has an a priori finite number of overlappings of the dilates by the factor* r_*/r *of the* $B^g_{\nu,r}$). *In addition, with*

$$g^\sigma_X(B-B') := \inf_{Y\in B,\,Y'\in B'} g^\sigma_X(Y-Y'),\ B,B' \subset \mathbb{R}^{2n},$$

and

$$\Delta_{\mu\nu}(r_*) := \max\{1, g^\sigma_{X_\mu}(B^g_{\mu,r_*} - B^g_{\nu,r_*}), g^\sigma_{X_\nu}(B^g_{\mu,r_*} - B^g_{\nu,r_*})\}^{1/2},$$

there exist constants N' *and* C' *such that*

$$\sup_\mu \sum_\nu \Delta_{\mu\nu}(r_*)^{-N'} < C'.$$

Moreover, given $0 < r' < C_0^{-1}$, *for all* $k \in \mathbb{Z}_+$ *there exist* $C > 0$ *and* $\ell \in \mathbb{Z}_+$ *such that for any given* $a \in S(m,g)$ *and* $b \in \operatorname{Conf}(1,g,X,r')$ *one has*

$$\|a\#b\|_{k,\operatorname{Conf}(1,g,X,r')} \le Cm(X)\|a\|_{\ell,S(m,g)}\|b\|_{\ell,\operatorname{Conf}(1,g,X,r')}. \tag{14.4}$$

Finally, given $0 < r' < C_0^{-1}$, *for all* $k,N \in \mathbb{Z}_+$ *there exist* $C > 0$ *and* $\ell \in \mathbb{Z}_+$ *such that for every* $\mu,\nu \in \mathbb{N}$, *and every* $a \in \operatorname{Conf}(1,g,X_\mu,r')$ *and* $b \in \operatorname{Conf}(1,g,X_\nu,r')$, *one has*

$$\begin{aligned}&\|a\#b\|_{k,\operatorname{Conf}(1,g,X_\mu,r')} + \|a\#b\|_{k,\operatorname{Conf}(1,g,X_\nu,r')} \le \\ &\qquad \le C\|a\|_{\ell,\operatorname{Conf}(1,g,X_\mu,r')}\|b\|_{\ell,\operatorname{Conf}(1,g,X_\nu,r')}\Delta_{\mu\nu}(r')^{-N}.\end{aligned} \tag{14.5}$$

As for bounds on the norm of $a^{\mathrm{w}}(x,D)$ when acting on L^2, with $a \in S(1,g)$ or $a \in \operatorname{Conf}(1,g,X,r)$, one has the following lemma (see [2]):

Lemma 14.2. *Let g be an amissible metric. Then there exist $C > 0$ and $k \in \mathbb{N}$ such that, according to whether $a \in \mathrm{Conf}(1,g,X,\tilde{r})$, $\tilde{r} \le 1$, or $a \in S(1,g)$, one has*

$$\|a^{\mathrm{w}}(x,D)\|_{L^2 \to L^2} \le C\|a\|_{k,\mathrm{Conf}(1,g,X,\tilde{r})}, \quad \|a^{\mathrm{w}}(x,D)\|_{L^2 \to L^2} \le C\|a\|_{k,S(1,g)}.$$

One has finally the following useful lemma, due to Bony and Lerner [2] (see also [13]).

Lemma 14.3. *Let g be an admissible metric, and let m be a g-admissible weight. Let B_ν be a g-ball as in Lemma 14.1. Let $g_\nu = g_{X_\nu}$ and $m_\nu = m(X_\nu)$. Let $\{a_\nu\}_{\nu \in \mathbb{Z}_+}$ be a sequence of symbols with $a_\nu \in S(m_\nu, g_\nu)$, such that for any given integer $k \in \mathbb{Z}_+$*

$$\sup_{\nu \in \mathbb{Z}_+} \|a_\nu\|_{k,\mathrm{Conf}(m_\nu,g_\nu,X_\nu,r)} < +\infty.$$

Then $a := \sum_{\nu \in \mathbb{Z}_+} a_\nu$ belongs to $S(m,g)$. The sequence $\{a_\nu\}_{\nu \in \mathbb{Z}_+}$ is said to be uniformly confined *in $S(m,g)$. When $m = 1$ we have from the Cotlar–Stein Lemma (see [12], Lemma 18.6.5) that $a^{\mathrm{w}} = \sum_\nu a_\nu^{\mathrm{w}}$ is a bounded operator in L^2.*

By $S(m,g;\mathsf{M}_N) = S(m,g) \otimes \mathsf{M}_N$, I shall denote the matrix-valued symbols. In the case of matrix-valued symbols, Definitions 14.2 and 14.3, and the composition formula (14.3), hold (being careful with the order of the terms). By $\Psi(m,g;\mathsf{M}_N)$, I shall denote the ψdos obtained by Weyl quantization of symbols in $S(m,g;\mathsf{M}_N)$.

Given $A, B > 0$, I shall throughout write $A \lesssim B$ when $A \le CB$ for some universal constant $C > 0$, and $A \approx B$ when $A \lesssim B$ and $B \lesssim A$.

14.3 The Basic Positivity Estimates in the Scalar Case

I now recall the *scalar* Sharp-Gårding inequality, due to Hörmander [12]. Hereafter, $(\cdot,\cdot)$ denotes the L^2-scalar product and $\|\cdot\|_s$ the norm in the L^2-based Sobolev space H^s.

Theorem 14.2. *If $0 \le p \in S(h^{-1},g)$, then there exists $C > 0$ such that*

$$(p^{\mathrm{w}}(x,D)u,u) \ge -C\|u\|_0^2, \quad \forall u \in \mathscr{S}(\mathbb{R}^n). \tag{14.6}$$

The theorem states that if p is *first-order* in the calculus and nonnegative, then one has control from below of the L^2-quadratic form associated with the operator.

Next, one has the scalar Fefferman–Phong inequality, proved by Fefferman and Phong in [7] (see also [12]), which states that a control from below is also possible for nonnegative symbols which are *second-order* in the calculus.

Theorem 14.3. *If $0 \le p \in S(h^{-2},g)$, then there exists $C > 0$ such that*

$$(p^{\mathrm{w}}(x,D)u,u) \ge -C\|u\|_0^2, \quad \forall u \in \mathscr{S}(\mathbb{R}^n). \tag{14.7}$$

Both proofs go through microlocally writing

$$p^{\mathrm{w}}(x,D) = q^{\mathrm{w}}(x,D)^2 + r^{\mathrm{w}}(x,D),$$

where $r^{\mathrm{w}}(x,D)$ is under control, that is, in the case of (14.6) it is L^2 bounded, and in the case of (14.7), one has $0 \leq r \in S(h^{-1},g)$ so that r^{w} satisfies (14.6). Furthermore, in the case of the Fefferman–Phong inequality, the proof is based on a *Calderón-Zygmund microlocalization*: One changes microlocally the metric (with uniform bounds on the structural constants and the seminorms of the symbol) to reduce matters to a "semiclassical" case on metric balls B_ν on which

- Either $p\big|_{B_\nu} \lesssim 1$, which gives rise, through a re-summation by a Cotlar–Stein argument, to an L^2-error term.
- Or $p\big|_{B_\nu}$ is elliptic, whence this piece gives a good microlocal control on u.
- Or $p\big|_{B_\nu}$ is microlocally (through a Fourier integral operator) conjugated to $e(y,\eta)\eta_1^2 + \tilde{p}_1(y_1,y',\eta')$, with $e > 0$ elliptic and $\tilde{p}_1 \geq 0$ (the *non-elliptic non-degenerate case*) depending on fewer variables, which makes it possible to use *an induction on the number of the variables*.
- Finally, a re-summation Cotlar–Stein argument gives the final estimate.

The constant in (14.7) depends on the structural constants of the metric g and on a number of seminorms of the symbol p which depends on the dimension n (see [7, 12, 13]).

Recently Tataru [31] has extended the Sharp-Gårding inequality to situations where the symbol has a reduced smoothness, and Ruzhansky and Turunen [29] have extended the Sharp-Gårding inequality to ψdos on Lie groups.

As for the Fefferman–Phong inequality, there is a paper by Bony [1] in which it is showed that the inequality holds, provided one has control *only* on derivatives of the symbol from the fourth order on, and papers by Tataru [31], Herau [9], Boulkhemair [3], and Lerner and Morimoto [14] in which the Fefferman–Phong inequality is extended to situations where the symbol has a limited smoothness. In [14], the authors work with the usual symbol space $S^m = S(h^{-m},g)$, where g and h are given by

$$g_X = |dx|^2 + \frac{|d\xi|^2}{\langle\xi\rangle^2}, \ h(X) = \langle\xi\rangle^{-1}, \ \text{with } \langle\xi\rangle = (1+|\xi|^2)^{1/2},$$

and succeed in giving a rather precise bound on the number of seminorms of the symbol which the constant appearing in the inequality depends on.

Let me mention in passing also Parenti and Parmeggiani [20] (Thm. 5.3), in which a sufficient condition is given in order to obtain a bound from below with a *positive* constant.

For understanding counterexamples in the case of systems, it will also be useful to consider:

- The *local Sharp-Gårding inequality* (local (SG), for short, that I directly state in the vector-valued case): *For a first-order symbol $p = p^* \geq 0$, for any given compact $K \subset \mathbb{R}^n$ there exists $C_K > 0$ such that*

$$(p^{\mathrm{w}}(x,D)u,u) \geq -C_K\|u\|_0^2, \ \forall u \in C_0^\infty(K;\mathbb{C}^N); \tag{14.8}$$

- The *local Fefferman–Phong inequality* (local (FP), for short, that I directly state in the vector-valued case): *For a second-order symbol $p = p^* \geq 0$, for any given compact $K \subset \mathbb{R}^n$, there exists $C_K > 0$ such that*

$$(p^{\mathrm{w}}(x,D)u,u) \geq -C_K\|u\|_0^2, \ \forall u \in C_0^\infty(K;\mathbb{C}^N). \tag{14.9}$$

This is due to the fact that some examples are *polynomial* in nature. (Of course, alternatively one may extend these examples outside a compact set in such a way that the coefficients of the resulting operator are bounded, along with all the derivatives to all orders.)

It is important to remark that

$$\text{(SG)} \Longrightarrow \text{local (SG)}$$

and that

$$\text{(FP)} \Longrightarrow \text{local (FP).}$$

Hence, any given time the local inequality does not hold, so is the case for the other.

I want also to mention that inequality (14.9) for the subclass of *classical* ψdos (i.e., ψdos whose symbol

$$p(x,\xi) \sim \sum_{j\geq 0} p_{2-j}(x,\xi), \ \ p_{2-j}(x,\xi) \text{ positively homogeneous of degree } 2-j \text{ in } \xi)$$

in the case of m-th order symbols with a control on the $(m-2)/2$-Sobolev norm is due to Hörmander [10]. This is the reason why Parenti and myself referred to inequality (14.9) in the case of classical symbols also as to *the Hörmander inequality*. In this regard, in the next subsection I will make a very short digression to discuss the validity of the local (FP) inequality, including the vector-valued case, in the more restrictive framework of *classical* ψdos.

14.3.1 The Case of Classical (Second-Order) ψdos

For second-order *classical* ψdos, Hörmander (see [10, 12]) proved that (14.9) holds even when the symbol goes to $-\infty$ in some (characteristic) directions, hence in directions along which a second-order symbol goes to $-\infty$ as fast as $|\xi|$. However, this is possible provided certain geometric assumptions hold, that is, the principal symbol p_2 has to vanish *exactly* to second order (transversal ellipticity) on the characteristic manifold Σ, which is assumed to be smooth and having $\mathrm{rk}\sigma\big|_\Sigma$ constant, and that the fundamental Melin *trace*$^+$ condition holds. The rank condition on the symplectic form has been slightly relaxed by Parenti and Parmeggiani [21] into a "stratified" condition. I remark that relaxing the transversal ellipticity condition is much more difficult (see [17]).

Extensions of the (local) Fefferman–Phong inequality to $N\times N$ systems of classical ψdos with *double characteristics*, that is, systems with Hermitian nonnegative principal symbol p_2 such that the characteristic manifold $\Sigma=\{\det p_2=0\}$ is smooth with $\mathrm{rk}\sigma\big|_\Sigma$=constant and such that

$$\det p_2(x,\xi)\approx|\xi|^{2N}\mathrm{dist}_\Sigma(x,\xi/|\xi|)^{2\ell},\ \dim\mathrm{Ker}\,p_2\big|_\Sigma=\ell,$$

is due to Hörmander [10] in the case $\ell=1$ and to Parenti and myself [19] when Σ is symplectic and $1\le\ell\le N$. In the latter paper, the technique of the *localized operator* (introduced by Boutet de Monvel) is used to reduce matters to a condition on the spectrum of the *localized system* P_ρ at each $\rho\in\Sigma$. (Although complicated, such a condition reduces to the Melin trace$^+$ condition in the scalar case $N=1$.) The localized system is a system obtained through Taylor expansion at each characteristic ρ in normal directions. It becomes a globally elliptic formally self-adjoint system with polynomial coefficients on the normal space $N_\rho\Sigma$ to Σ at each $\rho\in\Sigma$ when P has double characteristics and Σ is symplectic; see Parenti and Parmeggiani [19] (see also the survey Parmeggiani [22]).

I mention also the papers by Brummelhuis [5], Brummelhuis and Nourrigat [6], and Toft's thesis for extensions to the systems case of the fundamental *Melin inequality* (see [12]).

It is also interesting to remark that, as proved in Parenti and Parmeggiani [19] and in Nicola [18], for $N\times N$ systems with double characteristics and Σ symplectic, the sole nonnegativity condition $P_\rho\ge 0$ (on Schwartz vector-valued functions) on the localized system yields the inequality

$$(p^{\mathrm{w}}(x,D)u,u)\ge -C\|u\|_{1/4}^2,\ \forall u\in\mathscr{S}(\mathbb{R}^n;\mathbb{C}^N).$$

Thus, in this approach the problem of understanding the spectrum of a polynomial-coefficient system arises, a problem for which very few results are available (see [27]).

14.4 The Sharp-Gårding and the Fefferman–Phong Inequality for Systems

I now turn to discussing what is known about the Sharp-Gårding inequality and the Fefferman–Phong inequality for systems of ψdos.

As for (SG), one has the following very general statement, due to Hörmander (see [12]; the $N\times N$ case was shown to hold also by Lax and Nirenberg, and R. Beals), which holds in the infinite-dimensional case.

Theorem 14.4. *Let* H *be a Hilbert space, and let* $\mathscr{L}(\mathsf{H},\mathsf{H})$ *be the algebra of linear bounded operator from* H *into itself. If* $0\le p\in S(h^{-1},g;\mathscr{L}(\mathsf{H},\mathsf{H}))$ *(that is,* p *is a first-order symbol in the calculus which takes nonnegative values in* $\mathscr{L}(\mathsf{H},\mathsf{H})$*), then there exists* $C>0$ *such that*

$$(p^{\mathrm{w}}(x,D)u,u) \geq -C\|u\|_0^2, \;\; \forall u \in \mathscr{S}(\mathbb{R}^n;\mathsf{H}). \tag{14.10}$$

The proof (see [12]) goes through a semiclassical approximation and then the comparison of the operator with a system of harmonic oscillators.

As for (FP), the main assumption on the symbol $p \in S(h^{-1},g;\mathsf{M}_N)$ is again that

$$p(X) = p(X)^* \geq 0, \;\; \forall X \in \mathbb{R}^n \times \mathbb{R}^n. \tag{14.11}$$

Now the situation is drastically different: *In general condition (14.11) alone is* not sufficient *to grant the (FP) inequality. There are* counterexamples.

The main reason for this is the example due to Hörmander of a system with polynomial coefficients of degree 2 in x,ξ, whose Weyl quantization is *not* nonnegative.

Lemma 14.4 (Hörmander [11]). *For* $X = (x,\xi) \in \mathbb{R} \times \mathbb{R}$, *consider the following symbol*

$$A_H(X) = \begin{bmatrix} \xi^2 & x\xi \\ x\xi & x^2 \end{bmatrix} = \begin{bmatrix} \xi \\ x \end{bmatrix}^* \otimes \begin{bmatrix} \xi \\ x \end{bmatrix}. \tag{14.12}$$

Then A_H *satisfies (14.11), but* $A_H^{\mathrm{w}}(x,D)$ *is* not *nonnegative.*

(Recall that $v_1^* \otimes v_2$ is defined, for $v_1,v_2 \in \mathbb{C}^N$, as $(v_1^* \otimes v_2)\colon \mathbb{C}^n \ni w \longmapsto \langle w,v_1\rangle_{\mathbb{C}^N} v_2$.) The lemma allows me to give the perhaps simplest example of system for which (FP) cannot hold. Let $M \geq 1$ be a parameter. Let $g = |dX|^2/M$. Of course, g is an admissible metric (the semiclassical metric), with Plank function $h(X) = 1/M$ and structural constants independent of M. Choose $\chi \in C_0^\infty(\mathbb{R} \times \mathbb{R})$ with $0 \leq \chi \leq 1$ and such that $\chi \equiv 1$ for $|t| \leq 1$ and $\chi \equiv 0$ for $|t| \geq 2$.

Theorem 14.5. *Let*

$$p(X) = p_M(X) = M\chi(\frac{X}{M^{1/2}})A_H(X) = M^2\chi(\frac{X}{M^{1/2}})A_H(\frac{X}{M^{1/2}}). \tag{14.13}$$

Then $p \in S(h^{-2},g;\mathsf{M}_2)$ *for all* $M \geq 1$, *and* $p^{\mathrm{w}}(x,D)$ cannot *satisfy the Fefferman–Phong inequality, that is, there are* no *constants* $C \in \mathbb{R}$ *and* $M_0 \geq 1$, *depending only on the dimension and a finite number (dependent on the dimension) of seminorms of* p, *such that*

$$(p^{\mathrm{w}}(x,D)u,u) \geq -C\|u\|_0^2, \;\; \forall u \in \mathscr{S}(\mathbb{R};\mathbb{C}^2), \;\; \forall M \geq M_0. \tag{14.14}$$

Proof. That $p \in S(h^{-2},g;\mathsf{M}_2)$ for all $M \geq 1$ is clear, since $|\partial_X^\alpha p(X)| \leq C_\alpha M^{2-|\alpha|/2}$, for all $\alpha \in \mathbb{Z}_+^n$. Now, pick $u_0 \in \mathscr{S}(\mathbb{R};\mathbb{C}^2)$ such that $(A_H^{\mathrm{w}}(x,D)u_0,u_0) < 0$. Suppose we have (14.14) for some C and M_0. Then, using the fact that

$$\mathscr{S}' \ni a \longmapsto (a^{\mathrm{w}}(x,D)u,v) \text{ is continuous for all } u,v \in \mathscr{S},$$

on the one hand, we have

$$\lim_{M\to+\infty} (p^{\mathrm{w}}(x,D)u_0,u_0)/M \geq 0,$$

and on the other

$$(p^{\mathrm{w}}(x,D)u_0,u_0)/M \longrightarrow (A_H^{\mathrm{w}}(x,D)u_0,u_0) < 0, \text{ as } M \to +\infty,$$

a contradiction. □

Example (14.13), which appears here for the first time, is preceded by examples that came earlier on. The first one is due to Brummelhuis.

Theorem 14.6 (Brummelhuis [4]). *Consider the second-order symbol*

$$A_B(X) = \begin{bmatrix} \xi_1^2 & -ix_1\xi_1\xi_2 \\ ix_1\xi_1\xi_2 & x_1^2\xi_2^2 \end{bmatrix}. \tag{14.15}$$

Then $A_B^{\mathrm{w}}(x,D)$ does not satisfy the Fefferman–Phong inequality. Moreover, no intermediate inequality with the L^2-norm replaced by the s-Sobolev norm with $s \in [0,1/2)$ is possible.

Notice that the case $s = 1/2$ is the Sharp-Gårding inequality that in fact has to hold because the symbol is Hermitian nonnegative.

It was then observed in Parmeggiani [23] that the reason for the existence of such a counterexample is rooted into the "non-nonnegativity" of the localized operator at characteristic points. In fact, the nonnegativity of the localized operator is a necessary condition for (FP) (and for local (FP)) to hold, and, say at the characteristic point $\rho = (x,\xi) = (0,0,0,1)$, we have that the localized system of A_B is unitarily equivalent to A_H^{w} which is *not* nonnegative. This observation allowed me to construct more general families of counterexamples that are "robust," in the sense that they do not satisfy (FP) even if perturbed by first-order terms that are sufficiently small. (In the same paper, I also showed how to construct "anisotropic" counterexamples.)

Theorem 14.7 (Parmeggiani [23]). *Let Σ be a symplectic smooth manifold, let $p,q \in S^1$ be real, positively homogeneous of degree 1 in the fibers, such that $p|_\Sigma = q|_\Sigma = 0$ and $\{p,q\}(\rho_0) \neq 0$ for some $\rho_0 = (x_0,\xi_0) \in \Sigma$ with $|\xi_0| = 1$. Let $v_1, v_2 \in \mathbb{C}^N$ be orthonormal vectors and let $L(X) = p(X)v_1 + q(X)v_2$. Let A_2 be the second-order symbol given by $A_2 = L^* \otimes L$, and let $A = A_2 + A_1$, where A_1 is a first-order symbol such that, for some $\delta \in [0,1)$,*

$$\langle A_1(\rho_0)w,w\rangle_{\mathbb{C}^N} \leq \delta\{p,q\}(\rho_0)\langle \mathrm{Im}(v_2^* \otimes v_1)w,w\rangle_{\mathbb{C}^N}, \ \forall w \in \mathbb{C}^N.$$

Then $A^{\mathrm{w}}(x,D)$ does not satisfy the Fefferman–Phong inequality, and no intermediate estimate with $s \in [0,1/2)$ is possible (except of course for the Sharp-Gårding inequality).

How about positive results?

There are indeed classes of systems for which the Fefferman–Phong inequality holds. In the first place, we have the following result.

Theorem 14.8. *Consider the $N \times N$ system*

$$p(x,\xi) = A(x)e(\xi) + \sum_{j=1}^{n} B_j(x)\xi_j + C(x) = p(x,\xi)^* \geq -cI_N, \ \forall (x,\xi) \in \mathbb{R}^n \times \mathbb{R}^n, \tag{14.16}$$

where $c > 0$, $e \in S^2$ is a positive-definite quadratic form and where $A, B_1, \ldots, B_n, C \in C^\infty(\mathbb{R}^n; \mathsf{M}_N)$ are bounded along with all their derivatives to all orders. Then $p^{\mathrm{w}}(x,D)$ satisfies the Fefferman–Phong inequality.

Remark 14.1. Note that, the matrix A being only ≥ 0, the symbol $p(X)$ is *not* assumed to be elliptic. ◇

Theorem 14.8 was first proved by Sung [30] in the case $n = 1$. His proof goes through Fourier series and is, in my opinion, very complicated. So, I wanted to have a proof that was based on an induction on the size N of the system. Such a proof indeed exists and is given in Parmeggiani [24] (see also [25]). It holds for general n and goes through a Calderón-Zygmund microlocal decomposition which is achieved by means of the *Fefferman–Phong metric* (a "proper" metric, as introduced by Lerner and Nourrigat in [15]) essentially of the form

$$g_{x,\xi} = H(x,\xi)|dx|^2 + \frac{|d\xi|^2}{1+|\xi|^2},$$

where

$$H(x,\xi)^{-1} = \max\Big\{\frac{1}{(1+|\xi|^2)^{1/2}}, \sqrt{\mathrm{Tr}A(x)}\Big\},$$

where Tr denotes the matrix trace (notice that the hypotheses give that $A(x) = A(x)^* \geq 0$). Such a microlocalization allows a Schur reduction of the system which is then microlocally decoupled, modulo L^2 errors, into a 1×1 block for which we may use the Fefferman–Phong scalar result, and an $(N-1) \times (N-1)$ block for which the assumptions are fulfilled, and that we control by induction on N. Here the constants depend on an a priori bounded number of seminorms of the symbol, bound that depends on the dimension n, and on the size N.

One has also a semiclassical analog of Theorem 14.8; see Parmeggiani [26].

The system in Theorem 14.8 is of "factorized" form. It is then natural to ask the following question: *Are there results for systems of a more general form?* The answer is in the positive, at least for certain classes of 2×2 systems, which, however, do not exhaust the class of systems that should be considered.

To understand what kind of conditions one should impose, let us go back to the Brummelhuis system (14.15) and observe the following. Consider the system

$$A(X) = A_B(X) + \xi_2^2 \begin{bmatrix} 0 & 0 \\ 0 & 1 \end{bmatrix}, \ X \in \mathbb{R}^2 \times \mathbb{R}^2.$$

Then system A has an *elliptic* trace, since $\mathrm{Tr}A(X) = \xi_1^2 + (1+x_1^2)\xi_2^2 \approx |\xi|^2$. It is straightforward to see that

$$(A^{\mathrm{w}}(x,D)u,u) = \|D_1u_1 - ix_1D_2u_2\|_0^2 + \mathrm{Re}(D_2u_2,u_1) + \|D_2u_2\|_0^2 \geq -\frac{1}{2}\|u\|_0^2,$$

for all $u \in \mathscr{S}(\mathbb{R}^2;\mathbb{C}^2)$. Hence, *an elliptic trace* should be part of the "cure" for obtaining the Fefferman–Phong inequality.

Theorem 14.9 (Parmeggiani [28]). *Let g be an admissible metric. Let $a,b,c \in S(h^{-2},g)$, and let*

$$p(X) = \begin{bmatrix} a(X) & \overline{c(X)} \\ c(X) & b(X) \end{bmatrix} = p(X)^* \geq 0, \quad \forall X \in \mathbb{R}^n \times \mathbb{R}^n.$$

Suppose the following ellipticity assumption on the matrix trace of the symbol

$$\mathrm{Tr}\, p(X) \approx h(X)^{-2} \tag{14.17}$$

is fulfilled, and assume that

$$e_1 := a\{c,\bar{c}\} - 2i\mathrm{Im}(c\{a,\bar{c}\}),\ e_2 = b\{c,\bar{c}\} - 2i\mathrm{Im}(c\{b,\bar{c}\}) \in S(h^{-4},g). \tag{14.18}$$

Then the Fefferman–Phong inequality holds for $p^{\mathrm{w}}(x,D)$.

It is important to observe the following facts.

Remark 14.2.

- As for condition (14.18), a priori the symbols e_1 and e_2 have order 5 in the calculus, that is, $e_1, e_2 \in S(h^{-5},g)$. So the requirement is that they are actually fourth-order symbols in the calculus. This is certainly the case when c or ic is real-valued.
- The class of counterexamples fulfills condition (14.18), but *not* condition (14.17).
- By virtue of Lemma 14.5 below, condition (14.18) is *invariant* under conjugation of the symbol by *constant* unitary matrices. ◇

Lemma 14.5. *Let p be the symbol in Theorem 14.9. Let*

$${}^{\mathrm{co}}p(X) := \begin{bmatrix} b & -\bar{c} \\ -c & a \end{bmatrix}$$

be the cofactor matrix of p (that is, $p\,{}^{\mathrm{co}}p = {}^{\mathrm{co}}p\,p = \det p$). Define

$$p_-(X) := \begin{bmatrix} a & \bar{c} \\ c & -b \end{bmatrix}, \quad X \in \mathbb{R}^n \times \mathbb{R}^n.$$

Then

$$\mathrm{Tr}\Big(\{p,p\}\,{}^{\mathrm{co}}p\Big) = (a-b)\{c,\bar{c}\} - 2i\mathrm{Im}(c\{a-b,\bar{c}\}),$$

$$\mathrm{Tr}\Big(\{p_-,p_-\}\,{}^{\mathrm{co}}p_-\Big) = (a+b)\{c,\bar{c}\} - 2i\mathrm{Im}(c\{a+b,\bar{c}\}).$$

As a consequence, condition (14.18) is invariant under conjugation by constant unitary matrices.

Proof. The proof is a direct computation, taking into account that, for an $N \times N$ matrix A (with C^1 entries),

$$\{A,A\} = \sum_{j=1}^{n} \Big(\frac{\partial A}{\partial \xi_j} \frac{\partial A}{\partial x_j} - \frac{\partial A}{\partial x_j} \frac{\partial A}{\partial \xi_j} \Big) = \sum_{j=1}^{n} \Big[\frac{\partial A}{\partial \xi_j}, \frac{\partial A}{\partial x_j} \Big],$$

and that $\mathrm{Tr}(AB) = \mathrm{Tr}(BA)$, for any given $N \times N$ matrices A and B. □

Corollary 14.1. *Write $c = c_1 + ic_2$, and replace condition (14.18) in Theorem 14.9 by the following condition: There are constants $0 < \theta_1 < \theta_2 < 1$ and a symbol $\omega_1 \in S(1,g)$ with $\theta_1 \leq \omega_1(X) \leq \theta_2$ for all X, such that, with $\omega_2 := 1 - \omega_1$,*

$$c_j(X)^2 \leq \omega_j(X)^2 a(X) b(X), \ \ \forall X \in \mathbb{R}^n \times \mathbb{R}^n, \ j = 1,2. \tag{14.19}$$

Then the Fefferman–Phong inequality holds true for $p^{\mathrm{w}}(x,D)$.

It is interesting to notice that the proof uses *convexity*. One in fact writes

$$p = \begin{bmatrix} \omega_1 a & c_1 \\ c_1 & \omega_1 b \end{bmatrix} + \begin{bmatrix} \omega_2 a & -ic_2 \\ ic_2 & \omega_2 b \end{bmatrix} =: p_1 + p_2$$

and notices that p_1^{w} and p_2^{w} satisfy (FP) by virtue of Theorem 14.9.

The proof of Theorem 14.9 actually shows that one can relax condition (14.18). One has in fact the following further corollary.

Corollary 14.2. *In Theorem 14.9, condition (14.18) can be weakened to*

$$e_1(X)^2 + e_2(X)^2 \lesssim \big(\mathrm{Tr}\, p(X)\big)^3 \det p(X) \approx h(X)^{-6} \det p(X), \ \ \forall X \in \mathbb{R}^n \times \mathbb{R}^n. \tag{14.20}$$

One has also a statement that deals with operators of "arbitrary" order in the calculus.

Corollary 14.3. *Let g be an admissible metric, and let m be a g-admissible weight. Let $a,b,c \in S(m,g)$, and let*

$$p(X) = \begin{bmatrix} a(X) & \overline{c(X)} \\ c(X) & b(X) \end{bmatrix} = p(X)^* \geq 0, \ \ X \in \mathbb{R}^n \times \mathbb{R}^n.$$

Suppose that

$$\mathrm{Tr}\, p(X) = a(X) + b(X) \approx m(X), \ \ \forall X \in \mathbb{R}^n \times \mathbb{R}^n,$$

and that either

$$a\{c,\bar{c}\} - 2i\,\mathrm{Im}(c\{a,\bar{c}\}) \ \text{ and } \ b\{c,\bar{c}\} - 2i\,\mathrm{Im}(c\{b,\bar{c}\}) \in S(h^2 m^3, g)$$

or that there exist constants $\theta_1, \theta_2 \in (0,1)$ *and* $\omega_1 \in S(1,g)$, *with* $\theta_1 \leq \omega_1(X) \leq \theta_2$, *such that with* $c = c_1 + ic_2$ *and* $\omega_2 := 1 - \omega_1$, *one has*

$$c_j(X)^2 \leq \omega_j(X)^2 a(X) b(X), \ \forall X \in \mathbb{R}^n \times \mathbb{R}^n, \ j = 1, 2.$$

Then there exists a symbol $q = q^* \in S(h^2 m, g; \mathsf{M}_2)$ *such that*

$$p^{\mathrm{w}}(x,D) \geq q^{\mathrm{w}}(x,D).$$

It is interesting to mention that, as proved in Parmeggiani [28], there are systems satisfying the ellipticity condition (14.17), but *not* (14.18) and (14.19), for which the (local) Fefferman–Phong inequality *does not* hold. In fact, let

$$p(X) = \begin{bmatrix} \xi_1^2 & (1+ix_1)\xi_1\xi_2 \\ (1-ix_1)\xi_1\xi_2 & (1+x_1^2)\xi_2^2 \end{bmatrix}. \tag{14.21}$$

It is readily seen that (14.17) holds, but neither (14.18) nor (14.19) does. One may construct an elliptic zeroth-order ψdo B, with $Bf = B_- f_1 + B_+ f_2$, such that

$$B^* p^{\mathrm{w}}(x,D) B = \begin{bmatrix} B_-^* \Lambda_-(x,D) B_- & O_{L^2 \to L^2}(1) \\ O_{L^2 \to L^2}(1) & B_+^* \Lambda_+(x,D) B_+ \end{bmatrix},$$

where $O_{L^2 \to L^2}(1)$ denotes a bounded operator in L^2, the principal symbol of $\Lambda_+(x,D)$ is $\lambda_+(X) = \operatorname{Tr} p(X)$, so it is *elliptic*, and the principal symbol of $\Lambda_-(x,D)$ is $\lambda_-(X) = \xi_1 \xi_2^2 / \lambda_+(X)$. Hence $p^{\mathrm{w}}(x,D)$ satisfies the (local) Fefferman–Phong inequality iff $\Lambda_-(x,D)$ satisfies the (local) Sharp-Gårding inequality, which cannot be since λ_- changes sign. Hence the system cannot satisfy the Fefferman–Phong inequality.

There are a few comments to make regarding Theorem 14.9.

- In the first place, why should one take $N = 2$? The reason is the following: When p is $N \times N$ and is given in the block form

$$p(X) = \left[\begin{array}{c|c} a(X) & c(X)^* \\ \hline c(X) & b(X) \end{array}\right],$$

where $a(X)$ is scalar and (say) elliptic, and $b(X) = b(X)^*$ of size $(N-1) \times (N-1)$, in the size-reduction procedure, we get an $(N-1) \times (N-1)$ block

$$b'(X) = b(X) - \frac{c(X)^* \otimes c(X)}{a(X)} \geq 0,$$

for which we cannot automatically say that

$$\operatorname{Tr} b'(X) = \operatorname{Tr} b(X) - \frac{|c(X)|^2}{a(X)} \geq 0 \ \text{ is still elliptic}$$

(even on a different localization scale). However, when $N = 2$, we have that b' is *scalar* and nonnegative, and this suffices for obtaining (FP).

- In addition, consider the following example of 3×3 system:

$$\begin{bmatrix} \xi_1^2 & ix_1\xi_1\xi_2 & 0 \\ -ix_1\xi_1\xi_2 & x_1^2\xi_2^2 & 0 \\ 0 & 0 & \xi_2^2 \end{bmatrix} = \left[\begin{array}{c|c} A_B(x,\xi) & \begin{matrix} 0 \\ 0 \end{matrix} \\ \hline 0\ 0 & \xi_2^2 \end{array}\right].$$

  Its trace is elliptic, but it clearly cannot satisfy the (local) Fefferman–Phong inequality.
- As a concluding comment for this section, I have results also in case $\operatorname{Tr} p$ is *not* elliptic (but this is work in progress). As an example, consider the system

$$\begin{bmatrix} \alpha\xi_1^2 + \beta x_1^2\xi_2^2 & \bar{\gamma}x_1\xi_1\xi_2 \\ \gamma x_1\xi_1\xi_2 & \beta\xi_1^2 + \alpha x_1^2\xi_2^2 \end{bmatrix}, \ \alpha, \beta > 0, \ \gamma \in \mathbb{C}.$$

  When $|\gamma| \leq 2\min(\alpha,\beta)$ *one has that the Fefferman–Phong inequality holds.* Notice that the trace is no longer elliptic, but is *non-elliptic nondegenerate* (in the approach of Fefferman and Phong).

14.5 The Approximate Positive and Negative Parts of a First-Order System

I now turn to the problem of the negative/positive part of a first-order pseudodifferential system $P = p^{\mathrm{w}}(x,D)$, that is, to the problem of constructing, starting from the symbol $p = p^*$, L^2-bounded operators $\sqcap_\pm$ (the *approximate projections* to the positive and negative parts of P) such that the operators $\pm\sqcap_\pm P$ satisfy the Sharp-Gårding inequality and such that the sum of $\sqcap_\pm$ is the identity modulo an error R which is "under control."

Before stating the results, let me give the basic example. Let $Y(t)$ be the Heaviside function, equal to 1 for $t > 0$ and equal to 0 when $t < 0$. Then $Y(D_1)$, where

$$Y(D_1)u(x) = (2\pi)^{-n} \iint e^{i\langle x-y,\xi\rangle} Y(\xi_1)u(y)dyd\xi, \ u \in C_0^\infty(\mathbb{R}^n),$$

and $1 - Y(D_1)$ are the projections to the positive and negative parts, respectively, of D_1.

The idea, in a general case, is to localize in phase space to regions on which either the symbol has a definite sign, or else it changes sign across a smooth hypersurface, or else is just small on regions of phase space which have volume ≈ 1. In the regions on which the symbol changes sign across a smooth hypersurface, the operator is equivalent, through conjugation by a unitary FIO, to an elliptic multiple of D_1, whence one constructs the (approximate) projections through those of D_1.

The first result in this direction is due to Fujiwara [8].

Theorem 14.10 (Fujiwara [8]). *Let $a \in S^1$ be real-valued. Let $\varepsilon \in (0,1)$ be given, arbitrarily small. Then one can construct three bounded self-adjoint linear operators $\pi_\pm$, R: $L^2(\mathbb{R}^n) \longrightarrow L^2(\mathbb{R}^n)$ such that*

(i) $\pi_\pm \geq 0$.
(ii) $\pi_+ + \pi_- = I + R,\ \|R\|_{L^2 \to L^2} \leq \varepsilon$.
(iii) *There exists $C > 0$ such that*

$$\pm \mathrm{Re}\,(\pi_\pm a^{\mathrm{w}}(x,D)u, u) \geq -C\|u\|_0^2,\ \ \forall u \in \mathscr{S}(\mathbb{R}^n).$$

(iv) *One has*

$$\|Ra^{\mathrm{w}}(x,D)\|_{L^2\to L^2} + \|a^{\mathrm{w}}(x,D)R\|_{L^2\to L^2} + \sum_{\pm} \|\left[\pi_\pm, a^{\mathrm{w}}(x,D)\right]\|_{L^2\to L^2} \leq C.$$

The operators are constructed microlocally from the symbol a and are of the form

$$\pi_\pm = \sum_{\nu \in \mathbb{Z}_+} \phi_\nu^{\mathrm{w}}(x,D) T_\nu \pi_\pm^{(\nu)}(x,D) T_\nu^* \phi_\nu^{\mathrm{w}}(x,D),$$

(and likewise $R = \sum \phi_\nu^{\mathrm{w}} R_\nu^{\mathrm{w}} \phi_\nu^{\mathrm{w}}$) where $\phi_\nu \geq 0$ and compactly supported, T_ν are FIOs (each T_ν brings $a\big|_{B_\nu}$ to an elliptic multiple of D_1, when a changes sign in B_ν), and the symbols $\pi_\pm^{(\nu)}$ may happen to be discontinuous along a hypersurface.

Theorem 14.10 is actually a consequence of the following stronger result, also due to Fujiwara [8], that I shall actually use in the extension of his result to 2×2 systems.

Theorem 14.11. *Assume that $\{a_\lambda\}_{\lambda \in \Lambda}$ is a family of real symbols which is bounded in $S^1(\mathbb{R}^n \times \mathbb{R}^n)$ such that*

$$a_\lambda(X) a_{\lambda'}(X) \geq 0,\ \ \forall \lambda, \lambda' \in \Lambda,\ \forall X \in \mathbb{R}^n \times \mathbb{R}^n. \tag{14.22}$$

Let $\varepsilon \in (0,1)$ be given, arbitrarily small. Then one can construct three L^2-bounded self-adjoint linear operators $\pi_\pm$ and R such that (i) through (iv) in Theorem 14.10 hold for $a_\lambda^{\mathrm{w}}(x,D)$ in place of $a^{\mathrm{w}}(x,D)$, with constants independent of $\lambda \in \Lambda$.

The approach is tailored to the classical pseudodifferential metric and Planck's function

$$g_X = |dx|^2 + \frac{|d\xi|^2}{\langle \xi \rangle^2},\ \ h(X) = \frac{1}{\langle \xi \rangle},$$

and microlocalization through the Beals–Fefferman metric (*up to introducing a large parameter*) to reduce to a semiclassical case.

Let me mention that in the *second-order scalar* case we have only the following result, due to Mughetti and Nicola (which, however, leaves the extension of Fujiwara's construction to operators with multiple characteristics largely open).

Theorem 14.12 (Mughetti and Nicola [16]). *Given $A = A^* \in \mathrm{OPS}^2$, classical and transversally elliptic with respect to Σ, with Hamilton map F whose eigenvalues satisfy certain conditions, for any fixed $\rho_0 \in \Sigma$ there exists a conic neighborhood V on which one can construct bounded self-adjoint linear operators $\pi_\pm \colon L^2 \longrightarrow L^2$ and $R \in \mathrm{OPS}^{-1/2}_{1/2,1/2}$ such that $\pi_+ + \pi_- = I + R$, with PR, $RP \in \mathrm{OPS}^{1/2}_{1/2,1/2}$ and satisfying the following: For any given microlocalizer $\psi(x,D)$ supported in V and every compact K, there exists $C_K > 0$ such that*

$$\pm\mathrm{Re}\,(P\pi_\pm\psi(x,D)u, \psi(x,D)u) \geq -C_K\|u\|^2, \ \ \forall u \in C_0^\infty(K).$$

Unfortunately, the result is only microlocal. It nevertheless yields a local construction (which is, however, weaker than that of Fujiwara). In fact, let $\{\psi_j\}_{j=1}^m$ be zeroth-order classical (i.e., positively homogeneous of degree 0 in the fiber variable) microlocalizers supported in conic neighborhoods V_j on which one may construct the operators $\pi_\pm^{(j)}$ of the theorem, such that $\sum_j \psi_j^2 = 1$ near $K \times (\mathbb{R}^n \setminus \{0\})$. Putting $P_\pm := \sum_j \psi_j(x,D)^* P \pi_\pm^{(j)} \psi_j(x,D)$, one has $P_+ + P_- = P + R'$ with $R' \in \mathrm{OPS}^{1/2}_{1/2,1/2}$ and in this case

$$\pm\mathrm{Re}(P_\pm u, u) \geq -C_K\|u\|^2, \ \ \forall u \in C_0^\infty(K).$$

As for systems, I have the following result in a 2×2 case.

Theorem 14.13. *Let $p = p^* = \begin{bmatrix} a & \bar{c} \\ c & b \end{bmatrix} \in S^1(\mathbb{R}^n \times \mathbb{R}^n; \mathsf{M}_2)$ and suppose*

$$\|p(X)\|^2_{\mathsf{M}_2} = \mathrm{Tr}(p(X)^2) \approx \langle\xi\rangle^2, \ \ \forall X \in \mathbb{R}^n \times \mathbb{R}^n. \tag{14.23}$$

Then, given any $\varepsilon \in (0,1)$, there exist $L^2(\mathbb{R}^n;\mathbb{C}^2)$-bounded linear operators $\sqcap_\pm$ and R, such that

- $\sqcap_+ + \sqcap_- = I + \mathsf{R}$.
- *There exists $C > 0$ such that*

$$\pm\mathrm{Re}(\sqcap_\pm p^{\mathrm{w}}(x,D)u, u) \geq -C\|u\|^2, \ \ \forall u \in \mathscr{S}(\mathbb{R}^n;\mathbb{C}^2).$$

- $\mathsf{R} = \mathsf{R}_1 + \mathsf{R}_2$ *where* $\mathsf{R}_1 \in \Psi(h,g;\mathsf{M}_2)$ *(and therefore gains one derivative) and* $\|\mathsf{R}_2\|_{L^2 \to L^2} \leq \varepsilon$.

Proof. I start by stating the next crucial lemma, which is proved exactly as in Parmeggiani [28].

Lemma 14.6. *There exist $r > 0$, with $4r < C_0^{-1/2}$ (C_0 being the slowness constant of g), and $c_1 > 0$ such that for any given $Y \in \mathbb{R}^{2n}$, we have*

$$\textit{either } |a(X)| \geq c_1\, h(X)^{-1}, \textit{ or } |b(X)| \geq c_1\, h(X)^{-1}, \textit{ or } |c(X)| \geq c_1\, h(X)^{-1},$$

for all $X \in B^g_{Y,4r}$.

I shall also need the following lemma, which is a consequence of the Cotlar–Stein method.

Lemma 14.7. *Let* $\{\psi_\nu\}_{\nu\in\mathbb{Z}_+}$ *and* $\{\theta_\nu\}_{\nu\in\mathbb{Z}_+}\subset S(1,g)$ *be* uniformly confined *to the metric balls* $\{B^g_{X_\nu,r}\}_{\nu\in\mathbb{Z}_+}$ *forming a covering of* $\mathbb{R}^n\times\mathbb{R}^n$ *as in Lemma 14.1. Let* $\{T_\nu\}_{\nu\in\mathbb{Z}_+}\subset\mathscr{L}(L^2,L^2)$ *be such that* $\sup_\nu\|T_\nu\|_{L^2\to L^2}<+\infty$. *Consider*

$$Au:=\sum_{\nu\in\mathbb{Z}_+}\psi_\nu^{\mathrm{w}}T_\nu\theta_\nu^{\mathrm{w}}u,\ \forall u\in L^2.$$

Then $A\in\mathscr{L}(L^2,L^2)$.

Proof (of the lemma). By the Cotlar–Stein re-summation method, let us set $A_\nu:=\psi_\nu^{\mathrm{w}}T_\nu\theta_\nu^{\mathrm{w}}$. We then have to prove that

$$\sup_{\mu\in\mathbb{Z}_+}\sum_{\nu\in\mathbb{Z}_+}\|A_\mu^*A_\nu\|^{1/2}_{L^2\to L^2}<+\infty,\ \text{and}\ \sup_{\mu\in\mathbb{Z}_+}\sum_{\nu\in\mathbb{Z}_+}\|A_\mu A_\nu^*\|^{1/2}_{L^2\to L^2}<+\infty.$$

I shall prove only one of the two estimates, for the second follows in exactly the same way. Now,

$$A_\mu^*A_\nu=(\theta_\mu^{\mathrm{w}})^*T_\mu^*(\psi_\mu^{\mathrm{w}})^*\psi_\nu^{\mathrm{w}}T_\nu\theta_\nu^{\mathrm{w}}.$$

By (14.5) of Lemma 14.1 we have for all $k,M\in\mathbb{Z}_+$ the existence of $C>0$ and $\ell\in\mathbb{Z}_+$ such that for all $\mu,\nu\in\mathbb{Z}_+$

$$\|\bar{\psi}_\mu\#\psi_\nu\|_{k,\mathrm{Conf}(1,g,X_\mu,r)}\leq C\|\psi_\mu\|_{\ell,\mathrm{Conf}(1,g,X_\mu,r)}\|\psi_\nu\|_{\ell,\mathrm{Conf}(1,g,X_\nu,r)}\Delta_{\mu\nu}(r)^{-M},$$

where M is chosen so large that $\sup_\mu\sum_\nu\Delta_{\mu\nu}(r)^{-M/2}<\infty$. Hence, by Lemma 14.2,

$$\sup_{\mu\in\mathbb{Z}_+}\sum_{\nu\in\mathbb{Z}_+}\|A_\mu^*A_\nu\|^{1/2}_{L^2\to L^2}\leq C'\sup_\mu\sum_\nu\Delta_{\mu\nu}(r)^{-M/2}<\infty,$$

and this ends the proof of the lemma. □

Using Lemma 14.1, let hence $\{B_\nu\}_{\nu\in\mathbb{Z}_+}$ be a covering of $\mathbb{R}^n\times\mathbb{R}^n$ by metric g-balls $B_\nu=B^g_{X_\nu,r}$ such that on the g-balls $B_\nu^*:=B^g_{X_\nu,4r}$ we have the conclusion of Lemma 14.6, and let $\{\varphi_\nu\}_{\nu\in\mathbb{Z}_+}$ be a corresponding partition of unity uniformly in $S(1,g)$ such that $\sum_\nu\varphi_\nu^2=1$.

I next define the following sets of nonnegative integers:

- $\mathsf{N}(a):=\{\nu\in\mathbb{Z}_+;\ \pm a\big|_{B_\nu^*}\approx h^{-1}\}$.
- $\mathsf{N}(b):=\{\nu\in\mathbb{Z}_+;\ \nu\notin\mathsf{N}(a),\ \pm b\big|_{B_\nu^*}\approx h^{-1}\}$.
- $\mathsf{N}(c):=\{\nu\in\mathbb{Z}_+;\ \nu\notin\mathsf{N}(a)\cup\mathsf{N}(b),\ |c|\big|_{B_\nu^*}\approx h^{-1}\}$.

Then, for each ν belonging to $\mathsf{N}(a)$ or $\mathsf{N}(b)$, respectively, we may define on the corresponding B_ν^* the matrix-valued zeroth-order symbols

$$E_\nu := \begin{bmatrix} 1 & -\bar{c}/a \\ 0 & 1 \end{bmatrix}, \quad \tilde{E}_\nu := \begin{bmatrix} 1 & 0 \\ -\bar{c}/b & 1 \end{bmatrix},$$

which are invertible on B^*_ν with inverse

$$E_\nu^{-1} := \begin{bmatrix} 1 & \bar{c}/a \\ 0 & 1 \end{bmatrix}, \quad \tilde{E}_\nu^{-1} := \begin{bmatrix} 1 & 0 \\ \bar{c}/b & 1 \end{bmatrix}.$$

Moreover, one has on B^*_ν also the relations

$$E^*_\nu p E_\nu = \begin{bmatrix} a & 0 \\ 0 & b-|c|^2/a \end{bmatrix} =: p_\nu, \quad \tilde{E}^*_\nu p \tilde{E}_\nu = \begin{bmatrix} a-|c|^2/b & 0 \\ 0 & b \end{bmatrix} =: \tilde{p}_\nu.$$

When $\nu \in \mathsf{N}(c)$, one may instead consider on B^*_ν the *smooth* unitary symbols U_ν for which

$$U^*_\nu p U_\nu = \begin{bmatrix} \lambda_1 & 0 \\ 0 & \lambda_2 \end{bmatrix} =: p_{d,\nu},$$

where $\lambda_{1,2} = (a+b \pm \sqrt{(a-b)^2+4|c|^2})/2$ are the eigenvalues of p which are then *smooth* in B^*_ν. In fact, on B^*_ν the symbol $|c|$ is elliptic, so that for the eigenvalues of $p(X)$ one has

$$|\lambda_1-\lambda_2| \approx \langle\xi\rangle = h(X)^{-1}, \quad \forall X \in B^*_\nu,$$

and p can be smoothly decoupled there, through a unitary U_ν which is then smooth on B^*_ν. Let now $\chi_\nu \in C_0^\infty(\mathbb{R}^n \times \mathbb{R}^n)$ with $\mathrm{supp}(\chi_\nu) \subset B^g_{X_\nu,2r}$ and $0 \le \chi_\nu \le 1$, uniformly in $S(1,g)$ and such that $\chi_\nu \varphi_\nu = \varphi_\nu$ (so that $\chi^2_\nu \varphi_\nu = \varphi_\nu$ also). Denote next by $A^\pm_\nu$, $\tilde{A}^\pm_\nu$, and Q_ν the operators in $\Psi(1,g;\mathsf{M}_2)$ obtained by Weyl quantization of the symbols $E^{\pm 1}_\nu \chi_\nu$, $\tilde{E}^{\pm 1}_\nu \chi_\nu$, and $U_\nu \chi_\nu$, respectively, which are in $S(1,g;\mathsf{M}_2)$ and uniformly confined to the $B^g_{X_\nu,2r} \subset B^*_\nu$. I next use Fujiwara's Theorem 14.11. Consider the families (uniformly bounded in $S(h^{-1},g)$),

$$\{a\chi_\nu\}_{\nu\in\mathsf{N}(a)}, \quad \{(b-|c|^2/a)\chi_\nu\}_{\nu\in\mathsf{N}(a)},$$

$$\{b\chi_\nu\}_{\nu\in\mathsf{N}(b)}, \quad \{(a-|c|^2/b)\chi_\nu\}_{\nu\in\mathsf{N}(b)},$$

and

$$\{\lambda_j\chi_\nu\}_{\nu\in\mathsf{N}(c)}, \quad j=1,2.$$

It is readily seen that condition (14.22) of Fujiwara's Theorem 14.11 is fulfilled for all these families. Let then $\tilde{a} := a-|c|^2/b$, $\tilde{b} := b-|c|^2/a$, and let

$$\pi_\pm(a), \ \pi_\pm(b), \ \pi_\pm(\tilde{a}), \ \pi_+(\tilde{b}), \ \pi^{(j)}_\pm, \ j=1,2,$$

be the corresponding approximate projectors, granted by Thorem 14.11, such that for any given $\varepsilon > 0$,

$$\pi_+(a)+\pi_-(a) = I+R(a), \ \text{etc.},$$

where $\|R(a)\|_{L^2\to L^2} \le \varepsilon$, and likewise for the operators $\pi_\pm(b)$, $R(b)$ etc. Then define the operators $\sqcap_\pm$ by

$$\sqcap_\pm := \sum_{\nu\in \mathsf{N}(a)} \varphi_\nu^{\mathrm{w}}(A_\nu^-)^* \begin{bmatrix} \pi_\pm(a) & 0 \\ 0 & \pi_\pm(\tilde{b}) \end{bmatrix} (A_\nu^+)^* \varphi_\nu^{\mathrm{w}} +$$

$$+ \sum_{\nu\in \mathsf{N}(b)} \varphi_\nu^{\mathrm{w}}(\tilde{A}_\nu^-)^* \begin{bmatrix} \pi_\pm(\tilde{a}) & 0 \\ 0 & \pi_\pm(b) \end{bmatrix} (\tilde{A}_\nu^+)^* \varphi_\nu^{\mathrm{w}} + \sum_{\nu\in \mathsf{N}(c)} \varphi_\nu^{\mathrm{w}} Q_\nu \begin{bmatrix} \pi_\pm^{(1)} & 0 \\ 0 & \pi_\pm^{(2)} \end{bmatrix} Q_\nu^* \varphi_\nu^{\mathrm{w}}.$$

Hence

$$\sqcap_\pm = \sum_{\nu\in\mathbb{Z}_+} \varphi_\nu^{\mathrm{w}} \Pi_\pm^{(\nu)} \varphi_\nu^{\mathrm{w}},$$

where

$$\Pi_\pm^{(\nu)} \in \mathscr{L}(L^2,L^2), \text{ with } \sup_{\nu\in\mathbb{Z}_+} \|\Pi_\pm^{(\nu)}\|_{L^2\to L^2} \le C < +\infty.$$

Hence, by Lemma 14.7, the operators $\sqcap_\pm \in \mathscr{L}(L^2,L^2)$. We next have the following lemma:

Lemma 14.8. *For any given* $\varepsilon \in (0,1)$ *one has*

$$\sqcap_+ + \sqcap_- = I + \mathsf{R}, \text{ with } \mathsf{R} = \mathsf{R}_1 + \mathsf{R}_2,$$

where the L^2*-bounded operators* R_1 *and* R_2 *have the following properties:*

- $\mathsf{R}_1 \in \Psi(h,g;\mathsf{M}_2)$ *(hence,* R_1 *gains one derivative).*
- $\mathsf{R}_2 = \sum_{\nu\in\mathbb{Z}_+} \varphi_\nu^{\mathrm{w}} R_\nu \varphi_\nu^{\mathrm{w}}$*, and* $\|R_\nu\|_{L^2\to L^2} \le C\varepsilon$*, with* C *independent of* ν *(and of* ε*), so that also* $\|\mathsf{R}_2\|_{L^2\to L^2} \le C'\varepsilon$ *(for a new absolute constant* C'*).*

Proof. We have $(A_\nu^-)^*(A_\nu^+)^* = ((E_\nu^{*-1}\chi_\nu)\#(E_\nu^*\chi_\nu))^{\mathrm{w}} = (\chi_\nu^2 I + r_{1,\nu})^{\mathrm{w}}$ with $r_{1,\nu} \in S(h,g;\mathsf{M}_2)$ *uniformly confined* to B_ν^* and analogously for the other terms $\tilde{A}_\nu^\pm$ and Q_ν. It is also important to notice that $\varphi_\nu\#\chi_\nu^2\#\varphi_\nu = \varphi_\nu^2 + r_{2,\nu}$ where $r_{2,\nu} \in S(h^2,g)$, and that $r_{2,\nu}$ and $\varphi_\nu\#r_{1,\nu}\#\varphi_\nu \in S(h,g;\mathsf{M}_2)$ are uniformly confined to B_ν and B_ν^*, respectively. Since $\sum_\nu \varphi_\nu^2 = 1$, Theorem 14.11 applied to each individual $\pi_\pm(a)$, $\pi_\pm(b)$ etc., yields

$$\sqcap_+ + \sqcap_- = I + \mathsf{R},$$

with $\mathsf{R} = \mathsf{R}_1 + \mathsf{R}_2$, where

$$\mathsf{R}_1 = \sum_{\nu\in\mathbb{Z}_+} (r_{2,\nu} I + \varphi_\nu \# r_{1,\nu} \# \varphi_\nu)^{\mathrm{w}} \in \Psi(h,g;\mathsf{M}_2)$$

(by Lemma 14.3), and

$$\mathsf{R}_2 = \sum_{\nu\in \mathsf{N}(a)} \varphi_\nu^{\mathrm{w}}(A_\nu^-)^* \begin{bmatrix} R(a) & 0 \\ 0 & R(\tilde{b}) \end{bmatrix} (A_\nu^+)^* \varphi_\nu^{\mathrm{w}} +$$

$$+ \sum_{\nu\in \mathsf{N}(b)} \varphi_\nu^{\mathrm{w}}(\tilde{A}_\nu^-)^* \begin{bmatrix} R(\tilde{a}) & 0 \\ 0 & R(b) \end{bmatrix} (\tilde{A}_\nu^+)^* \varphi_\nu^{\mathrm{w}} + \sum_{\nu\in \mathsf{N}(c)} \varphi_\nu^{\mathrm{w}} Q_\nu \begin{bmatrix} R^{(1)} & 0 \\ 0 & R^{(2)} \end{bmatrix} Q_\nu^* \varphi_\nu^{\mathrm{w}}$$

having the required properties. This ends the proof of the lemma. □

I next consider $\Pi_+ p$ (the conclusion for $-\Pi_- p$ follows in exactly the same way). Since

$$(A_\nu^+)^* \varphi_\nu^{\mathrm{w}} p^{\mathrm{w}} = \Big(E_\nu^* p \chi_\nu E_\nu (E_\nu^{-1}\chi_\nu)\varphi_\nu \Big)^{\mathrm{w}} + \Psi(1,g;\mathsf{M}_2) =$$

$$= \Big(E_\nu^* p \chi_\nu E_\nu \Big)^{\mathrm{w}} A_\nu^- \varphi_\nu^{\mathrm{w}} + r_\nu^{\mathrm{w}},$$

where $r_\nu \in S(1,g;\mathsf{M}_2)$ is uniformly confined, and analogously for the other terms, we have

$$\mathrm{Re}(\Pi_+ p^{\mathrm{w}} u, u) = \sum_{\nu\in \mathsf{N}(a)} \mathrm{Re}\Big(\begin{bmatrix} \pi_+(a) & 0 \\ 0 & \pi_+(\tilde{b}) \end{bmatrix} p_\nu^{\mathrm{w}} A_\nu^- \varphi_\nu^{\mathrm{w}} u, A_\nu^- \varphi_\nu^{\mathrm{w}} u) +$$

$$+ \sum_{\nu\in \mathsf{N}(b)} \mathrm{Re}\Big(\begin{bmatrix} \pi_+(\tilde{a}) & 0 \\ 0 & \pi_+(b) \end{bmatrix} \tilde{p}_\nu^{\mathrm{w}} \tilde{A}_\nu^- \varphi_\nu^{\mathrm{w}} u, \tilde{A}_\nu^- \varphi_\nu^{\mathrm{w}} u) +$$

$$+ \sum_{\nu\in \mathsf{N}(c)} \mathrm{Re}\Big(\begin{bmatrix} \pi_+^{(1)} & 0 \\ 0 & \pi_+^{(2)} \end{bmatrix} p_{d,\nu}^{\mathrm{w}} Q_\nu^* \varphi_\nu^{\mathrm{w}} u, Q_\nu^* \varphi_\nu^{\mathrm{w}} u) + O(\|u\|_0^2) =$$

$$= I(a) + I(b) + I(c) + O(\|u\|_0^2).$$

By Theorem 14.11 we get that there exists $C > 0$, independent of ν, such that

$$I(a) + I(b) + I(c) \geq$$

$$\geq -C\Big[\sum_{\nu\in \mathsf{N}(a)} (A_\nu^- \varphi_\nu^{\mathrm{w}} u, A_\nu^- \varphi_\nu^{\mathrm{w}} u) + \sum_{\nu\in \mathsf{N}(b)} (\tilde{A}_\nu^- \varphi_\nu^{\mathrm{w}} u, \tilde{A}_\nu^- \varphi_\nu^{\mathrm{w}} u) + \sum_{\nu\in \mathsf{N}(c)} (Q_\nu^* \varphi_\nu^{\mathrm{w}} u, Q_\nu^* \varphi_\nu^{\mathrm{w}} u) \Big]$$

$$\geq -C' \|u\|_0^2, \text{ by Lemma 14.7.}$$

This concludes the proof of the theorem. □

Remark 14.3. Notice that the theorem is not so satisfactory, in that it does not state that $\Pi_\pm$ are self-adjoint. This does not follow from the present construction (I am just able to construct them out of conjugation orbits of diagonal matrices made of the scalar pieces $\pi_\pm(a)$, $\pi_\pm(b)$, etc.). An analogous "defect" of the proof is that we cannot conclude that the norm of R is as small as we wish, but only that R is made of two pieces, one as small in norm as we wish, and the other gaining one derivative.

However, I believe the smallness of R can be recovered, but this is work in progress, which goes along with an extension of Fujiwara's results Theorems 14.10 and 14.11 to symbols in $S(h^{-1}, g)$ for more general Hörmander metrics g. The Bony FIO calculus in the context of Hörmander metrics will play a crucial role. It seems that not all admissible metrics are going to be allowed. ◇

Remark 14.4. As a final comment, notice that the ellipticity assumption (14.23) is *not* useful for obtaining the vector-valued Fefferman–Phong inequality. This is shown by example (14.21), the reason being that the first-order terms that arise cannot be controlled just by L^2-norms. ◇

Acknowledgements I wish to thank the organizers of the conference, Massimo Cicognani, Ferruccio Colombini, and Daniele Del Santo, for the great work they have done in the organization of the workshop. I wish also to thank Sandro Coriasco for the very useful remarks.

References

1. Bony, J.-M.: Sur l'inégalité de Fefferman–Phong, Sém. EDP, Ecole Polytechnique (1998–99), Exposé 3
2. Bony, J.-M., Lerner, N.: Quantification asymptotique et microlocalisations d'ordre supérieur. I. Ann . Sci. Éc. Norm. Sup. **22**, 377–433 (1989)
3. Boulkhemair, A.: On the Fefferman–Phong inequality. Ann. Inst. Fourier **58**, 1093–1115 (2008)
4. Brummelhuis, R.: Sur les inégalités de Gårding pour les systèmes d'opérateurs pseudo-différentiels. C.R. Acad. Sci. Paris, Série I, **315**, 149–152 (1992)
5. Brummelhuis, R.: On Melin's inequality for systems. Comm. Part. Differ. Equat. **26**, 1559–1606 (2001)
6. Brummelhuis, R., Nourrigat, J.: A necessary and sufficient condition for Melin's inequality for a class of systems. J. Anal. Math. **85**, 195–211 (2001)
7. Fefferman, C.L., Phong, D.H.: On positivity of pseudo-differential operators. Proc. Natl. Acad. Sci. USA **75** 4673–4674 (1978)
8. Fujiwara, D.: A construction of approximate positive parts of essentially self-adjoint pseudo-differential operators. Comm. Pure Appl. Math. **37**, 101–147 (1984)
9. Herau, H.: Melin-Hörmander inequality in a Wiener type pseudo-differential algebra. Ark. Mat. **39**, 311–338 (2001)
10. Hörmander, L.: The Cauchy problem for differential equations with double characteristics. J. Anal. Math. **32**, 118–196 (1977)
11. Hörmander, L.: The Weyl calculus of pseudodifferential operators. Comm. Pure Appl. Math. **32**, 359–443 (1979)
12. Hörmander, L.: The analysis of linear partial differential operators, vol. I–IV. Springer, Berlin (1983/85)
13. Lerner, L.: Metrics on the phase space and non-self-adjoint pseudo-differential operators. In: Pseudo-Differential Operators. Theory and Applications, vol. 3, p. xii+397. Birkhäuser Verlag, Basel (2010)
14. Lerner, L., Morimoto, Y.: On the Fefferman–Phong inequality and a Wiener-type algebra of pseudodifferential operators. Publ. Res. Inst. Math. Sci. **43**, 329–371 (2007)
15. Lerner, L., Nourrigat, J.: Lower bounds for pseudo-differential operators. Ann. Inst. Fourier (Grenoble) **40**, 657–682 (1990)
16. Mughetti, M., Nicola, F.: On the positive parts of second order symmetric pseudodifferential operators. Integr. Equ. Oper. Theor. **34**, 553–572 (2009)

17. Mughetti, M., Parenti, C., Parmeggiani, A.: Lower bound estimates without transversal ellipticity. Comm. Part. Differ. Equat. **32**, 1399–1438 (2007)
18. Nicola, F.: A lower bound for systems with double characteristics. J. Anal. Math. **96**, 297–311 (2005)
19. Parenti, C., Parmeggiani, A.: Lower bounds for systems with double characteristics. J. Anal. Math. **86**, 49–91 (2002)
20. Parenti, C., Parmeggiani, A.: On the hypoellipticity with a big loss of derivatives. Kyushu J. Math. **59**, 155–230 (2005)
21. Parenti, C., Parmeggiani, A.: A remark on the Hörmander inequality. Comm. Part. Differ. Equat. **31**, 1071–1084 (2006)
22. Parmeggiani, A.: On lower bounds of pseudodifferential systems. In: Colombini, F., Nishitani, T. (eds.) Hyperbolic Problems and Related Topics Somerville. Graduate Series in Analysis, pp. 269–293. International Press, Somerville (2003)
23. Parmeggiani, A.: A class of counterexamples to the Fefferman–Phong inequality for systems. Comm. Part. Differ. Equat. **29**, 1281–1303 (2004)
24. Parmeggiani, A.: On the Fefferman–Phong inequality for systems of PDEs. In: Bove, A., et al. (eds.) Phase Space Analysis of Partial Differential Equations. Progress in Nonlinear Differential Equations and their Applications, vol. 69, pp. 247–266. Birkhäuser Verlag, Boston (2006)
25. Parmeggiani, A.: On positivity of certain systems of partial differential equations. Proc. Nat. Acad. Sci. U.S.A. **104**, 723–726 (2007)
26. Parmeggiani, A.: A remark on the semiclassical Fefferman–Phong inequality for certain systems od PDEs. Rend. Lincei Mat. Appl. **19**, 339–359 (2008)
27. Parmeggiani, A.: In: Spectral Theory of Non-commutative Harmonic Oscillators: An Introduction. Lecture Notes in Mathematics, vol. 1992, pp. xii+254. Springer, Berlin (2010)
28. Parmeggiani, A.: A remark on the Fefferman–Phong inequality for 2×2 systems. Pure Appl. Math. Q. **6**, Special Issue: In honor of Joseph J. Kohn. Part 2, 1081–1103 (2010)
29. Ruzhansky, M., Turunen, V: Sharp-Gårding inequality on compact Lie groups. J. Funct. Anal. **260**, 2881–2901 (2011)
30. Sung, L.-Y.: Semi-boundedness of Systems of differential operators. J. Differ. Equat. **65**, 427–434 (1986)
31. Tataru, D.: On the Fefferman–Phong inequality and related problems, Comm. Part. Differ. Equat. **27**, 2101–2138 (2002)

Chapter 15
Scattering Problems for Symmetric Systems with Dissipative Boundary Conditions

Vesselin Petkov

Abstract We study symmetric systems with dissipative boundary conditions. The solutions of the mixed problems for such systems are given by a contraction semi-group $V(t)f = e^{tG_b}f$, $t \geq 0$, and the solutions $u = e^{tG_b}f$ with eigenfunctions f of the generator G_b with eigenvalues λ, $\operatorname{Re}\lambda < 0$, are called asymptotically disappearing (ADS). We prove that the wave operators are not complete if there exist (ADS). This is the case for Maxwell system with special boundary conditions in the exterior of the sphere. We obtain a representation of the scattering kernel, and we examine the inverse backscattering problem related to the leading term of the scattering kernel.

Key words: Asymptotically disappearing solutions, Backscattering inverse problem, Dissipative boundary conditions, Scattering kernel

2010 Mathematics Subject Classification: Primary: 35P25; Secondary: 47A40, 35L50, 81U40.

15.1 Introduction

Let $K \subset \{x \in \mathbb{R}^n, |x| \leq \rho\}, n \geq 3$, n odd, be an open bounded domain with C^∞ boundary ∂K. Consider in $\Omega = \mathbb{R}^n \setminus \bar{K}$ the operator $G = \sum_{j=1}^n A_j(x)\partial_{x_j} + B(x)$, where $A_j(x)$, $j = 1, \dots, n$, are smooth symmetric $(r \times r)$ matrices and $B(x)$ is a smooth $(r \times r)$ matrix. For simplicity in this paper we will assume that A_j are constant matrices and $B = 0$, but our results remain true for operators with variable coefficients under some conditions on the decay of $A_j(x)$ and $B(x)$ as $|x| \to \infty$.

We assume that the eigenvalues of $A(\xi) = \sum_{j=1}^n A_j\xi_j$ for $\xi = (\xi_1, \dots, \xi_n) \in \mathbb{R}^n \setminus \{0\}$ have constant multiplicity independent of ξ. Denote by $\nu(x)=(\nu_1(x), \dots, \nu_n(x))$

V. Petkov (✉)
Institut de Mathématiques de Bordeaux, 351 Cours de la Libération, 33405 Talence, France
e-mail: petkov@math.u-bordeaux1.fr

M. Cicognani et al. (eds.), *Studies in Phase Space Analysis with Applications to PDEs*, Progress in Nonlinear Differential Equations and Their Applications 84,
DOI 10.1007/978-1-4614-6348-1_15,

the unit normal at $x \in \partial\Omega$ pointing into K and set $A(\nu(x)) = \sum_{j=1}^{n} A_j \nu_j(x)$. Let $\mathscr{N}(x) \subset \mathbb{C}^r$ be a linear space depending smoothly on $x \in \partial\Omega$ such that

(*i*) $\langle A(\nu(x))u(x), u(x)\rangle \leq 0$ for all $u(x) \in \mathscr{N}(x)$.

(*ii*) $\mathscr{N}(x)$ is maximal with respect to (i).

Consider the boundary problem

$$\begin{cases}(\partial_t - G)u = 0 \,\text{in}\, \mathbb{R}^+ \times \Omega, \\ u(t,x) \in \mathscr{N}(x) \,\text{for}\, t \geq 0,\, x \in \partial\Omega, \\ u(0,x) = f(x) \,\text{in}\, \Omega.\end{cases} \tag{15.1}$$

The conditions (i) and (ii) make it possible to introduce a contraction semigroup $V(t) = e^{tG_b}$, $t \geq 0$ in $H = L^2(\Omega : \mathbb{C}^r)$ related to the problem (15.1) with generator G_b. The domain $D(G_b)$ of G_b is the closure with respect to the graph norm $(\|g\|^2 + \|Gg\|^2)^{1/2}$ of functions $g(x) \in C^1_{(0)}(\bar{\Omega} : \mathbb{C}^r)$ satisfying the boundary condition $g(x)|_{\partial\Omega} \in \mathscr{N}(x)$.

Next consider the unitary group $U_0(t) = e^{tG_0}$ in $H_0 = L^2(\mathbb{R}^n : \mathbb{C}^r)$ related to the Cauchy problem

$$\begin{cases}(\partial_t - G)u = 0 \,\text{in}\, \mathbb{R} \times \mathbb{R}^n, \\ u(0,x) = f(x) \,\text{in}\, \mathbb{R}^n,\end{cases} \tag{15.2}$$

where G_0 with domain $D(G_0) = \{f \in H_0 : \; Gf \in H_0\}$ is the generator of $U_0(t)$. Let $H_b \subset H$ be the space generated by the eigenvectors of G_b with eigenvalues $\mu \in \mathbf{i}\mathbb{R}$ and let $H_b^{\perp}$ be the orthogonal complement of H_b in H. The generator $G_0 = \sum_{j=1}^{n} A_j \partial_{x_j}$ is skew self-adjoint in H_0, and the spectrum of G_0 on the space $H_0^{ac} = (\text{Ker}\, G_0)^{\perp} \subset H_0$ is absolutely continuous (see Chap. IV in [18]).

For dissipative symmetric systems some solutions can have global energy decreasing exponentially as $t \to \infty$, and it is possible also to have disappearing solutions. The precise definitions are given below.

Definition 15.1. We say that $u = V(t)f$ is a disappearing solution (DS), if there exists $T > 0$ such that $V(t)f = 0$ for $t \geq T$.

Definition 15.2. We say that $u = V(t)f$ is asymptotically disappearing solution (ADS), if there exists $\lambda \in \mathbb{C}$ with $\text{Re}\,\lambda < 0$ and $f \neq 0$ such that $V(t)f = e^{\lambda t} f$.

Notice that if $V(t)f = e^{\lambda t} f$, then $f \in D(G_b)$ and $G_b f = \lambda f$. The existence of disappearing solutions perturbs strongly the inverse backscattering problem since the leading term of the backscattering matrix vanishes for all directions (see Sect. 15.5). This phenomenon is well known for the wave equations with dissipative boundary conditions [8, 15, 18]. For symmetric systems with dissipative boundary conditions the situation is much more complicated. It seems rather difficult to construct disappearing solutions, and even for Maxwell system the problem of the existence of disappearing solutions remains open (see [1, 18] and the references given there). In this paper we present a survey of some results related to the existence of (ADS). First in Sect. 15.2 we show that the completeness of the wave operators $W_\pm$ related to $V(t)$ and $U_0(t)$ fails if (ADS) exist. Therefore we must define the scattering operator by using another operator W. Secondly, we describe in Sect. 15.3 some recent results obtained in [1], where (ADS) for Maxwell system have been constructed. We study maximally dissipative boundary conditions for which there are no disappearing solutions, but (ADS) exist. This shows the importance of (ADS) which are stable under perturbations. Next in Sect. 15.4, we establish a representation of the scattering kernel of the scattering operator in the case of characteristics of constant multiplicity following the arguments in [16, 17] for strictly hyperbolic systems. Finally, in Sect. 15.5 we study the inverse backscattering problem connected with the leading singularity of the scattering matrix $\left(S^{jk}(s,-\omega,\omega)\right)_{j,k=1}^{d}$. Here the boundary condition plays a crucial role, and we investigate the problem assuming that

$$\mathscr{N}(x) \ominus \mathrm{Ker}\,(A(\nu(x))) \neq \Sigma_-(\nu(x)),$$

$\Sigma_-(\nu(x))$ being the space spanned by the eigenvectors of $A(\nu(x))$ with negative eigenvalues. This condition guarantees that at least one element of the scattering matrix $S^{jk}(s,-\omega,\omega)_{j,k=1}^{d}$ has a nonvanishing leading singularity related to the support function $\rho(\omega) = \min_{x\in\partial\Omega}\langle x,\omega\rangle$ in direction $\omega \in \mathbb{S}^{n-1}$.

15.2 Wave Operators

To introduce the wave operators, consider the operator $J : H \longrightarrow H_0$ extending $f \in H$ as 0 for $x \in K$ and let $J^* : H_0 \longrightarrow H$ be the adjoint of J. Let P_{ac} be the orthogonal projection on the space H_0^{ac}. The wave operators related to perturbed and non-perturbed problems have the form

$$W_- f = \lim_{t\to+\infty} V(t)J^*U_0(-t)P_{ac}f,\ f \in H_0,$$

$$W_+ f = \lim_{t\to+\infty} V^*(t)J^*U_0(t)P_{ac}f,\ f \in H_0.$$

Under the above hypothesis it is not difficult to prove the existence of $W_\pm$ and to show that (see, for instance, [9] and Chap. III in [18])

$$\mathrm{Ran}\,W_\pm \subset H_b^\perp.$$

To obtain more precise results for Ran $W_\perp$, we need to impose the following *coercive conditions*:

(H) : For each $f \in D(G_b) \cap (\mathrm{Ker}\,G_b)^\perp$ we have

$$\sum_{j=1}^{n} \|\partial_{x_j} f\| \le C(\|f\| + \|G_b f\|)$$

with a constant $C > 0$ independent of f.

(H^*) : For each $f \in D(G_b^*) \cap (\mathrm{Ker}\, G_b^*)^{\perp}$ we have

$$\sum_{j=1}^{n} \|\partial_{x_j} f\| \leq C(\|f\| + \|G_b^* f\|)$$

with a constant $C > 0$ independent of f.

Remark 15.1. The conditions (H) and (H^*) are satisfied for a large class of non-elliptic symmetric systems (see [13]) for which it is possible to construct a first order $(l \times r)$ matrix operator $Q = \sum_{j=1}^{n} Q_j \partial_{x_j}$ so that

$$Q(\xi)A(\xi) = 0, \ \mathrm{Ker}\, Q(\xi) = \mathrm{Im}\, A(\xi),$$

where $Q(\xi) = \sum_{j=1}^{n} Q_j \xi$. In our case we need coercive estimates for $f \in D(G_b) \cap H_b^{\perp}$. On the other hand, the space $\mathrm{Ker}\, G_b$ is infinite dimensional if Q with the properties above exist. Notice also that for the generator G_b we have $\mathrm{Ker}\, G_g = \mathrm{Ker}\, G_b^*$.

Introduce the spaces

$$\mathscr{H}_{\infty}^{+} = \{f \in H : \lim_{t \to +\infty} V(t)f = 0\}, \ \mathscr{H}_{\infty}^{-} = \{f \in H : \lim_{t \to +\infty} V^*(t)f = 0\}.$$

We have the following

Theorem 15.1 ([4]). *Assume the conditions* (H) *and* (H^*) *fulfilled. Then*

$$\overline{\mathrm{Ran}\, W_{\pm}} = H_b^{\perp} \ominus \mathscr{H}_{\infty}^{\pm}.$$

To obtain a completeness of the wave operators we must have $\mathscr{H}_{\infty}^{+} = \mathscr{H}_{\infty}^{-}$. On the other hand, in general, it is not clear if these subspaces are not empty.

For our analysis we use the translation representation $\mathscr{R}_n : H_0^{ac} \longrightarrow (L^2(\mathbb{R} \times \mathbb{S}^{n-1}))^d$ of $U_0(t)$, where Rank $A(\xi) = r - d_0 = 2d > 0$ for $\xi \neq 0$. Let $\tau_j(\xi)$, $j = 1, \ldots, 2d$, be the nonvanishing eigenvalues of $A(-\xi)$ ordered as follows:

$$\tau_1(\xi) > \ldots > \tau_d(\xi) > 0 > \tau_{d+1}(\xi) > \ldots > \tau_{2d}(\xi), \ \xi \neq 0.$$

Denote by $r_j(\xi)$, $j = 1, \ldots, 2d$, the normalized eigenvectors of $A(-\xi)$ related to $\tau_j(\xi)$. Then $\mathscr{R}_n$ has the form (see Chap. IV in [18])

$$(\mathscr{R}_n f)(s, \omega) = \sum_{j=1}^{d} \tilde{k}_j(s, \omega) r_j(\omega),$$

where

$$\tilde{k}_j(s, \omega) = \tau_j(\omega)^{1/2} k_j(s\tau_j(\omega), \omega), \ j = 1, \ldots, d,$$

and $k_j(s,\omega) = 2^{-(n-1)/2}D_s^{(n-1)/2}\langle (Rf)(s,\omega), r_j(\omega)\rangle$, $(Rf)(s,\omega)$ being the Radon transform of $f(x)$. $\mathscr{R}_n$ map H_0^{ac} isometrically into $(L^2(\mathbb{R}\times\mathbb{S}^{n-1}))^d$ and $\mathscr{R}_n U_0(t) = T_t\mathscr{R}_n$, $\forall t\in\mathbb{R}$, where $T_t g = g(s-t,\omega)$. Let $0 < v_0 = \min_{\omega\in\mathbb{S}^{n-1}} \tau_d(\omega)$. To introduce the Lax-Phillips spaces (see Chap. VI in [11]) , we need the following

Definition 15.3. We say that $f \in D_\pm$ if

$$U_0(t)f = 0 \,\text{for}\, |x| < \pm v_0 t, \ \pm t > 0.$$

We have $f\in D_\pm$ if and only if $\mathscr{R}_n f(s,\omega) = 0$ for $\mp s > 0$. Set $D_\pm^b = U_0(\pm b/v_0)D_\pm$, $b > 0$. For $t \geq 0$ it is easy to prove (see Lemma 4.1.5 in [18]) the following equalities:

$$U_0(-t)V(t)f = V^*(t)U_0(t)f = f,\, f\in D_+^\rho,$$

$$U_0(t)V^*(t)f = V(t)U_0(-t)f = f,\, f\in D_-^\rho.$$

The next result is similar to that in [5] established for strictly hyperbolic systems.

Theorem 15.2. *If $f\in D(G_b^j)\cap \mathscr{H}_\infty^+ \cap (D_-^\rho)^\perp, \forall j\in\mathbb{N}, f = G_b f_0, f_0\in H_b^\perp$, then we have $(V(t)f)(t,x) = 0$ for $|x| > 2\rho$.*

This yields the following

Corollary 15.1. *Assume that there exists an asymptotically disappearing solution $u(t,x) = e^{\lambda t}f(x)$ such that $f(x)\neq 0$ does not have a compact support. Then $\mathscr{H}_\infty^+ \neq \mathscr{H}_\infty^-$, and the wave operators are not complete.*

Proof. If $G_b f = \lambda f$, $\operatorname{Re}\lambda < 0$, we have $f\in D(G_b^j)\cap\mathscr{H}_\infty^+$, $\forall j\in\mathbb{N}$. Assuming $\mathscr{H}_\infty^- = \mathscr{H}_\infty^+$, for $g\in D_-^\rho$ we get

$$(f,g) = (f, V(t)U_0(-t)g) = (V^*(t)f, U_0(-t)g) \to_{t\to\infty} 0.$$

Thus $f\in (D_-^\rho)^\perp$ and we can apply Theorem 15.2 □

Proof (Theorem 15.2). Given $g\in\mathscr{H}_\infty^+\cap(D_-^\rho)^\perp$ and $f\in D_+^\rho$, for every fixed $t\geq 0$ we get

$$(V(t)g,f) = (V(t)g, V^*(s)U_0(s)f) = (V(t+s)g, U_0(s)f) \longrightarrow 0 \,\text{as}\, s\to+\infty.$$

Thus, $V(t)g\perp D_+^\rho$, $t\geq 0$. Also we obtain easily $V(t)g\perp D_-^\rho$, $t\geq 0$. Indeed, for $h\in D_-^\rho$ we have

$$(V(t)g,h) = (g, V^*(t)h) = (g, U_0(-t)h) = 0$$

since $U_0(-t)h\in D_-^\rho$ for $t\geq 0$. R. Phillips called the solutions with data $h\in\mathscr{D}(G_b)$ with $V(t)h\perp(D_+^\rho\oplus D_-^\rho)$, $t\geq 0$ *incontrollable* (see [12]).

We modify the argument of Lemma 2.2 in [5] (see also Proposition 4.2.5 in [18]) established for strictly hyperbolic systems in order to cover the situation when $A(\xi)$

has eigenvalues of finite multiplicity and $\operatorname{Ker} A(\xi)$ is not trivial. Let $\varphi(x) \in C^\infty(\mathbb{R}^n)$ be such that $\varphi(x) = 1$ for $|x| \geq 2\rho$, $\varphi(x) = 0$ for $|x| \leq \rho$. Set $w(t,x) = \varphi(x)V(t)f$. We have $w(t,x) = G\varphi(x)V(t)f_0 + [\varphi, G]V(t)f_0 = v(t,x) + [\varphi, G]V(t)f_0$. Thus $v(t,.) \in H_0^{ac}$, and we may consider the translation representation of $v(t,.)$. On the other hand, $v(t,.) \in \bigcap_{j=1}^\infty D(G_0^j)$ since $G_0 v = G_b v$. Next we have

$$(\partial_t - G)v(t,x) = \sum_{j=1}^n (A_j \varphi_{x_j})V(t)f = g(t,x).$$

Applying the transformation $\mathscr{R}_n$ to both sides of the above equality and setting

$$h_j(s,\omega,t) = \langle \mathscr{R}_n(v)(s,\omega,t), r_j(\omega)\rangle,\ l_j(s,\omega,t) = \langle \mathscr{R}_n(g)(s,\omega,t), r_j(\omega)\rangle,\ j = 1,\dots,d,$$

we obtain the equations

$$(\partial_t + \tau_j(\omega)\partial_s)h_j(s,\omega,t) = l_j(s,\omega,t),\ j = 1,\dots,d.$$

Next we repeat the argument of Lemma 2.2 in [5] based on the following

Lemma 15.1 ([5]). *Let $g \in \bigcap_{j=1}^\infty (G_0^j) \cap H_0^{ac}$ and let $\mathscr{R}_n g = 0$ for $|s| \geq b$. Then $g = 0$ for $|x| \geq b$ if and only if*

$$\int\int_{\mathbb{R}\times\mathbb{S}^{n-1}} [A(\omega)]^k \big[(\mathscr{R}_n g)(s,\omega) + (-1)^{(n-1)/2}(\mathscr{R}_n g)(-s,-\omega)\big] s^a Y_j(\omega)\, ds\, d\omega = 0$$

for $a = 0,1,2,\dots$ and any spherical harmonic function $Y_m(\omega)$ of order $m \geq a + k + (3-n)/2$.

Thus we conclude that $v(t,x) = 0$ for $|x| \geq 2\rho$, hence $V(t)f = 0$ for $|x| > 2\rho$, and this completes the proof of Theorem 15.2. □

Remark 15.2. For $d_0 = 0$ we may obtain a stronger version of Theorem 15.2 assuming that $f \in \mathscr{H}_\infty^+ \cap (D_-^\rho)^\perp$. For this purpose we must construct, as in [6], a sequence φ_ε such that

$$\varphi_\varepsilon \in \Big(\bigcap_{j=1}^\infty D(G_b^j)\Big) \cap D_-^\rho,\ \varphi_\varepsilon \in \mathscr{H}_\infty^+$$

with $\|\varphi_\varepsilon - f\| < \varepsilon$ and take the limit $\varepsilon \to 0$.

Remark 15.3. For systems with $\operatorname{Ker} A(\xi) = \{0\}$ V. Georgiev proved in [6] that if $f \in \mathscr{H}_\infty^+ \cap (D_-^\rho)^\perp$, then $V(t)f$ is a disappearing solution. On the other hand, the assumption $f \in (D_-^\rho)^\perp$ cannot be relaxed. In fact, in Sect. 15.3 we construct an example of (ADS) for which $f \in \bigcap_{j=1}^\infty D(G_b^j) \cap \mathscr{H}_\infty^+$, but $f(x)$ has not compact support with respect to x.

The space D_+^ρ is invariant with respect to the semigroup $V(t)$. Thus the generator G_b is the extension of the generator G_+ of the group $V(t)|_{D_+^\rho} = U_0(t)|_{D_+^\rho}$. By using the translation representation $\mathscr{R}_n$, it is easy to see that G_+ has spectrum in $\{z \in \mathbb{C} : \mathrm{Re}\, z \leq 0\}$, so $\mathbf{i}\mathbb{R}$ is in $\sigma(G_b)$ (see [10]). On the other hand, it was proved in [9] that the eigenvalues of G_b on $\mathbf{i}\mathbb{R}$ have finite multiplicity, and the only accumulating point of these eigenvalues could be 0. Thus the $\sigma(G_b)$ is continuous on $\mathbf{i}\mathbb{R}$. The analysis of $\sigma(G_b)$ in $\mathrm{Re}\, z < 0$ is more complicated. In the next section we show for the Maxwell system that there exist eigenvalues $\lambda < 0$ of G_b.

To define a scattering operator we prove the existence of the operator:

$$Wf = \lim_{t \to \infty} U_0(-t) J V(t) f,\ f \in H_b^\perp$$

assuming the hypothesis (H) fulfilled (see [9, 18]). We define the scattering operator $S = W \circ W_-$ by using the diagram on Fig. 15.1.

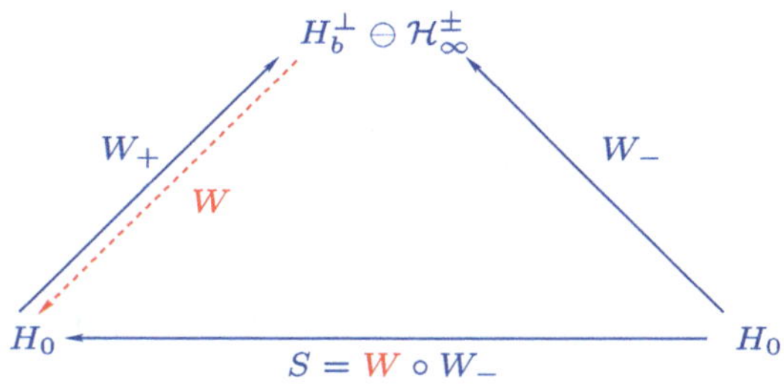

Fig. 15.1 Scattering operator

15.3 Asymptotically Disappearing Solutions for Maxwell System

In this section we show that for the Maxwell system with maximal dissipative boundary conditions there exist (ADS). The Maxwell system in $\mathbb{R}^3$ is given by the equations

$$\partial_t E - \mathrm{curl}\, B = 0,\ \partial_t B + \mathrm{curl}\, E = 0.$$

$$\mathrm{div}\, E = 0,\ \mathrm{div}\, B = 0.$$

It is well known that the wave equation $u_{tt} - \Delta u = 0$ in $\mathbb{R}_t \times \mathbb{R}^3$ admits almost spherical solutions $\frac{f(|x|+t)}{|x|}$ defined outside $x = 0$. It was proved in [1] that for Maxwell system there are no such almost spherical solutions with the exception of functions $g(x) + ct$ linear in t. To find a family of incoming divergence-free solutions of Maxwell systems depending on $|x|$ and t, we apply the following:

Theorem 15.3 ([1]). *Let $h \in C^\infty(\mathbb{R})$ and let $h^{(k)} = \partial_s^k h(s) \in L^1([0,\infty[)$, for all $k \in \mathbb{N}$. Then*

$$E := \left(\frac{h''(|x|+t)}{|x|} - \frac{h'(|x|+t)}{|x|^2} \right) \frac{x}{|x|} \wedge (1,0,0), \tag{15.3}$$

$$B := -\left(\frac{h''}{|x|} - \frac{3h'}{|x|^2} + \frac{3h}{|x|^3} \right) \frac{x}{|x|} \wedge \left(\frac{x}{|x|} \wedge (1,0,0) \right) + 2\left(\frac{h'}{|x|^2} - \frac{h}{|x|^3} \right) (1,0,0), \tag{15.4}$$

where the argument of the functions $h^{(k)}$ is $|x|+t$, define smooth divergence-free incoming solutions of Maxwell system in $\mathbb{R}_t^+ \times (\mathbb{R}^3 \setminus 0)$.

The proof of this theorem is technical. The starting point is to search E in the form $E = \operatorname{curl}(g,0,0)$ with $g(t,x) = \frac{f(|x|+t)}{|x|}$. This guarantees $\operatorname{div} E = 0$ and $\Box E = 0$. Next we determine B from the equation $B_t = -\operatorname{curl} E$ so that $\operatorname{div} B = 0$. Finally,

$$\partial_t (E_t - \operatorname{curl} B) = E_{tt} + \operatorname{curl}\operatorname{curl} E = E_{tt} - \Delta E = 0$$

and since $E_t - \operatorname{curl} B \to 0$ as $t \to \infty$, we obtain $E_t - \operatorname{curl} B = 0$. We can exchange the role of E and B and start with $B = \operatorname{curl}(g,0,0)$, but this leads to similar expressions.

We wish to construct asymptotically disappearing solutions in $|x| > 1$ that satisfy a homogeneous boundary condition

$$u = (E,B) \in \mathcal{N}(x), \quad \text{on} \quad |x| = 1. \tag{15.5}$$

Here $\mathcal{N}(x)$ is a four dimensional linear subspace of $\mathbb{C}^6$ depending smoothly of x.

Write the Maxwell equations in matrix form

$$u_t - \sum_{j=1}^{3} A_j \partial_j u = 0.$$

The matrices A_j are real symmetric, and $A(\xi) := \sum_{j=1}^3 A_j \xi_j$ for $\xi \neq 0$ has rank equal to 4 and eigenvalues $0, \pm|\xi|$ of multiplicity 2.

A sufficient condition for $\mathcal{N}(x)$ to be maximally dissipative and to obtain a well-posed mixed initial boundary value problem generating a contraction semigroup $V(t)$ on $(L^2(\{|x| \geq 1\}))^6$ is that

$$\dim \mathcal{N}(x) = 4, \quad \text{and,} \quad \langle A(\nu(x))u, u \rangle \leq 0, \forall u \in \mathcal{N}(x), \forall |x| = 1.$$

It follows that $\mathcal{N}(x) \supset \operatorname{Ker} A(\nu(x))$ for all boundary points $x \in \partial\Omega$.

For any unit vector ν the eigenvalues of $A(\nu)$ are $-1, 0, 1$. The kernel of $A(\nu)$ is the set of (E,B) so that both E and B are parallel to ν. The condition that $\mathcal{N}(x)$ contain the kernel is equivalent to say that (E,B) belonging to $\mathcal{N}(x)$ if (E,B) is determined entirely by the tangential components (E_{tan}, B_{tan}). The eigenspace $\Sigma_\pm(\nu)$ of $A(\nu)$ corresponding to eigenvalue ± 1 is equal to

$$\Sigma_\pm(\nu) := \left\{ (E,B) : E_{tan} = \mp \nu \wedge B_{tan} \right\}.$$

The span of eigenspaces $\Sigma_-(\nu) \oplus \operatorname{Ker} A(\nu)$ with nonpositive eigenvalues is strictly dissipative, that is, for all $u \in \Sigma_-(\nu) \oplus \operatorname{Ker} A(\nu)$, we have

$$\langle A(\nu)u\,,u\rangle \;=\; -\|u_{tan}\|^2 \;=\; -\big\|(E_{tan},B_{tan})\big\|^2.$$

To construct asymptotically disappearing solutions we choose h in a special way in Theorem 15.3.

Theorem 15.4 ([1]). *Let $\varepsilon_0 > 0$ be sufficiently small and for $0 < \varepsilon < \varepsilon_0$ set $2r = 1-\sqrt{1+4/\varepsilon} < 0$ and $h(s) = e^{rs}$. Then $(E,B)(t,x)$ defined by (15.3) and (15.4) yield a divergence-free solution of boundary value problem defined by the Maxwell equations in $|x| > 1$ with maximal dissipative boundary condition*

$$(1+\varepsilon)E_{tan} - \nu \wedge B_{tan} = 0, \quad \text{on} \quad |x| = 1. \tag{15.6}$$

For each α there is a constant $C(\varepsilon,\alpha) > 0$ so that $\big|\partial^\alpha(E,B)(t,x)\big| \le C(\varepsilon,\alpha)\, h(t+|x|)$. In particular, the energy decays exponentially as $t \to \infty$, and G_b has an eigenvalues $r < 0$.

It is important to note that for the mixed problem with boundary conditions (15.6) and $\varepsilon > 0$, there are no disappearing solutions. This follows from Theorem 3 in [6] saying that if $\mathscr{N}(x) \cap \Sigma_-(\nu(x)) = \{0\}$, $\forall x \in \partial\Omega$, for system with real analytic boundary conditions, then there are no disappearing solutions. In our situation, if $(E,B) \in \mathscr{N}(x) \cap \Sigma_-(\nu(x))$, we have $\varepsilon E_{tan} = 0$, and this yields $B_{tan} = 0$, so $(E,B) \in \operatorname{Ker} A(\nu(x))$. On the other hand, it is clear that for the sphere $|x| = 1$ the boundary condition (15.6) is real analytic with respect to x. For $\varepsilon = 0$ the boundary condition

$$E_{tan} - \nu \wedge B_{tan} = 0, \quad \text{on} \quad |x| = 1 \tag{15.7}$$

satisfies $E_-(\nu(x)) \subset \mathscr{N}(x)$, $\forall x \in \partial\Omega$, and (15.7) is the analog of the condition $\Big((\partial_\nu + \partial_t)u\Big)|_{\partial\Omega} = 0$, for the wave equation $\partial_{tt} - \Delta u = 0$ (see [12] for the results concerning this mixed problem).

It is interesting to see that for the Maxwell system with dissipative boundary condition, if $G_b g = \lambda g$ with $\operatorname{Re}\lambda < 0$, then $g(x) = \mathscr{O}(e^{\operatorname{Re}\lambda|x|})$ as $|x| \to \infty$. To see this, consider the function $\varphi(x)$ introduced in Sect. 15.2 and set $w(x) = \varphi(x)g(x)$. We have

$$G\begin{pmatrix}E\\B\end{pmatrix} = \begin{pmatrix}0 & \operatorname{curl}\\ -\operatorname{curl} & 0\end{pmatrix}\begin{pmatrix}E\\B\end{pmatrix}.$$

Taking

$$Q\begin{pmatrix}E\\B\end{pmatrix} = \begin{pmatrix}\operatorname{div}E\\ \operatorname{div}B\end{pmatrix},$$

we have $Q(\xi)A(\xi) = 0$, $\operatorname{Ker} Q(\xi) = \operatorname{Im} A(\xi)$, and we obtain $(G^2 - Q^*Q)\begin{pmatrix}E\\B\end{pmatrix} = \Delta\begin{pmatrix}E\\B\end{pmatrix}$. Let $g = \begin{pmatrix}E\\B\end{pmatrix}$ be an eigenfunction of G_b with eigenvalue $\lambda, \operatorname{Re}\lambda < 0$. Obviously, $G_b g = \lambda g$ implies $Qg = 0$. Thus we get

$$(\Delta + (\mathbf{i}\lambda)^2)w = [(G^2 - Q^*Q) - \lambda^2]w = (G_b^2 - \lambda^2)w - Q^*\begin{pmatrix}\langle \operatorname{grad}\varphi, E\rangle\\ \langle \operatorname{grad}\varphi, B\rangle\end{pmatrix}$$

$$= (G_b + \lambda)[G_b, \varphi]g - Q^* \begin{pmatrix} \langle \operatorname{grad}\varphi, E\rangle \\ \langle \operatorname{grad}\varphi, B\rangle \end{pmatrix} = F_\varphi(g).$$

Here we have used the fact that $G^2 w = G_b^2 w$. The right-hand side $F_\varphi(g)$ has compact support and $\|F_\varphi(g)\| \leq C\|g\|$ with constant depending on φ. The (incoming) resolvent of the free Laplacian $R_-(\mu) = (\Delta + \mu^2)^{-1}$ for $\operatorname{Im}\mu < 0$ has kernel

$$R_-(x, y; \mu) = -\frac{e^{-\mathbf{i}\mu|x-y|}}{4\pi|x-y|}$$

By using the above equation for w and the kernel of $R_-(\mathbf{i}\lambda)$, we obtain

$$(\varphi g)(x) = -\frac{1}{4\pi}\int \frac{e^{\lambda|x-y|}}{|x-y|} F_\varphi(g)(y)dy$$

and this yields

$$|g(x)| \leq C_0 e^{(\operatorname{Re}\lambda)|x|}\|g\|, \; |x| > 3\rho.$$

For the Maxwell system with strictly dissipative boundary conditions it is natural to conjecture that the spectrum of G_b in $\{z \in \mathbb{C} : \operatorname{Re} z < 0\}$ is formed by isolated eigenvalues with finite multiplicity. This conjecture has been proved recently [2] for boundary problems satisfying the coercive conditions (H) and (H^*).

15.4 Representation of the Scattering Kernel

In this section we obtain a representation of the scattering kernel. Such a representation has been obtained in [16, 17] for strictly hyperbolic systems with, respectively, conservative and dissipative boundary conditions. Here for completeness we sketch the proof of a representation in the case when G has characteristics of constant multiplicity. By using the translation representation $\mathscr{R}_n$ of $U_0(t)$, consider the scattering operator

$$\tilde{S}k = \mathscr{R}_n S \mathscr{R}_n^{-1} k = \lim_{t' \to +\infty} \mathscr{R}_n U_0(-t') J V(2t') J^* U_0(-t') \mathscr{R}_n^{-1} k.$$

Let $k(s, \omega) \in (C_0^\infty(\mathbb{R} \times \mathbb{S}^{n-1}))^d$ with $k = 0$ for $|s| > R_1$. Next for simplicity we write $\mathscr{R}$ instead of $\mathscr{R}_n$. Set $f = \mathscr{R}^{-1}k$. We must study

$$\lim_{t' \to +\infty} T_{-t'} \mathscr{R}\Big(JV(2t')J^* U_0(-t')f\Big).$$

Choose $p > R_1 + \frac{\rho}{v_0}$. It is easy to see that for $t' > p$ we have $J^* U_0(-t')f = U_0(-t')f \in D_-^\rho$. Set $u_0(t,x) = U_0(t)f$ and denote by $u(t,x;t') = V(t)U_0(-t')f$ the solution of the problem

$$\begin{cases} (\partial_t - G)u = 0 \,\text{in}\, \mathbb{R}^+ \times \Omega, \\ u(t,x) \in \mathscr{N}(x) \,\text{on}\, \mathbb{R}^+ \times \partial\Omega, \\ u|_{t \le t'-p} = u_0(t-t',x). \end{cases} \tag{15.8}$$

Consequently,

$$\tilde{S}k = \lim_{t' \to +\infty} T_{-t'} \mathscr{R} J \tilde{u}(t',x), \tag{15.9}$$

where $\tilde{u}(t,x)$ is the solution of the problem

$$\begin{cases} (\partial_t - G)\tilde{u} = 0 \,\text{in}\, \mathbb{R} \times \Omega, \\ \tilde{u}(t,x) \in \mathscr{N}(x) \,\text{on}\, \mathbb{R} \times \partial\Omega, \\ \tilde{u}|_{t \le -p} = u_0(t,x). \end{cases} \tag{15.10}$$

Next we repeat the argument of Sect. 4 in [17]. Setting

$$v = \begin{cases} \tilde{u} \text{ in } \mathbb{R} \times (\mathbb{R}^n \setminus K), \\ 0 \text{ in } \mathbb{R} \times K, \end{cases}$$

we have

$$\langle R((\partial_t - G)v), r_j(\omega)\rangle = (\partial_t + \tau_j(\omega)\partial_s)\langle Rv, r_j(\omega)\rangle$$

and we get

$$\langle Rv, r_j(\omega)\rangle = \langle Ru_0, r_j(\omega)\rangle + \int_{-\infty}^{t} \langle R((\partial_t - G)v)(\tau, s + \tau_j(\omega)(\tau - t), \omega), r_j(\omega)\rangle d\tau. \tag{15.11}$$

On the other hand, for every vector-valued function $\psi \in (C_0^\infty(\mathbb{R}^{n+1}))^d$ we obtain

$$((\partial_t - G)v, \psi) = \int\int_{\mathbb{R} \times \partial\Omega} \langle v, A(\nu(x))\psi\rangle d\tau dS_x$$

and since $(\partial_t - G)v$ has compact support with respect to x, we can apply the above formula combined with (15.11) for the calculus of $(\mathscr{R}v)(s,\omega)$. Taking the limit $t' \to +\infty$ in (15.9), we deduce

$$(\tilde{S}k)(s,\theta) = k(s,\theta) + d_n \sum_{j=1}^{d} \tau_j(\theta)^{1/2} r_j(\theta)$$

$$\times \int\int_{\mathbb{R} \times \partial\Omega} \delta^{(n-1)/2}(\langle x,\theta\rangle - \tau_j(\theta)(s+\tau)) \big\langle r_j(\theta), A(\nu(x))\tilde{u}(\tau,x)\big\rangle d\tau dS_x,$$

where d_n is a constant depending only on n and the integral is taken in the sense of distributions.

Next we repeat the argument of Sect. 3, [16, 17]. Set $\tilde{u}(t,x) = u_0(t,x) + u^s(t,x)$,

$$w_k^o(t,x,\omega) = \tau_k(\omega)^{1/2} \delta^{(n-1)/2}(\langle x,\omega\rangle - \tau_k(\omega)t) r_k(\omega),$$

and consider the (outgoing) solution $w_k^s(t,x,\omega)$ of the problem

$$\begin{cases} (\partial_t - G)w_k^s = 0 \,\mathrm{in}\, \mathbb{R}\times\Omega, \\ w_k^s + w_k^o \in \mathscr{N}(x) \,\mathrm{on}\, \mathbb{R}\times\partial\Omega, \\ w_k^s|_{t\leq -\frac{\rho}{v_0}} = 0. \end{cases} \tag{15.12}$$

called disturbed plane wave.

To justify the existence of w_k^s, we set $w_k^s = z_k - w_k^o$ and consider the mixed problem

$$\begin{cases} (\partial_t - G)\tilde{z}_k = 0 \,\mathrm{in}\, \mathbb{R}\times\Omega, \\ \tilde{z}_k \in \mathscr{N}(x) \,\mathrm{on}\, \mathbb{R}\times\partial\Omega, \\ \tilde{z}_k|_{t\leq -\frac{\rho}{v_0}} = (-\tau_k(\omega))^{-(n+2)/2}H(\langle x,\omega\rangle - \tau_k(\omega)t)r_k(\omega), \end{cases} \tag{15.13}$$

where

$$H(\eta) = \begin{cases} 0, \ \eta > 0, \\ \eta, \ \eta \leq 0 \end{cases}.$$

It is easy to show the existence of $\tilde{z}_k$ and we get $z_k = \mathbf{i}\partial_t^{(n+3)/2}\tilde{z}_k$. By using $w_k^s + w_k^o$, $k = 1,\ldots,d$, we can express $\tilde{u}(\tau,x)$ by $k(s,\omega)$, and we obtain a representation

$$(\tilde{S}k)(s,\theta) = k(s,\theta) + \int\int_{\mathbb{R}\times\mathbb{S}^{n-1}} K^{\#}(s-\tau,\theta,\omega)k(\tau,\omega)d\tau d\omega.$$

The distribution $K^{\#}(s,\theta,\omega) = \left(S^{jk}(s,\theta,\omega)\right)_{j,k=1}^{d}$ is called scattering kernel. Thus we have the following:

Theorem 15.5. *The scattering kernel $K^{\#}(s,\theta,\omega)$ computed with respect to the basis $\{r_j(\omega)\}_{j=1}^d$ has elements*

$$S^{jk}(s,\theta,\omega) = d_n^2\tau_j(\theta)^{1/2}$$

$$\times \int\int_{\mathbb{R}\times\partial\Omega} \delta^{(n-1)/2}(\langle x,\theta\rangle - \tau_j(\theta)(s+t))\langle r_j(\theta), A(\nu(x))(w_k^o + w_k^s)(t,x,\omega)\rangle dt dS_x. \tag{15.14}$$

15.5 Back Scattering Inverse Problem for the Scattering Kernel

Consider the scattering operator S. It is easy to see that $(S - Id)(D_-^\rho) \subset (D_+^\rho)_0^\perp$, where $(.)_0^\perp$ denotes the orthogonal complement in H_0^{ac}. Then taking $k(\sigma,\omega) = 0$ for $\sigma > -\rho/v_0$, we deduce $((\tilde{S} - I)k)(s,\theta) = 0$ for $s > \rho/v_0$, and we get

$$K^{\#}(s,\theta,\omega) = 0 \,\mathrm{for}\, s > 2\rho/v_0.$$

This property implies that the Fourier transform $\hat{K}^{\#}(\lambda,\theta,\omega)$ of $K^{\#}(s,\theta,\omega)$ admits an analytic continuation for $\operatorname{Im}\lambda < 0$, and the same is true for the operator-valued function $\tilde{S}(s) : (L^2(\mathbb{S}^{n-1}))^d \to (L^2(\mathbb{S}^{n-1}))^d$.

In fact, a more precise result holds for the backscattering matrix $S^{jk}(s,-\omega,\omega)$.

Theorem 15.6. *We have*

$$\max_{s\in\mathbb{R}} \operatorname{supp} S^{jk}(s,-\omega,\omega) \leq -\Big(\frac{\tau_j(-\omega)+\tau_k(\omega)}{\tau_j(-\omega)\tau_k(\omega)}\Big)\rho(\omega), \tag{15.15}$$

where $\rho(\omega) = \min_{y\in\partial\Omega}\langle y,\omega\rangle$ *is the support function of* $\partial\Omega$ *in direction* ω.

Applying (15.14), the proof of the above inequality is the same as that in Theorem 3.2 in [16].

Before going to the analysis of an equality in (15.15), consider the problem for the wave equation with dissipative boundary conditions

$$\begin{cases} (\partial_t^2 - \Delta)w = 0 \text{ in } \mathbb{R}^+ \times \Omega, \\ \partial_\nu w + \gamma(x)\partial_t w = 0 \text{ on } \mathbb{R}^+ \times \partial\Omega, \\ (w(0,x), w_t(0,x)) = (f_1, f_2), \end{cases} \tag{15.16}$$

where $\gamma(x) \geq 0, x \in \partial\Omega$ is a smooth function. We can introduce a scattering operator $S \in \mathscr{L}(L^2(\mathbb{R}\times\mathbb{S}^{n-1}))$ related to (15.16), and the kernel $K^{\#}(s-s',\theta,\omega)$ of $S - Id$ is called scattering kernel. The singularities of $K^{\#}(s,\theta,\omega)$ with respect to s are closely related to the geometry of the obstacle. Thus if $\gamma(x) \neq 1$, $\forall x \in \partial\Omega$, and $\theta \neq \omega$, we have (see [14]) for strictly convex obstacles

$$\max \operatorname{singsupp}_s K^{\#}(s,\theta,\omega)\} = \max_{y\in\partial\Omega}\langle y, \theta-\omega\rangle.$$

For backscattering $\theta = -\omega$ this yields

$$\max \operatorname{singsupp}_s K^{\#}(s,-\omega,\omega) = -2\rho(\omega) = -T_\gamma. \tag{15.17}$$

Here T_γ is the shortest sojourn time of a ray incoming with direction ω and outgoing with direction $-\omega$ (see [19] for the definition of sojourn time). To obtain (15.17), it is necessary to study the asymptotics of the *filtered* scattering amplitude

$$a_\varphi(\lambda,-\omega,\omega) = \int e^{-\mathrm{i}\lambda s} K^{\#}(s,-\omega,\omega)\varphi(s)ds,$$

where $\varphi(s) \in C_0^\infty(\mathbb{R})$ has small support around $-T_\gamma$, $\varphi(-T_\gamma) = 1$, and the integral is taken in the sense of distributions. By using the propagation of the wave front sets of the solutions of the mixed problem (15.16) in the diffraction region and a microlocal parametrix, Majda [14] showed that if the support of φ is sufficiently small, then

$$a_\varphi(\lambda,-\omega,\omega) = c_n\lambda^{(n-1)/2}\mathscr{K}^{-1/2}(x_+)e^{\mathrm{i}\lambda T_\gamma}\Big(\frac{1-\gamma(x_+)}{1+\gamma(x_+)} + \mathscr{O}(|\lambda|^{-1})\Big). \tag{15.18}$$

Here $x_+ \in \partial\Omega$ is the unique point on $\partial\Omega$ with $\nu(x_+) = \omega$, and $\mathscr{K}(x_+)$ is the Gauss curvature at x_+. For $\gamma(x) \neq 1$, similar result holds for arbitrary (non-convex) obstacles [15] (see also [18]), and there exists an open dense subset $\Sigma \subset \mathbb{S}^{n-1}$ such that for every $\omega \in \Sigma$ we have

$$\max \operatorname{singsupp}_s K^{\#}(s, -\omega, \omega) = -2\rho(\omega). \tag{15.19}$$

Thus from the leading singularities of the backscattering kernel we can determine the convex hull of the obstacle. If $\gamma(x) = 1, \forall x \in \partial\Omega$, the leading term in (15.18) vanishes for all directions $\omega \in \mathbb{S}^{n-1}$. In this direction V. Georgiev and J. Arnaoudov proved the following:

Theorem 15.7 ([8]). *Let $\gamma(x) = 1$ for all $x \in \partial\Omega$. Then for $n = 3$ if K is strictly convex, we have*

$$a_\varphi(\lambda, -\omega, \omega) = \frac{c}{16\pi}\mathscr{K}^{-1/2}(x_+)e^{\mathrm{i}\lambda T_\gamma} + \mathscr{O}(|\lambda|^{-1})$$

with $c \neq 0$. Moreover, for an arbitrary smooth obstacle, there exists an open dense subset $\Sigma \subset \mathbb{S}^{n-1}$ such that for every $\omega \in \Sigma$, we have (15.19).

Remark 15.4. Let us note that it is not proved that the problem (15.16) with $\gamma(x) = 1$, $\forall x \in \partial\Omega$, has disappearing solutions, and this explains the inverse scattering result in the above theorem. In [12] Majda established the existence of disappearing solutions for mixed problem outside the sphere $|x| = 1$ with boundary condition $\partial_\nu w + \partial_t w + w = 0$ on $\mathbb{R} \times \mathbb{S}^2$.

Passing to the case of symmetric systems, consider first strictly hyperbolic systems with $\operatorname{Ker} A(\xi) = \{0\}$ and maximal energy preserving boundary condition

$$\langle A(\nu(x))u, u\rangle = 0,\ u \in \mathscr{N}(x).$$

In this case $r/2 = d$ and there exists (see Chap. VI in [11]) an orthonormal basis $(p_j(x))_{j=1}^d$ of the positive eigenspace $\Sigma_+(\nu(x))$ of $A(\nu(x))$ with respect to the inner product $< A(\nu(x))u, u >$ and an orthonormal basis $(n_j((x))_{j=1}^d$ of the negative eigenspace $\Sigma_-(\nu(x))$ of $A(\nu(x))$ with respect to the inner product $- < A(\nu(x))u, u >$. Then a maximal energy preserving space $\mathscr{N}(x)$ is spanned by the vectors $(p_j(x) + n_j(x))_{j=1}^d$, and we have $u(x) \in \mathscr{N}(x)$ if and only if

$$\langle p_j(x) + n_j(x), A(\nu(x))u(x)\rangle = 0,\ 1 \leq j \leq d.$$

When we change x on $\partial\Omega$, this special basis changes. For strictly hyperbolic systems with maximal energy preserving boundary conditions, Majda and Taylor [16], assuming that (p_j) and (n_j) depend implicitly through the unit normal $\nu(x)$, proved that for every given $k, 1 \leq k \leq d$, there exists an open dense set $\Sigma \subset \mathbb{S}^{n-1}$ such that for every $\omega \in \Sigma$ for some j, $1 \leq j \leq d$, we have

$$\max \operatorname{singsupp}_s S^{jk}(s, -\omega, \omega) = -\Big(\frac{\tau_j(-\omega) + \tau_k(\omega)}{\tau_j(-\omega)\tau_k(\omega)}\Big)\rho(\omega). \tag{15.20}$$

We should mention that for strictly convex obstacles Majda and Taylor showed [16] that the leading term of the filtered (j,k) term of the scattering matrix has the form

$$c_n \lambda^{(n-1)/2} \exp\Big(\mathbf{i}\lambda \Big(\frac{\tau_j(-\omega)+\tau_k(\omega)}{\tau_j(-\omega)\tau_k(\omega)}\Big)\rho(\omega)\Big)\mathscr{K}(x_+)^{-1/2}$$

$$\times \sum_{\mu=1}^{d} \langle r_k(\omega), p_\mu(\omega)\rangle \langle r_j(-\omega), n_\mu(\omega)\rangle.$$

When we fix k, $1 \le k \le d$, the vector $\sum_{\mu=1}^{d} \langle r_k(\omega), p_\mu(\omega)\rangle n_\mu(\omega)$ is in the eigenspace $\Sigma_-(\omega)$, and it is easy to see that for some j, $1 \le j \le d$, we have

$$\sum_{\mu=1}^{d} \langle r_k(\omega), p_\mu(\omega)\rangle \langle r_j(-\omega), n_\mu(\omega)\rangle \neq 0.$$

In the case of maximal dissipative boundary conditions, the structure of the boundary space $\mathscr{N}(x)$ is more complicated. It was shown [7, 17] that we have the following

Lemma 15.2. *Let $\mathscr{N}(x)$ be a maximal dissipative linear space depending smoothly on $x \in \partial\Omega$. Then for every $\hat{x} \in \partial\Omega$ there exists a neighborhood $\mathscr{V}$ of $\hat{x}$ and smooth in $\mathscr{V} \cap \partial\Omega$ vectors $p_1(x), \dots, p_d(x), n_1(x), \dots, n_d(x)$ satisfying*

$$\langle p_i(x), A(\nu(x))p_j(x)\rangle = \delta_{i,j},\ \langle n_i(x), A(\nu(x))n_j(x)\rangle = -\delta_{i,j},$$

$$\langle n_i(x), A(\nu(x))p_j(x)\rangle = 0,\ i,j = 1,\dots,d,$$

and $0 \le \mu(x) \le d$ so that $\mathscr{N}(x) \ominus (\mathrm{Ker} A(\nu(x)))$ is spanned by the vectors

$$\{p_1(x)+n_1(x), \dots, p_{\mu(x)}(x)+n_{\mu(x)}(x), n_{\mu(x)+1}(x), \dots, n_d(x)\}.$$

In the strictly dissipative case we have $\mu(x) = 0$. It is important to note that in general $(p_j(x))_{j=1}^d$ and $(n_j(x))_{j=1}^d$ do not form a basis, respectively, in the spaces $\Sigma_+(\nu(x))$ and $\Sigma_-(\nu(x))$. On the other hand, $u(x) \in \mathscr{N}(x)$ if and only if

$$\begin{cases} \langle p_j(x)+n_j(x), A(\nu(x))u(x)\rangle = 0,\ j = 1,\dots,\mu(x), \\ \langle p_j(x), A(\nu(x))u(x)\rangle = 0,\ j = \mu(x)+1,\dots,d. \end{cases} \tag{15.21}$$

It is clear that for strictly dissipative boundary conditions if

$$\mathscr{N}(x) \ominus (\mathrm{Ker} A(\nu(x)) = \Sigma_-(\nu(x)), \tag{15.22}$$

then $(n_j(x))_{j=1}^d$ span $\Sigma_-(\nu(x))$ and $(p_j(x))_{j=1}^d$ span $\Sigma_+(\nu(x))$. The condition (15.22) is the analog of the boundary condition (15.16) with $\gamma(x) = 1$, $\forall x \in \partial\Omega$.

For strictly hyperbolic systems we have the following

Theorem 15.8 ([17]). *Let $A(\xi)$ have simply characteristic roots for $\xi \neq 0$ and let* $\mathrm{Ker} A(\xi) = \{0\}$. *Consider the problem (15.1) and assume that the vectors (p_j) and*

(n_j) depend implicitly through the unit normal $\nu(x)$ and (p_j) and (n_j) are smooth functions of this normal with μ independent on $\nu(x)$. Moreover, assume that for every $x \in \partial\Omega$ in Lemma 15.2 $\mu(x) = const$ and if $\mu = 0$, we have

$$\mathscr{N}(x) \ominus (\mathrm{Ker} A(\nu(x)) \neq \Sigma_-(\nu(x)). \tag{15.23}$$

Then there exists an open dense set $\Sigma \subset \mathbb{S}^{n-1}$ such that for every $\omega \in \Sigma$, there exist (k, j) depending on ω so that we have

$$\max \operatorname{singsupp}_s S^{jk}(s, -\omega, \omega) = -\Big(\frac{\tau_j(-\omega) + \tau_k(\omega)}{\tau_j(-\omega)\tau_k(\omega)}\Big)\rho(\omega). \tag{15.24}$$

Remark 15.5. In contrast to the problem (15.16) with $\gamma(x) \neq 1$, $\forall x \in \partial\Omega$, the condition (15.22) does not exclude the existence of disappearing solutions. On the other hand, we are interesting to have at least one term in the scattering matrix $S^{jk}(s, -\omega, \omega)$ with leading singularity given by (15.24). This means that all other terms could be regular or vanishing.

In the proof of Theorem 15.8 the crucial point is the construction of a microlocal outgoing parametrix with boundary data distribution $\tau_k(\omega)\delta^{(n-1)/2}(\langle x, \omega\rangle - \tau_k(\omega)t)r_k(\omega)$. More precisely, we assume that locally the boundary is given by $x = 0$ with $(x, y) \in \mathbb{R}^+ \times \mathbb{R}^{n-1}$. After a microlocalization around the point $(0, y, t, 0, -1)$ with $\nu(y) = \omega$ and an application of the decoupling procedure of the matrix symbol [20], we are going to study the problem (see [16, 17])

$$\frac{\partial}{\partial x}\tilde{v}_k - \begin{bmatrix} \lambda_1 & & 0 \\ & \ddots & \\ 0 & & \lambda_r \end{bmatrix} \tilde{v}_k = F(x, y, t),$$

$$\Lambda\Big((V\tilde{v}_k)|_{x=0} - \Big(\tau_k(\omega)^{1/2}\delta^{(n-1)/2}(\langle x, \omega\rangle - \tau_k(\omega)t)r_k(\omega)\Big)|_{\partial\Omega}\Big) = g(y, t), \tag{15.25}$$

$$\tilde{v}_k \text{ is smooth for } t \leq -T_0 < 0.$$

Here $F(x, y, t)$, $g(y, t)$ are smooth functions, $\lambda_j(x, y, t, \eta, \tau)$, $j = 1, \ldots, r$, are different first order pseudodifferential operators, V is a classical matrix pseudodifferential elliptic operator of order 0, and $\Lambda W(x) = 0$ means that $W \in \mathscr{N}(x)$. For this construction, we must determine the boundary data of $(V\tilde{v}_k)$ on $x = 0$ from (15.25). For this purpose, by using (15.21) and the condition (15.23), we obtain an elliptic pseudodifferential system $\mathscr{E}\Big((V\tilde{v}_k)|_{x=0}\Big) = Y_k$ on the boundary modulo smooth terms, and choosing suitably k, we arrange $Y_k \neq 0$. Thus we determine the boundary data for $V\tilde{v}_k$, and we construct an outgoing parametrix in a standard way.

Remark 15.6. Notice that in the case of the boundary condition (15.22), we get $Y_k = 0, k = 1, \ldots, d$, and the leading term of the backscattering matrix vanishes for all j, k.

We conjecture that the statement of Theorem 15.8 holds for symmetric systems with characteristics of constant multiplicity, but this needs extra work concerning the microlocal matrix reduction. For other inverse scattering problems for symmetric systems with dissipative boundary conditions, we refer to [3], where the case of directions $(\theta, \omega) \in \mathbb{S}^{n-1} \times \mathbb{S}^{n-1}$ satisfying $\|\theta + \omega\| < \delta$ with sufficiently small $\delta > 0$ has been studied.

References

1. Colombini, F., Petkov, V., Rauch, J.: Incoming and disappearing solutions of Maxwell's equations. Proc. AMS **139**, 2163–2173 (2011)
2. Colombini, F., Petkov, V., Rauch, J.: Spectral problems for non elliptic symmetric systems with dissipative boundary conditions, Preprint 2012.
3. Georgiev, V.: High frequency asymptotics of the filtered scattering amplitudes and the inverse scattering problem for dissipative hyperbolic systems. Math. Nachr. I, **117**, 111–128 (1984); II, **122**, 267–275 (1985)
4. Georgiev, V.: Existence and completeness of the wave operators for dissipative hyperbolic systems. J. Oper. Theor. **14**, 291–310 (1985)
5. Georgiev, V.: Controllability of the scattering operator for dissipative hyperbolic systems. Math. Nachr. **122**, 339–346 (1985)
6. Georgiev, V.: Disappearing solutions for dissipative hyperbolic systems of constant multiplicity. Hokkaido Math. J. **15**, 357–385 (1986)
7. Georgiev, V., Arnaoudov, Ja.: Construction of a smooth basis for a matrix depending on a parameter. Higher School Appl. Math. **20**(2), 27–30 (1984)
8. Georgiev, V., Arnaoudov, Ja.: Inverse scattering problem for dissipative wave equation. Math. Appl. Comput. **9**(1), 59–78 (1990)
9. Georgiev, V., Stefanov, P.: Existence of the scattering operator for dissipative hyperbolic systems of variable multiplicity. J. Oper. Theor. **19**, 217–241 (1988)
10. Lax, P.D., Phillips, R.S.: Scattering theory for dissipative systems. J. Funct. Anal. **14**, 172–235 (1973)
11. Lax, P.D., Phillips, R.S.: Scattering Theory, 2nd edn. Academic, New York (1989)
12. Majda, A.: Disappearing solutions for the dissipative wave equation. Indiana Math. J. **24**, 1119–1133 (1975)
13. Majda, A.: Coercive inequalities for nonelliptic symmetric systems. Comm. Pure Appl. Math. **28**, 49–89 (1975)
14. Majda, A.: High frequency asymptotics for the scattering matrix and the inverse problem of acoustical scattering. Comm. Pure Appl. Math. **29**, 261–291 (1976)
15. Majda, A.: A representation formula for the scattering operator and the inverse problem for arbitrary bodies. Comm. Pure Appl. Math. **30**, 165–194 (1977)
16. Majda, A., Taylor, M.: Inverse scattering problems for transparent obstacles, electromagnetic waves and hyperbolic systems. Comm. PDE **2**, 395–438 (1977)
17. Petkov, V.: Representation of the scattering operator for dissipative hyperbolic systems. Comm. PDE **6**, 993–1022 (1981)
18. Petkov, V.: Scattering Theory for Hyperbolic Operators. North Holland, Amsterdam (1989)
19. Petkov, V., Stoyanov, L.: Sojourn times, singularities of the scattering kernel and inverse problems. In: Uhlmann, G. (ed.) Inside Out: Inverse Problems and Applications. MSRI Proceedings, vol. 47, pp. 297–332. Cambridge University Press, Cambridge (2003)
20. Taylor, M.: Reflection of singularities of solutions to systems of differential equations. Comm. Pure Appl. Math. **28**, 547–478 (1975)

Chapter 16
Kato Smoothing Effect for Schrödinger Operator

Luc Robbiano

Abstract In this paper we give a survey of results on smoothing effect for Schrödinger operators. Several phenomena can be called smoothing effect. Here we limit us on the Kato or one half smoothing effect. We shall give the new and old results in different contexts, global in time and local in time in all spaces, in exterior domain. In this last case we shall give the recent result where the geometrical control condition is not satisfied and replaced by a damping condition. This kind of assumption originates in the control theory. We shall give also some references on the other smoothing effect, Strichartz estimate and related problem for wave equation, and KdV equation without exhaustiveness.

Key words: Schrödinger equation, Smoothing effect

2010 Mathematics Subject Classification: 35Q41, 35B65.

16.1 Introduction

If the smoothing effect is well-known for heat equation, for dispersive equation this phenomenon was been found rather recently. The first result was probably the one of Kato [38] on KdV equation. After this pioneer work, Constantin and Saut [19], Vega [69], Sjölin [65], and Yajima [74] give results on a large class of dispersive equations including Schrödinger equation. These results are obtained under the assumption that the principal symbol does not depend on space variables.

L. Robbiano (✉)
Laboratoire de Mathématiques, Université de Versailles Saint-Quentin,
45, avenue des États-Unis, F-78035 Versailles Cedex, France
e-mail: luc.robbiano@uvsq.fr

M. Cicognani et al. (eds.), *Studies in Phase Space Analysis with Applications to PDEs*, Progress in Nonlinear Differential Equations and Their Applications 84,
DOI 10.1007/978-1-4614-6348-1_16,

For operators with variable coefficients, Doi [25] proved a Kato smoothing effect for operator of Schrödinger type. One of the main tools of the Doi's method is the construction of an escape function associated with an energy method. To prove the smoothing effect, he assumed that the metric is not trapping and he proved the smoothing effect cannot occur if this assumption is not fulfilled. After this result, Doi [27] obtained the same result for unbounded potential. Typically, the potential is assumed estimated by $|x|^2$ at infinity. For potential superquadratic some weaker smoothing effect was found by Yajima and Zhang [75] and by Robbiano and Zuily [58] following the Doi's approach. In case of superquadratic potential, the smoothing effect is weaker and related to the size of the potential at infinity. For exterior domains Robbiano and Zuily [59] proved the smoothing effect in non-trapping geometry. In this geometrical context the geodesic can be trapping by the metric or by the boundary. Even for the flat metric a geodesic can be trapped. As in nonboundary case, Burq [16] proved that the nontrapping assumption is necessary. In the trapped case, in a work in preparation, we prove in [7] a smoothing effect under the assumption that the trapped geodesic hit a damping zone. The damping is an adding first-order term in the Schrödinger equation. The assumption on the geodesic flow is called geometric control condition and was introduced by Bardos et al. [9] in context of control and stabilization.

Another approach to prove smoothing effect was initiated by Ben-Artzi and Devinatz [10] and Ben-Artzi and Klainerman[11]. This method relates the smoothing effect and a property on the resolvent. This method was exploited by Burq et al. [18] who used the resolvent estimate obtained by Burq [14]. The advantage of this method is one can prove global smoothing effect in time. With the Doi's method we can only prove local smoothing effect in time, but this method allows us to prove smoothing for operators which have not the global smoothing effect property in time. For instance, $-\Delta+|x|^2$ has eigenfunctions and cannot have global smoothing effect but have the local smoothing effect property in time. The Doi's method can treat operators as $-\Delta+|x|^2$. In this case the solution is locally in time and in space more regular than the datum, but the integral in time does not converge.

In the sequel we give some smoothing results, global in time and local in time in $\mathbb{R}^n$; next, we give results for exterior domains, with geometrical control condition, with damping term. In a last section we give some references on related subjects, C^∞ smoothing, analytic smoothing, Strichartz estimate, and dispersive estimate.

16.2 Results Global in Time

The result given by Vega [69] was global in time but concerns Schrödinger operator, i.e., $i\frac{\partial u}{\partial t}=\Delta u$ with a lost on the smoothing effect. He gave an estimate with a weight in space variables as in Theorem 16.1 below. We give the theorem proved by Ben-Artzi and Klainerman [11] which is global in time and optimal with respect to the smoothing and the decay. The method of proof is rather general and consists to deduce the smoothing effect from a resolvent estimate. This method appears already in Kato and Yajima [40]. We give in the Proposition 16.1 the kind of resolvent estimate used to prove the smoothing effect. The same kind of results was proved by Watanabe [71] for $(-\Delta)^\alpha$ using the same method.

Here and in the sequel we denote by $\langle x\rangle = \sqrt{1+|x|^2}$ and by $\langle D\rangle$ the operator associated with symbol $\langle \xi \rangle$.

Theorem 16.1 ([11]). *For all $\mu > 1/2$, there exists $C > 0$ such that for all $u_0 \in L^2(\mathbb{R}^n)$, the solution u of the following problem*

$$
\begin{aligned}
&i\frac{\partial u}{\partial t} = \Delta u \text{ in } \mathbb{R}\times\mathbb{R}^n\\
&u(0,\cdot) = u_0 \text{ on } \mathbb{R}^n,
\end{aligned}
$$

satisfies

$$\int_{\mathbb{R}^{n+1}} |\langle x\rangle^{-\mu}\langle D\rangle^{1/2}u(t,x)|^2 dt dx \le C\int_{\mathbb{R}^n} |u_0(x)|^2 dx.$$

Remark 16.1. In Ruzhanski and Sugimoto [62, 63] we can find an alternative proof of this result and some generalization, using the Operator Integral Fourier expression of $u(t,x)$ which is explicit for constant coefficient operators.

In the case of exterior domains, Burq et al. [18] proved a result following an analog method. More precisely, let K be a compact set with smooth boundary, we denote by $\Omega = \mathbb{R}^n \setminus K$. We assume that the generalized bicharacteristic flow is non-trapping (see Sect. 16.4, Definition 16.2 for a description of this assumption). Let Δ_D be the Laplacian with Dirichlet boundary condition.

Proposition 16.1. *Assume $K \neq \emptyset$. Then for every $\chi \in C_0^\infty(\mathbb{R}^n)$, $n \ge 2$, $\chi \ge 0$, every $s \ge -1$, there exist $C > 0$, $\varepsilon_0 > 0$ such that for every $\lambda \in \mathbb{R}$ and $\varepsilon \in (0,\varepsilon_0]$, one has*

$$\|\chi\left(-\Delta_D - (\lambda \pm i\varepsilon)^2\right)^{-1}\chi\|_{H_D^s(\Omega)\to H_D^{s+1}(\Omega)} \le C,$$

where $H_D^s(\Omega)$ is the domain of $(1-\Delta_D)^{s/2}$ equipped with natural norm.

This proposition allows to prove with Ben-Artzi and Klainerman method the following smoothing result:

Theorem 16.2 ([18]). *Assume $K \neq \emptyset$. Then for every $\chi \in C_0^\infty(\mathbb{R}^n)$, $n \ge 2$, there exists $C > 0$ such that*

$$\|\chi u\|_{L^2(\mathbb{R},H_D^{s+1}(\Omega))} \le C\|\chi f\|_{L^2(\mathbb{R},H_D^s(\Omega))},$$

where $s \in [-1,1]$ and $u(t) = \int_0^t e^{i(t-\tau)\Delta_D}\chi f(\tau)d\tau$.

$$\|\chi v\|_{L^2(\mathbb{R},H_D^{s+1/2}(\Omega))} \le C\|v_0\|_{H_D^s(\Omega)},$$

where $s \in [0,1]$ and $v(t) = e^{it\Delta_D}v_0$.

Remark 16.2. Using the result obtained by Vasy and Zworski [70], we can replace the cutoff χ by the weight $\langle x\rangle^{-\mu}$ at the left-hand side of inequalities in Theorem 16.2 to obtain a generalization of Theorem 16.1.

Remark 16.3. The Proposition 16.1 has a long story in literature. We can cite the works of Lax and Philips [43], Vainberg [68], and the microlocal approach following the works of Melrose and Sjöstrand [49, 50]. We may find the references on the subject in Vasy and Zworski [70] and Burq [14].

Remark 16.4. Burq, Gérard, and Tzvetkov proved the second inequality of Theorem 16.2 using a $TT^\star$ method and the first inequality. This method is rather general and can be used, for instance, under other assumptions on operators or locally in time.

16.3 Results Local in Time in $\mathbb{R}^n$

Here we give results on smoothing locally in time. It is in this context that we have the most general result for Schrödinger operators. Nevertheless for general dispersive equations with constant coefficients we have rather general results. We give here the result of Constantin and Saut [19] which is one of the most complete for general dispersive equations with constant coefficients.

We assume

$$p(\xi) \text{ is } L^\infty_{\text{loc}} \text{ and there exists } R > 0 \text{ such that } p(\xi) \text{ is } C^1 \text{ if } |\xi| \geq R. \tag{16.1}$$

There exist $m > 1$, $R > 0$, and $C > 0$ such that

$$\left|\frac{\partial p}{\partial \xi_j}(\xi)\right| \geq C\langle\xi\rangle^{m-2}|\xi_j|, \text{ for all } \xi \in \mathbb{R}^n \text{ such that } |\xi| \geq R \tag{16.2}$$

Theorem 16.3 ([19]). *Assume* (16.1) *and* (16.2). *Let* $\chi \in C_0^\infty(\mathbb{R}^n)$, $\chi \geq 0$, *and* $T > 0$, *there exists* $C > 0$ *such that*

$$\int_{-T}^{T}\int_{\mathbb{R}^n}\left|\chi(x)\langle D\rangle^{s+(m-1)/2}u(t,x)\right|^2 dxdt \leq C\int_{\mathbb{R}^n}|\langle D\rangle^s u_0(x)|^2 dx,$$

where $u(t,\cdot) = e^{itp(D)}u_0$.

Remark 16.5. The assumptions (16.1) and (16.2) allow symbol as $|\xi|^m$. It is the case treated by Sjölin [65].

Remark 16.6. The assumption (16.2) implies that the symbol p is of principal type, i.e., $\partial_\xi p \neq 0$.

Remark 16.7. Constantin and Saut [20] gave smoothing results for Schrödinger operator perturbed by a potential. In Koch and Saut [42] we can find a result with a forcing term, analogous to the one on Schrödinger equation. The important fact is we have the double smoothing effect with respect to the smoothing on initial data.

Remark 16.8. We give a simple version of the result of Constantin and Saut [19]. The precise result allows to replace χ and the interval of integration $[-T,T]$ by functions with enough decrease in space and in time.

For Schrödinger operator we have better results. We give here a simple version of the results of Doi [25, 27]. Doi considers operators with following form:

$$H(t) = \sum_{j,k=1}^{n} (D_j - a_j(t,x))g^{jk}(x)(D_k - a_k(t,x)) + V(t,x).$$

We assume the metric elliptic, more precisely,
g^{jk} is symmetric, and there exist $C_2 > C_1 > 0$ such that

$$C_1 Id \leq (g^{jk})_{1\leq j,k\leq n} \leq C_2 Id, \text{ for all } x \in \mathbb{R}^n. \tag{16.3}$$

Let $T > 0$, we assume for all $\alpha \in \mathbb{N}^n$, there exists $C_\alpha > 0$ such that for all j,k,

$$|\partial_x^\alpha g^{jk}(t,x)| \leq C_\alpha \langle x\rangle^{-|\alpha|}, \text{ for all } (t,x) \in [0,T]\times\mathbb{R}^n. \tag{16.4}$$

We assume a_j real valued, and for all $\alpha \in \mathbb{N}^n$, there exists $C_\alpha > 0$ such that for all j,

$$|\partial_x^\alpha a_j(t,x)| \leq C_\alpha \langle x\rangle^{1-|\alpha|}, \text{ for all } (t,x) \in [0,T]\times\mathbb{R}^n. \tag{16.5}$$

We assume V real valued, and for all $\alpha \in \mathbb{N}^n$, there exists $C_\alpha > 0$ such that

$$|\partial_x^\alpha V(t,x)| \leq C_\alpha \langle x\rangle^{2-|\alpha|} \text{ for all } (t,x) \in [0,T]\times\mathbb{R}^n. \tag{16.6}$$

We need also a convexity condition on the metric at infinity. Doi gives a general condition (see [27, assumption H5]). This condition is not so obvious. We do not know if it is necessary to obtain the result. For simplicity we give only a stronger condition.

There exist $\nu > 0$ and $C > 0$, such that for all j,k,p,

$$|\partial_{x_p} g^{jk}(x)| \leq C\langle x\rangle^{-1-\nu}, \text{ for all } x \in \mathbb{R}^n. \tag{16.7}$$

We need a geometrical assumption on the bicharacteristic flow. If we denote by $p(x,\xi) = \sum_{j,k=1}^n g^{jk}(x)\xi_j\xi_k$, the bicharacteristic issues from $(x,\xi) \in T^\star(\mathbb{R}^n)$ is the solution of differential equation $\dot{x}(t) = \partial_\xi p(x,\xi)$, $\dot{\xi}(t) = -\partial_x p(x,\xi)$ satisfying $x(0) = x$, $\xi(0) = \xi$.

Remark 16.9. The solution $(x(t),\xi(t))$ exists globally in t. Indeed, $p(x,\xi)$ is a conserved quantity; this implies that $\xi(t)$ stays bounded for all t in existence domain. Now $\partial_\xi p(x,\xi)$ is bounded; this implies that $x(t)$ is bounded if it is not defined globally in time.

Definition 16.1. We say that the metric is nontrapped if for all $(x,\xi) \in T^\star(\mathbb{R}^n)$, the bicharacteristic issued from (x,ξ) satisfies $|x(t)| \to +\infty$ when $t \to +\infty$ or when $t \to -\infty$.

Theorem 16.4 ([27]). *We assume* (16.3)–(16.7), *and the metric* $(g^{jk})_{1\le j,k\le n}$ *is nontrapped. Let* $\mu > 1/2$, *there exists* $C > 0$ *such that for all* $u \in C^1([0,T], S(\mathbb{R}^n))$ *satisfying* $(i\partial_t + H(t))u(t) = f(t)$ *for* $t \in [0,T]$, *we have for all* $t \in [0,T]$,

$$\|E_s u(t)\|^2_{L^2(\mathbb{R}^n)} + \int_0^T \|\langle x\rangle^{-\mu} E_{s+1/2} u(t)\|^2_{L^2(\mathbb{R}^n)} dt$$
$$\le C\|E_s u(0)\|^2_{L^2(\mathbb{R}^n)} + C\int_0^T \|\langle x\rangle^{\mu} E_{s-1/2} f(t)\|^2_{L^2(\mathbb{R}^n)} dt, \quad (16.8)$$

where $E_s = (\langle x\rangle + \langle D\rangle)^s$.

Remark 16.10. If $f = 0$, a part of inequality (16.8) is the classical energy estimate. If $f \neq 0$, the estimate on $\|E_s u(t)\|^2$ is not the energy estimate. The norm on f is better with respect regularity but worst with respect decreasing.

Remark 16.11. In the assumption we take $u(t,\cdot)$ and $f(t,\cdot)$ in $S(\mathbb{R}^n)$, but we can extend the estimate (16.8) if the right-hand side is finite. The solution u is in the space defined with the norm given by the left-hand side of (16.8).

We give now a result if the increases of a_j and V are more faster. The first result was given by Yajima and Zhang [75], and we give here the result obtained in [58].

We keep the general form of H taken by Doi, the assumptions (16.3), (16.4), and (16.7). We replace the assumption (16.5) and (16.6) by the following.

Let $m \ge 2$, we assume a_j real valued and for all $\alpha \in \mathbb{N}^n$, there exists $C_\alpha > 0$ such that for all j,

$$|\partial_x^\alpha a_j(t,x)| \le C_\alpha \langle x\rangle^{-|\alpha|+m/2} \text{ for all } (t,x) \in [0,T]\times\mathbb{R}^n. \quad (16.9)$$

We assume V real valued and for all $\alpha \in \mathbb{N}^n$, there exists $C_\alpha > 0$ such that

$$|\partial_x^\alpha V(t,x)| \le C_\alpha \langle x\rangle^{m-|\alpha|} \text{ for all } (t,x) \in [0,T]\times\mathbb{R}^n. \quad (16.10)$$

We need also that P is essentially self-adjoint for each $t \in [0,T]$. This assumption is fulfilled if we assume there exists $C > 0$ such that for all $x \in \mathbb{R}^n$, $V(t,x) \ge -C\langle x\rangle^2$.

We denote by F_s the operator with symbol $(1+|\xi|^2+|x|^m)^{s/2}$ in Weyl quantification.[1]

Theorem 16.5 ([58]). *We assume* (16.3), (16.4), (16.7), (16.9), (16.10), *and the metric is nontrapped. Let* $\mu > 1/2$, *there exists* $C > 0$ *such that for all* $u \in C^1([0,T], S(\mathbb{R}^n))$ *satisfying* $(i\partial_t + H(t))u(t) = f(t)$ *for* $t \in [0,T]$, *we have for all* $t \in [0,T]$,

[1] For a function on $q(x,\xi)$, there are several manner to associate an operator. For instance if $q(x,\xi) = a(x)\xi$, the classical quantification of q is $a(x)D_j$. Other possibility is $D_j a(x)$. The Weyl quantification is the mix of the previous one, i.e., $(1/2)(a(x)D_j + D_j a(x))$. For a general symbol, the formula is more complicated; see Hörmander [32, Chap. 18, ğ5].

$$\|u(t)\|^2_{L^2(\mathbb{R}^n)} + \int_0^T \|\langle x\rangle^{-\mu} F_{1/m} u(t)\|^2_{L^2(\mathbb{R}^n)} dt$$
$$\leq C\|u(0)\|^2_{L^2(\mathbb{R}^n)} + C\int_0^T \|\langle x\rangle^{\mu} F_{-1/m} f(t)\|^2_{L^2(\mathbb{R}^n)} dt. \quad (16.11)$$

Remark 16.12. For $m = 2$ we find the Doi's estimate (16.8) with $s = 0$. Actually to prove (16.11) we adapt the Doi's proof in our assumptions. The main trick is to construct a symbol to obtain the estimate by a commutator method.

Remark 16.13. In [58, Theorem 1.2] we give a smoothing estimate more precise where we distinguish the tangential derivative to the sphere and the normal derivative. For the tangential derivative we can have an estimate with $\mu = 1/2$ at the left-hand side.

16.4 Results in Exterior Domains

In this section we keep the assumptions (16.3) and (16.5) where V does not depend on t, (16.7) and the operator will be $H = \sum_{j,k=1}^n D_j g^{jk}(x) D_k + V(x)$. We assume there exists $C_0 > 0$ such that for all $x \in \mathbb{R}^n$, $V(x) \geq -C_0$. We recall the notation given for the result of Burq, Gérard, and Tzvetkov. Let K be a compact set with smooth boundary, we denote by $\Omega = \mathbb{R}^n \setminus K$. Like the problem in $\mathbb{R}^n$, we need a geometrical condition. We define the generalized bicharacteristic flow. In Ω, this flow is the usual bicharacteristic flow defined before Definition 16.1. When the bicharacteristic hit the boundary of Ω, if the bicharacteristic is transverse to the boundary, the generalized flow is continued with respect to the geometrical optical law; if the bicharacteristic is tangential, the generalized bicharacteristic can glide on the boundary or continue in Ω if the usual flow hit the boundary only in one point. The precise description of this flow can be found in Hörmander [32, Definition 24.3.7] and a description in local coordinates near the boundary in [59, Appendix 1]. To assure the uniqueness of the flow, we also assume that the flow has no contact of infinity order with the boundary; see [59, Definition A.3].

The operator H is essentially self-adjoint and for simplicity we denote by H the unique self-adjoint extension with Dirichlet boundary condition. With the assumption above, $(1 + C_0)Id + H$ is positive; by functional calculus, we can define the power of this operator, and we denote by $\Lambda^s = ((1 + C_0)Id + H)^{s/2}$.

We denote the norm in $L^2(\Omega)$ by $\|\cdot\|$.

16.4.1 Geometrical Control Condition

Definition 16.2. We say that the metric is nontrapping in Ω if for all $(x, \xi) \in T^\star(\Omega)$, the generalized bicharacteristic satisfied $|x(t)| \to +\infty$ if $t \to +\infty$ or for $t \to -\infty$.

Remark 16.14. Even if the metric is flat, we can have trapped bicharacteristics, for instance, if K is the union of two disjoint balls.

In this context we prove with Claude Zuily [59] the following result:

Theorem 16.6 ([59]). *We assume that the metric is nontrapping in Ω. Let $T > 0$ and $\chi \in C_0^\infty(\overline{\Omega})$. There exists $C > 0$ such that for all $u_0 \in C_0^\infty(\Omega)$, we have*

$$\int_{-T}^{T} \|\chi \Lambda^{s+1/2} u(t)\|^2_{L^2(\Omega)} dt \leq C \|\Lambda^s u_0\|^2_{L^2(\Omega)}, \tag{16.12}$$

where $u(t) = e^{itH} u_0$.

Remark 16.15. The operator Λ^s plays the role of the H^s norm in the space. Here we must take account the boundary condition. The operator Λ^s commutes with H, and it is sufficient to prove (16.12) for $s = 0$.

Remark 16.16. The proof of Theorem 16.6 uses the technic of microlocal defect measure. In this context this method was initiated by Burq [14] to prove an estimate on the resolvent. We have adapted this method in nonstationary case. We can find more information on microlocal defect measure in Gérard and Leichtnam [28], Lebeau [45], Burq [13], Burq and Gérard [17], and Miller [51]. These works concern wave equation, but we can adapt the technics for Schrödinger equation.

16.4.2 Damping Condition

Recently with Lassaad Aloui and Moez Khenissi [7], we study the smoothing effect for the following problem:

$$\begin{cases} (D_t + H)u - iaH^{1/2}au = f \text{ in } (0,\infty) \times \Omega \\ u = 0 \text{ on } (0,\infty) \times \partial\Omega, \\ u_{|t=0} = u_0, \end{cases} \tag{16.13}$$

Here for simplicity we assume $H \geq 1$, and we remark that this can be realized working with $v = e^{itC}u$ with C large enough. Of course $H^{1/2}$ is changed in $(H+C)^{1/2}$. We assume $a \in C_0^\infty(\overline{\Omega})$ and $a \geq 0$. We denote by $\omega = \{x \in \Omega,\ a(x) > 0\}$.

Definition 16.3 (E.G.C.). We say that ω verifies the exterior geometric control condition if all generalized bicharacteristics $\gamma(t) = (x(t), \xi(t))$ is such that:

- $|x(t)|$ goes to $+\infty$ when t goes to $+\infty$ or $-\infty$.
- $x(t)$ meets ω.

Theorem 16.7 ([7]). *Let $T > 0$, $s \in (-1/2, 1/2)$, and $\mu \in (1/2, 1]$. We assume that ω satisfies the exterior geometric control condition and then there exists $C > 0$ such that*

$$\int_0^T \left\| \Lambda^{s+1/2} \langle x \rangle^{-\mu} u \right\|^2_{L^2(\Omega)} dt + \sup_{t \in [0,T]} \| \Lambda^s u(t) \|^2_{L^2(\Omega)}$$
$$\leq C \left(\| \Lambda^s u_0 \|^2_{L^2(\Omega)} + \int_0^T \left\| \Lambda^{s-1/2} \langle x \rangle^{\mu} f \right\|^2_{L^2(\Omega)} dt \right) \quad (16.14)$$

for all u_0 *in* $C_0^\infty(\Omega)$, f *in* $C_0^\infty(\mathbb{R}_+ \times \Omega)$, *where* u *denotes the solution of* (16.13).

Remark 16.17. The estimate (16.14) is analog of the estimate (16.8), and (16.14) can be extended if the right-hand side is finite. The range of s and μ is different. On μ the result is sharp. On s, probably we can improve the result. The difficulty to do this is to commutate Λ^s with a or $\langle x \rangle^{\pm\mu}$. The technical Lemmas exist in [7] to do this, but the number of commutator to compute and to estimate increases with s, and we have no applications to motivate this tedious computation. This difficulty does not appear in Theorem 16.6.

Remark 16.18. We can prove that the norms on u, $\|\Lambda^s \langle x \rangle^\mu u\|$ and $\|\langle x \rangle^\mu \Lambda^s u\|$, are equivalent.

Remark 16.19. This result improves Theorem 16.6 with respect to the forcing term f.

Remark 16.20. The proof of Theorem 16.7 follows the strategy of proof of Theorem 16.6. The first step is to reduce the estimate to a semiclassical estimate. We need to compute commutator between $\chi(h^2 H)$ and smooth function compactly supported. The main tool is the Helffer-Sjöstrand formula [31] (see also Davies [22]). The semiclassical estimate is proved by contradiction. This is decomposed in few steps. We constructed a semiclassical defect measure. On the one hand, we prove that this measure is not identically null, and on the other, we prove that the measure has no support over ω and on the incoming point; the support is propagated by the generalized bicharacteristic flow. These properties imply that the measure is identically null. Both parts of the proof yield the contradiction.

Remark 16.21. The term $iaH^{1/2}a$ in (16.13) appears in the works on stabilization. For Schrödinger operator, this term appear in the work of Aloui [1] and in Aloui et al. [6]. Dehman et al. [24] used the term $ia(1+H)^{-1}aD_t$ instead of $-iaH^{1/2}a$ for stabilization. We can remark that on the characteristic set, we have $D_t \sim -H$, and this damping term is of order 0. Aloui and Khenissi [4] used $-ia$ instead of $-iaH^{1/2}a$. In these both works, the damping is of order 0 and too weak to obtain smoothing effect. This term was also used in compact manifold by Aloui [1] and in [2] for a bounded domain of $\mathbb{R}^n$ with boundary condition to prove smoothing effect if all the bicharacteristic hit the damping term. The stabilization had studied more extensively for wave equation since [9] where the terminology "geometric control" was introduced. The control problem was studied for Schrödinger equation by microlocal method by Lebeau [44]. Aloui and Khenissi [3, 5] had used this notion for the stabilization in a exterior domain for wave equation.

16.5 Related Problems

16.5.1 Analytic and C^∞ Smoothing Effect

The $1/2$-smoothing effect is related to the L^2 norm (or H^s norm) on the datum u_0. If the datum is more decreasing, we can obtain a C^∞ or analytic smoothing.

Kapitanski and Safarov [35] studied the singularities of fundamental solution of Schrödinger equation comparing the Schrödinger fundamental solution and the one of wave. After, Craig et al. [21] proved that if u_0 is quickly decreasing at infinity, then the solution of Schrödinger equation is C^∞ for all time $t \neq 0$. Their result is more precise and can be formulated microlocally that if u_0 decreases quickly in a cone Γ at infinity, (x,ξ) is not in the wave front of the solution at time $t \neq 0$ if the bicharacteristic issue of (x,ξ) goes in the cone Γ.

This result was precised by Wunsch [72] who defined a wave front set at infinity and proved a propagation result in this context. This allows to find the result of Craig, Kappeler, and Strauss by another way, and we can understand why, at some time, the solution is not smooth if the datum oscillates at infinity (see also a pioneer work by Shananin [64]).

We can find also other result in Nakamura [54] where it is precised the wave front set at a fixed time. The previous works giving results for a uniform wave front set in time.

In the context of compact manifold or in bounded domain of $\mathbb{R}^n$, we may not expect smoothing effect, but with damping term analog to the one of (16.13), Aloui in [1, 2] proved C^∞ smoothing effect in this case.

These works were proved in the analytic context first by Hayashi and Saito [30] in one dimension. In general dimension, we proved in [55] with Claude Zuily, under the same assumption than the one of Craig, Kappeler, and Strauss, that (x,ξ) is not in the analytic wave front set if u_0 decrease exponentially in the cone Γ. In [56, 57], we gave results for oscillating data in the first paper and in the spirit of the Wunsch results in second paper. In this framework other results can be found in Martinez et al. [46–48] and Ito [33].

Results for Schrödinger operator with potential as x^2 at infinity were proven by Wunsch [73] in C^∞ case and by Atallah-Baraket and Mechergui [8] in the analytic case. New phenomenon arrives in this case; periodically the singularities appear from infinity.

Other results were proven in Gevrey context; see De Bouard et al. [23]; Kato and Taniguchi [39]; Morimoto et al. [53]; Kajitani and Taglialatela [36]; Taniguchi [67]; and Mizuhara [52].

Some of previous works concern also other operators, as KdV, general dispersive equation. For Schrödinger operators where the Laplacian is replaced by a power of Laplacian, we can see the work of Kamoun and Mechergui [37].

To give an example of result in this field, we give the one of Craig et al. [21]. To explain the more recent results, we should need to introduce a lot of notations and geometrical notions.

Let u solution of

$$i\partial_t u(t,x) = \sum_{1\le j,k\le n} D_{x_j} g_{jk}(x) D_{x_k} u(t,x),\ u(0,x) = u_0(x),$$

where g_{jk} is symmetric and elliptic, i.e., there exist $C_1 > 0$ and $C_2 > 0$ such that

$$C_1|\xi|^2 \le \sum_{1\le j,k\le n} g_{jk}(x)\xi_j\xi_k \le C_2|\xi|^2$$

for all $x \in \mathbb{R}^n$.

We assume for all $\alpha \in \mathbb{N}^n$, there exists $C_\alpha > 0$ such that

$$|\partial^\alpha(g_{jk}(x) - \delta_{jk})| \le \frac{C_\alpha}{\langle x\rangle^{1+\nu+|\alpha|}}$$

where $\nu > 0$.

Theorem 16.8 ([21]). *Let $(x_0,\xi_0) \in T^\star\mathbb{R}^n$, we assume that (x_0,ξ_0) is not backward trapped, if*

$$\int_{\mathbb{R}^n} |x|^k |u_0(x)|^2 dx < +\infty$$

for all $k \in \mathbb{N}^n$, then $(x_0,\xi_0) \notin WF(u(t,\cdot))$.

Remark 16.22. We remind that a bicharacteristic is not backward trapped if $|x(t)|$ goes to $+\infty$ when t goes to $-\infty$.

Remark 16.23. Under the assumption of Theorem 16.8, we can prove that $\frac{x(t)}{|x(t)|}$ goes to some $\eta \in \mathbb{R}^n$ when t goes to $-\infty$ where $x(t)$ is the bicharacteristic issue of (x_0,ξ_0). We can precise the theorem. We define

$$\Gamma = \left\{x \in \mathbb{R}^n,\ \left|\frac{x}{|x|} - \frac{\eta}{|\eta|}\right| < \varepsilon,\ |x| > R\right\}.$$

If we assume

$$\int_\Gamma |x|^k |u_0(x)|^2 dx < +\infty,$$

for all $k \in \mathbb{N}^n$, then $(x_0,\xi_0) \notin WF(u(t,\cdot))$.

Remark 16.24. With analytic assumptions we can prove same results for the analytic wave front set. More precisely, we assume there exist $C > 0$ such that for all $\alpha \in \mathbb{N}^n$,

$$|\partial^\alpha(g_{jk}(x) - \delta_{jk})| \le \frac{C^{1+|\alpha|}}{\langle x\rangle^{1+\nu+|\alpha|}}$$

where $\nu > 0$.

We assume (x_0,ξ_0) is not backward trapped, and

$$\int_\Gamma e^{\delta|x|} |u_0(x)|^2 dx < +\infty,$$

for a $\delta > 0$, then $(x_0,\xi_0) \notin WF_A(u(t,\cdot))$.

Remark 16.25. Theorem 16.8 and the Kato smoothing effect suggest that we can have intermediate results. If the datum decreases in a cone with a rate, the solution is in a Sobolev space (microlocally) with an index of regularity related with the rate of decreasing. To our knowledge this kind of results are not proved in literature. In Doi [26] we may find an intermediate smoothing effect with an assumption on the datum which is not related with decreasing.

16.5.2 Strichartz Estimates and Dispersive Estimates

The Strichartz estimates can be considered as smoothing effect in L^p space. A way to obtain Strichartz estimates is to apply the $TT^\star$ method and dispersive estimate. We cannot give the huge literature on the subject. We can consult the introduction of Bouclet and Tzvetkov [12] and the references therein for more information. The Kato smoothing effect plays a role in the proof of Strichartz estimate as in Staffilani and Tataru [66] and Burq [15]. Strichartz estimates are also related to dispersive estimate. This relation can be found in Ginibre and Velo [29] and Keel and Tao [41]. We can find dispersive results global in time in Journé et al. [34]; Rodnianski and Schlag [60]; and Rodnianski et al. [61]. The three subjects, Kato smoothing effect, dispersive estimate, and Strichartz estimate, are related. These estimates are proved for linear operators; they play an important role in the proofs of local or global existence for nonlinear equations.

References

1. Aloui, L.: Smoothing effect for regularized Schrödinger equation on compact manifolds. Collect. Math. **59**, 53–62 (2008)
2. Aloui, L.: Smoothing effect for regularized Schrödinger equation on bounded domains. Asympt. Anal. **59**, 179–193 (2008)
3. Aloui, L., Khenissi, M.: Stabilisation de l'équation des ondes dans un domaine extérieur. Rev. Math. Iberoamericana, **28**, 1–16 (2002)
4. Aloui, L., Khenissi, M.: Stabilization of Schrödinger equation in exterior domains. Contr. Optim. Calculus Variat. ESAIM **13**, 570–579 (2007)
5. Aloui, L., Khenissi, M.: Boundary stabilization of the wave and Schrödinger equations in exterior domains. Discrete Contin. Dyn. Syst. Ser. A **27**, 919–934 (2010)
6. Aloui, L., Khenissi, M., Vodev, G.: Smoothing effect for the regularized Schrödinger equation with non controlled orbits. Comm. Partial Diff. Eq. **38**, 265–275 (2013)
7. Aloui, L., Khenissi, M., Robbiano, L.: The Kato smoothing effect for regularized Schrödinger equations in exterior domains. Submitted
8. Atallah-Baraket, A., Mechergui, C.: Analytic smoothing effect for the Schrödinger equation relative to the harmonic oscillator equation. Asymp. Anal. **63**, 29–48 (2009)
9. Bardos, C., Lebeau, G., Rauch, J.: Sharp sufficient conditions for the observation, control, and stabilization of waves from the boundary. SIAM J. Contr. Optim. **30**, 1024–1065 (1992)
10. Ben-Artzi, M., Devinatz, A.: Local smoothing and convergence properties of Schrödinger type equations. J. Funct. Anal. **101**, 231–254 (1991)

11. Ben-Artzi, M., Klainerman, S.: Decay and regularity for the Schrödinger equation. J. Anal. Math. **58**, 25–37 (1992)
12. Bouclet, J-M., Tzvetkov, N.: Strichartz estimates for long range perturbations. Am. J. Math. **129**, 1565–1609 (2007)
13. Burq, N.: In: Mesures semi-classiques et mesures de défaut. Séminaire Bourbaki, vol. 1996/97. Astérisque **245**, 167–195 (1997)
14. Burq, N.: Semi-classical estimates for the resolvent in nontrapping geometries. Int. Math. Res. Not. **5**, 221–241 (2002)
15. Burq, N.: Estimations de Strichartz pour des perturbations à longue portée de l'opérateur de Schrödinger. Séminaire Équations aux Dérivées Partielles, 2001–2002, exposé n^{0}11, École Polytechnique
16. Burq, N.: Smoothing effect for Schrödinger boundary value problems. Duke Math. J. **123**, 403–427 (2004)
17. Burq, N., Gérard, P.: Condition nécessaire et suffisante pour la contrôlabilité exacte des ondes. C. R. Acad. Sci. Paris Sér. I Math. **325**, 749–752 (1997)
18. Burq, N., Gérard, P., Tzvetkov, N.: On nonlinear Schrödinger equations in exterior domains. Ann. Inst. H. Poincaré Anal. Non Linéaire **21**, 295–318 (2004)
19. Constantin, P., Saut, J.-C.: Local smoothing properties of dispersive equations. J. Am. Math. Soc. **1**, 413–439 (1988)
20. Constantin, P., Saut, J.-C.: Local smoothing properties of Schrödinger equations. Indiana Univ. Math. J. **38**, 791–810 (1989)
21. Craig, W., Kappeler, T., Strauss, W.: Microlocal dispersive smoothing for the Schrödinger equation. Comm. Pure Appl. Math. **48**, 769–860 (1995)
22. Davies, E.B.: In: Spectral Theory and Differential Operators. Cambridge Studies in Advanced Mathematics, vol. 42. Cambridge University Press, Cambridge (1995)
23. De Bouard, A., Hayashi, N., Kato, K.: Gevrey regularizing effect for the (generalized) Korteweg-de Vries equation and nonlinear Schrödinger equations. Ann. Inst. H. Poincaré Anal. Non Linéaire **12**, 673–725 (1995)
24. Dehman, B., Gérard, P., Lebeau, G.: Stabilization and control for the nonlinear Schrödinger equation on a compact surface. Math. Z. **254**, 729–749 (2006)
25. Doi, S.: Smoothing effects for Schrödinger evolution group on Riemannian manifolds. Duke Math. J. **82**, 679–706 (1996)
26. Doi, S.: Smoothing effects for Schrödinger evolution equation and global behavior of geodesic flow. Math. Ann. **318**, 355–389 (2000)
27. Doi, S.: Smoothing of solutions for Schrödinger equations with unbounded potentials. Publ. Res. Inst. Math. Sci. **41**, 175–221 (2005)
28. Gérard, P., Leichtnam, E.: Ergodic properties of eigenfunctions for the Dirichlet problem. Duke Math. J. **71**, 559–607 (1993)
29. Ginibre, J., Velo, G.: Smoothing properties and retarded estimates for some dispersive evolution equations. Comm. Math. Phys. **144**, 163–188 (1992)
30. Hayashi, N., Saitoh, S.: Analyticity and smoothing effect for the Schrödinger equation. Ann. Inst. H. Poincaré Phys. Théor. **52**, 163–173 (1990)
31. Helffer, B., Sjöstrand, J.: Équation de Schrödinger avec champ magnétique et équation de Harper. In: Schrödinger Operators (Sønderborg (1988)). Lecture Notes in Physics, vol. 345, pp. 118–197. Springer, Berlin (1989)
32. Hörmander, L.: The Analysis of Linear Partial Differential Operators, vol. III. Springer, Berlin (1994)
33. Ito, K.: Propagation of singularities for Schrödinger equations on the Euclidean space with a scattering metric. Comm. Part. Differ. Equat. **31**, 1735–1777 (2006)
34. Journé, J.-L., Soffer, A., Sogge, C.D.: Decay estimates for Schrödinger operators. Comm. Pure Appl. Math. **44**, 573–604 (1991)
35. Kapitanski, L., Safarov, Y.: Dispersive smoothing for Schrödinger equations. Math. Res. Lett. **3**, 77–91 (1996)
36. Kajitani, K., Taglialatela, G.: Microlocal smoothing effect for Schrödinger equations in Gevrey spaces. J. Math. Soc. Jpn. **55**, 855–896 (2003)

37. Kamoun, I., Mechergui, C.: Generalization of the analytic wavefront set and application. Indiana Univ. Math. J. **56**, 2569–2600 (2007)
38. Kato, T.: On the Cauchy problem for the (generalized) KdV equation. In: Guillemin, V. (ed.) Studies in Applied Mathematics. Adv. Math. Suppl. Stud., vol. 8, pp. 93–128. Academic, New York (1983)
39. Kato, K., Taniguchi, K.: Gevrey regularizing effect for nonlinear Schrödinger equations. Osaka J. Math. **33**, 863–880 (1996)
40. Kato, T., Yajima, K.: Some examples of smooth operators and the associated smoothing effect. Rev. Math. Phys. **1**, 481–496 (1989)
41. Keel, M., Tao, T.: Endpoint Strichartz estimates. Am. J. Math. **120**, 955–980 (1998)
42. Koch, H., Saut, J.-C.: Local smoothing and local solvability for third order dispersive equations. SIAM J. Math. Anal. **38**, 1528–1541 (2006/07)
43. Lax, P., Phillips, R.: Scattering Theory. Academic, Boston (1989)
44. Lebeau, G.: Contrôle de l'équation de Schrödinger. J. Math. Pures Appl. **71**, 267–291 (1992)
45. Lebeau, G.: Équation des ondes amorties. Algebraic and Geometric Methods in Mathematical Physics (Kaciveli, 1993), pp. 73–109. Kluwer, Dordrecht (1996)
46. Martinez, A., Nakamura, S., Sordoni, V.: Analytic smoothing effect for the Schrödinger equation with long-range perturbation. Comm. Pure Appl. Math. **59**, 1330–1351 (2006)
47. Martinez, A., Nakamura, S., Sordoni, V.: Analytic wave front set for solutions to Schrödinger equations. Adv. Math. **222**, 1277–1307 (2009)
48. Martinez, A., Nakamura, S., Sordoni, V.: Analytic wave front set for solutions to Schrödinger equations II—long range perturbations. Comm. Part. Differ. Equat. **35**, 2279–2309 (2010)
49. Melrose, R.B., Sjöstrand, J.: Singularities of boundary value problems. I. Comm. Pure Appl. Math. **31**, 593–617 (1978)
50. Melrose, R.B., Sjöstrand, J.: Singularities of boundary value problems. II. Comm. Pure Appl. Math. **35**, 129–168 (1982)
51. Miller, L.: Refraction of high-frequency waves density by sharp interfaces and semiclassical measures at the boundary. J. Math. Pures Appl. **79**, 227–269 (2000)
52. Mizuhara, R.: Microlocal smoothing effect for the Schrödinger evolution equation in a Gevrey class. J. Math. Pures Appl. **91**, 115–136 (2009)
53. Morimoto, Y., Robbiano, L., Zuily, C.: Remark on the analytic smoothing for the Schrödinger equation. Indiana Univ. Math. J. **49**, 1563–1579 (2000)
54. Nakamura, S.: Wave front set for solutions to Schrödinger equations. J. Funct. Anal. **256**, 1299–1309 (2009)
55. Robbiano, L., Zuily, C.: Microlocal analytic smoothing effect for the Schrödinger equation. Duke Math. J. **100**, 93–129 (1999)
56. Robbiano, L., Zuily, C.: Effet régularisant microlocal analytique pour l'équation de Schrödinger: le cas des données oscillantes. Comm. Part. Differ. Equat. **25**, 1891–1906 (2000)
57. Robbiano, L., Zuily, C.: Analytic theory for the quadratic scattering wave front set and application to the Schrödinger equation. Astérisque **283** (2002)
58. Robbiano, L., Zuily, C.: Remark on the Kato smoothing effect for Schrödinger equation with superquadratic potentials. Comm. Part. Differ. Equat. **33**, 718–727 (2008)
59. Robbiano, L., Zuily, C.: The Kato smoothing effect for Schödinger equations with unbounded potentials in exterior domains. Int. Math. Res. Notices **9**,1636–1698 (2009)
60. Rodnianski, I., Schlag, W.: Time decay for solutions of Schrödinger equations with rough and time-dependent potentials. Invent. Math. **155**, 451–513 (2004)
61. Rodnianski, I., Schlag, W., Soffer, A.: Dispersive analysis of charge transfer models. Comm. Pure Appl. Math. **58**, 149–216 (2005)
62. Ruzhansky, M., Sugimoto, M.: A new proof of global smoothing estimates for dispersive equations. In: Advances in Pseudo-differential Operators. Oper. Theory Adv. Appl., vol. 155, pp. 65–75. Birkhäuser, Basel (2004)
63. Ruzhansky, M., Sugimoto, M.: A smoothing property of Schrödinger equations in the critical case. Math. Ann. **335**, 645–673 (2006)
64. Shananin, N.A.: Singularities of the solutions of the Schrödinger equation for a free particle. Math. Notes **55**, 626–631(1994)

65. Sjölin, P.: Regularity of solutions to the Schrödinger equation. Duke Math. J. **55**, 699–715 (1987)
66. Staffilani, G., Tataru, D.: Strichartz estimates for a Schrödinger operator with nonsmooth coefficients. Comm. Part. Differ. Equat. **27**, 1337–1372 (2002)
67. Taniguchi, K.: Gevrey regularizing effect for the initial value problem for a dispersive operator. Osaka J. Math. **41**, 911–932 (2004)
68. Vaĭnberg, B.R.: Asymptotic Methods in Equations of Mathematical Physics. Gordon & Breach Science Publishers, New York (1989)
69. Vega, L.: Schrödinger equations : pointwise convergence to the initial data. Proc. A.M.S. **102**, 874–878 (1988)
70. Vasy, A., Zworski, M.: Semiclassical estimates in asymptotically Euclidean scattering. Comm. Math. Phys. **212**, 205–217 (2000)
71. Watanabe, K.: Smooth perturbations of the selfadjoint operator $|\Delta|^{\alpha/2}$. Tokyo J. Math. **14**, 239–250 (1991)
72. Wunsch, J.: Propagation of singularities and growth for Schrödinger operators. Duke Math. J. **98**, 137–186 (1999)
73. Wunsch, J.: The trace of the generalized harmonic oscillator. Ann. Inst. Fourier (Grenoble) **49**, 351–373 (1999)
74. Yajima, K.: In: On Smoothing Property of Schrödinger Propagators. Lecture Notes in Mathematics, vol. 1450, pp. 20–35. Springer, Berlin (1990)
75. Yajima, K., Zhang, G.: Local smoothing property and Strichartz inequality for Schrödinger equations with potentials superquadratic at infinity. J. Differ. Equat. **202**, 81–110 (2004)

Chapter 17
On the Cauchy Problem for NLS with Randomized Initial Data

Nicola Visciglia

Abstract We show that a general family of Cauchy problems associated to NLS with randomized initial data is well posed locally in time. As initial data we consider the random vector which is associated to the Gibbs measure.

Key words: Cauchy problem, NLS, Randomized initial data

2010 Mathematics Subject Classification: 35Q40, 35Q55.

17.1 Introduction

In this paper we consider the following family of Cauchy problems with randomized Cauchy data:

$$i\partial_t u - Au \pm u|u|^{\alpha-1} = 0, (t,x) \in \mathbb{R} \times \mathbb{R}, \tag{17.1}$$
$$u(0,x) = f^{\omega}(x).$$

Since now on we mean by solutions to (17.1) as the solutions to the following associated integral equations:

$$u(t) = e^{itA}\varphi \pm i \int_0^t e^{i(t-s)A}(u(s)|u(s)|^{\alpha-1})ds. \tag{17.2}$$

In the sequel we shall use the following notations:

$$L_x^r = L^r(\mathbb{R}), L_T^q L_x^r = L^q((-T,T);L_x^r)$$

N. Visciglia (✉)
Dipartimento di Matematica “L. Tonelli”, Università di Pisa,
largo Bruno Pontecorvo 5, I-56127 Pisa, Italy
e-mail: viscigli@dm.unipi.it

M. Cicognani et al. (eds.), *Studies in Phase Space Analysis with Applications to PDEs*, Progress in Nonlinear Differential Equations and Their Applications 84, DOI 10.1007/978-1-4614-6348-1_17,

and for any $1 \le p \le \infty$ we denote by p' the dual Hölder exponent, i.e. $1/p + 1/p' = 1$. Next we assume that

$$A : L^2_x \supset D(A) \to L^2_x$$

is a self-adjoint operator with discrete spectrum, $\{\lambda_n\}_n$ is the corresponding sequence of eigenvalues counted with its multiplicity, and $\{e_n\}_n$ is the associated basis of eigenvectors.
We shall assume that the initial data are parametrized by the following map:

$$(\Omega, \mathscr{F}, \mathbf{p}) \ni \omega \to f^{\omega}(x) = \sum_{n \in \mathbb{N}} \gamma_n g_n(\omega) e_n(x) \tag{17.3}$$

where $\gamma_n \in \mathbb{C}$ is a suitable sequence of complex numbers, $(\Omega, \mathscr{F}, \mathbf{p})$ is a probability space, and $g_n(\omega)$ are Gaussian independent random variables.
The main interest in this paper is to consider the case γ_n is almost in $l^2(\mathbb{N})$, that is,

$$\sum_{n \in \mathbb{N}} |\gamma_n|^2 \lambda_n^{-\delta} < \infty \; \forall \delta > 0. \tag{17.4}$$

In [1] it is considered (17.1) with $A = -\partial_x^2 + |x|^2$ and with a randomized initial data that satisfies the assumptions as above. In fact the analysis of the local Cauchy theory with randomized initial data is the basic step in order to get a probabilistic global well-posedness result via the invariance of the Gibbs measure.

The main point in this paper is that we treat a general family of operators A and we provide a simple proof of the local well-posedness of (17.1) for almost every $\omega \in \Omega$. More specifically we shall assume that

$$A : L^2_x \supset D(A) \to L^2_x \text{ is self-adjoint;} \tag{17.5}$$

$$\exists\, \bar{t} > 0 \text{ s.t. } \sup_{t \in (0,\bar{t})} \sqrt{t} \|e^{itA}\|_{\mathscr{L}(L^1_x, L^\infty_x)} < \infty \tag{17.6}$$

$$\exists\, (e_n, \lambda_n) \in L^2_x \times \mathbb{R} \text{ s.t. } e_n \text{ is a basis of } L^2(\mathbb{R}) \text{ and } Ae_n = \lambda_n e_n; \tag{17.7}$$

$$\forall r > 2\; \exists C(r), \varepsilon(r) > 0 \text{ s.t. } \|e_n\|_{L^r_x} \le C(r) \lambda_n^{-\varepsilon(r)}. \tag{17.8}$$

Our first result concerns the local well-posedness of (17.2) in the case $1 < \alpha < 5$.

Theorem 17.1. *Assume A satisfies* (17.5)–(17.8)*; $1 < \alpha < 5$ and $\gamma_n \in \mathbb{C}$ satisfy* (17.4)*. Then for almost every $\omega \in \Omega$ there exists $T(\omega) > 0$ such that the integral equation* (17.2) *(in both focusing and defocusing case) has one unique solution*

$$u(t,x) \in L^q_{T(\omega)} L^r_x \; \forall (q,r) \in [1,\infty] \times [1,\infty] \text{ s.t. } \frac{2}{q} + \frac{1}{r} = \frac{1}{2}. \tag{17.9}$$

More precisely there exist $a, b, c = a(\alpha), b(\alpha), c(\alpha) > 0$ such that

$$\forall M > 0\; \exists\, \Omega_M \subset \Omega \text{ measurable s.t.} \tag{17.10}$$

$$\mathbf{p}(\Omega_M) \ge 1 - e^{-cM^2} \text{ and } T(\omega) \ge \min\{\bar{t}, aM^{-b}\} \; \forall \omega \in \Omega_M.$$

In order to state our result in the L^2-supercritical regime, first we introduce for every $\alpha \geq 5$ the following set of $\mathbb{R}^2$:

$$\mathscr{A}_\alpha \equiv Conv\left\{(0,\frac{1}{\alpha+1}),\left(0,\frac{1}{\alpha}\right),\left(\frac{\alpha+3}{2\alpha(\alpha-1)},\frac{\alpha-3}{\alpha(\alpha-1)}\right),\left(\frac{\alpha+1}{2\alpha(\alpha-1)},\frac{1}{\alpha}\right)\right\}$$

where $Conv\{P_1,P_2,P_3,P_4\}$ denotes the convex envelope of P_1,P_2,P_3,P_4.

Theorem 17.2. *Assume that A satisfies* (17.5)–(17.8)*;* $\alpha \geq 5$ *and* $\gamma_n \in \mathbb{C}$ *satisfy* (17.4)*. Let us fix* $q(\alpha), r(\alpha) \in [1,\infty]$ *such that* $\left(\frac{1}{q(\alpha)},\frac{1}{r(\alpha)}\right) \in \mathscr{A}_\alpha$*. Then for almost every* $\omega \in \Omega$*, there exists* $T(\omega) > 0$ *such that the integral equation* (17.2) *has one unique solution*

$$u(t,x) \in L^{q(\alpha)}_{T(\omega)} L^{r(\alpha)}_x. \tag{17.11}$$

More precisely there exist $a,b,c = a(q(\alpha),r(\alpha)), b(q(\alpha),r(\alpha)), c(q(\alpha),r(\alpha)) > 0$ *such that*

$$\forall M > 0 \; \exists \; \Omega_M \subset \Omega \text{ measurable s.t.} \tag{17.12}$$

$$\mathbf{p}(\Omega_M) \geq 1 - e^{-cM^2}$$

and

$$T(\omega) \geq min\{\bar{t}, aM^{-b}\} \; \forall \omega \in \Omega_M. \tag{17.13}$$

Remark 17.1. Next we compare Theorems 17.1 and 17.2 with the result in [1]. In this paper we treat a general class of NLS and not only the NLS associated to the harmonic oscillator; moreover, we do not impose α to be an integer. On the other hand in [1], it is also presented as a probabilistic global well-posedness result, while in this paper we focus only on the local Cauchy theory. We plan to pursue elsewhere the globalization argument.

Remark 17.2. Following [6] the assumptions in Theorems 17.1 and 17.2 are satisfied in the case $A = -\partial_x^2 + V(x)$ where $V(x)$ satisfies suitable growth condition at infinity. In particular the case $V(x) = |x|^2$ enters in our analysis.

Remark 17.3. We underline that in the statement of Theorems 17.1 and 17.2, we don't say nothing about the regularity of the solution along the time of existence, unless a space–time integrability property.

Remark 17.4. A huge literature has been devoted to the deterministic analysis of (17.1), that is, the case that initial data do not depend on any randomization and belong to suitable Sobolev spaces. We cannot be exhaustive with the literature in this direction; however, we quote at least the pioneering papers [2, 4], where the Cauchy problem (17.1) is studied for $A = -\Delta$ and in any dimension.

17.2 Probabilistic A-Priori Estimates

Along this section we shall assume that f^ω is the random vector (17.3), γ_n satisfies (17.4), and λ_n are the eigenvalues of A with eigenfunctions e_n.

Proposition 17.1. *Assume* (17.8)*, then for every given* $T > 0$, $2 < r < \infty$ *and* $1 \leq q < \infty$*, there exists* $C = C(T,q,r) > 0$ *such that*

$$\mathbf{p}\{\omega \in \Omega \text{ s.t. } \|e^{itA} f^\omega\|_{L^q_T L^r_x} > \lambda\} \leq e^{-C\lambda^2}. \tag{17.14}$$

Proof. By using the Minkowski inequality we get the following estimate:

$$\|e^{itA} f^\omega\|_{L^p_\omega L^q_T L^r_x} \leq \|e^{itA} f^\omega\|_{L^q_T L^r_x L^p_\omega}$$

$$= \|\sum_n \gamma_n g_n(\omega) e^{it\lambda_n} e_n(x)\|_{L^q_T L^r_x L^p_\omega} \ \forall p > max\{q,r\}.$$

On the other hand we have

$$\|\sum_n \gamma_n g_n(\omega) e^{it\lambda_n} e_n(x)\|_{L^p_\omega} \leq \sqrt{p} \|(\sum_n |\gamma_n e_n(x)|^2)^{\frac{1}{2}}\|_{L^q_T L^r_x}$$

$$= \sqrt{p} T^{\frac{1}{q}} \|\gamma_n e_n(x)\|_{L^r_x l^2_N} \leq C\sqrt{p} T^{\frac{1}{q}} \|(\gamma_n e_n(x)\|_{l^2_N L^r_x}$$

where at the last step we have used the assumption $r > 2$. By using the hypothesis (17.8) we finally deduce

$$\|e^{itA} f^\omega\|_{L^p_\omega L^q_T L^r_x} \leq C\sqrt{p} T^{\frac{1}{q}} (\sum_n |\gamma_n|^2 \lambda_n^{-2\varepsilon(r)})^{\frac{1}{2}} = C(r)\sqrt{p} T^{\frac{1}{q}}.$$

This estimate implies (17.14) via the classical Chebichev inequality. □

17.3 Proof of Theorem 17.1

The proof of Theorem 17.1 is simpler than the proof of Theorem 17.2 (and hence we shall skip it). In fact the proof of Theorem 17.1 can be done following step by step the proof of Theorem 17.2 except that in the context of theorem 17.1 (i.e., in the L^2 subcritical regime), the fixed point argument can be done by using the standard Strichartz estimates, and it is not necessary to exploit the inhomogeneous ones.

17.4 Deterministic Theory via Inhomogeneous Strichartz Estimates for L^2 Supercritical NLS

Following [3, 5] we recall the inhomogeneous Strichartz estimates satisfied by the propagator e^{itA}.

Proposition 17.2. *Assume A satisfies* (17.5) *and* (17.6)*. Then for every* $(q,r,\tilde{q},\tilde{r}) \in [1,\infty]$ *there exists* $C = C(q,r,\tilde{q},\tilde{r}) > 0$ *such that*

$$\left\| \int_0^t e^{i(t-s)A} F(s) ds \right\|_{L^q_{\bar{t}} L^r_x} \leq C \|F\|_{L^{\tilde{q}'}_{\bar{t}}; L^{\tilde{r}'}_x}$$

provided that

$$\frac{1}{q} + \frac{1}{r} < \frac{1}{2}, \frac{1}{\tilde{q}} + \frac{1}{\tilde{r}} < \frac{1}{2}, 0 < \frac{1}{q}, \frac{1}{\tilde{q}} \leq 1, 0 \leq \frac{1}{r}, \frac{1}{\tilde{r}} \leq \frac{1}{2}, \frac{1}{q} + \frac{1}{\tilde{q}} = \frac{1}{2}\left(1 - \frac{1}{r} - \frac{1}{\tilde{r}}\right). \tag{17.15}$$

Proof. Notice that the family of operators $T(t) = \chi_{(0,\bar{t})}(t) e^{itA}$ satisfies

$$\int_0^t T(t) T^*(s) ds = \int_0^t e^{i(t-s)A} F(s) ds \; \forall t \in (0,\bar{t}). \tag{17.16}$$

Hence the result follows as an application, to the family $T(t)$, of the general result proved in [3]. □

Next we prove the existence of fixed points for the integral equation associated with (17.2):

$$u(t) = e^{itA} \varphi \pm i \int_0^t e^{i(t-s)A} (u(s)|u(s)|^{\alpha-1}) ds \tag{17.17}$$

by using the inhomogeneous estimates in Proposition 17.2. First we prove the following result, where $\mathscr{A}_\alpha$ is the set that appears in Theorem 17.2:

Proposition 17.3. *Let* $\alpha \geq 5$*. Then* $\left(\frac{1}{q(\alpha)}, \frac{1}{r(\alpha)}\right) \in \mathscr{A}_\alpha$ *if and only if there exist* $q(\alpha)$, $r(\alpha)$, $\tilde{q}(\alpha)$, $\tilde{r}(\alpha) \in [1,\infty]$ *that satisfy the following condition:*

$$\frac{1}{q(\alpha)} + \frac{1}{r(\alpha)} < \frac{1}{2} \tag{17.18}$$

$$\frac{1}{\tilde{q}(\alpha)} + \frac{1}{\tilde{r}(\alpha)} < \frac{1}{2} \tag{17.19}$$

$$0 < \frac{1}{q(\alpha)}, \frac{1}{\tilde{q}(\alpha)} \leq 1 \tag{17.20}$$

$$0 \leq \frac{1}{r(\alpha)}, \frac{1}{\tilde{r}(\alpha)} \leq \frac{1}{2} \tag{17.21}$$

$$\frac{1}{q(\alpha)} + \frac{1}{\tilde{q}(\alpha)} = \frac{1}{2}\left(1 - \frac{1}{r(\alpha)} - \frac{1}{\tilde{r}(\alpha)}\right) \tag{17.22}$$

$$\frac{\alpha}{r(\alpha)} = 1 - \frac{1}{\tilde{r}(\alpha)} \tag{17.23}$$

$$\frac{\alpha}{q(\alpha)} < 1 - \frac{1}{\tilde{q}(\alpha)}. \tag{17.24}$$

Proof. For simplicity we shall write $q(\alpha), r(\alpha), \tilde{q}(\alpha), \tilde{r}(\alpha) = q, r, \tilde{q}, \tilde{r}$. First we eliminate the variable $\tilde{r}$ by using the identity (17.23); hence, the conditions (17.18)–(17.24) are equivalent to

$$\frac{1}{q} + \frac{1}{r} < \frac{1}{2} \tag{17.25}$$

$$\frac{1}{q} + \frac{1}{\tilde{q}} = \frac{\alpha - 1}{2r} \tag{17.26}$$

$$\frac{\alpha - 1}{2r} - \frac{1}{q} + 1 - \frac{\alpha}{r} < \frac{1}{2} \tag{17.27}$$

$$0 < \frac{1}{q}, \frac{1}{\tilde{q}} \leq 1 \tag{17.28}$$

$$0 \leq 1 - \frac{\alpha}{r}, \frac{1}{r} \leq \frac{1}{2} \tag{17.29}$$

$$1 - \frac{\alpha}{q} > \frac{\alpha - 1}{2r} - \frac{1}{q} \tag{17.30}$$

(more precisely (17.26) is equivalent to (17.22) due to (17.23), (17.27) is equivalent to (17.19) due to (17.23) and (17.26), (17.30) is equivalent to (17.24) due to (17.26)). Next we eliminate $\tilde{q}$ by using (17.26) and we get

$$\frac{1}{q} + \frac{1}{r} < \frac{1}{2} \tag{17.31}$$

$$\frac{1}{q} + \frac{\alpha + 1}{2r} > \frac{1}{2} \tag{17.32}$$

$$0 < -\frac{1}{q} + \frac{\alpha - 1}{2r} \leq 1 \tag{17.33}$$

$$0 < \frac{1}{q} \leq 1 \tag{17.34}$$

$$\frac{1}{2} \leq \frac{\alpha}{r} \leq 1 \tag{17.35}$$

$$0 \leq \frac{1}{r} \leq \frac{1}{2} \tag{17.36}$$

$$\frac{\alpha - 1}{q} + \frac{\alpha - 1}{2r} < 1 \tag{17.37}$$

Actually since we are assuming $\alpha \geq 5$ it is easy to check that condition (17.43) is stronger than (17.31); Hence, the above conditions reduce to

$$\frac{1}{q} + \frac{\alpha+1}{2r} > \frac{1}{2} \tag{17.38}$$

$$0 < -\frac{1}{q} + \frac{\alpha-1}{2r} \leq 1 \tag{17.39}$$

$$0 < \frac{1}{q} \leq 1 \tag{17.40}$$

$$\frac{1}{2} \leq \frac{\alpha}{r} \leq 1 \tag{17.41}$$

$$0 \leq \frac{1}{r} \leq \frac{1}{2} \tag{17.42}$$

$$\frac{\alpha-1}{q} + \frac{\alpha-1}{2r} < 1. \tag{17.43}$$

It is easy to check that $(\frac{1}{q}, \frac{1}{r})$ satisfy the above conditions if and only if $(\frac{1}{q}, \frac{1}{r}) \in \mathscr{A}_\alpha$. □

Proposition 17.4. *Assume the operator A satisfies* (17.5) *and* (17.6)*. Let $\alpha \geq 5$ and $q(\alpha)$, $r(\alpha) \in [1,\infty]$ be any fixed couple such that $\left(\frac{1}{q(\alpha)}, \frac{1}{r(\alpha)}\right) \in \mathscr{A}_\alpha$. Assume moreover that $\varphi(x)$ is such that*

$$\|e^{itA}\varphi\|_{L_{\bar{t}}^{q(\alpha)} L_x^{r(\alpha)}} \leq K < \infty. \tag{17.44}$$

Then there exists $T > 0$ and one unique local solution $u(t,x)$ to (17.17) *such that*

$$u(t,x) \in L_T^{q(\alpha)} L_x^{r(\alpha)}.$$

Moreover there exist $a, b := a(q(\alpha), r(\alpha)), b(q(\alpha), r(\alpha)) > 0$ such that

$$T \geq min\{\bar{t}, aK^{-b}\}.$$

Proof. We introduce the nonlinear operator

$$S_\varphi(u) = e^{itA}\varphi \pm i \int_0^t e^{i(t-s)A} u(s)|u(s)|^{\alpha-1} ds.$$

For simplicity we shall write $q(\alpha), r(\alpha) = q, r$. First we prove the following:

Claim 1

$$\exists\, c, d = c(q(\alpha), r(\alpha)), d(q(\alpha), r(\alpha)) > 0 \text{ s.t. } S_\varphi(B_{2K}(L_T^q L_x^r) \subset B_{2K}(L_T^q L_x^r)$$

$$\forall T \leq min\{\bar{t}, cK^{-d}\}.$$

Notice that by combining the inhomogeneous Strichartz estimates with the hypothesis on φ we get

$$\|S_\varphi(u)\|_{L_T^q L_x^r} \leq K + C\|u|u|^{\alpha-1}\|_{L_T^{\tilde{q}'} L_x^{\tilde{r}'}} \; \forall 0 < T < \bar{t}$$

provided that $q, r, \tilde{q}, \tilde{r}$ satisfy the hypothesis (17.15) in Proposition 17.2. By using the Hölder inequality, we get the existence of $k = k(q,r) > 0$ such that

$$\|S_\varphi(u)\|_{L_T^q L_x^r} \leq K + \|u\|^\alpha_{L_T^{\alpha\tilde{q}'} L_x^{\alpha\tilde{r}'}} \leq K + CT^k \|u\|^\alpha_{L_T^q L_x^r} \tag{17.45}$$

provided that we can choose $\tilde{q}, \tilde{r}$ such that

$$\alpha\tilde{q}' < q \text{ and } \alpha\tilde{r}' = r. \tag{17.46}$$

Summarizing we get the estimate (17.45) for a suitable $k = a(\alpha) > 0$ provided that we can find $\tilde{q}, \tilde{r}$ that satisfy (17.15) and (17.46). Indeed this is possible due to Proposition 17.3 and since we are assuming $\left(\frac{1}{q}, \frac{1}{r}\right) \in \mathscr{A}_\alpha$. Finally notice that by (17.45) we easily deduce the claim provided that we choose $CT^k(2K)^\alpha \leq K$.

Next we prove that S_φ is a contraction on $B_{2K}(L_T^q L_x^r)$ for $T > 0$ small enough. More precisely we have the following:

Claim 2

$$\exists \tilde{c}, \tilde{d} = \tilde{c}(q,r), \tilde{d}(q,r) > 0 \quad \textit{such that}$$

$$\|S_\varphi(v) - S_\varphi(w)\|_{L_T^q L_x^r} \leq \frac{1}{2}\|v - w\|_{L_T^q L_x^r} \quad \forall v, w \in B_{2K}(L_T^q L_x^r)$$

$$\textit{with } T < \textit{min}\{\bar{t}, \tilde{c}K^{-\tilde{d}}\}.$$

By fixing $q, r, \tilde{q}, \tilde{r}$ as above we get

$$\|S_\varphi(v) - S_\varphi(w)\|_{L_T^q L_x^r} \leq C\|v|v|^{\alpha-1} - w|w|^{\alpha-1}\|_{L_T^{\tilde{q}'} L_x^{\tilde{r}'}}$$

$$\leq C\alpha\|(v - w)(|v|^{\alpha-1} + |w|^{\alpha-1})\|_{L_T^{\tilde{q}'} L_x^{\tilde{r}'}} \quad \forall T < \bar{t}$$

where we have used the pointwise estimate

$$|s|s|^{\alpha-1} - t|t|^{\alpha-1}| \leq \alpha|t - s|(|t|^{\alpha-1} + |s|^{\alpha-1}).$$

By the Hölder inequality we continue the estimate above and we get

$$\|S_\varphi(v) - S_\varphi(w)\|_{L_T^q L_x^r}$$

$$\leq 2^{\alpha-1} C T^k \alpha \|v - w\|_{L_T^q L_x^r} Max\{\|v\|_{L_T^q L_x^r}, \|w\|_{L_T^q L_x^r}\}^{\alpha-1}$$

$$\leq 2^{\alpha-1}\alpha C(2K)^{\alpha-1} T^k \|u - w\|_{L_T^q L_x^r} \; \forall T < \bar{t}$$

where $k > 0$ is the same number as in (17.45). Hence it is sufficient to fix

$$2^{\alpha-1}\alpha C(2K)^{\alpha-1}T^k < \frac{1}{2}.$$

In conclusion we have that S_φ is a contraction on $B_{2K}(L_T^q L_x^r)$ provided that

$$T < min\{\bar{t}, min\{c,\tilde{c}\}K^{-max\{d,\tilde{d}\}}\}. \qquad \square$$

17.5 Proof of Theorem 17.2

We introduce the following sets:

$$\Omega_M = \{\omega \in \Omega \text{ s.t. } \|e^{itA}f^\omega\|_{L_{\bar{t}}^{q(\alpha)}L_x^{r(\alpha)}} \leq M\} \ \forall M > 0.$$

Then by using Proposition 17.1, we deduce that

$$\mathbf{p}(\Omega_M) \geq (1 - e^{-CM^2}).$$

Moreover for every $\omega \in \Omega_M$ we deduce by Proposition 17.4 the existence of a unique solution to (17.2) that satisfies (17.11) and (17.13).

References

1. Burq, N., Thomann, L., Tzvetkov, N.: Long time dynamics for the one dimensional non linear Schrödinger equation, arXiv:1002.4054
2. Cazenave, T., Weissler, F.B.: The Cauchy problem for the critical nonlinear Schrödinger equation in H^s. Nonlinear Anal. **14**(10), 807–836 (1990)
3. Foschi, D.: Inhomogeneous Strichartz estimates. J. Hyperbol. Differ. Equat. **2**(1), 1–24 (2005)
4. Ginibre, J., Velo, G.: On a class of nonlinear Schrödinger equations. I. The Cauchy problem, general case. J. Funct. Anal. **32**(1), 1–32 (1979)
5. Vilela, M.C.: Inhomogeneous Strichartz estimates for the Schrödinger equation. Trans. Am. Math. Soc. **359**(5), 2123–2136 (2007)
6. Yajima, K., Zhang, G.: Smoothing property for Schrödinger equations with potential superquadratic at infinity. Comm. Math. Phys. **221**(3), 573–590 (2001)

MIX
Papier aus verantwortungsvollen Quellen
Paper from responsible sources
FSC® C105338

If you have any concerns about our products,
you can contact us on
ProductSafety@springernature.com

In case Publisher is established outside the EU,
the EU authorized representative is:
Springer Nature Customer Service Center GmbH
Europaplatz 3, 69115 Heidelberg, Germany

Printed by Libri Plureos GmbH
in Hamburg, Germany